EXPERTISE OUT OF CONTEXT

Proceedings of the Sixth International Conference on Naturalistic Decision Making

EXPERTISE OUT OF CONTEXT

Proceedings of the Sixth International Conference on Naturalistic Decision Making

Edited by

Robert R. Hoffman
Institute for Human and Machine Cognition
Pensacola, Florida

Lawrence Erlbaum Associates
Taylor & Francis Group

New York   London

Cover design by Tomai Maridou

Lawrence Erlbaum Associates
Taylor & Francis Group
270 Madison Avenue
New York, NY 10016

Lawrence Erlbaum Associates
Taylor & Francis Group
2 Park Square
Milton Park, Abingdon
Oxon OX14 4RN

Visit the Taylor & Francis Web site at
http://www.taylorandfrancis.com

and the LEA Web site at
http://www.erlbaum.com

I would like to dedicate this volume to my loving family,
Robin, Rachel, and Eric.
I love you very very much.

I would also like to dedicate this book
to Larry and Sylvia Erlbaum,
supporters and loving friends for many years.

Contents

PART V: TEAMS OUT OF CONTEXT

Series Editor's Preface

Early studies of expertise addressed well-defined tasks with highly circum-scribed contexts such as those associated with the game of chess, physics problem solving, or memory for digits. Much was learned about different levels of performance on those tasks along with some more general "principles" of expertise, such as the domain-specific nature of expertise. That is, for example, the skills of the chess master do not transfer to skilled memory for digits.

There are many instances, however, in which real expertise is not so easily confined and appears to stretch beyond its routine boundaries. The medical doctor may be called on to diagnose a rare disease or perform emergency surgery outside of his or her area of specialization because other experts are not available. And in some cases, the context for expertise is in a constant state of flux such that no one case is exactly like another. Experts need to continually adapt to the changing environment. Researchers in the fields of cognitive systems engineering and naturalistic decision making have contributed to the understanding of this extreme expertise or expertise out of context. This volume of the series documents much of this exciting work.

—*Nancy J. Cooke*

Editor's Preface

Many of the chapters in this volume were selected from the presentations at the Sixth International Conference on Naturalistic Decision Making, held at Pensacola Beach, Florida, May 15–17, 2003. The international Conferences on Naturalistic Decision Making are held every other year, alternating between Europe and North America. The Conferences are a gathering of researchers who attempt to study cognition, perception, and reasoning not in the traditional laboratory, but in "real-world" domains that are of importance to business, government, industry, and society at large. Naturalistic decision making (NDM) researchers have investigated such things as the knowledge and decision-making skills of expert firefighters, critical-care nurses, military commanders, and aircraft pilots. The goal of NDM research is to increase our understanding of expertise and its development, so that people can be accelerated on the learning curve, so that the knowledge, skills, and wisdom of top experts can be preserved for future generations, and so that new and better forms of human-computer interaction can be developed.

NDM6 had a number of topical sessions, including "The Transfer and Generalization of Expertise," "How to Train People to Become Experts," and "NDM Applications to World Needs." It was fairly clear before the meeting was over that a theme had emerged. Thereafter it shaped this volume.

The editor would like to thank the agencies and organizations that sponsored NDM6: The Army Research Institute and the Army Research Laboratory; the U.S. Air Force, the Air Force Office of Scientific Research, the AFRL Information Institute, and the Air Force Rome Laboratory; the U.S. Navy and the Office of Naval Research; and finally, the National Aeronautics and Space Administration. Corporate sponsors were Lawrence Erlbaum Associates, Klein Associates, and Micro Analysis and Design.

Thanks to the Army Research Laboratory, and Mike Strub in particular, and thanks also to the people at Micro Analysis and Design who, through their leadership and participation in the Advanced Decision Architectures Collaborative Alliance, provided a mechanism to support conference organization and management.

The editor would like to thank the Institute for Human and Machine Cognition for its support, encouragement, and guidance, and for permitting him the privilege of working with the amazing people who inhabit it.

The editor would like to thank all the reviewers of draft chapter submissions: Laurel Allender, John Annett, John Coffey, Cindy Dominguez, Rhona Flin, Rob Hutton, Richard Kaste, Raanan Lipshitz, Ric Lowe, David Malek, Laura Militello, Brian Moon, Judith Orasanu, Debbie Peluso, Karol Ross, Eduardo Salas, Jan Maarten Schraagen, Lawrence Shattuck, Paul Sirett, and Peter Thunholm.

The editor would like to thank Jeff Yerkes, Mary Berger, Brooke Layton, and Eric Hoffman for helping in the preparation of the indexes.

—*Robert R. Hoffman*

Part I

OVERVIEWS AND PERSPECTIVES

Chapter 1

Introduction: A Context for "Out of Context"

Robert R. Hoffman
Institute for Human and Machine Cognition, Pensacola, FL

The main theme of this volume is captured in its title. The authors were encouraged to discuss topics in Naturalistic Decision Making (NDM), expertise studies, and cognitive systems engineering (CSE) with an eye toward questions of what happens when domain practitioners are forced, for one reason or another, to work outside of their comfort zone.

As shown in the titles and the organization of the chapters in this volume into parts, "out-of" contexts can occur in a variety of ways and for a variety of reasons. Domain practitioners in any domain, at any time, can always be knocked out of joint by rare, tough, or unusual cases. Researchers who have used methods of Cognitive Task Analysis (CTA) have noted for some time that probing experts about their experiences with tough cases can be a rich source of information about expert knowledge and reasoning. A good case study is provided in chapter 5 by Lipshitz, Omodei, McClellan and Wearing, who discuss their research on a model of the heuristic strategies that firefighters use as they cope with difficult cases that are rife with uncertainty.

Another set of "out-of" contexts involves changes in the workplace. Many contexts for cognitive work—sociotechnical systems—are themselves always changing. There can be changes in the domain, in the sense that as more is learned about the domain on the basis of continuing research, more can be done in the cognitive work There are also changes in the workplace—new software, new workstations, and so forth. As chapter 13 by Ballas illustrates, cognitive work is always a "moving target."

Another class of "out-of" contexts involves changes in team or organizational structure. A domain practitioner might have to work in a new team, or even have a job description that says that their task is to work in ad hoc teams, ever shifting. Another class of "out-of" contexts involves having people work in new locations, that is, new sociocultural venues where cultural differences and cognitive styles play a key role in shaping how expertise will be exercised. A number of chapters deal with these topic areas.

Another class of "out-of" contexts is a special one, one that applies to us as researchers. Specifically, whenever we as NDM-ers or CSE-ers have taken on the challenge of conducting research on some new domain of practice or expertise, we ourselves are forced to work out of context. We would like to think that we have the experienced-based skills and toolkits sufficient to allow us to do this.

Research and issues for "out-of" contexts that confront teams are discussed in a number of chapters, including Maier and Taber's study of "oppositional expertise" (high-speed tactical decision making; chap. 9), Harville et al.'s chapter (chap. 21) on the tough tasks of sustained military operations, Klein and Steele-Johnson's discussion of their work on training people for work in multicultural teams (chap. 22), Graham, Gonzalez, and Schneider's chapter on the analysis of "netcentric" team organizations in defense operations (chap. 18), and Burke et al.'s discussions, including multi-cultural teams, of the challenges of training teams to be flexible and adaptive (chaps. 19, 23). These chapters touch on topics that are of immediate concern in the military, as explained clearly in chapter 17 by Lt.Gen. Frederick Brown.

Cognitive task analysis research in the domain of weather forecasting is also discussed in a number of chapters (by Ballas, chap. 13; by Kirschenbaum, Trafton, and Pratt, chap. 14; Trafton and Hoffman, chap. 15; and Klinger et al., chap. 16). By virtue of constant change in the workplace (workstations, software, new data types and displays), weather forecasters are almost always having to work "out of context."

Problems that challenge practitioners in analytical domains are discussed in chapters by Alison, Barrett, and Crego (the domain of criminal investigation; chap. 4) and by Hutchins, Pirolli, and Card (the domain of intelligence analysis; chap. 12).

The volume also has a few excellent chapters on broader theoretical and methodological issues. Hunt and Joslyn (chap. 2) discuss how laboratory research and NDM research complement and feed into one another in the investigation of practical problems. Gary Klein and colleagues (chap. 6) have contributed a thorough treatment of the macrocognitive process of "sensemaking," a notion that lies at the heart of any explanation of how domain practitioners are able to work "out of context." One way of defining when a practitioner is working out of context is when their ability to engage in "recognition-primed decision making" is short-circuited. Warwick and Hutton (chap. 20) discuss their attempt to develop a computational model of recognition-primed decision making, and wrap their discussion in a clear treatment of the challenges of developing a microcognitive model of what is essentially a macrocognitive process.

Some chapters discuss "tools for thinking out of context." Van den Bosch and De Beer (chap. 8) summarize their research aimed at developing a training package to support analytical reasoning, based on exercises that were crafted with the assistance of domain experts. Cohen et al. (chap. 10) discuss their research aimed at developing a tool to support analytical reasoning, based on a notion of "critical thinking through dialogue." Burns (chap. 11) discusses an approach to aid practitioners in tackling challenging problems that involve uncertainty and probability. Bonaceto and Burns (chap. 3) present a scheme for how we CTA researchers ourselves might make sense of, and extend, all of the various methods and methodologies that are in the CTA toolkit. Militello and Quill (chap. 7) ask how we ourselves as researchers develop an ability to become "expert apprentices."

It might be said that all of the chapters represent some maturation in the field of expertise studies and in the community of practice known as NDM. No longer concerned with demonstrating the defining features of experts versus novices (e.g., knowledge base, reasoning strategies, etc.), we are now looking at challenges to expertise, both for domain practitioners and for ourselves as CTA/NDM researchers. The field of expertise studies remains a fascinating one, rich with challenges for empirical investigation.

Chapter 2

The Dynamics of the Relation Between Applied and Basic Research

Earl Hunt
Susan Joslyn
The University of Washington

In the late 1950s, the senior author of this chapter did some moonlighting from his graduate studies to work for an applied research organization, a "Beltway Bandit." The projects were not all memorable. He spent most of this time doing research assistant chores building what may have been the worst ground-to-air control panel ever designed. Fortunately the Air Force canceled the entire system, thus improving our national defense.

Either because of or in spite of this experience, he became convinced that there were purposes to psychological research other than publication in *Psychological Review.* He was also bothered by the totally empirical, almost guild mentality of human factors researchers of the time. Theory was of little interest to them. The important thing was to have had experience, and to have read handbooks.

Graduate school at Yale during that era had almost exactly the opposite bias. The important thing was to find ways to test theories. To be fair, there was concern for theoretical connections to clinical psychology and the biomedical community, and social psychologists (notably Carl Hovland) were concerned about understanding communications, advertising, and propaganda. Technology, and especially computer programming, was beginning to have an impact on thinking about cognition. Nevertheless, with the sole exception of Donald Taylor (later dean of the Graduate School), the faculty had only mild interest in the study of issues that had practical importance but were not of particular theoretical relevance. Or at least, that was the way that it appeared to this graduate student.

Of course there were some exceptions to these broad statements. In the United Kingdom, the Applied Psychology Unit at Cambridge was doing applied research that would, eventually, lead to major advances in theories of cognition. The same remark could be made of a few U.S. universities. The relation between Carnegie Institute of Technology, and the RAND Corporation and MIT's Lincoln Laboratories led to notable achievements. Likewise, researchers at universities such as the University of Connecticut were making significant inroads in the area of speech perception and synthesis, through collaborations with the Haskins Laboratory in New Haven Connecticut. Nevertheless, the human factors engineers that EH encountered in his foray into industry had remarkably little interest in theory, and the institutions that dominated the headlines in experimental psychology had relatively little interest in applications, except perhaps in education. People working on basic research and on applied psychology in what we would, today, call the field of cognitive systems engineering had relatively little to do with each other. This is somewhat surprising, because much of the genesis of human information-processing studies had come out of research at U.S. Air Force and Navy laboratories during World War II (Hunt, 2002).

Today we take the more sophisticated view that the interests of basic and applied researchers are positively correlated although they are not identical. Hoffman and Deffenbacher (1993) summarized the present conceptual situation, and offered several suggestions for further conceptualization. Two of their points are central here. The first is that there are, in fact, multiple, distinguishable criteria for research, reliant on either epistemological (theory-based) or ecological (application-based) criteria. The different criteria can be lumped together by saying that epistemological criteria emphasize logical coherence and precision of definition. In other words, explanatory accuracy, having a good explanation, is the crucial thing. By contrast, ecological criteria emphasize relevance to some extralaboratory situation. Both epistemological coherence and ecological validity have several dimensions. What some might regard as an "ideal" study would be high on every dimension within each criterion, although such an ideal study may not exist. An important conclusion to draw from Hoffman and Deffenbacher's analysis that should be drummed into every psychology graduate student is that applied and basic research are linked by many, continuous and interacting dimensions. Information from a study can flow either way along these dimensions. The contribution is not a one-way street.

The present chapter complements and extends Hoffman and Deffenbacher's (1993) approach. Our argument is based on a conceptual point that was not emphasized in their analysis. The distinction between epistemological and ecological criteria applies to theories as well as empirical studies and methodology. We provide examples by tracing the development of several prominent theories. Moreover, the theoretical balance can shift from an emphasis on epistemological considerations to ecological considerations within a single project. We illustrate this point by describing some projects in which we have been involved, not necessarily because they are the best projects but rather because we know them best.

ELABORATIONS ON THE HOFFMAN
AND DEFFENBACHER VIEW

Hoffman and Deffenbacher (1993) were concerned primarily with the ecological and epistemological qualities of empirical studies. The same arguments can be applied to theoretical models. A theory can be evaluated for its internal coherence, that is, its epistemological validity and also its ecological relevance. Different theories have different degrees of formal statement and internal coherence. They also vary in their ecological validity.

Example: Models of Memory

Consider, for instance, the justly famous Atkinson and Shiffrin (1968) model of short-term and long-term memory. This theory was a mathematical tour de force. It provided surprisingly accurate predictions of the data from experiments that were designed to evaluate the theory. You can actually work out the predictions with mathematical precision, providing that the experimental situation maps precisely onto the model. On the other hand, whereas the general principle of Atkinson and Shiffrin's model—a distinction between short-term and long-term memory—can be applied in many situations, their model could be evaluated only by using experimental paradigms designed for that purpose. The precision of the model disappears when tested in most applied situations because one can no longer always or easily measure with precision a number of variables, such as the items to be recalled, the number of intervening items, the strategy being used by the participant, and the degree of interference between items.

We may contrast this to Alan Baddeley's (1986, 1992) equally famous working memory model. Baddeley's model, which has been stated solely in terms of diagrams and discourse, postulates passive "notebook" memories for auditory and visual/spatial information. A somewhat mysterious "executive process in working memory" keeps track of and manipulates information in the notebooks. Baddeley's model is not nearly as high in internal coherence as the Atkinson and Shiffrin model, for it does not have the same degree of mathematical precision. Nevertheless, it does draw our attention to important variables of a situation, such as the number of streams of information that have to be monitored and maintained in order to perform a task. Numerous investigators have been able to apply Baddeley's model to the sorts of information available for the analysis of everyday performance. We provide an example later. Furthermore, there is now substantial physiological evidence for the concepts contained in Baddeley's model. We can even identify locations in the brain that are active during the performance of a task that places heavy demands on either the central executive or one of the passive memory systems (Smith & Jonides, 1997).

From the viewpoint of an applied psychologist, Atkinson and Shiffrin's model represents an unobtainable ideal. If it could be mapped onto a practical situation, the human engineer could generate equations that were as precise predictions of human behavior as an aeronautical engineer's equations are predictions about airframe behavior. Alas, the required mapping can seldom be achieved. On the other hand the Baddeley model, although less precise, can be used as, if not a blueprint, at least a set of architects' notes to generate predictions about human performance in an applied setting. Indeed, Meyer and Kieras (1997) have written a computer program that does just this. The numbers produced by the program are suspect, but the functional relations it uncovers are reliable.

THE SITUATION IN THE DECISION SCIENCES

The decision sciences (also known as judgment and decision making) have analogues to the contrast between the Atkinson–Shiffrin and Baddeley models. Modern research on decision making began with an epistemologically outstanding theory, utility theory (Von Neumann & Morgenstern, 1947). One of the great advantages of the coherence of utility theory was that it provided clear directions for the translation of theory into action, via the concepts of subjective utility and subjective probability. In theory, if you could determine a decision maker's utility functions and what the decision maker thought the probabilities were, you should be able to tell the decision maker what to do. In work for which he won a Nobel award, Herbert Simon (1955) criticized the Von Neumann and Morgenstern approach for failing to consider two crucial variables that constrain actual decision making: time pressure on the decision maker and the effects of limited computing (or memory) capacity. Subsequently, Daniel Kahneman, in work for which *he* won a Nobel award, criticized the Von Neumann theory on the grounds that one has to consider relative rather than absolute changes in wealth, and that the utility function varies depending on whether one is concerned about gains or losses (Kahneman & Tversky, 1979).

To complete the story, Klein (1998), Hammond (1996), and Wagner (1991) have pointed out that in many everyday situations the concept of a rational decision maker should be replaced by the concept of a pattern recognizer, who tries to take action based largely on knowledge of what has worked before. Wagner went to the heart of the matter when he pointed out that managerial decision makers reject the concept of a gamble. This stance makes a fundamental difference because the idea of a "gamble" is basic both to the Von Neumann and Morgenstern approach and to Kahneman and Tversky's modification of it. Emphasis shifts from choosing between alternatives to situation assessment. We point out later how important this is in the discussion of our own second project.

This brief recitation should not be seen as an account of a failure of a theory. It is an account of a success. A theory was developed with sufficient epistemological coherence and ecological validity. In practice, however, utility theory had few naturally occurring instances, except after the theory/method were forced on people. When it was applied it was found wanting on ecological grounds. This lead to further experimentation (especially in the case of Kahneman and Tversky's research), additional modification of the theory, and additional comparisons of theory to everyday behavior. People such as Hammond, Klein, Wagner, and Eric Hollnagel (2005) are now challenging the entire rationalist-normative approach on the grounds that the pattern recognition approach has more ecological validity than the utility maximization approach. However the pattern recognition approach, as yet, lacks theoretical coherence. It functions more as a description after the fact than as a prescription that tells us what patterns an individual is likely to recognize. Hopefully we will soon see some sort of combination of the two, and indeed Hammond's (1996) book points the way toward such reconciliation. That is the way that things should happen.

The description just given is of a macroscopic interplay between theory and practical observation that took place over some 50 years. Our claim is that this grand scope of interplay between theory, experiment, and naturalistic observation also occurs on a micro scale, within the scope of a single project. We now illustrate this with three stories of studies in our own laboratory.

ILLUSTRATION 1: GOING BACK AND FORTH FROM FIELD TO LABORATORY: THE TASK DEMANDS ON PUBLIC SAFETY DISPATCHERS

Our first illustration deals with public safety dispatching, popularly referred to as 911 dispatching. Public safety dispatchers are responsible for assigning personnel to emergency incidents and monitoring the status of each incident until it is resolved. Our work and results have been described previously (Joslyn & Hunt, 1998). Here we concentrate on how we went about conducting the studies.

The task that we set ourselves was to develop a procedure that could be used to evaluate people's ability to deal with situations in which they have to cope with, and make decisions about, a number of difficult, and sometimes novel and unexpected situations occurring more or less at the same time. Problems of this sort arise in air traffic control and combat information centers, as well as public safety dispatching, indeed whenever experts are required to work "out of context."

Previous research on attention switching both in our own laboratory and in many others indicated that there is an ability to switch back and forth between several things. This ability is apparently related to working memory and appears to be largely a matter of being able to control attention (Engle, 2001). The evidence in support of this assertion rests almost entirely on laboratory-based studies, using laboratory paradigms that were designed to test theories of working memory and the control of attention. In Hoffman and Deffenbacher's (1993) terms, this work has a high level of epistemological validity but low ecological validity or relevance. We wanted to establish ecological validity of the theory, rather than of any particular experimental paradigm.

A colleague suggested to us that we examine public safety dispatching. (Outside the United States this is sometimes referred to as emergency responding.) At that time, however, we had only a television viewer's knowledge of what dispatchers do. Accordingly, we spent considerable time sitting in public safety dispatch centers. We decided to concentrate on the operations of a center that served several midsize suburban cities in a mixed residential and industrial area just south of Seattle. In this center, two-person teams did most dispatching. The *call receiver* talked to the caller while the *dispatcher* coordinated the movements of public safety units in the field. Dispatchers typically deal with several emergencies at once, and also have responsibility for control of a good deal of nonemergency communication, such as keeping track of whether or not an officer has left a police vehicle on a break, or informing an officer to pick up a warrant from a court and serve it at the relevant address.

Though we did study call receiving, we concentrated our analysis on the dispatcher, as this job appeared to place more demands on memory and the coordination of attention. We believe that it would be clear to anyone who has observed a public safety dispatch center that dispatching is an example of time-pressured decision making.

Armed with some knowledge of the dispatcher's job, plus our previous knowledge of psychological theory, we then developed a laboratory task that we believed abstracted the domain-general memory, attention control, and decision-making aspects of a dispatcher's task. Details can be found in Joslyn and Hunt (1998). The important thing here is that this was an artificial task involving the sorting of geometric shapes into bins. It intentionally made no contact with domain-specific knowledge, on which we knew that dispatchers relied. This was appropriate because: (a) We wanted to determine the information-processing aspects of a dispatcher's job rather than the knowledge aspects and (b) our goal was to develop a procedure that could identify people who were capable of meeting these information-processing demands, before they had actually learned a dispatcher's job. This would be one step toward satisfying Hoffman and Deffenbacher's (1993) criterion of ecological validity. The second goal is important because the use of a simple, albeit artificial task in employment screening avoids the trouble (and expense) of teaching someone to be a dispatcher to find out he or she can't handle the job. The use of an artificial task, which can be taught to a person in a few minutes, is intermediate between on-the-job evaluation and the use of employment tests designed to cover classes of jobs, which may have very different information-processing demands.

The resulting task was epistemologically valid, in the sense that we had generated it from a research tradition and a theoretical model of time-pressured decision making. The next step was to determine whether or not the task might be ecologically valid despite its superficial lack of ecological relevance. The straightforward way to do this would be to evaluate the task by comparing it to the performance of actual dispatchers, on the job. But this approach immediately ran into trouble.

Dispatchers are routinely evaluated, and we were able to sit in on a supervisor's evaluation conference. It was immediately clear that all the working dispatchers were considered competent in the narrow sense of dealing with incidents. Although there were a few mistakes, these were rare and usually not serious. The evaluation centers as much, if not more, on their performance as employees than it does on their performance as dispatchers, in the narrow sense. For instance, there was a good deal of discussion about whether or not certain dispatchers could be counted on to relieve the watch on time. Other discussions centered on tactful communication with the police and fire department administrations. These are all important aspects of a job, but they were not the aspects we were studying.

A second possibility would be to observe dispatchers on the job and create our own rating and use that to evaluate our task. This alternative was not feasible because, on the one hand, it might intrude on the handling of what can literally be life-and-death situations, and, on the other hand, performance is driven at least as much by the situation as the abilities of a person in the situation. We had no control over who would be on duty on a dull or busy night. Barring unanticipated cooperation by the criminal elements in the relevant cities (and even more cooperation from careless drivers), we had no way of creating comparable situations for different dispatchers.

Accordingly, we turned to a second laboratory task to use as a comparison. We created a simulation of dispatching. Actual dispatching requires knowledge of over 200 specialized incident classifications (e.g., robbery with/without a weapon). It takes a candidate dispatcher a week or more to learn all these codes. However, the common and most important subset can be learned rather quickly. We used this fact to create a computer simulation of a reduced set of the 15 most common dispatching tasks (Franklin & Hunt, 1993).

The interface in our simulated task was constructed to look very much like a dispatcher's display screen. The task itself involved responding to a simulated sequence of emergency messages. In a sense, our computer program re-created "Sin City," for things never went right in our simulated world. More seriously, we now had a situation that satisfied two important criteria. The first was that we could control the occurrence of emergency incidents in accord with an experimental context. The second was that the reduced dispatching task could be taught to naive participants (ideally, with educational backgrounds similar to the typical dispatcher's backgrounds). This made it possible to do correlational studies relating performance in dispatching to performance on our abstract decision-making task, and to covariate measures, such as measures of information integration and updating of working memory.

Before we could trust such studies, however, one more step remained to be taken. We had to show that our Sin City simulation had ecological validity. We did this by asking experienced dispatchers to work through and evaluate the simulation. They uniformly approved. In fact, they required almost no instruction in how to operate the simulation; they just sat down and went to work. One even used the simulation to assist him in tutoring a candidate dispatcher who was receiving on-the-job training.

Having accomplished these steps, we were then prepared to conduct the correlational studies, intended to establish ecological validity of the theory behind the abstract decision-making task. Several studies were needed, relating performance on the abstract task to dispatching, and determining the relation between dispatching, the abstract task, and other information-processing and intelligence measures. We could not perform these studies with dispatchers, for the simple reason that there were not enough dispatchers, nor were they available for a long enough time, to conduct multivariate studies. However we now had a tool that we could use, our dispatcher simulation.

We trained college and community college students (who matched the dispatchers in education) to use the simulation. We then conducted several studies showing that the decision-making task was indeed a substantial predictor of performance on the simulation ($r = .68$ in the first validation study, similar results in later studies). We also showed that the predictive power of the test could not be accounted for by other, simpler information-processing tasks not involving decision making. Moreover, it was only moderately correlated with a standard measure of intelligence (a matrix reasoning task similar to Raven's progressive matrices).

In sum, we began with a theoretical model and a vague idea that it related to a class of tasks outside of the laboratory. We then examined one such task in detail, and designed an artificial task, based on the theory, that we felt captured the relevant parts of the real-world task. Being unable to determine the correlation between the main workplace performance measure and the theoretically oriented measure directly, we developed a simulation of a reduced part of the workplace task and determined its ecological validity, in the sense convergent validity. We then examined the correlations between the theoretical task and the simulation, and on that basis asserted the ecological validity of the original theoretical analysis.

We now turn to a second illustration, which shows much the same logic in a quite different decision situation, one that proceeds more slowly than public safety dispatching (although it is not without time pressure), and that requires far more analysis on the part of the decision maker.

ILLUSTRATION 2: NAVAL WEATHER FORECASTING

Our next example is intended to show how an applied problem in decision making can draw attention to new theoretical problems, and suggest new basic research. It also shows that an initial analysis of the situation may have limited applicability.

We were asked to join a team of meteorologists, statisticians, and display designers working on a project to improve Naval weather forecasting at the mesoscale level. Mesoscale forecasting is the task of predicting a few hours to a day or so ahead in a localized region (roughly, the size of a county). It is a forecast at a higher spatial and temporal resolution than many local TV forecasts. Crucial Navy operations, such as aircraft landings (our principal concern), at-sea refueling, and amphibious landings are events very sensitive to different aspects of the mesoscale weather. A number of civilian activities, including aircraft landings, building construction, and firefighting, are also contingent on mesoscale weather.

Modern forecasters rely on many sources of data, including but not limited to satellite and radar images, direct observations, and the predictions generated from mathematical models of the weather, referred to as *numerical forecast models*. Historically, numerical model forecasts were made at the synoptic scale, that is, over distances that can span a continent or sizeable part of an ocean and over time periods of days to weeks. Only recently have meteorologists had the computing power and sufficient data to extend modeling to the mesoscale (see Ballas, chap. 13, this volume)

The typical Naval forecaster will have available predictions from four or five different numerical models, each of which may be at least slightly different. Different predictions arise from a number of things, including the fact that each model can start with a slightly different *initialization*. For our purposes, the initialization can be thought of as a measurement of today's weather that the model's equations use as a starting point. In theory, initializations are based on a fine grid of actual observations of the weather. In practice, however, there are areas where observations are sparse. For example, we do not have sensors everywhere in the Northeast Pacific Ocean, although we do have buoys, surface ship readings, and airplane observations from some points on that vast and stormy sea. Therefore in practice, initializations rely on extrapolation across an area, based on the limited data that are available. Depending on the data available and the extrapolation method used, different initializations may be calculated. Because weather processes can be chaotic, in the mathematical sense, the initialization that one chooses can make major differences in the output of the numerical forecasting program.

To make matters still more confusing, a numerical model may show varying accuracy and precision of prediction, depending on the time and parameter being forecast. The statement "ETA [a model] was doing pretty well, but recently MM5 [another model] has been more accurate" makes sense in some contexts (region, climate, time period). This statement could even be qualified to say that "ETA has been more accurate in predicting dew point, but MM5 does a better job with winds." Such statements about "model biases" change over time, depending on such things as shifts in ocean currents and oscillations such as the El Niño/La Niña. For instance, models may underestimate wind speed in situations where there are funnels in the terrain, something that happens quite a bit in the mountainous terrain of the forecast area that we studied. And finally, a model's prediction may have to be qualified to take into account local topographical features that can influence the weather. Forecasting can be a formidable exercise in data manipulation, keeping track of many variables, perceiving patterns, and making decisions.

Quite a few different governmental agencies and private-sector companies deal with weather forecasting. These include the National Weather Service, civil aviation forecasting centers, commercial forecast groups, and Air Force and Navy forecasting operations. Early in our visits to forecasting offices we noticed some differences between the Navy forecasters with whom we were to work and civilian agencies. The most obvious difference was in the display equipment available. The National Weather Service has programs that permit them to overlay different forecasts on the same screen, whereas, at the time, the Naval Air Station forecasters had to split their attention over several screens. In addition, there are important differences in the people and conditions of work. Naval forecasters generally have much less formal training and on-the-job experience than civilian weather forecasters. In particular, the military forecasters have less experience with the forecast area, because Naval personnel are moved from post to post more than are civilian personnel. There is also an important difference in education. Naval forecasters are usually petty officers, first class, and have less than a year of formal meteorological training. Civilian forecasters, on the other hand, almost all hold relevant college degrees in meteorology or a related discipline, and many have graduate degrees as well.

The third major difference between Naval and civilian forecasters has to do with the conditions of work. The Naval forecasters whom we observed, unlike National Weather Service forecasters, were constantly being interrupted as they prepared new forecasts. In one case that we observed, the forecaster was interrupted 13 times over an hour as he attempted to prepare a forecast. The most common interruptions were requests by pilots for forecasts related to their own oncoming flight. Such interruptions are high-priority tasks, and must be responded to quickly, even if it means suspending work on the local forecast.

In order to take a closer look at the forecasting operation, as conducted at a Naval Air Station, we supplemented our informal observations with a formal protocol analysis of "think aloud" records obtained from forecasters as they prepared the terminal aerodrome forecast (TAF). This is the forecast covering the mesoscale region that includes the air field itself. TAFs are relied on by both incoming and outgoing flights. And what did we find?

Previous research on weather forecasting suggested that the expert forecaster progresses through a series of steps, although there is minor disagreement about how to describe them. Trafton et al. (2000), in an analysis of Navy and Marine forecasters, describe the first step as initializing one's understanding of the large-scale weather. This is the approach recommended in Naval forecaster training. Forecasters are taught to follow the "forecast funnel." They begin by assessing the global patterns that are driving the weather and then work down the funnel to ascertain the local conditions. Pilske et al. (1997), in an analysis of Air Force forecasters, and Hoffman, Trafton, and Roebber (2007) in an analysis of Naval forecasting, describe the first step as "diagnosing the current weather situation" by extracting and matching patterns in the available data to those stored in memory. Pilske et al. (1997) say that expert forecasters begin by identifying a weather problem of the day. Again, this corresponds with forecaster training. Focusing on a primary weather problem serves to narrow the scope of information gathering in a domain in which information is plentiful.

The second step involves building an integrated mental model of the atmosphere (Hoffman, 1991; Pilske et al., 1997; Trafton et al., 2000). Trafton et al. refer to the mental model as a representation similar to a four-dimensional picture in which features are located relative to one another. The representation is qualitative, although much of the information from which it was derived is quantitative and it will be translated back into quantitative information when the final forecast is made. Half of those interviewed by Pilske et al. describe a picturelike model as well. For expert forecasters, the model incorporates the physical cause-and-effect relationships in the atmosphere as well as the weather features themselves, and hence integrates each component causally with the rest (Pilske et al., 1997).

Hoffman, Coffee, and Ford (2000) suggest that the initial mental model is largely hypothetical. It is then tested against various information sources and adjusted when necessary. Trafton et al. (2000; Hoffman et al., 2007) describe this step as "verifying and adjusting the mental model." According to Trafton et al., most of information gathering takes place at this point in an effort to check the first hypothesis. This is the first time, according to Pilske et al. (1997), that some expert forecasters begin to use the output of numerical models.

The preceding description is somewhat idealized. All of the researchers cited would agree that the process is iterative, that there are individual differences and differences due to levels of experience. In fact, some argue that there is no "one" description of forecaster reasoning, but hundreds, depending on climate, locale, experience level, and a host of other factors (Hoffman et al., 2007).

By analyzing forecasters' statements about what they were doing and by analyzing the information on the computer screens they were looking at, we found that the four forecasters that we studied evaluated relatively few information sources before arriving at their decisions. In addition, we observed rather more reliance on the numerical models than some researchers have reported.[1] This tendency has been noted by other researchers who attribute heavy reliance on forecasting models to lack of experience (Hoffman et al., 2007). Indeed, the most experienced forecaster that we observed, a civilian who had a great deal of local experience, made less use of the models than did the other forecasters and looked at them later in the process. We also found that although information search was narrowed to some extent by focusing on a "weather problem of the day," a surprising amount of the search was determined by a simple variable—how easy it was

---

[1]This discrepancy from previous reports should not be exaggerated. The previous studies span roughly a dozen years, and over that time the models have improved, and a new method has been developed to combine the outputs of more than one numerical model into an "ensemble forecast." As the predicting power of the models has improved, it is rational for a forecaster to rely on a numerical model more today than he or she might have in, say, 1993.

to access the information in question. This is important because, as we have noted, there are several numerical models that can give different predictions. However, the forecasters we observed tended to focus on a single model, rather than trying to strike a compromise between model estimates. Sometimes a second model was used for a particular parameter, but this appeared to motivated largely by the ease with which the desired information could be extracted from the interface.

The forecasters statements suggested that they all had some understanding of how models functioned. They all understood that model initialization times are important in estimating model inaccuracy. They all had some understanding of model biases. Moreover, they all made some attempt to evaluate model predictions. For some parameters, for example, pressure, forecasters compared the numerical value given by the model to the observation and then adjusted the predicted value up or down as indicated. For some parameters, adjustments were made based on knowledge of model biases. Thus, there was strong evidence that each forecaster observed had some understanding of the numerical models themselves and used that information in generating their own forecasts.

Another interesting thing is what we did not observe. Based on their verbal descriptions, only one of the four forecasters appeared to build an integrated mental model of the atmosphere and use it to forecast the weather. Such a goal was inferred when a forecaster described a coherent, qualitative, four-dimensional (spatial/temporal), causally interrelated representation of the weather that explained two or more parameters and did not refer directly to some external information source. This definition was derived largely from prior research (Hoffman, 1991; Pilske et al., 1997; Trafton et al., 2000). A complete mental model, by this definition, was not detected in the majority of the verbal protocols of the U.S. Navy forecasters observed in this study. Although some referred to what might be considered partial mental models, for three of the forecasters there was no evidence of a representation with all of the features in our definition. Instead, decision making for these three seemed to be much more localized. The reasoning we observed might just as well be described as application of the if–then rules of a production system. Clearly such an approach is much less demanding of working memory capacity and perhaps advantageous under the circumstances of the weather office we observed.

These results can be generalized in two ways. First, from the viewpoint of applied psychology, the Navy forecasters appear to have made an adaptive response to their particular situation, where forecasts have to be made under time pressure, in the face of interruptions, and with limited direct experience. The forecasting process has been streamlined to one that relies heavily on rules of thumb and the numerical models. Given the increasing accuracy and precision of current models, it makes sense to respond to the demands of their situation by starting with and then adjusting a model.

We then asked a general question: Was this tendency to fix on the recommendation from a single-model characteristic of forecasters (and perhaps forecasters' training), was it a response to the constraints of the situation we observed, or was it a more general characteristic? The issue was particularly urgent for us because meteorologists and statisticians interested in forecasting have proposed that the mathematical models display probabilistic information, that is, a range of forecast possibilities, rather than a point forecast. On the one hand, this seems like a good idea. On the other, forecasting is already very much a process of winnowing down information, so presenting yet more information may not be that informative, and may actually interfere!

To answer this question, we again turned to both the literature and the laboratory. In the literature, we found support for the proposition that people can combine and weight information from multiple advisers (Fisher & Harvey, 1999), although there is evidence for "bias" (or the use of heuristics) in the process (Yaniv, 1997; Yaniv & Kleinberger, 2000). However, these studies were conducted in what seems to us to be rather too sterile a laboratory setting, mostly using artificial materials, artificial-probability juggling tasks, and college freshmen as participants. The investigators sometimes assert that their laboratory situation mimics a real decision-making situation, but work in this laboratory paradigm generally does not provide any bridging evidence, beyond the investigator's assertion, that the analogy is a good one. (In fairness to the investigators, they sometimes acknowledge this.)

On the other hand, we do not see how these issues could be investigated solely by observations in the field, as there are too many uncontrolled variables. Accordingly, we are now constructing a series of laboratory experiments in which the participants are meteorology students and the stimuli are "controlled" observations and model outputs, constructed from historical records of actual weather situations. Depending on the condition, various kinds of probabilistic information about model performance will or will not be displayed. We believe that these studies will move our work toward both greater ecological and epistemological validity.

For the present purposes, the important point is not the outcome of our studies (which waits to be determined) but the process that we went through. Once again we began with a theoretical approach to an applied question. Observation caused us to shift our theoretical approach, and suggested further questions. The questions themselves required controlled experimentation, but with simulations of actual tasks rather than further field explorations.

We now look at another and very different situation that involves a quite different technique for decision making—the choice of instructional strategies for teaching high school science.

ILLUSTRATION 3: THE SYSTEMS CONTEXT OF APPLICATIONS: RESEARCH ON SCIENCE EDUCATION

The third example raises issues that go beyond the interaction between psychological theory and application. We look at the need to understand theories of social behavior outside of psychology. Why? At the outset of this chapter, we argued that psychological research should result in something beyond articles in academic journals. We want our research to be used outside of the context of scientific psychology itself. In cases where the investigator is essentially serving a client, as in the design of interfaces for specific tasks, such an outcome is built in. If the client adopts your recommendation, a limited application is ensured. Whether or not the research leads to a general change in practice depends on whether or not others in the client's industry will follow suit. Adoption issues are also relevant when research is conducted under the aegis of a general funding agency, such as the Department of Education or the Division of Science and Engineering Education of the National Science Foundation. Here the purpose of having the research program is to affect current practice as much or more than doing research to advance scientific knowledge.

Our example is a research program on science education that has been conducted by EH and his colleague Jim Minstrell. The driving idea is that students enter instruction with some systematic initial ideas about a topic, which Minstrell has referred to as *facets* (Minstrell, 2001). An example is the belief that static objects cannot exert a force. Therefore, according to this facet, if you run into a wall you exert a force on the wall but the wall does not exert a force on you.

We hypothesized that education will be effective if it is based on identifying facets and addressing them directly, rather than repeatedly presenting "the right answer." Accordingly, the technique is called *facet-based* instruction. Note that it is a decision-making process, the instructor infers facets from student behavior, and chooses acts of instruction accordingly. This can be done on either a class or an individual basis.

In the spirit of the previous discussions, could psychological theory help here? The conclusion was, basically, no. Minstrell's ideas can be reworded into discussions of schema theory, but all this does is to translate educational jargon into psychological jargon. The psychological concept of a schema is so underspecified that it was not useful.

The next, more empirical step was to find out whether the method worked at all. Minstrell himself had used it to great effect, but of course such a demonstration confounds the personal effect of the teacher and possible task demand characteristics (such as a possible Hawthorne effect) with the educational effect of the method. During the early 1990s we conducted a series of experiments comparing the class performance of science classes that were matched on relevant variables and that were taught the same material using either facet-based instruction or conventional instruction.

These studies involved several teachers, who were given instruction in the teaching technique. We also evaluated the performance of students taught by the same instructor, before and after this instructor adopted the technique. Details can be found in Hunt and Minstrell (1994, 1996). The evidence favoring facet-based instruction was very strong.

The problem is that this particular technique, like many teaching innovations based on the experience of a single teacher, does not scale up well. The technique depends on an instructor's knowing what facets of thought are likely to occur and how to react to them. Instructors trained in traditional teaching methods know, often very well, how to deliver the accepted answer in an effective manner. Most of them do not have a catalog of facets, distilled from years of teaching in a way that involves a lot of listening to students.

We attempted to approach this problem with technology. We developed an interactive computer program, the DIAGNOSER, that contains questions designed to reveal student facets, provide immediate feedback for students, and provide a report for teachers of the facets revealed (on either an individual or class basis). It also includes a teacher's guide that contains further descriptions of facets and suggestions for classroom exercises that can be used to address particular facets. The program's success was demonstrated by applications in what might best be called "controlled classroom" situations, where the teachers and researchers had a fairly close interaction. This is the typical educational intervention model. The question is whether or not it would scale up to adoption by teachers who had not had personal interactions with the researchers.

In order to pursue this, we placed the program on the World Wide Web (WWW), where it has been in operation for the last three academic years. As of March 2003, there were over 6,000 students registered, and more than 2,000 of them had made substantial use of the program. Preliminary comparisons of the performance of students who used and did not use the system during the 2001–2002 school year were moderately encouraging. Subsequent studies have confirmed that there is an effect of DIAGNOSER use combined with the teacher's adoption of at least some of the principles of facet-based instruction. However, the effect can be shown only by examinations that probe students' knowledge fairly deeply. There is essentially no effect if knowledge is evaluated using conventional summative assessments, such as those usually adopted for evaluation as required by the No Child Left Behind Act.

One the whole, we believe that this work has shown that the provision of a WWW version of facet-based instruction does help somewhat in achieving understanding of scientific topics, compared to traditional study and review methods. The benefits received are not nearly as great as the benefits of having a teacher who is trained in this admittedly time-consuming technique of instruction. On the other hand, the intervention is far less costly. We also point out, as this may be understood, that the students typically utilize the program during class time, rather than as homework. Therefore they are not receiving *more* instruction; they are receiving *different* instruction.

Perhaps most important, our research uncovered a problem that appears to be inherent in many summative educational assessments. Facet-based instruction is not compatible with traditional psychometric analyses of student performance. In conventional testing, answers can be graded as "right" or "wrong," and test scores can be thought of as an attempt to locate a student on a unidimensional gradation of knowledge. Facet-based instruction does, of course, recognize correct reasoning. However, incorrect answers are not treated uniformly as "wrong," they are treated as manifestations of a particular idea, for example, an Aristotelian rather than a Newtonian view of motion. To make things even more complicated, we also have to be concerned with the context in which beliefs are expressed. For example, we find people who are essentially Newtonian when they think about objects in space, and Aristotelian when they think about motion on Earth. (Parenthetically, we point out that this is not at all unreasonable. Aristotle's rules do describe how objects move in a friction-filled world.)

Accordingly, in order to handle the large amounts of data generated by the WWW program, and to remain true to the ideas behind facet-based instruction, we had to move away from the psychometric view that dimensions represented abilities. Instead we have taken a multidimensional scaling approach, in which people are located in space, at some distance from an "ideal" point. This representation allows us to depict the extent to which people hold consistent ideas, and the way in which they move from one idea to another.

However, conventional multidimensional scaling alone was not enough, because we needed some indication of the reliability of our identifications of responses in different contexts as indicators of a coherent facet of thought. Therefore we had to take a brief foray into basic statistics in order to solve this problem. In this foray, we place facet expressions in spaces defined by people, rather than placing people in a space defined by facets. By looking at the extent to which the same students express particular facets in different contexts, we gain some notion of the generality of a particular facet. This is important information for people designing instruction.

The final phase of this project introduces a new consideration. Will the innovation spread through the educational community? Here the results are, at best, mixed. The jury is still out on whether or not we succeeded. Some of our usage statistics, such as the 6,000-plus registered students, sound impressive. On the other hand, usage records indicate a very high attrition rate, especially among teachers. Somewhere around 10% of the teachers contacted will actually make use of the program. Our usage records also indicate that some teachers who use the program assign DIAGNOSER exercises to students but never look at the student results. This, of course, defeats the purpose of facet-based instruction, tailoring further instruction to a student's answers. These results are discouraging because it is hard to see how programs could be made more available, at low cost, than by placing them on the WWW.

Of course, this could be just because we designed a bad system. However, we do not think so. We have talked with other people who have put similar systems in place, and they have found similar results. The history of educational reform efforts is littered with ideas that could be shown to work, but simply were not adopted. Why? The answer to this question may be found in another discipline: sociological and anthropological studies of the decision to adopt innovative practices (Rogers, 1995). These studies deal primarily with the adoption of agricultural practices, although health practices and other types of innovation have also been studied. In the case of our work, the literature on innovation highlights three variables that, in retrospect, we failed to attend to sufficiently. They are:

1. *Innovation depends on a network over which successful innovators can spread the message.* Although the educational system has designated periods for continuing education, this is oriented toward receiving top-down messages. Informal teacher–teacher communication is much harder to achieve. This is particularly true for specialty disciplines, such as science and mathematics, where there are relatively few teachers for that discipline within a single school.

2. *A successful innovator must be central in the network, the innovation must be a public success, and the innovator must be perceived to be like the people who are considering adoption of the innovation.* Several of the teachers with whom we worked were clearly central in their school district's network of science teachers. They were senior, and were often assigned duties as mentors or resource teachers. On the other hand, because they were assigned these duties they were relieved of others. A teacher who has received a (justly needed) reduction in teaching load in order to participate in a research program is not perceived as being the same as a teacher who is considering

innovations while dealing with a full teaching load. Public success in education is hard to achieve. In the case of our own program, success would be measured by student performance on an examination toward the end of the school year. Teachers typically do not monitor the performance of other teachers' students.

3. *The short-term costs and benefits are extremely important.* The costs of taking on a new and expensive teaching method are clear. The benefits, whether immediate or long-term, are less clear. Continuing the analogy to farming, farmers compete in a public marketplace. Teachers do not. Therefore teachers (and anyone else in a noncompetitive situation) can afford to be behind the cutting edge in production methods. Competitive marketers do not have this luxury.

We do not claim to fully understand spread of innovation in teaching. (Our colleague, Jim Minstrell, currently is working on a project that investigates this topic.) The speculations here may be wrong, and in any case the details of the particular episode are of little interest to this audience. The general point is important. If the goal of a research project includes spreading an innovation through a field, it is important to consider the communication channels and resource allocation procedures of that field. In most cases, if you just build it, they probably will not come.

CONCLUDING REMARKS

Hoffman and Deffenbacher (1993) correctly pointed out that there are multiple continua relating basic and applied research. They showed that the criteria for basic and applied studies differ, and they made the important point that individual studies vary in the extent to which they meet the criteria deemed important from a basic or an applied viewpoint. We have tried to expand on this point by looking at the interplay between applied and basic research during the process of conducting research. We did so using examples from our own research, not because these are the best examples but because they are the ones with which we are most familiar.

At the outset of a project, the investigator approaches an applied topic with a worldview. This is largely based on conceptual formulations, although experienced investigators probably develop idiosyncratic varieties based on the literature. The worldview leads the investigator to make observations that may or may not reinforce it. At this point we think it extremely important to be willing to admit that one's worldview may be wrong, or irrelevant to the particular problem.

Though field observations may lead to hunches about why inconsistencies occur, the field situation will probably be too uncontrolled to provide acceptable scientific evidence for a new theory, evidence that permits the refutation of hypotheses. Therefore, if possible, it may be a good idea to return to the laboratory in order to modify one's initial ideas. How far the laboratory can be from the field situation depends on one's purposes. Are you trying to solve the current applied problem or are you trying to advance knowledge in general? If the former, the laboratory must mimic the field as closely as possible. If the latter, the field observations will have suggested new models, and the laboratory studies should be designed to evaluate those models, even if they do not directly address the original applied problem.

Finally, a solution to the applied project is proposed. The extent to which it is accepted depends on how one defines acceptance. If the goal of the study is a technical report to a sponsor or a publication in a journal, success is easy to achieve. If the goal is to introduce innovations across a wide field, then it is necessary to publicize results in a way that is consistent with the network over which innovations spread in that field. Often we must acknowledge that the natural networks of a field, all of which exist for a good reason, vary greatly in the extent to which they facilitate or inhibit the adoption of new ideas.

ACKNOWLEDGMENTS

The projects described in this chapter were supported by the Office of Naval Research, Cognitive Science Program, project contract N00014-92-C-0062, a Department of Defense Multidisciplinary University Research Initiative (MURI) program administered by the Office of Naval Research under Grant N00014-01-10745, the National Science Foundation (Project REC 9972999), and grants from the James S. McDonnell Foundation. In addition to this financial support, we are happy to acknowledge the contributions of A. Lynne Franklin, Karla Schweitzer, John Pyles, David Jones, Jim Minstrell, Aurora Graf, Bjorn Levidow, Anne Thissen-Roe, the staffs of the South King County (ValleyCom) public safety dispatch center, the forecasting unit of the Naval Air Station, Whidbey Island Washington, and the teachers who have participated in the DIAGNOSER project.

The research reported as "Illustration 3" was conducted by EH in collaboration with James Minstrell. The "we" in this section should be understood to refer to the two of them and their collaborators. EH is solely responsible for the preparation of this part of the chapter.

REFERENCES

Atkinson, R. C., & Shiffrin, R. M. (1968). Human memory: A proposed system and its control processes. In K. W. Spence & J. T. Spence (Eds.), *The psychology of learning and motivation: Advances in research and theory* (Vol. 2; pp. 89–195). New York: Academic Press.

Baddeley, A. D. (1986). *Working memory*. Oxford, England: Oxford University Press.

Baddeley, A. D. (1992). Working memory. *Science, 255,* 556–559.

Engle, R. W. (2001). What is working memory capacity? In H. L. Roediger III, J. S. Nairne, I. Neath, & A. M. Suprenant (Eds.), *The nature of remembering: Essays in honor of Robert G. Crowder* (pp. 297–314). Washington, DC: American Psychological Association.

Fisher, I., & Harvey, N. (1999). Combining forecasts: What information do judges need to outperform the simple average? *International Journal of Forecasting, 15,* 227–246.

Franklin, A. L., & Hunt, E. B. (1993). An emergency situation simulator for examining time-pressured decision making. *Behavior Research Methods, Instruments & Computers, 25,* 143–147.

Hammond, K. R. (1996). *Human judgment and social policy: Irreducible uncertainty, inevitable error, unavoidable injustice.* Oxford, England: Oxford University Press.

Hoffman, R. R. (1991). Human factors psychology in the support of forecasting: The design of advance meteorological workstations. *Weather and Forecasting, 6,* 98–110.

Hoffman, R. R., Coffey, J. W., & Ford, K. M. (2000). *A case study in the research paradigm of human-centered computing: Local expertise in weather forecasting* (Report). Pensacola, FL: Institute for Human and Machine Cognition.

Hoffman, R. R., & Deffenbacher, K. A. (1993). An analysis of the relations between basic and applied psychology. *Ecological Psychology, 5,* 315–352.

Hoffman, R. R., Trafton, G., & Roebber, P. (2007). *Minding the weather: How expert forecasters think.* Cambridge MA: MIT Press.

Hollnagel, E. (2005, June). *The natural unnaturalness of rationality.* Paper presented at the Seventh International Conference on Naturalistic Decision Making, Amsterdam.

Hunt, E., & Minstrell, J. (1994). A collaborative classroom for teaching conceptual physics. In K. McGilly (Ed.), *Classroom lessons: Integrating cognitive theory and the classroom* (pp. 51–74). Cambridge, MA: MIT Press.

Hunt E., & Minstrell, J. (1996). Effective instruction in science and mathematics: Psychological principles and social constraints. *Issues in Education: Contributions from Educational Psychology, 2,* 123–162.

Joslyn, S., & Hunt, E. (1998). Evaluating individual differences in response to time-pressure situations. *Journal of Experimental Psychology: Applied, 4,* 16–43.

Kahneman, D., & Tversky, A. (1979). Prospect theory: An analysis of decisions under risk. *Econometrica, 47,* 263–291.

Klein, G. (1998). *Sources of power: How people make decisions.* Cambridge, MA: MIT Press.

Meyer, D. E., & Kieras, D. E. (1997). A computational theory of executive cognition processes and multi-task performance. Part 1: Basic mechanisms. *Psychological Review, 104,* 3–65.

Minstrell, J. (2001). The role of the teacher in making sense of classroom experience and effecting better learning. In S. M. Carver & D. Klahr (Eds.), *Cognition and instruction: 25 years of progress* (pp. 121–150). Mahwah, NJ: Lawrence Erlbaum Associates.

Pliske, R., Klinger, D., Hutton, R., Crandall, B., Knight, B., & Klein, G. (1997). *Understanding skilled weather forecasting: Implications for training and the design of forecasting tools.* (Tech. Rpt. No. AL/HR-CR-1997-0003). Brooks AFB, TX: U.S. Air Force, Armstrong Laboratory.

Rogers, E. M. (1995). *The diffusion of innovations* (4th ed.). New York: The Free Press.

Simon, H. A. (1955). A behavioral model of rational choice. *Quarterly Journal of Economics, 69,* 99–118.

Smith, E. E., & Jonides, J. J. (1997). Working memory: A view from neuroimaging. *Cognitive Psychology, 33,* 5–42.

Trafton, J. G., Kirschenbaum, S. S., Tsui, T. L., Miyamoto, R. T., Ballas, J. A., & Raymond, P. A. (2000). Turning pictures into numbers: extracting and generating information from complex visualizations. *International Journal of Human–Computer Studies, 53,* 827–850.

von Neumann, J., & Morgenstern, O. (1947). *Theory of games and economic behavior* (2nd ed.). Princeton, NJ: Princeton University Press.

Wagner, R. K. (1991). Managerial problem solving. In R. J. Sternberg & P. A. Frensch (Eds.), *Complex problem solving: Principles and mechanisms* (pp. 159–184). Hillsdale, NJ: Lawrence Erlbaum Associates.

Yaniv, I. (1997). Weighting and trimming: Heuristics for aggregating judgments under uncertainty. *Organizational Behavior and Human Decision Processes, 69,* 237–249.

Yaniv, I., & Kleinberger, E. (2000). Advice taking in decision making: Egocentric discounting and reputation formation. *Organizational Behavior and Human Decision Processes, 83,* 260–281.

Chapter 3

A Survey of the Methods and Uses
of Cognitive Engineering

Craig Bonaceto
Kevin Burns
The MITRE Corporation, Bedford, MA

Cognitive engineering has produced a variety of methods for a variety of uses. This chapter provides a survey of the methods and uses, in order to help us and others apply cognitive systems engineering to human–systems integration in the design of new information-processing technologies. We begin by enumerating and classifying over a hundred specific methods of cognitive systems engineering. Our classification scheme is intended to help both cognitive systems engineers as well as systems engineers who may or may not be familiar with the field of cognitive systems engineering to understanding the main point of each method. This is not a trivial task because there are so many methods and because most claims about what a method can do are made (by their authors and others) in isolated papers with parochial interests. Figure 3.1 shows our classification scheme as a concept map.

After classifying each method according to its major purpose, we examine the potential uses of each method in the systems engineering design process. To do this, we outline a dozen major "phases" (uses) in the systems engineering life cycle, including "concept definition," "requirements analysis," "function allocation," "training development," and "performance assurance."

For each method, we identify those uses in systems engineering to which the method is particularly well suited. Our assessment is based on an analysis of the design activities conducted during each phase and, based on our review of the literature, the suitability of a particular method in addressing

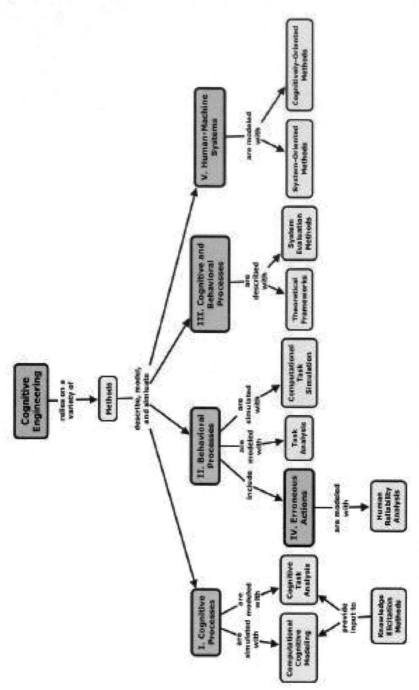

FIG. 3.1.   A concept map showing our classification of cognitive systems engineering methods.

the design activities. Our results are displayed in Table 3.1 in the form a matrix of methods (rows) versus uses (columns). In this matrix, the cells that are filled in black represent those pairings that, based on our review, offer the best opportunities for applying cognitive systems engineering to systems engineering. Cells that are filled in gray represent pairings that are less than optimal but still, based on our review, have potential. Armed with this matrix as a reference guideline, we believe that systems engineers in various domains will be better equipped to understand where and how they can apply the "human-centered" techniques of cognitive systems engineering to their problem set.

Our methods matrix relates to this volume's theme of *expertise out of context* at two levels. At an operational level, systems designed with the aid of cognitive systems engineering methods can better help system users cope with various challenges of decision making. At an engineering level, the matrix provides a road map for experts in one discipline, that is, systems engineering, who wish to apply the methods of another discipline, namely cognitive systems engineering. Because many systems engineers are not cognitive engineers, the mapping provided by our matrix might help inject cognitive systems engineering considerations and methods into systems engineering and thereby close the gap between these two mutually dependent fields of engineering expertise.

COGNITIVE SYSTEMS ENGINEERING METHODS

Each method of cognitive systems engineering that we identified has been organized into one of five primary categories. Figure 3.1 shows the categories and how they are related. The categories are: (I) Modeling Cognitive Processes, (II) Modeling Behavioral Processes, (III) Describing Cognitive and Behavioral Processes, (IV) Modeling Erroneous Actions, and (V) Modeling Human–Machine Systems. Each method that is linked under the categories is placed in a single category/subcategory. Though some (perhaps many) methods might be placed in several categories, we chose what we believe is the one "primary" category for each. In this regard, our purpose is not to establish definitive boundaries (because we doubt that such exist). Rather, our purpose is to provide a high-level road map.

Cognitive Processes

The methods in this category primarily concern themselves with modeling the knowledge and cognitive activities necessary to perform tasks. Cognitive task analysis seeks to model workers' knowledge and cognition descriptively, whereas computational cognitive models are more detailed at a

TABLE 3.1
Matrix of Cognitive Systems Engineering Methods and Systems Engineering Phases.
(Available online at http://mentalmodels.mitre.org/cog_eng/ce_sys_eng_phases_matrix.htm)

Cognitive Task Analysis — Method		Concept Definition	Requirements Analysis	Function Analysis	Function Allocation	Task Design	Interface & Team Development	Performance, Workload, Training Estimation	Requirements Review	Personnel Selection	Training Development	Performance Assurance	Problem Investigation
I.A.1	Applied Cognitive Task Analysis	█	█				█				█	▒	▒
I.A.2	Critical Decision Method		▒	▒			▒	▒		▒	█	█	█
I.A.3	PARI										█		
I.A.4	Skill-Based CTA Framework	▒				█	▒	▒		▒	█	▒	
I.A.5	Decompose, Network, and Asses									▒	█		
I.A.6	Task-Knowledge Structures		█	█			█	█	▒		█		
I.A.7	Goal-Directed Task Analysis	▒		█	█	█	█	█	▒		█	█	▒
I.A.8	Cognitive Function Model				█		▒	█			█		
I.A.9	Cognitively Oriented Task Analysis						█	▒			█		
I.A.10	Hierarchical Task Analysis	▒	█						▒		█		▒
I.A.11	Interacting Cognitive Subsystems				█						█		

	Code	Method
	I.A.12	Knowledge Analysis and Documentation System
Interview/Observe	I.B.1	Unstructured Interviews
	I.B.2	Structured Interviews
	I.B.3	Step Listing
	I.B.4	Group Interview
	I.B.5	Questionnaires
	I.B.6	Teachback
	I.B.7	Field Observations/ Ethnographic Methods
	I.B.8	Twenty Questions
Process Tracing	I.B.9	Discourse/ Conversation/ Interaction Analysis
	I.B.10	Activity Sampling
	I.B.11	Think-Aloud Problem-Solving/ Protocol Analysis
	I.B.12	Retrospective/ Aided Recall
	I.B.13	Interruption Analysis
	I.B.14	Shadowing Another

TABLE 3.1
(Continued)

Method		Concept Definition	Requirements Analysis	Function Analysis	Function Allocation	Task Design	Interface & Team Development	Performance, Workload, Training Estimation	Requirements Review	Personnel Selection	Training Development	Performance Assurance	Problem Investigation
I.B.15	Simulators/ Mockups					■	■	■	■	■	■	■	
I.B.16	Exploratory Sequential Data Analysis					■	▨	▨	■	▨			
I.B.17	Minimal Scenario Technique	▨	▨	▨		■	■	▨					
I.B.18	Critical Incident Technique					■	■	▨			■	■	■
I.B.19	Cloze Technique	▨	▨	▨	▨	■	■	▨		▨			
I.B.20	Critiquing					■	■	▨	■			▨	▨
I.B.21	Crystal Ball/ Stumbling Block				▨	■	■	▨	▨	■	■	▨	
I.B.22	Table-Top Analysis	■	■	▨	▨	■	■	▨	■	▨	■		
I.B.23	Wizard of Oz									■	■		
I.B.24	Decision Analysis	▨	▨		▨		▨				■		
I.B.25	Rating and Sorting Tasks	■	▨							■	■		

34

	I.B.26	Magnitude Estimation
	I.B.27	Repertory Grid
	I.B.28	P Sort
	I.B.29	Q Sort
	I.B.30	Hierarchical Sort
Conceptual Methods	I.B.31	Cluster Analysis
	I.B.32	Multidimensional Scaling
	I.B.33	Likert Scale Elicitation
	I.B.34	Structural Analysis
	I.B.35	Conceptual Graph Construction
	I.B.36	Concept Mapping
	I.B.37	Diagramming
	I.B.38	Laddering
	I.B.39	Influence Diagram Construction
	I.C.1	Keystroke-Level Model
Computational	I.C.2	CMN-GOMS
Cognitive Modeling	I.C.3	NGOMSL
	I.C.4	CPM-GOMS
	I.C.5	Cognitive Analysis Tool
	I.C.6	COGNET

(Continued)

TABLE 3.1
(Continued)

Engineering Phase / Method		Concept Definition	Requirements Analysis	Function Analysis	Function Allocation	Task Design	Interface & Team Development	Performance, Workload, Training Estimation	Requirements Review	Personnel Selection	Training Development	Performance Assurance	Problem Investigation
	I.C.7	COGENT											
	I.C.8	ACT-R											
	I.C.9	Soar											
	I.C.10	EPIC											
	I.C.11	Apex											
	I.C.12	MIDAS											
	I.C.13	SAMPLE											
	I.C.14	OMAR											
Task Analysis	II.A.1	Behavioral Task Analysis											
	II.A.2	Operational Sequence Diagrams											
	II.A.3	Timeline Analysis											
	II.A.4	Operator Function Model											
	II.A.5	Link Analysis											

(Continued)

Task Simulation	II.B.1	IMPRINT
	II.B.2	CART
	II.B.3	Micro Saint
	II.B.4	WinCrew
	II.B.5	IPME
System Evaluation Methods	III.A.1	Heuristic Evaluation
	III.A.2	Walk-Throughs/Cognitive Walk-Throughs/Talk-Throughs
	III.A.3	Formal Usability Studies
	III.A.4	Rapid Prototyping
	III.A.5	Storyboarding
	III.A.6	Interface Evaluation Surveys
	III.A.7	Ergonomic Checklists
	III.A.8	Contextual Inquiry
Theoretical Frameworks	III.B.1	Activity Theory
	III.B.2	Situated Cognition
	III.B.3	Distributed Cognition
	III.B.4	Naturalistic Decision Making

TABLE 3.1
(Continued)

Method	Engineering Phase											
	Concept Definition	Requirements Analysis	Function Analysis	Function Allocation	Task Design	Interface & Team Development	Performance, Workload, Training Estimation	Requirements Review	Personnel Selection	Training Development	Performance Assurance	Problem Investigation
IV.1 Event Tree Analysis											■	■
IV.2 Fault Tree Analysis											■	■
IV.3 Failure Modes and Effects Analysis											■	■
IV.4 Barrier Analysis					■	▨					■	
IV.5 Hazard and Operability Analysis											■	
IV.6 Management Oversight Risk Tree											■	
IV.7 Work Safety Analysis										▨	■	
IV.8 Confusion Matrices											■	
IV.9 Operator Action Event Tree											■	

Human Reliability Analysis

	Concept Definition	Requirements Analysis	Function Analysis	Function Allocation	Task Design	Interface & Team Development	Performance, Workload, Training Estimation	Requirements Review	Personnel Selection	Training Development	Performance Assurance	Problem Investigation
Human Reliability Analysis												
IV.1 Event Tree Analysis											■	■
IV.2 Fault Tree Analysis											■	■
IV.3 Failure Modes and Effects Analysis											■	■
IV.4 Barrier Analysis					■	▨						
IV.5 Hazard and Operability Analysis												
IV.6 Management Oversight Risk Tree												
IV.7 Work Safety Analysis										▨	■	
IV.8 Confusion Matrices											■	
IV.9 Operator Action Event Tree											■	

Note. Available at http://mentalmodels.mitre.org/cog_eng/ce_sys_eng_phases_matrix.htm

microcognitive level, and can be run on a computer. Knowledge elicitation techniques provide input to both cognitive task analyses and computational cognitive modeling.

Cognitive Task Analysis (CTA).   These are methods to analyze and represent the knowledge and cognitive activities workers utilize to perform complex tasks in a work domain (Gordon & Gill, 1997; Schraagen, Chipman, & Shalin, 2000). They are most useful in developing training programs and performance measures, establishing criteria to select people for certain jobs, and providing insight into the types of support systems that people may need.

Applied cognitive task analysis (ACTA) is a series of three structured interviews that aim to identify aspects of expertise that underlie proficient task performance. The first interview generates the task diagram, which provides a broad overview of the task. Next, the knowledge audit interview surveys the aspects of expertise required for a specific task. Finally, the cognitive processes of experts are probed within the context of a specific scenario in the simulation interview (Crandall, Klein, & Hoffman, 2005; Hardinge, 2000; Militello & Hutton, 1998; Militello, Hutton, Pliske, Knight, & Klein, 1997; Miller, Copeland, Phillips, & McCloskey, 1999; Phillips, McDermott, Thordsen, McCloskey, & Klein, 1998).

In the critical decision method (CDM), the domain expert recounts a nonroutine incident, and a set of cognitive probe questions is used to determine the bases for situation assessment and decision making (Crandall et al., 2005; Freeman & Cohen, 1998; Grassi, 2000; Hoffman, Crandall, & Shadbolt, 1998; Klein, 1996; Klein, Calderwood, & MacGregor, 1989; Thordsen, 1991).

The PARI (precursor [reason for action], action, result, interpretation [of result]). method consists of a structured interview in which novice and expert troubleshooters diagnose a fault in a problem scenario posed by another expert. The products are PARI diagrams of both expert and novice solutions to a representative set of troubleshooting problems (Gott, 1998; Hall, Gott, & Pokorny, 1995; Marsh, 1999).

The skill-based CTA framework is an approach to CTA that attempts to identify the hierarchy of skills needed to operate in a domain. This hierarchy, starting at the most complex skill type, includes: strategies, decision-making skills, representational skills, procedural skills, and automated skills (Redding, 1992; Redding et al., 1991; Seamster, Redding, & Kaempf, 1997, 2000).

The decompose, network, and assess (DNA) method is a computer-based method that helps to decompose a domain into its constituent elements, then network the elements into an inheritance hierarchy, and then assess this knowledge structure for validity and reliability. The knowledge structure combines aspects of production systems and conceptual graphs (Shute, Sugrue, & Willis, 1997; Shute & Torreano, 2002; Shute, Torreano, & Willis, 2000).

Task-knowledge structures (TKS) model conceptual, declarative, and procedural knowledge in terms of the roles, goals, objects, and actions associated with work tasks. A set of methods has been developed to collect and analyze data to create TKS models and use them in system design (Hamilton, 1996; Hourizi & Johnson, 2001; Johnson & Johnson, 1991; Johnson, Johnson, & Hamilton, 2000; Keep & Johnson, 1997).

Goal-directed task analysis (GDTA) is a technique that focuses on uncovering the situation awareness (SA) requirements associated with a job. The method identifies the major goals and subgoals of the domain, as well as the primary decision needed for each subgoal and the SA information requirements for making those decisions and performing each subgoal (Bolstad & Endsley, 1998; Endsley, 1999; Endsley, Bolte, & Jones, 2003; Endsley, Farley, Jones, & Midkiff, 1998; Endsley & Garland, 2000; Endsley & Hoffman, 2002; Endsley & Jones, 1997; Endsley & Robertson, 2000; Endsley & Rodgers, 1994; Hanson, Harper, Endsley, & Rezsonya, 2002).

Cognitive function modeling (CFM) is a technique that bridges the gap between operator function models and CTA. The intent is to identify nodes in an operator function model that involve highly challenging cognitive tasks and that should therefore be pursued more deeply with CTA (Chrenk, Hutton, Klinger, & Anastasi, 2001).

Cognitively oriented task analysis (COTA) is a collection of methods to improve the assessment of job expertise and performance. A verbal protocol analysis technique is used to determine the standard methods used for accomplishing tasks; to determine how these methods are selected, initiated, and completed; and to determine how these methods are adapted to novel situations (DuBois & Shalin, 2000; DuBois, Shalin, & Borman, 1995).

Hierarchical task analysis (HTA) breaks down a task into a hierarchy of goals and supporting subgoals, as well as the actions performed to accomplish those goals (Merkelbach & Schraagen, 1994; Shepherd, 2000).

Interacting cognitive subsystems (ICS) is a theory of cognition comprising a multiprocessor architecture made up of nine subsystems related to sensory, central, and effector processing activity. Aspects of the theory underpin a set of methods to conduct CTA (Barnard & May, 1993, 2000; Medical Research Council, n.d.).

The knowledge analysis and documentation system (KADS) is a knowledge elicitation method that models knowledge in terms of generic tasks (e.g., classification, diagnosis, planning, etc.) and standard organizations for declarative knowledge. By identifying the category of task being analyzed, the analyst is able to make use of established patterns of knowledge

(Hickman, Killin, Land, & Mulhall, 1990; Merkelback & Schraagen, 1994; Schreiber, Wielinga, & Breuker, 1993).

Knowledge Elicitation.   These are methods to determine the knowledge required to perform tasks. These techniques provide input to cognitive task analysis and computational cognitive models, as well as to traditional task analysis (Cooke, 1994; Diaper, 1989; Hoffman, Shadbolt, Burton, & Klein, 1995; Wilson, Barnard, Green, & MacLean, 1987).

This category includes observational methods of watching workers and talking with them. These methods are well suited to the initial phases of knowledge elicitation, and can provide direction on where to focus further knowledge elicitation activities. Observation methods used in cognitive task analysis are similar to, and sometimes borrow ideas from ethnographic methods. Workers are observed and interviewed in their actual work environments as they perform regular work activities. These methods can be useful for gathering user requirements, understanding and developing user models, and evaluating new systems and iterating their design (Bentley et al., 1992; Hughes, King, Rodden, & Andersen, 1994; Hughes, O'Brien, Rodden, Rouncefield, & Sommerville, 1995; Millen, 2000; Patterson, Woods, Sarter, & Watts-Perotti, 1999; Roth, Malsh, & Multer, 2001; Roth & O'Hara, 2002; Roth & Patterson, 2005; Shuler & Namioka, 1993; Woods, 1993).

The category of observational methods also includes process tracing, which immerses the worker in an actual task (i.e., problem-solving context), and is used to make inferences about the cognitive processes and knowledge structure that underlie task performance (Woods, 1993). Discourse/conversation/interaction analysis examines messages exchanged among group members to discover the systematic processes by which the members communicate, including standard sequences of interaction (Belkin, Brooks, & Daniels, 1987; Olson & Biolsi, 1991). In activity sampling, observations are made of a group of machines, processes, or workers over some period. Each observation records what is happening at that instant. The percentage of observations recorded for a particular activity enables the total time during which that activity is occurring to be predicted (Kirwan & Ainsworth, 1992; Tijerinia et al., 1996).

In think-aloud problem-solving/protocol analysis, the expert thinks aloud while actually performing some task or solving a problem. The procedure generates a protocol (i.e., a recording of the expert's deliberations) that can be transcribed and analyzed to uncover information about the expert's goal structure and reasoning sequence (Conrad, Blair, & Tracey, 1999; Ericsson & Simon, 1993; Gordon, Coovert, Riddle, & Miles, 2001; Lees, Manton, & Triggs, 1999; Woods, 1993). In a variation of think-aloud problem solving called the minimal scenario technique, the expert is given a problem to solve, but only the minimal amount of information required to state the problem is provided. The expert describes the domain by requesting the information that they need to solve the problem (Geiwitz, Klatzky, & McCloskey, 1988). A similar method is the cloze technique, in which experts are presented with textual material, typically a sentence or a passage of text, in which important information has been omitted. They are then asked to fill in the missing information according to their knowledge of the task domain (Leddo & Cohen, 1988). In another variation, called the Wizard of Oz technique, the expert is asked to simulate the behavior of a future system. One approach is to have the expert play the role of a future expert system, responding to users' queries (Cordingley, 1989; Maulsby, Greenberg, & Mander, 1993).

Additional observational methods are:

• Interruption analysis: While performing a task or solving a problem, experts are periodically interrupted with probe questions, such as "What were you just doing?" or "What would you have done just then if . . . ?" (Olson & Reuter, 1987).

• Shadowing: One expert observes the performance of another expert, either live or on tape, and provides real-time commentary (e.g., what the expert is doing, why the expert is doing it, etc.). Shadowing another is useful in process-control situations in which the expert performing the task may not be able to comment (Clarke, 1987).

• Simulators/mockups and microworld simulation: Equipment or information that is representative of what will be used during the task is constructed. Task activity while using this equipment is then observed and recorded. Microworld simulations are simplified versions of a more complex (real-world) task that simulate key challenges for testing/training human operators (Cooke, Shope, & Kiekel, 2001; Entin, Serfaty, Elliot, & Schiflett, 2001; Fahey, Rowe, Dunlap, & deBoom, 2001; Gray, 2002; Hess, MacMillan, Serfaty, & Elliot, 1999; Stammers, 1986).

The analysis of observation-derived data sometimes involves the use of exploratory sequential data analysis (ESDA). This is a family of observational analysis techniques that aim to uncover common sequences of interactions. For example, the analyst may look for repeated sequences of keystrokes that suggest poor design (Fisher & Sanderson, 1996; Sanderson & Fisher, 1994; Sanderson et al., 1994).

Interviews of a number of different types are also used for knowledge elicitation. The unstructured interview is an open dialogue in which the interviewer asks open-ended questions about the expert's knowledge and reasoning (Scott, Clayton, & Gibson, 1991; Weiss & Kulikowski, 1984). In a structured interview, the range and content of the questions are carefully planned prior to meeting with experts (Roth, Malsh, & Multer, 2001; Wood, 1997; Wood & Ford, 1993). A type of structured interview is step listing, in which the expert lists the steps involved in performing a specific task in his or her domain (Schvaneveldt et al., 1985). In teachback interviews, the expert explains a concept to the analyst. The analyst then explains this concept back to the expert. This process continues until the expert is satisfied with the analyst's understanding of the concept (Johnson & Johnson, 1987). In a twenty questions interview, the expert is provided with little or no information about a particular problem to be solved and must ask the analyst yes–no questions for information needed to solve the problem. The information that the expert requests, along with the order in which it is requested, can provide the analyst with insight into the expert's problem-solving strategies (Breuker & Wielinga, 1987; Grover, 1983; Shadbolt & Burton, 1990; Welbank, 1990). In a critiquing interview, the expert discusses both positive and negative aspects of a particular problem-solving strategy (possibly a novice strategy) in comparison to alternatives that might be reasonable or preferred (Miller, Patterson, & Woods, 2001). In the crystal ball/stumbling block technique, the expert describes a challenging assessment or decision. The analysis insists that the assessment is wrong, and the expert generates alternative interpretations of events, missing information, or assumptions (Cohen, Freeman, & Thompson, 1998). In decision analysis, the decision maker lists all the optional decision paths and all the possible outcomes and then rates them in terms of likelihoods and utilities. From the data, the researcher attempts to construct mathematical models of utilities (i.e., the values associated with particular events) and probabilities (i.e., the likelihoods of particular events) based on inputs provided by the expert. The models include decision trees and inference networks, which are sometimes implemented as expert systems (Bradshaw & Boose, 1990; Fischhoff, 1989; Goldstein & Hogarth, 1997; Hart, 1986; Keeney, 1982; Wright & Bolger, 1992).

Interviews can also be structured around the process of retrospective/aided recall. The expert performs some task or solves a problem in an uninterrupted manner. During this process, all actions of the expert may be recorded so that the task execution can be replayed. On completion of the task, the expert is guided through her task behavior and asked to report on her thoughts (Wood & Ford, 1993; Zachary, Ryder, & Hicinbothom, 2000). In another retrospection method called critical-incident technique (CIT), the expert is asked to recall and discuss a specific incident that was of particular importance in some context of task performance (Black, Dimarki, van Esselstyn, & Flanagan, 1995; Cohen et al., 1995; Flanagan, 1954; Woods, 1993).

Sometimes, group interviews are conducted if experts in the domain have differing areas of expertise, or if reliance on more than one expert is necessary to assess the reliability or importance of particular aspects of knowledge or reasoning (Boose, 1986; Boy, 1997b; Grabowski, Massy, & Wallace, 1992; McGraw & Harbison-Briggs, 1989; Peterson, Stine, & Darken, 2000). Questionnaires are useful when there is value in collecting a large number of viewpoints (Schweigert, 1994; Sinclair, 1990). In table-top analysis, a group of experts meets to discuss a problem perspective of the task, using task scenarios to explore the problem and derive a solution (Gustafson, Shulka, Delbecq, & Walister, 1975).

Conceptual Methods.   These are techniques that aim to produce representations of the relevant concepts in a domain and the relationships between these representations. Representations include diagrams, ratings, categorizations, and mathematical models.

In rating tasks, experts are asked to rate domain elements and provide a basis for their ratings. The sorting task is a variation of the rating task in which experts sort domain elements into various categories (Cordingley, 1989; Geiwitz, Kornell, & McCloskey, 1990; Zachary, Ryder, & Purcell, 1990). Magnitude estimation is a rating task in which experts are asked to assign a magnitude to a particular concept along some scale. The technique is often combined with other techniques, such as the repertory grid method or card sorting, which elicit attributes that are then scaled with magnitude estimation (Zachary et al., 1990). The repertory grid technique is a rating task in which experts are asked to generate important aspects of a domain and then provide dimensions on which those aspects may be rated (Bhatia & Yao, 1993; Boose, 1990; Bradshaw, Ford, Adams-Webber, & Boose, 1993; Cordingley, 1989; Gaines & Shaw, 1993; Zacharias, Illgen, Asdigha, Yara, & Hudlicka, 1996; Zachary et al., 1990). In Likert scale elicitation, experts are presented with assertions regarding some aspect of the domain. They are then asked to indicate their degree of agreement on a Likert scale. Likert scales are typically 5-point rating scales ranging from "strongly agree" through "neither agree nor disagree" to "strongly disagree" (Geiwitz & Schwartz, 1987; Geiwitz et al., 1988).

The P-sort is a sorting technique in which the expert sorts domain concepts into a fixed number of categories with limitations on the number of concepts per category (Zachary et al., 1990). The Q-sort is a sorting technique in which the analyst presents the expert with domain concepts that are then sorted into piles based on relatedness (Zachary et al., 1990). The hierarchical sort is a variation of the sorting task in which the expert sorts concepts in an increasing number of piles on each pass (Shadbolt & Burton, 1990).

Data from rating and sorting tasks can be analyzed using methods of cluster analysis. Clusters are groups of concepts in which the similarity is high for concepts within a group and low for concepts between groups

(Corter & Tversky, 1986; Leddo et al., 1988; Lewis, 1991; Zacharias et al., 1996). In analyzing data using multidimensional scaling, the analyst presents all combinations of pairs of concepts from the domain to an expert who is asked to rate them in terms of proximity or similarity. The data are used to create a k-dimensional space in which the concepts are located (Cooke, 1992; Cooke, Durso, & Schvaneveldt, 1986; Schenkman, 2002; Zacharias et al., 1996).

In many cognitive task analysis projects, the domain expert is simply asked to draw pictures of the domain's basic constructs and their interrelationships. Examples of diagrams include maps showing spatial relationships, flowcharts showing temporal relationships, and network or tree diagrams showing hierarchical relationships (Geiwitz et al., 1988). Graphical methods include structural analysis, in which a mathematical algorithm is used to transform the expert's concept-relatedness measures into a graphical representation of the domain (Boose, 1990; Cooke & McDonald, 1987; Schvaneveldt et al., 1985). In conceptual graph construction, the expert and analyst work together to draw a graphical representation of the domain in terms of the relationships (links in the graph) between the domain elements (nodes in the graph) (Gordon & Gill, 1997; Gordon, Schmierer, & Gill, 1993; Reichert & Dadam, 1997; Thordsen, 1991). In concept mapping, diagrams represent domain concepts and their relationships using nodes and labeled links that describe the relationships (Carnot, Dunn, & Canas, 2001; Coffey, Hoffman, Canas, & Ford, 2002). Laddering is a diagramming technique in which the analyst asks the expert questions to systematically build a hierarchy of domain concepts (Diederich, Ruhmann, & May, 1987; Shadbolt & Burton, 1990). In influence diagram construction, targeted events (e.g., system failures) are defined, and then the analyst produces a directed graph representing the influences (e.g., procedure adequacy) that determine the outcome (success or failure) of each event together with any dependencies between nodes (Howard & Matheson, 1984).

Computational Cognitive Modeling.   This subcategory includes methods that produce detailed models of how humans perform complex cognitive tasks. These models, which are run on a computer, can provide predictions of how well a proposed system will support its users by assessing factors such as how easy the system will be to learn and use, the workload it imposes, and the propensity for errors (Byrne, 2002; Campbell & Cannon-Bowers, n.d.; Gray & Altmann, 1999; Pew & Mavor, 1998; Polk & Seifert, 2002). The various model approaches are described in Table 3.2.

GOMS (Goals, Operators, Methods, Selection rules) is a task-modeling language that decomposes tasks into goals, operators, methods, and selection rules (Card, Moran, & Newell, 1983; John & Kieras, 1994). Goals are an

TABLE 3.2
Variants of GOMS Modeling

Keystroke-level model (KLM) Haunold & Kuhn, 1994; John & Kieras, 1994, 1996	A simplified version of GOMS that models a task as a specific sequence of keystroke-level primitive operators.
CMN-GOMS (Card Moran Newell GOMS) John & Kieras, 1994, 1996	Describes a task in terms of a hierarchical goal structure and set of methods in program form, each of which consists of a series of steps executed in a strictly sequential manner.
NGOMSL (Natural GOMS Language) Kieras, 1994	Refines the CMN-GOMS model by connecting it to a simple cognitive architecture. It provides a structured natural language to describe tasks in practical applications.
GLEAN3 Kieras, 1997a, 1998; Kieras, Wood, Abotel, & Hornof, 1995; Santoro, Kieras, & Campbell, 2000	A tool that generates quantitative predictions of human performance from a supplied NGOMSL model.
CPM-GOMS (Critical-path method/cognitive-perceptual-motor GOMS) Gray, John, & Atwood, 1992; John & Kieras, 1994, 1996; John, Vera, Matessa, Freed, & Remington, 2002	Is based directly on a parallel multiprocessor stage model of human information processing. CPM-GOMS uses a schedule chart (Pert chart) to represent the operators and dependencies between operators. The critical path through a schedule chart provides a simple prediction of total task time.
CAT (cognitive analysis tool) Baumeister, John, & Byrne, 2000; Voigt, 2000; Williams, 2000	A software application that uses a structured interview process to elicit from workers descriptions of how they would perform a task. It can structure those descriptions as production rules that provide the basis for a GOMS model.

end state that must be achieved to accomplish a task. Operators are the task actions that must be performed to attain a goal or subgoal. Methods are sequences of operators used to accomplish a specific goal. Selection rules are sets of conditions or decision rules that are used to determine which specific method should be used to accomplish a goal if more than one method is applicable. Variations of the GOMS approach are described in Table 3.2.

Integrated cognitive architectures allow for the memory, cognition, motor behavior, sensation, and perception necessary to perform complex tasks to be modeled at a high level of detail (Byrne, 2002). It is the analyst's job to describe a task or problem solving activity using the modeling framework provided by the architecture. Modeling tools in this subcategory are described in Table 3.3.

Behavioral Processes

This family consists of techniques that describe the sequences of behaviors necessary to perform work tasks, as well as rule-based decisions that determine when particular sequences are activated and how sequences interact. These methods are not suited for analyzing highly cognitive tasks, but they can be used as starting points by representing the overt tasks that workers perform and then focusing on aspects of those tasks that are cognitively demanding and require further analysis. Subcategories are task analysis and computational task simulation.

Task Analysis. This includes methods for producing detailed descriptions of the way a task is currently performed or could be performed (Kieras, 1997b; Kirwan & Ainsworth, 1992; Lewis & Rieman, 1994). In be-

TABLE 3.3
Integrated Cognitive Architectures

COGNET (cognition as a network of tasks) Ryder, Weiland, Szcepkowski, & Zachary, 1998; Weiland, Campbell, Zachary, & Cannon-Bowers, 1998; Zachary, Ryder, & Hicinbothom, 1998, 2000	Provides a language for describing cognitive tasks in information processing, similar to the GOMS notation, but includes additional features to allow for the modeling of cognitive operators. COGNET models may be implemented using the iGEN software system.
COGENT (cognitive objects within a graphical environment) Cooper, Yule, Fox, & Sutton, 1998; Cooper & Fox, 1997	Is a software environment for developing cognitive models. It provides a visual programming interface to allow users to build cognitive models using memory buffers, rule-based processes (similar to production rules), connectionist networks, and I/O sources and sinks.
ACT-R (Adaptive control of thought—rational) Anderson & Lebiere, 1998; Byrne, 2001; Gray, Schoelles, & Fu, 2000; Salvucci, 2001; Schollen & Gray, 2000)	This is a "unified theory of cognition" that allows modeling of the declarative and procedural knowledge in a variety of cognitive tasks. Procedural knowledge is modeled as if–then-like rules called productions that direct behavior. ACT–R/PM includes perceptual and motor extensions.

(Continued)

TABLE 3.3
(Continued)

Soar Conati & Lehman, 1993; Darkow & Marshak, 1998	This is also said to be a "unified theory of cognition" that implements goal-oriented behavior as a search through a problem space and learns the results of its problem solving
Kieras & Meyer, 1998, 2000; Kieras, Wood, & Meyer, 1995	A symbolic cognitive architecture that emphasizes the modeling of perceptual and motor processes, but also includes a simple production rule cognitive processor for modeling task performance.
Apex Freed & Remington, 2000; Freed, Shaflo, & Remington, 2000; John et al., 2002; NASA Ames Research Center, n.d.-a	A software architecture that allows for the modeling of behaviors ranging from perceptual motor actions to intelligent management of multiple, long-duration tasks in complex, dynamic environments. The language used to describe cognition is PDL, which incorporates and extends the capabilities provided by GOMS.
MIDAS (man–machine integrated design and analysis system) Corker, Pisanich, & Bunzo, 1997; NASA Ames Research Center, n.d.-b; Smith & Tyler, 1997	This was developed by NASA to model pilot behavior. It consists of an agent-based operator model, with modules for representing perceptual, cognitive, and motor processing, as well as a model of the proximal environment (e.g., displays and controls) and the distal environment (e.g., other aircraft, terrain)
SAMPLE (Situation awareness model for pilot-in-the-loop evaluation) Hanson et al., 2002; Zacharias, Miao, Illgen, & Yara, 1995	A stage-based cognitive architecture that decomposes decision making into information processing, situation assessment, and procedure execution. It assumes that behavior is guided by standard procedures and driven by detected events and assessed situations.
OMAR (operator model architecture) Deutsch, Cramer, Keith, & Freeman, 1999; Deutsch & Pew, 2002	This assumes that human behavior is goal-directed, and agents in OMAR are capable of executing goals, plans, and tasks. It has provisions for executing multiple tasks concurrently, as well as modeling multiple, communicating operators using D-OMAR (Distributed Operator Model Architecture).

havioral task analysis, tasks are decomposed as series of observable behaviors that are required to complete the task (Kirwan & Ainsworth, 1992). In operational sequence diagramming, the behavioral operations, inputs, outputs, and simple rule-based decisions necessary to perform a task are diagrammed in a flowchart in the order in which they are to be carried out. The diagram may also be supported by a text description (Dugger, Parker, Winters, & Lackie, 1999; Kirwan & Ainsworth, 1992; Stammers, Carey, & Astley, 1990). In timeline analysis, a temporally ordered sequence of actions necessary to achieve a task is constructed, along with duration estimates of each action (Kirwan & Ainsworth, 1992). Using the method called operator function modeling (OFM), a hierarchic/heterarchic network of nodes is created to describe/prescribe the role of the operator in a complex system. At the topmost (heterarchic) level of an OFM are high-level functions that are further broken down into a supporting hierarchy of subfunctions (Chu, Mitchell, & Jones, 1995; Degani, Mitchell, & Chappell, 1995; Mitchell, 1999; Rubin, Jones, & Mitchell, 1988). In link analysis, one creates an annotated diagram showing the frequency of operator movements between system components. The diagram is then used to analyze the relationships between system components to optimize their arrangement by minimizing movement times and distances (Hendy, Edwards, & Beevis, 1999; Kirwan & Ainsworth, 1992).

Computational Task Simulation. These techniques are the analogue of computational cognitive modeling, but model only the observable actions necessary to perform tasks rather than the underlying cognitive activities that drive task performance. The simulations can dynamically "run" tasks in real or fast time as a way of estimating complete cycle times, error likelihoods, workload, accuracy, and so on (Kirwan & Ainsworth, 1992). Specific simulation systems are described in Table 3.4.

Cognitive and Behavioral Processes

These approaches "describe" how people actually perform work tasks, so they are generally less formal than the "models" considered in the first two sections. The methods examine how workers use the tools they currently have available to them to perform tasks in the work domain. Typically, these methods are used when there is already a system (or prototype) in place that is to be evaluated and improved. Subcategories in this family are system evaluation methods and theoretical frameworks.

System Evaluation Methods. These methods evaluate how workers interact with existing or proposed systems. They aim to assess how easy a particular system is to learn and use, and how well the system supports the tasks

TABLE 3.4
Computational Task Simulation Systems

IMPRINT (**im**proved **p**erformance **r**esearch **in**tegration **t**ool) Army Research Laboratory, n.d.; Kelley & Allender, 1996; Mitchell, 2000	An event-based task network in which a mission is decomposed into functions that are further decomposed into tasks. The tasks are linked together in a network that represents the flow of events. Task performance time and accuracy, as well as expected error rates, are entered for each task.
CART (combat automation requirements testbed) Brett, Doyal, Malek, Martin, & Hoagland, 2000; Brett et al., 2002; Caretta, Doyal, & Craig, 2001	CART is built upon IMPRINT, so it is also a task networking modeling environment that allows for worker behavior to be represented in terms of the tasks and functions. CART adds functionality for integration with military simulations.
Micro Saint (system analysis of integrated network of tasks) See & Vidulich, 1997a, 1997b	A general-purpose, discrete-event simulation software tool. Micro Saint models of task execution yield estimates of times to complete tasks and task accuracies, as well as estimates of workload and task load (i.e., the number of tasks an operator has to perform over time).
WinCrew Mitchell, 2000	This is primarily concerned with determining whether operators will be able to handle the workload in existing or conceptual systems. It can predict operator workload for a crew given a design concept, as well as simulate how operators dynamically alter their behavior under high-workload conditions.
Integrated Performance Modeling Environment (IPME) Hendy, Beevis, Lichacz, & Edwards, 2002	Provides a means to model performance-shaping factors (which are environmental factors such as temperature and time that impact task performance) and performance-shaping functions (which define how performance-shaping factors affect performance), as well as a scheduler to simulate operator loading and an algorithm for estimating operator workload.

that workers perform. They are typically used in an iterative fashion to test and refine a proposed system design, evolving it from a prototype to a final design.

In contextual inquiry, field observations and interviews are used to collect detailed information about how people work, the context in which the work takes place, which tools are more important and less important, and which design changes to existing tools would help make their work easier (Beyer & Hotzblatt, 2002; Raven & Flanders, 1996). In storyboarding, pictures or display screens of candidate design concepts and how they would operate in a simulated scenario are constructed. Storyboarding enables system developers to receive feedback early in the concept development phase (Halskov Madsen, & Aiken, 1993).

Moving a step closer toward a finished system, in rapid prototyping a prototype of a proposed system is presented to workers for critical comments. Revisions are made to the original prototype, producing a second version that is again presented to workers for critical analysis. The process of revising and submitting to workers continues until some criterion for acceptability is reached (Mayhew, 1999; Thompson & Wishbow, 1992; Wilson, Jonassen, & Cole, 1993). Also at the prototype stage, in heuristic evaluation the analyst assesses the usability of a computer interface using rules of thumb, such as "speak the user's language," "provide feedback," and "be consistent." The analyst evaluates how well the proposed interface follows the rules of thumb and provides feedback as to how it could be improved (Nielsen, 1993; Nielsen & Mack, 1994; Nielsen & Phillips, 1993; Polson, Rieman, & Wharton, 1992).

In walk-throughs, cognitive walk-throughs, and talk-throughs, workers who know a system perform a task using an actual system or a realistic mock-up for analysis (Kirwan & Ainsworth, 1992; Nielsen & Mack, 1994; Polson et al., 1992). Cognitive walk-throughs attempt to evaluate the state of the worker's thought processes at each step of task performance, with emphasis on identifying aspects of the interface that are confusing (Jacobsen & John, 2000; John & Packer, 1995; Wharton, Bradford, Jeffries, & Franzke, 1992).

In formal usability studies, workers are observed or videotaped while they perform tasks using a proposed system in a controlled environment. By observing different workers performing the same tasks under such controlled conditions, it is possible to identify aspects of the interface that require improvement (Mayhew, 1999; Nielsen, 1993; Nielsen & Phillips, 1993). Formal analysis of finished or nearly finished systems can involve conducting interface evaluation surveys, a group of information collection methods used to identify specific ergonomics problems or deficiencies in interfaces. They address issues such as the labeling and consistency of controls and how well the system works within its environment (Brykczynski, 1999; Kirwan & Ainsworth, 1992; Mahemoff & Johnston, 1998). Formal analysis can also include ergonomics checklists that an analyst can use to ascertain whether particular ergonomic criteria are being met by a system. The items within these checklists can range from overall subjective opinions to very specific objective checks (Kirwan & Ainsworth, 1992).

Theoretical Frameworks.   These are perspectives about how people perform cognitive work. Although there are a host of theoretical frameworks, those discussed here are most relevant to cognitive systems engineering. They can help focus knowledge elicitation efforts by positing the important aspects of worker–technology interaction that one should take into account.

Activity theory is a psychological theory that characterizes human practices as developmental processes. The fundamental unit of analysis is the human activity, which is directed toward an object, mediated by artifacts, and occurs within a social context. Activity theory focuses primarily on the goals workers are trying to satisfy with artifacts (Bardram, 1997; Nardi, 1996).

Situated cognition is an approach to studying cognition that emphasizes the fact that cognition must be studied in its natural context. It is based on the fact that the same set of actions will have different effects at different times. Each situation is a unique context, and, as a result of the situated context, a different set of actions will be required to achieve the same task goal (Bardram, 1997; Wilson & Madsen-Myers, 1999).

Distributed cognition is an approach to studying cognition that attempts to understand how cognitive phenomena are distributed across individuals and artifacts. It emphasizes the fact that cognition does not lie strictly within the individual, but instead is an emergent, distributed activity, performed by people with tools within the context of a team or organization, in an evolved cultural context (Hollan, Hutchins, & Kirsh, 2000; Hutchins, 1995; Rogers, 1997; Rogers & Ellis, 1994; Wright, Fields, & Harrison, 2000).

We would be remiss not to mention naturalistic decision making (NDM), the study of how people make judgments and decisions in environments that involve high stakes, time pressure, multiple players, ill-defined goals, and uncertain and dynamic conditions. A major theory of NDM is the recognition-primed decision (RPD) model, which asserts that experts assess the current situation and draw on their experience to relate it to previous situations and successful courses of action (Klein, 1998; Klein, Orasanu, & Calderwood, 1993; Salas & Klein, 2001; Zsambok & Klein, 1996).

Erroneous Actions: Human Reliability Analysis (HRA)

These methods are used for analyzing situations in which errors have happened, or might happen. The goal of these techniques is to determine whether human errors will have serious consequences, and to quantify how likely the errors are to occur. The design of the system or the interface can then be modified to reduce the likelihood of human errors, or mitigate the consequences of them (Henley & Kumamoto, 1981; Kirwan & Ainsworth, 1992). Specific methods are described in Table 3.5.

Human–Machine Systems: Analysis at the Whole-System Level

Methods in this family consider how the entire system, consisting of machines and humans, works as a whole in order to accomplish the overall system goal. Subcategories are cognitively oriented methods, which model the goals of the work domain and the cognitive demands that are imposed on humans operating in those domains, and system-oriented methods, which are primarily focused on determining what role individual human operators will play in the system.

Cognitively Oriented Methods. These include cognitive work analysis (CWA), which is an integrated approach to human-centered system design

TABLE 3.5
Methods for Studying Error

Event Tree Analysis Henley & Kumamoto, 1981; Neogy, Hanson, Davis, & Fenstermacher, 1996; Vesely, Goldberg, Roberts, & Haasl, 1981	An event tree is a graphical method for identifying the various possible outcomes of an initiating event. The course of events from the occurrence of the initiating event until its final consequence is determined by the operation or nonoperation of various human and physical systems.
Fault Tree Analysis Henly & Kumamoto, 1981, Vesely et al., 1981	Fault trees show failures that would have to occur to cause an undesired event (an accident). They are constructed as a series of logic gates descending through subsidiary events to basic events at the bottom of the tree. The basic events may be human errors, equipment failures, or environmental events.
Failure Modes and Effects Analysis (FMEA) Henley & Kumamoto, 1981, Vesely et al., 1981; Zalosh, Beller, & Till, 1996	The analyst determines what errors might occur during the execution of a task and their likely consequences for the system.
Barrier Analysis Neogy et al., 1996; Trost & Nertney, 1985	This approach aims to identify hazards that could lead to accidents. For each hazard, any barriers that could prevent the accident are recorded along with their method of functioning and modes of failure (including human error).

(Continued)

TABLE 3.5
Methods for Studying Error

Hazard and Operability Analysis (HAZOP) Neogy et al., 1996; Whalley, 1988	An interdisciplinary group is assembled to identify potential hazards, possible consequences, and preventive mechanisms at each step in a process.
Management Oversight Risk Tree (MORT) Kirwan & Ainsworth, 1992; Whitaker-Sheppard & Wendel, 1996	A technique used to investigate the adequacy of safety management structures, either to ensure that they exist, or, if an incident has occurred, to determine which safety management functions have failed.
Work Safety Analysis Bell & Swain, 1983	A systematic analysis of a chosen work situation for all possible occupational accidents, plus the measures that may be adopted to reduce or eliminate their likelihood.
Confusion Matrices Kirwan & Ainsworth, 1992	A tabular plot of a set of stimuli (e.g., displays) against a set of responses is constructed. The frequencies of actual responses are recorded, with the diagonal showing the frequency of correct responses. The grid can be used to identify which responses may be made mistakenly for a given stimulus.
Operator Action Event Tree Kirwan & Ainsworth, 1992	A representation of success and failure routes through a sequence of actions necessary to perform a task is constructed. Each stage in the route can be given a failure probability resulting in an overall probability of failure or success for the complete event sequence.
Generic Error Modeling System (GEMS) Alm, 1992; Reason, 1990	An error classification scheme based on Rasmussen's skills, rules, knowledge (SRK) taxonomy, which describes the competencies needed by workers to perform their roles in complex systems. GEMS describes three major categories of errors: skill-based slips and lapses, rule-based mistakes, and knowledge-based mistakes.
Cognitive Reliability and Error Analysis Method (CREAM) Hollnagel, 1998	A comprehensive approach to HRA that includes a method that can both search for the causes of errors and predict performance, plus an error classification scheme that describes person-related, technology-related, and organization-related errors. CREAM also includes an underlying model of operators that describes how actions are chosen based on the result of the interaction between competence and context.

that comprises five stages: work domain analysis, control task analysis, strategies analysis, social organization and cooperation analysis, and worker competencies analysis. The work domain is modeled as a function abstraction hierarchy (FAH), which shows goal–means relationships on different levels of the hierarchy, including functional purpose, abstract function, generalized function, physical function, and physical form (Bisantz et al., 2003; Burns, Bryant, & Chalmers, 2001; Burns & Vicente, 2001; Chalmers, Easter, & Potter, 2000; Chin, Sanderson, & Watson, 1999; Flach, Eggleston, Kuperman, & Dominguez, 1998; Flach & Kuperman, 1998; Martinez, Talcott, Bennett, Stansifer, & Shattuk, 2001; Naikar & Sanderson, 2001; Rasmussen, Pejtersen, & Goodstein, 1994; Rasmussen, Pejtersen, & Schmidt, 1990; Vicente, 1999, 2000).

Applied cognitive work analysis (ACWA) represents the results of knowledge elicitation using a goal–means decomposition. The goal–means decomposition focuses explicitly on the goals to be accomplished in the work domain, the relationships between goals, and the means to achieve goals (including decisions and the information required to make those decisions) (Gualtieri, Elm, Potter, & Roth, 2001; Potter, Elm, Gualtieri, Easter, & Roth, 2002; Potter, Elm, Roth, & Woods, 2001; Potter, Roth, Woods, & Elm, 2000; Roth, Patterson, & Mumaw, 2001).

In cognitive function analysis (CFA), cognitive functions are defined as mappings from tasks to activities. The phases of this approach include design of a set of primitive cognitive functions through the use of participatory design and domain analysis, definition of evaluation criteria to guide the distribution of cognitive functions among agents, and incremental design and assessment of cognitive functions (by designers, users, and usability specialists) to build active design documents during the life cycle of an artifact (Boy, 1997a, 1998a, 1998b, 2000).

The COADE (**co**gnitive **a**nalysis **d**esign and **e**valuation) framework is an approach to the development of cognitively centered systems comprising a set of activities for cognitive analysis, design, and evaluation. In the analyze phase, a behavior model representing the goal structure of the task and a cognitive model representing the knowledge and mental processes applied in tasks are created. In the design phase, the cognitive requirements are translated into system design specifications (Essens, Fallesen, McCann, Cannon-Bowers, & Dorfel, 1995).

The perceptual control theory (PCT) approach to system design claims that all human behavior occurs as a result of a perceptually driven, goal-referenced feedback system. The approach produces a hierarchical goal analysis based on PCT (Chery & Farrell, 1998; Hendy et al., 2002).

System Oriented Methods. These include:

• Information flow analysis: A flow chart of the information and decisions required to carry out the functions of a complex system is constructed (Davis et al., n.d.; Meister, 1989; Randel, Pugh, & Wyman, 1996).

• Functional flow analysis: The system is decomposed into the functions it must support. Function-flow diagrams are constructed to show the sequential or information-flow relationships between system functions. Petri nets may be used as a modeling formalism to implement function-flow diagrams (Kirwan & Ainsworth, 1992; Meister, 1989).

• Function allocation: A set of informal techniques for determining which system functions people should perform and which machines should perform. A human factors technique to conduct function allocation involves the use of a Fitts list, which describes tasks that humans perform well and tasks that machines perform well (Campbell et al., n.d.; Dearden, Harrison, & Wright, 2000; Fitts, 1951; Merisol & Saidane, 2000; Price, 1985).

• Mission and scenario analysis: An approach to designing a system based on what the system has to do (the mission), especially using concrete examples, or scenarios (Carroll, 1995; McDermott, Klein, Thordsen, Ransom, & Paley, 2000; Rosson, Carroll, & Cerra, 2001).

• Signal flow graph analysis: This technique identifies the important variables within a system and enables their relationship to be detailed. An output variable from the system is selected and all the variables that can influence this output are identified (Kirwan & Ainsworth, 1992).

SYSTEM ENGINEERING USES

To establish the columns of our matrix (Table 3.1), which define the potential uses of various methods, we adopt the 12 "phases" of systems engineering that are proposed by Dugger, Parker, and Winters (1999). Below we summarize each phase and highlight the cognitive systems engineering activities that are typically performed in each phase:

1. Concept definition: At the outset of system design, the objective is to identify the system's mission and required capabilities, that is, the reason for the system to exist. Cognitive tasks that are particularly challenging and that may require support systems may also be identified at this stage.

2. Requirements analysis: After the system concept is defined, more detailed system requirements and specifications are then developed. The cognitive engineer develops human performance requirements, including us-

ability and learnability requirements, and identifies human information needs and decision points.

3. Function analysis: Next, the system functions needed to meet the mission requirements are defined. The cognitive engineer ensures that the function analysis includes all aspects relevant to inclusion of people in the system, especially the human functions that are needed to allow the system to function.

4. Function allocation: At this stage, design decisions are made to effectively distribute functions between humans and systems. The cognitive engineer conducts performance and workload studies to determine optimal allocations, possibly through the use of simulation.

5. Task design: The goal of this stage is to analyze how people would (and should) carry out the functions that have been assigned to them. The cognitive engineer identifies task characteristics (interactions and sequences) and possible strategies that people may employ.

6. Interface and team development: Once the roles and tasks of people (with respect to the system) have been determined, general concepts and specific designs for interfaces between these people, their system(s), and other people/systems are developed.

7. Performance, workload, and training estimation: Given a proposed system design, the physical and cognitive workloads of individuals and teams are assessed. Small-scale or full-scale simulations may be particularly useful at this stage.

8. Requirements review: Throughout the development process, the system design is reviewed with respect to its requirements (i.e., operational needs). The role of the cognitive engineer is to evaluate the system with respect to its impact on human performance, including usability, learnability, and decision making.

9. Personnel selection: The goal of this phase is to establish the required human competencies.

10. Training development: The goal of this phase is to impart and assess knowledge and skills.

11. Performance assurance: Once the system has been deployed, the goal of this phase is to ensure that it continues to function as intended. Capabilities and deficiencies of the operational system are examined and may lead to new system requirements. Cognitive systems engineering activities focus on how well the system and people work together.

12. Problem investigation: This phase refers to any accidents and incidents that occur after the system is deployed. The focus is on modifying the system itself and/or human training or procedures to prevent problem recurrence.

MAPPING METHODS TO USES

For each phase of the systems engineering process (columns of Table 3.1), we assessed the extent to which the various methods of cognitive systems engineering (rows in Table 3.1) could help. Though our assessment is both subjective and based on experience, we considered both the actual match between method and uses (as documented in available references), as well as the potential matches that are suggested by the distinguishing features of each method (as described earlier in our summaries). The results of our assessment appear as filled cells in a "methods matrix" (Table 3.1), where gray fill represents "some match" (between row and column) and black fill represents a "good match." Next we provide an example to illustrate the rationale we applied in making our assessments. This example considers the various uses (columns) for the method (row) of cognitive work analysis (CWA).

The applicability of CWA begins during concept definition, where the work domain analysis phase of CWA can help answer questions about why a new system should exist (based on an analysis of the high-level goals of the work domain). As the system moves into the requirements analysis and function analysis stages, the function abstraction hierarchy (FAH) of CWA can show the lower level functions and physical systems, such as databases and sensors, that are needed to support the higher level goals. Moreover, the FAH can provide additional information by showing how multiple systems support the same higher level goals (Bisantz et. al., 2003).

Results from CWA's control task analysis (of what needs to be done to achieve the functions of the domain) and strategies analysis (of specific ways that the control tasks may be carried out) can inform the task design and human interface and team development phases of the system engineering process. These analyses can also provide insight into the information needs of workers, and they can suggest potential designs for visual displays. In addition, the models generated from these analyses can provide measures of the complexity of specific tasks (and the likelihood of errors in these tasks), thus informing the performance, workload, and training estimation phase of systems engineering.

The worker competencies analysis phase of CWA identifies the requisite skills, rules, and knowledge to perform the work of the system, and can therefore inform the personnel selection and training development phases. The social organization and cooperation analysis of CWA addresses function allocation and team organization issues. Throughout the system design process, the FAH of CWA can be used to evaluate how well the proposed system design satisfies the goals and constraints of the work domain, thus providing insight into requirements review (Naikar, Lintern, & Sanderson, 2002). And even after the operational system is deployed, the

CWA's FAH can be refined and evolved to determine whether the system continues to meet the goals and constraints of the work domain during the performance assurance and problem investigation phases.

This example shows that the CWA framework addresses a broad range of systems engineering challenges, which is why all the columns of the matrix (Table 3.1) are filled in black (meaning "good match") for this method (row). Other methods were assessed in a similar manner, with some having widespread applicability to various problems (many black columns) and some having more narrow applicability (fewer black columns).

For a systems engineer with a specific problem (column), the matrix can be read the other way around to provide a guide to the methods. That is, reading down a column, the cells filled black denote those cognitive systems engineering methods (rows) that are best suited to addressing the system engineering problem (column). Given the candidate methods (black or gray) for addressing the problem, as indicated by the methods matrix (Table 3.1), the systems engineer can then use the method summaries (see Cognitive Systems Engineering Methods section) along with their more detailed references to compare the relative merits of each method. As such, we believe that our matrix provides a useful tool for matching methods to uses, and we are currently using the matrix for this purpose in systems engineering efforts at MITRE's Command and Control Center.

CONCLUSION

We have presented a survey of cognitive systems engineering methods and system engineering uses. The survey is summarized in the form of a matrix of methods (rows) versus uses (columns). Each method is categorized, summarized, and referenced with specific citations. The matrix and supporting text provide a practical road map for applying cognitive systems engineering methods to systems engineering uses. As such, this work can help researchers and system developers, and helps put cognitive systems engineering *in the context* of systems engineering activities, where such expertise may not otherwise reside.

REFERENCES

Alm, I. (1992). *Cognitive tasks: A meta analysis of methods* (FOA Rep. No. C 500985.5). Sundbyberg, Sweden: National Defense Research Establishment.

Anderson, J. R., & Lebiere, C. (1998). *The atomic components of thought.* Mahwah, NJ: Lawrence Erlbaum Associates.

Army Research Laboratory. (n.d.). *IMPRINT home page.* Retrieved May 24, 2004, from http://www.arl.army.mil/ARL-Directorates/HRED/imb/imprint/imprint.htm

Bardram, J. E. (1997). Plans as situated action: An activity theory approach to workflow systems. In J. A. Hughes, W. Prinz, T. Rodden, & K. Schmidt (Eds.), *Proceedings of the 5th Euro-*

pean Conference on Computer Supported Cooperative Work (pp. 17–32). Lancaster, England: Kluwer Academic Publishers.

Barnard, P. J., & May, J. (1993). Cognitive modeling for user requirements. In P. F. Byerley, P. J. Barnard, & J. May (Eds.), *Computers, communication and usability: Design issues, research, and methods for integrated services* (pp. 101–145). Amsterdam: Elsevier.

Barnard, P. J., & May, J. (2000). Toward a theory-based form of cognitive task analysis of broad scope and applicability. In J. M. Schraagen, S. F. Chipman, & V. L. Shalin (Eds.), *Cognitive task analysis* (pp. 147–164). Mahwah, NJ: Lawrence Erlbaum Associates.

Baumeister, L. K., John, B. E., & Byrne, M. D. (2000). A comparison of tools for building GOMS models. In *Proceedings of the SIGCHI Conference on Human Factors in Computing Systems* (pp. 502–509). New York: ACM Press.

Belkin, N. J., Brooks, H. M., & Daniels, P. J. (1987). Knowledge elicitation using discourse analysis. *International Journal of Man–Machine Studies, 27,* 127–144.

Bell, B. J., & Swain, A. D. (1983). *A procedure for conducting a human reliability analysis for nuclear power plants* (Rep. No. NUREG/CR-2254). Washington, DC: U.S. Nuclear Regulatory Commission.

Bentley, R., Hughes, J. A., Randall, D., Rodden, T., Sawyer, P., Shapiro, D., et al. (1992). Ethnographically-informed systems design for air traffic control. In *Proceedings of the ACM conference on computer supported cooperative work* (pp. 123–129). New York: ACM Press.

Beyer, H., & Holtzblatt, K. (2002). *Contextual design: A customer-centered approach to systems design.* New York: Morgan Kaufmann.

Bhatia, S. K., & Yao, Q. (1993). A new approach to knowledge acquisition by repertory grids. In *CIKM 93: Proceedings of the 2nd International Conference on Information and Knowledge Management* (pp. 738–740). New York: ACM Press.

Bisantz, A. M., Roth E., Brickman, B., Gosbee, L. L., Hettinger, L., & McKinney, J. (2003). Integrating cognitive analyses in a large-scale system design process. *International Journal of Human–Computer Studies, 58(2),* 177–206.

Black, J. B., Dimarki, E., van Esselstyn, D., & Flanagan, R. (1995). Using a knowledge representation approach to cognitive task analysis. In *Proceedings of the 1995 Annual National Convention of the Association for Educational Communications and Technology (AECT)* (ERIC Document Reproduction Service No. ED383287). Washington, DC: AECT.

Bolstad, C. A., & Endsley, M. R (1998). *Information dissonance, shared mental models, and shared displays: An empirical evaluation of information dominance techniques* (Tech. Rep. No. AFRL-HE-WP-TR-1999-0213). Wright-Patterson Air Force Base, OH: U.S. Air Force Research Laboratory, Human Effectiveness Directorate.

Boose, J. H. (1986). *Expertise transfer for expert system design.* Amsterdam: Elsevier.

Boose, J. H. (1990). Uses of repertory grid-centered knowledge acquisition tools for knowledge-bases systems. In J. H. Boose & B. R. Gaines (Eds.), *The foundations of knowledge acquisition* (pp. 61–84). San Diego, CA: Academic Press.

Boy, G. A. (1997a). Active design documents. In S. Coles (Ed.), *Proceedings of the Conference on Designing Interactive Systems: Processes, Practices, Methods, and Techniques* (pp. 31–36). New York: ACM Press.

Boy, G. A. (1997b). The group elicitation method for participatory design and usability testing. *Interactions, 4(2),* 27–33.

Boy, G. A. (1998a). *Cognitive function analysis.* Stamford, CT: Ablex.

Boy, G. A. (1998b). Cognitive function analysis for human-centered automation of safety-critical systems. In *Proceedings of the SIGCHI Conference on Human Factors in Computing Systems* (pp. 265–272). New York: ACM Press

Boy, G. A. (2000). Active design documents as software agents that mediate participatory design and traceability. In J. M. Schraagen, S. F. Chipman, & V. L. Shalin (Eds.), *Cognitive task analysis* (pp. 291–301). Mahwah, NJ: Lawrence Erlbaum Associates.

Bradshaw, J. M., & Boose, J. H. (1990). Decision analysis techniques for knowledge acquisition: combining information and preferences using Acquinas and Axotl. *International Journal of Man–Machine Studies, 32,* 121–186.

Bradshaw, J. M., Ford, K. M., Adams-Webber, J. R., & Boose, J. H. (1993). Beyond the repertory grid: new approaches to constructivist knowledge acquisition tools development. *International Journal of Intelligent Systems, 8,* 287–333.

Breuker, J. A., & Wielinga, B. J. (1987). Use of models in the interpretation of verbal data. In A. Kidd (Ed.), *Knowledge acquisition for expert systems* (pp. 17–43). New York: Plenum.

Brett, B. E., Doyal, J. A., Malek, D. A., Martin, E. A., & Hoagland, D. G. (2000). *The combat automation requirements testbed (CART) task 1 final report: Implementation concepts and an example* (Rep. No. AFRL-HE-WP-TR-2000-0009). Wright-Patterson Air Force Base, OH: U.S. Air Force Research Laboratory, Human Effectiveness Directorate.

Brett, B. E., Doyal, J. A., Malek, D. A., Martin, E. A., Hoagland, D. G., & Anesgart, M. N. (2002). The combat automation requirements testbed (CART) task 5 interim report: Modeling a strike fighter pilot conducting a time critical target mission (Rep. No. AFRL-HE-WP-TR-2002-0018). Wright-Patterson Air Force Base, OH: U.S. Air Force Research Laboratory, Human Effectiveness Directorate.

Brykczynski, B. (1999). A survey of software inspection checklists. *Software Engineering Notes, 24*(1), 82–90.

Burns, C. M., Bryant, D. J., & Chalmers, B. A. (2001). Scenario mapping with work domain analysis. In *Proceedings of the Human Factors and Ergonomics Society 45th Annual Meeting* (pp. 424–429). Santa Monica, CA: Human Factors and Ergonomics Society.

Burns, C. M., & Vicente, K. J. (2001). Model-based approaches for analyzing cognitive work: A comparison of abstraction hierarchy, multilevel flow modeling, and decision modeling. *International Journal of Cognitive Ergonomics, 5*(3), 357–366.

Byrne, M. D. (2001). ACT-R/PM and menu selection: Applying a cognitive architecture to HCI. *International Journal of Human–Computer Studies, 55,* 41–84.

Byrne, M. D. (2002). Cognitive architecture. In J. A. Jacko & A. Sears (Eds.), *The human–computer interaction handbook: Fundamentals, evolving technologies and emerging applications* (pp. 97–117). Mahwah, NJ: Lawrence Erlbaum Associates.

Campbell, G. E., & Cannon-Bowers, J. A. (n.d.). *The application of human performance models in the design and operation of complex systems.* Retrieved May 24, 2004, from the Office of Naval Research, SC-21 Science and Technology Manning Affordability Initiative Web site at http://www.manningaffordability.com/S&tweb/Index_main.htm

Campbell, G. E., Cannon-Bowers, J., Glenn, F., Zachary, W., Laughery, R., & Klein, G. (n.d.). *Dynamic function allocation in the SC-21 manning initiative program.* Retrieved May 24, 2004, from the Office of Naval Research, SC-21 Science and Technology Manning Affordability Initiative Web site at http://www.manningaffordability.com/s&tweb/PUBS/Dfapaper.htm

Card, S. K., Moran, T. P., & Newell, A. (1983). *The psychology of human–computer interaction.* Mahwah, NJ: Lawrence Erlbaum Associates.

Carnot, M. J., Dunn, B., & Canas, A. J. (2001). *Concept maps vs. web pages for information searching and browsing.* Retrieved May 23, 2004, from the Institute for Human and Machine Cognition, CMap Tools Web site at http://www.ihmc.us/users/acanas/Publications/CMapsVSWebPagesExp1/CMapsVSWebPagesExp1.htm

Carretta, T. R., Doyal, J. A., & Craig, K. A. (2001). *Combat automation requirements testbed (CART): An example application* (Rep. No. AFRL-HE-WP-TR-2001-0151). Dayton, OH: U.S. Air Force Research Laboratory, Human Effectiveness Directorate.

Carroll, J. M. (Ed.). (1995). *Scenario-based design: Envisioning work and technology in system development.* New York: Wiley.

Chalmers, B. A., Easter, J. R., & Potter, S. S. (2000). Decision-centred visualizations for tactical decision support on a modern frigate. In *Proceedings of the 2000 Command and Control Research and Technology Symposium* (CD-ROM), Monterey, CA.

Chery, S., & Farrell, P. S. E. (1998). *A look at behaviourism and perceptual control theory in interface design* (Rep. No. DCIEM-98-R-12). Ontario, Canada; Defence and Civil Institute of Environmental Medicine.

Chin, M., Sanderson, P., & Watson, M. (1999). Cognitive work analysis of the command and control work domain. *Proceedings of the 1999 Command and Control Research and Technology Symposium* (CD-ROM), Monterey, CA.

Chrenk, J., Hutton, R. J. B., Klinger, D. W., & Anastasi, D. (2001). The cognimeter: Focusing cognitive task analysis in the cognitive function model. In *Proceedings of the Human Factors and Ergonomics Society 45th Annual Meeting* (pp. 1738–1742). Santa Monica, CA: Human Factors and Ergonomics Society.

Chu, R. W., Mitchell, C. M., & Jones, P. M. (1995). Using the operator function model/OFMspert as the basis for an intelligent tutoring system: Towards a tutor-aid paradigm for operators for supervisory control systems. *IEEE Transactions on Systems, Man, and Cybernetics, 25*(7), 1054–1075.

Clarke, B. (1987). Knowledge acquisition for real-time knowledge-based systems. In *Proceedings of the First European Workshop on Knowledge Acquisition for Knowledge Based Systems* (pp. C2.1–C2.7). Reading, England: Reading University.

Coffey, J. W., Hoffman, R. R., Canas, A. J., & Ford, K. M. (2002). A concept map-based knowledge modeling approach to expert knowledge sharing. In M. Bourmedine (Ed.), *Proceedings of the IASTED International Conference on Information and Knowledge Sharing* (pp. 212–217). Calgary, Canada: ACTA Press.

Cohen, M. S., Freeman, J. T., & Thompson, B. T. (1998). Critical thinking skills in tactical decision making: A model and a training method. In J. Canon-Bowers & E. Salas (Eds.), *Decision-making under stress: Implications for training and simulation* (pp. 155–189). Washington, DC: American Psychological Association.

Cohen, M. S., Thompson, B. B., Adelman, L., Bresnick, T. A., Tolcott, M. A., & Freeman, J. T. (1995). *Rapid capturing of battlefield mental models* (Tech. Rep. 95-3). Arlington, VA: Cognitive Technologies.

Conati, C., & Lehman, J. F. (1993). EFH-Soar: Modeling education in highly interactive microworlds. In P. Torasso (Ed.), *Proceedings of the 3rd Congress for the Italian Association for Artificial Intelligence: Advances in Artificial Intelligence* (pp. 47–48). London: Springer-Verlag.

Conrad, F., Blair, J., & Tracy, E. (1999). Verbal reports are data! A theoretical approach to cognitive interviews. In *Proceedings of the Federal Committee on Statistical Methodology Research Conference* (pp. 11–20). Washington, DC: U.S. Bureau of Labor Statistics Office of Survey Methods Research.

Cooke, N. J. (1992). Predicting judgment time from measures of psychological proximity. *Journal of Experimental Psychology: Learning, Memory, and Cognition, 18*, 640–653.

Cooke, N. J. (1994). Varieties of knowledge elicitation techniques. *International Journal of Human–Computer Studies, 41*(6), 801–849.

Cooke, N. J., Shope, S. M., & Kiekel, P. A. (2001). *Shared-knowledge and team performance: A cognitive systems engineering approach to measurement* (Final Tech. Rep. No. AFRL-SR-BL-TR-01-0370). Arlington, VA: U.S. Air Force Office of Scientific Research.

Cooke, N. J., Durso, F. T., & Schvaneveldt, R. W. (1986). Recall and measures of memory organization. *Journal of Experimental Psychology: Learning, Memory, and Cognition, 12*, 538–549.

Cooke, N. J., & McDonald, J. E. (1987). The application of psychological scaling techniques to knowledge elicitation for knowledge-based systems. *International Journal of Man–Machine Studies, 26*, 533–550.

Cooper, R., & Fox, J. (1997). Learning to make decisions under uncertainty: The contribution of qualitative reasoning. In R. W. Scholz & A. C. Zimmer (Eds.), *Qualitative aspects of decision making* (pp. 83–106). Lengerich, Germany: Pabst Science.

Cooper, R., Yule, P., Fox, J. & Sutton, D. (1998). COGENT: An environment for the development of cognitive models. In U. Schmid, J. F. Krems, & F. Wysotzki (Eds.), *A cognitive science approach to reasoning, learning and discovery* (pp. 55–82). Lengerich, Germany: Pabst Science.

Cordingley, E. S. (1989). Knowledge elicitation techniques for knowledge-based systems. In D. Diaper (Ed.), *Knowledge elicitation: Principles, techniques, and applications* (pp. 89–175). New York: Wiley.

Corker, K., Pisanich, G., & Bunzo, M. (1997). A cognitive system model for human/automation dynamics in airspace management. In *Proceedings of the First European/US Symposium on Air Traffic Management,* Saclay, France.

Corter, J. E., & Tversky, A. (1986). Extended similarity trees. *Psychometrika, 51,* 429–451.

Crandall, B., Klein, G., & Hoffman, R. (2005). *A practitioner's handbook of cognitive task analysis.* Cambridge MA: MIT Press.

Darkow, D. J., & Marshak, W. P. (1998). *Low-level cognitive modeling of aircrew function using the SOAR artificial intelligence architecture* (Tech. Rep. No. AFRL-HE-WP-TR-1998-0056). Dayton, OH: U.S. Air Force Research Laboratory, Human Effectiveness Directorate, Crew System Interface Division.

Davis, J. G., Subrahmanian, E., Konda, S., Granger, H., Collins, M., & Westerberg, A. W. (n.d.). *Creating shared information spaces to support collaborative design work.* Retrieved May 24, 2004, from the Carnegie Mellon University, n-dim group Web site at http://www.ndim .edrc.cmu.edu/papers/creating.pdf

Dearden, A., Harrison, M., & Wright, P. (2000). Allocation of function: Scenarios, context, and the economics of effort. *International Journal of Human Computer Studies, 52*(2), 289–318.

Degani, A., Mitchell, C. R., & Chappell, A. R. (1995). Task models to guide analysis: Use of the operator function model to represent mode transitions. In R. S. Jensen (Ed.), *Proceedings of the 8th International Aviation Psychology Symposium Conference* (pp. 210–215). Columbus: Ohio State University.

Deutsch, S. E., Cramer, N., Keith, G., & Freeman, B. (1999). *The distributed operator model architecture* (Tech. Rep. No. AFRL-HE-WP-TR-1999-0023). Wright-Patterson Air Force Base, OH: U.S. Air Force Research Laboratory, Human Effectiveness Directorate, Deployment and Sustainment Division.

Deutsch, S., & Pew, R. (2002). Modeling human error in a real-world teamwork environment. In W. D. Gray & C. D. Schunn (Eds.), *Proceedings of the 24th Annual Meeting of the Cognitive Science Society* (pp. 274–279). Mahwah, NJ: Lawrence Erlbaum Associates.

Diaper, D. (Ed.). (1989). *Knowledge elicitation: Principles, techniques, and applications.* New York: Wiley.

Diederich, J., Ruhmann, I., & May, M. (1987). KRITON: A knowledge acquisition tool for expert systems. *International Journal of Man–Machine Studies, 26,* 29–40.

DuBois, D. A., & Shalin, V. L. (2000). Describing job expertise using cognitively oriented task analysis (COTA). In J. M. Schraagen, S. F. Chipman, & V. L. Shalin (Eds.), *Cognitive task analysis* (pp. 41–56). Mahwah, NJ: Lawrence Erlbaum Associates.

DuBois, D. A., Shalin, V. L., & Borman, W. C. (1995). *A cognitively-oriented approach to task analysis and test development.* Minneapolis, MN: Personnel Decision Research Institutes, Inc.

Dugger, M., Parker, C., & Winters, J. (1999). *Interactions between systems engineering and human engineering.* Retrieved May 24, 2004 from the Office of Naval Research, SC-21 Science and Technology Manning Affordability Initiative Web site at http://www.manningaffordability .com/s&tweb/PUBS/SE_HE/SE_HE_Inter.htm

Dugger, M., Parker, C., Winters, J., & Lackie, J. (1999). *Systems engineering task analysis: Operational sequence diagrams (OSDs).* Retrieved May 24, 2004, from the Office of Naval Research

SC-21 Science and Technology Manning Affordability Initiative Web site at http://www.manningaffordability.com/S&tweb/Index_main.htm

Endsley, M. R. (1999). Situation awareness and human error: Designing to support human performance. Paper presented at the 1999 High Consequences Systems Surety Conference. Retrieved May 24, 2004, from http://www.satechnologies.com/papers/pdf/sandia99-safety.pdf

Endsley, M. R., Bolte, B., & Jones, D. G. (2003). *Designing for situation awareness: An approach to human-centered design.* New York: Taylor & Francis.

Endsley, M. R., Farley, T. C., Jones, W. M., & Midkiff, A. H. (1998). *Situation awareness information requirements for commercial airline pilots* (Rep. No. ICAT-98-1). Cambridge, MA: MIT, Department of Aeronautics and Astronautics, International Center for Air Transportation.

Endsley, M. R., & Garland, D. J. (Eds.). (2000). *Situation awareness analysis and measurement.* Mahwah, NJ: Lawrence Erlbaum Associates.

Endsley, M., & Hoffman, R. R. (2002). The Sacagawea principle. *IEEE Intelligent Systems, 17*(6), 80–85.

Endsley, M. R. & Jones, W. M. (1997). *Situation awareness information dominance and information warfare* (Tech. Rep. No. AL/CF-TR-1997-0156). Wright-Patterson Air Force Base, OH: U.S. Air Force Armstrong Laboratory.

Endsley, M. R., & Robertson, M. M. (2000). Training for situation awareness. In M. R. Endsley & D. J. Garland (Eds.), *Situation awareness analysis and measurement* (pp. 349–365). Mahwah, NJ: Lawrence Erlbaum Associates.

Endsley, M. R. & Rodgers, M. D. (1994). *Situation awareness information requirements for en route air traffic control* (Rep. No. DOT/FAA/AM-94/27). Washington, DC: Federal Aviation Administration, Office of Aviation Medicine.

Entin, E., Serfaty, D., Elliot, L., & Schiflett, S. G. (2001). DMT-Rnet: An Internet-based infrastructure for distributed multidisciplinary investigations of C2 performance. Retrieved May 24, 2004, from http://www.dodccrp.org/events/2001/6th_ICCRTS/Cd/Tracks/Papers/Track4/22_Tr4.pdf

Ericsson, K. A., & Simon, H. A. (1993). *Protocol analysis: Verbal reports as data* (Rev. ed.). Cambridge, MA: MIT Press.

Essens, P., Fallesen, J., McCann, C., Cannon-Bowers, J., & Dorfel, G. (1995). *COADE: A framework for cognitive Analysis, design, and evaluation* (Final Rep. No. AC/243 [Panel 8] TR/17). Brussels, Belgium: NATO Defence Research Group.

Fahey, R. P., Rowe, A. L., Dunlap, K. L., & deBoom, D. O. (2001). *Synthetic task design: Cognitive task analysis of AWACS weapons director teams* (Tech. Rep. No. AFRL-HE-AZ-TR-2000-0159). Mesa, AZ: U.S. Air Force Research Laboratory, Human Effectiveness Directorate, Warfighter Training Research Division.

Fischhoff, B. (1989). Eliciting knowledge for analytical representation. *IEEE Transactions on Systems, Man, and Cybernetics, 19*(3), 448–461.

Fisher, C., & Sanderson, P. (1996). Exploratory sequential data analysis: exploring continuous observational data. *Interactions, 3*(2), 25–34.

Fitts, P. M. (Ed.). (1951). *Human engineering for an effective air navigation and traffic control system.* Washington, DC: National Research Council.

Flach, J. M., Eggleston, R. G., Kuperman, G. G., & Dominguez, C. O. (1998). *SEAD and the UCAV: A preliminary cognitive systems analysis* (Rep. No. AFRL-HE-WP-TR-1998-0013). Wright-Patterson AFB, OH: U.S. Air Force Research Laboratory, Human Effectiveness Directorate, Crew System Interface Division.

Flach, J., & Kuperman, G. G. (1998) *Victory by design: War, information, and cognitive systems engineering* (Rep. No. AFRL-HE-WP-TR-1998-0074). Wright-Patterson AFB, OH: U.S. Air Force Research Laboratory, Human Effectiveness Directorate, Crew System Interface Division.

Flanagan, J. C. (1954). The critical incident technique. *Psychological Bulletin, 51*(4), 327–358.

Freed, M. A., & Remington, R. W. (2000). Making human–machine system simulation a practical engineering tool: An Apex overview. In N. Taatgen & J. Aasman (Eds.), *Proceedings of the 2000 International Conference on Cognitive Modeling* (pp. 110–117). Veenendaal, Netherlands: Universal Press.

Freed, M. A., Shaflo, M. G., & Remington, R. W. (2000). Employing simulation to evaluate designs: The Apex approach. In N. Taatgen & J. Aasman (Eds.), *Proceedings of the 2000 International Conference on Cognitive Modeling* (pp. 110–117). Veenendaal, Netherlands: Universal Press.

Freeman, J. T., & Cohen, M. S. (1998). *A critical decision analysis of aspects of naval anti-air warfare* (Tech. Rep. No. 98-2). Arlington, VA: Cognitive Technologies.

Gaines, B. R., & Shaw, M. L. G. (1993). Knowledge acquisition tools based on personal construct theory. *Knowledge Engineering Review, 8*(1), 49–85.

Geiwitz, J., Klatzky, R., & McCloskey, B. (1988). *Knowledge acquisition techniques for expert systems: Conceptual and empirical comparisons.* Santa Barbara, CA: Anacapa Sciences.

Geiwitz, J., Kornell, J., & McCloskey, B. P. (1990). *An expert system for the selection of knowledge acquisition techniques* (Tech. Rep. No. 785-2). Santa Barbara, CA: Anacapa Sciences.

Geiwitz, J., & Schwartz, D. R. (1987) . *Delco soldier–machine interface: Demonstration and evaluation* (Tech. Rep. No. 714-3). Santa Barbara, CA: Anacapa Sciences.

Goldstein, W. M., & Hogarth, R. M. (Eds.). (1997). *Research on judgment and decision making: Currents, connections, and controversies.* Cambridge, England: Cambridge University Press.

Gordon, S. E., & Gill, R. T. (1997). Cognitive task analysis. In C. E. Zsambok & G. Klein (Eds.), *Naturalistic decision making* (pp. 131–140). Mahwah, NJ: Lawrence Erlbaum Associates.

Gordon, S. E., Schmierer, K. A., & Gill, R. T. (1993). Conceptual graph analysis: Knowledge acquisition for instructional system design. *Human Factors, 35*(3), 459–481.

Gordon, T. R., Coovert, M. D., Riddle, D. L., & Miles, D. E. (2001). Classifying C2 decision making jobs using cognitive task analyses and verbal protocol analysis. Retrieved May 24, 2004, from http://www.dodccrp.org/events/2001/6th_ICCRTS/Cd/Tracks/Papers/Track4/096_tr4.pdf

Gott, S. P. (1998). *Rediscovering learning: Acquiring expertise in real world problem solving tasks* (Rep. No. AL/HR-TP-1997-0009). Brooks Air Force Base, TX: Armstrong Research Laboratory.

Grabowski, M., Massey, A. P., & Wallace, W. A. (1992). Focus groups as a group knowledge elicitation technique. *Knowledge Acquisition, 4,* 407–425.

Grassi, C. R. (2000). *A task analysis of pier side ship-handling for virtual environment ship-handling simulator scenario development.* Unpublished master's thesis, Naval Postgraduate School, Monterey, CA. (Defense Technical Information Center Scientific and Technical Information Network Service 2001128 089)

Gray, W. D. (2002). Simulated task environments: the Role of high-fidelity simulations, scaled worlds, synthetic environments, and laboratory tasks in basic and applied cognitive research. *Cognitive Science Quarterly, 2,* 205–227.

Gray, W. D., & Altmann, E. M. (1999). Cognitive modeling and human-computer interaction. In W. Karwowski (Ed.), *International encyclopedia of ergonomics and human factors* (Vol. 1, pp. 387–391). New York: Taylor & Francis.

Gray, W. D., John, B. E., & Atwood, M. E. (1992). The precis of Project Ernestine or an overview of a validation of GOMS. In *Proceedings of the SIGCHI Conference on Human Factors in Computing Systems* (pp. 307–312). New York: ACM Press.

Gray, W. D., Schoelles, M. J., & Fu, W. (2000). Modeling a continuous dynamic task. In N. Taatgen & J. Aasman (Eds.), *Proceedings of the Third International Conference on Cognitive Modeling* (pp. 158–168). Veenendal, Netherlands: Universal Press.

Grover, M. D. (1983). A pragmatic knowledge acquisition methodology. In A. Bundy (Ed.), *Proceedings of the 8th International Joint Conference on Artificial Intelligence* (pp. 436–438). Los Altos, CA: Kaufmann.

Gualtieri, J. W., Elm, W. C., Potter, S. S., & Roth, E. (2001). Analysis with a purpose: Narrowing the gap with a pragmatic approach. In *Proceedings of the Human Factors and Ergonomics Society 45th Annual Meeting* (pp. 444–448). Santa Monica, CA: Human Factors and Ergonomics Society.

Gustafson, D., Shulka, R., Delbecq, A., & Walister, G. (1975). A comparative study of the differences in subjective likelihood estimates made by individuals, interacting groups, Delphi groups, and nominal groups. *Organizational Behavior and Human Performance, 9,* 280–291.

Hall, E. P., Gott, S. P., & Pokorny, R. A. (1995). *A procedural guide to cognitive task analysis: The PARI methodology* (Tech. Rep. No. 1995-0108). Brooks Air Force Base, TX: Armstrong Research Laboratory.

Halskov-Madsen, K., & Aiken, P. H. (1993). Experiences using cooperative interactive storyboard prototyping. *Communications of the ACM, 36*(4), 57–67.

Hamilton, F. (1996). Predictive evaluation using task knowledge structures. In M. J. Tauber (Ed.), *Conference companion on human factors in computing systems: Common ground* (pp. 261–262). New York: ACM Press.

Hanson, M. L., Harper, K. A., Endsley, M., & Rezsonya, L. (2002). Developing cognitively congruent HBR models via SAMPLE: A case study in airline operations modeling. In *Proceedings of the 11th Computer Generated Forces (CGF) Conference* (pp. 49–57). Orlando, FL: Simulation Interoperability Standards Organization.

Hardinge, N. (2000). *Cognitive task analysis of decision strategies of submarine sonar and target motion analysis operators: Phase 1* (Tech. Rep. Contract No. N00014-99-1044). Office of Naval Research.

Hart, A. (1986). *Knowledge acquisition for expert systems.* London: Kogan Page.

Haunold, P., & Kuhn, W. (1994). A keystroke level analysis of manual map digitizing. In *Proceedings of the SIGCHI Conference on Human Factors in Computing Systems: Celebrating Interdependence* (pp. 337–343). New York: ACM Press.

Hendy, K. C., Beevis, D., Lichacz, F., & Edwards, J. L. (2002). Analyzing the cognitive system from a perceptual control theory point of view. In M. D. McNeese & M. A. Vidulich (Eds.), *Cognitive systems engineering in military aviation environments: Avoiding cogminutia fragmentosa!* (pp. 201–252). Wright-Patterson Air Force Base, OH: Human Systems Information Analysis Center.

Hendy, K. C., Edwards, J. L., & Beevis, D. (1999). *Analysing advanced concepts for operations room layouts.* Retrieved May 24, 2004, from the Office of Naval Research SC-21 Science and Technology Manning Affordability Initiative Web site at http://www.manningaffordability .com/s&tweb/PUBS/AdvConcOperRoomLayout/AdvConcOperRoomLayout.htm

Henley, J., & Kumamoto, H. (1981). *Reliability engineering and risk assessment.* New York: Prentice-Hall.

Hess, S. M., MacMillan, J., Serfaty, D., & Elliot, L. (1999). From cognitive task analysis to simulation: Developing a synthetic team task for AWACS weapons directors. In *Proceedings of the 1999 Command and Control Research and Technology Symposium.* Retrieved May 24, 2006, from http://www.dodccrp.org/events/1999/1999ccrts/pdf-files/track_3/078hess.pdf

Hickman, F. R., Killin, J. L., Land, L., & Mulhall, T. (1990). *Analysis for knowledge-based systems: A practical guide to the kads methodology.* New York: Ellis Horwood.

Hoffman, R. R., Crandall, B., & Shadbolt, N. (1998). Use of the critical decision method to elicit expert knowledge: A case study in the methodology of cognitive task analysis. *Human Factors, 40*(2), 254–276.

Hoffman, R. R., Shadbolt, N. R., Burton, A. M., & Klein, G. (1995). Eliciting knowledge from experts: A methodological analysis. *Organizational Behavioral and Human Decision Processes, 62*(2), 129–158.

Hollan, J., Hutchins, E., & Kirsh, D. (2000). Distributed cognition: Toward a new foundation for human-computer interaction research. *ACM Transactions on Computer Human Interaction, 7*(2), 174–196.

Hollnagel, E. (1998). *Cognitive reliability and error analysis method.* New York: Elsevier Science.

Hourizi, R., & Johnson, P. (2001). Unmasking mode errors: A new application of task knowledge principles to the knowledge gaps in cockpit design. In M. Hirose (Ed.), *Proceedings of Human Computer Interaction: Interaction 2001* (pp. 255–262). Amsterdam: IOS Press.

Howard, R. A., & Matheson, J. E. (1984). Influence diagrams. In R. A. Howard and J. E. Matheson (Eds.), *The principles and applications of decision analysis* (pp. 721–762). Menlo Park, CA: Strategic Decisions Group.

Hughes, J., King, V., Rodden, T., & Andersen, H. (1994). *Moving out from the control room: Ethnography in system design* (Rep. No. CSCW/9/1994). Lancaster, England: Lancaster University, Center for Research in Computer Supported Cooperative Work.

Hughes, J., O'Brien, J., Rodden, T., Rouncefield, M., & Sommerville, I. (1995). Presenting ethnography in the requirements process. In *Proceedings of the Second IEEE International Symposium on Requirements Engineering* (pp. 27–34). Washington, DC: IEEE Computer Society.

Hutchins, E. (1995). *Cognition in the wild.* Cambridge, MA: MIT Press.

Institute for Human and Machine Cognition. (n.d.). *Cmap tools: Concept mapping software toolkit.* Retrieved May 24, 2004, from http://cmap.ihmc.us

Jacobsen, N. E., & John, B. E. (2000). *Two case studies in using cognitive walkthrough for interface evaluation* (Rep. No. CMU-CHII-00-100). Pittsburgh, PA: Carnegie Mellon University, Human–Computer Interaction Institute.

John, B. E., & Kieras, D. E. (1994). *The GOMS family of analysis techniques: Tools for design and evaluation* (Tech. Rep. No. CMU-HCII-94-106). Pittsburgh, PA: Carnegie Mellon University, Human–Computer Interaction Institute.

John, B. E., & Kieras, D. E. (1996). Using GOMS for user interface design and evaluation: Which technique? *ACM Transactions on Computer–Human Interaction, 3*(4), 287–319.

John, B. E., & Packer, H. (1995). Learning and using the cognitive walkthrough method: A case study approach. In *Proceedings of the SIGCHI Conference on Human Factors in Computing Systems* (pp. 429–436). New York: ACM Press.

John, B., Vera, A., Matessa, M., Freed, M., & Remington, R. (2002). Controlling complexity: Automating CPM-GOMS. *Proceeding of the SIGCHI Conference on Human Factors in Computing Systems: Changing Our Word, Changing Ourselves* (pp.147–154). New York: ACM Press.

Johnson, H., & Johnson, P. (1991). Task knowledge structures: psychological basis and integration into system design. *Acta Psychologica, 78,* 3–26.

Johnson L., & Johnson, N. (1987). Knowledge elicitation involving teachback interviewing. In A. Kidd (Ed.), *Knowledge elicitation for expert systems: A practical handbook* (pp. 91–108). New York: Plenum.

Johnson, P., Johnson, H., & Hamilton, F. (2000). Getting the knowledge into HCI: Theoretical and practical aspects of task knowledge structures. In J. M. Schraagen, S. F. Chipman, & V. L. Shalin (Eds.), *Cognitive task analysis* (pp. 201–214). Mahwah, NJ: Lawrence Erlbaum Associates.

Keeney, R. L. (1982). *Decision analysis: State of the field* (Tech. Rep. No. 82-2). San Francisco: Woodward Clyde Consultants.

Keep, J., & Johnson, H. (1997). HCI and requirements engineering: Generating requirements in a courier despatch management system. *SIGCH Bulletin, 29*(1).

Kelley, T., & Allender, L. (1996). *A process approach to usability testing for IMPRINT* (Rep. No. ARL-TR-1171). Aberdeen Proving Ground, MD: Army Research Laboratory.

Kieras, D. E. (1994). GOMS modeling of user interfaces using NGOMSL. In C. Plaisant (Ed.), *Conference companion on human factors in computing systems* (pp. 371–372). New York: ACM Press.

Kieras, D. E. (1997a). A guide to GOMS model usability evaluation using NGOMSL. In M. G. Helander, T. K. Landauer, & P. V. Prabhu (Eds.), *The handbook of human–computer interaction* (2nd ed., pp. 733–766). New York: North-Holland.

Kieras, D. E. (1997). Task analysis and the design of functionality. In A. Tucker (Ed.), *The computer science and engineering handbook* (pp. 1401–1423). Boca Raton, FL: CRC Press.

Kieras, D. E. (1998). *A guide to GOMS model usability evaluation using GOMSL and GLEAN3* (Tech Rep. No. 38, TR-98/ARPA-2). Ann Arbor: University of Michigan, Electrical Engineering and Computer Science Department.

Kieras, D. E., & Meyer, D. E. (1998). *The EPIC architecture: Principles of operation.* Retrieved May 24, 2004, from University of Michigan, Electrical Engineering and Computer Science Department Web site at ftp://www.eecs.umich.edu/people/kieras/EPIC/EPICArch.pdf

Kieras, D. E., & Meyer, D. E. (2000). The role of cognitive task analysis in the application of predictive models of human performance. In J. M. Schraagen, S. F. Chipman, & V. L. Shalin (Eds.), *Cognitive task analysis* (pp. 237–260). Mahwah, NJ: Lawrence Erlbaum Associates.

Kieras, D. E., Wood, S. D., Abotel, K., & Hornof, A. (1995). GLEAN: A computer-based tool for rapid GOMS model usability evaluation of user interface designs. In *Proceedings of the 8th Annual ACM Symposium on User Interface and Software Technology* (pp. 91–100). New York: ACM Press.

Kieras, D. E., Wood, S. D., & Meyer, D. E. (1995). Predictive engineering models based on the EPIC architecture for a multimodal high-performance human–computer interaction task. In *ACM transactions on computer–human interaction* (Vol. 4, pp. 230–275). New York: ACM Press.

Kirwan, B., & Ainsworth, L. K. (1992). *A guide to task analysis.* New York: Taylor & Francis.

Klein, G. (1996). *The development of knowledge elicitation methods for capturing military expertise* (ARI Research Note No. 96-14). Alexandria, VA: U.S. Army Research Institute for the Behavioral and Social Sciences.

Klein, G. (1998). *Sources of power: How people make decisions.* Cambridge, MA: MIT Press.

Klein, G., Calderwood, R., & MacGregor, D. (1989). Critical decision method for eliciting knowledge. *IEEE Transactions on Systems, Man, and Cybernetics, 19*(3), 462–472.

Klein, G., Orasanu, J., & Calderwood, R. (1993). *Decision making in action: Models and methods.* Norwood, NJ: Ablex.

Leddo, J., & Cohen, M. S. (1988). *Objective structure analysis: A method of eliciting expert knowledge.* Alexandria, VA: Army Research Institute.

Leddo, J. M., Mullin, T. M., Cohen, M. S., Bresnick, T. A., Marvin, F. F., & O'Connor, M. F. (1988). *Knowledge elicitation: Phase I final report* (Vol. 1, Tech. Rep. No. 87-15). Alexandria, VA: Army Research Institute.

Lees, C., Manton, J., & Triggs, T. (1999). *Protocol analysis as a tool in function and task analysis* (Rep. No. DSTO-TR-0883). Salisbury, Australia: Defence Science and Technology Organisation, Electronics and Surveillance Research Laboratory.

Lewis, C., & Rieman, J. (1994). *Task-centered user interface design: A practical introduction.* Retrieved May 24, 2004, from http://hcibib.org/tcuid/

Lewis, S. (1991). Cluster analysis as a technique to guide interface design. *International Journal of Man–Machine Studies, 35,* 251–265.

Mahemoff, M. J., & Johnston, L. J. (1998). Principles for a usability-oriented pattern language. In P. Calder & B. Thomas (Eds.), *Proceedings of the Australian Conference on Computer Human Interaction* (pp. 132–139). Washington, DC: IEEE Computer Society.

Marsh, C. (1999). The F-16 maintenance skills tutor. *The Edge: MITRE's Advanced Technology Newsletter, 3*(1). Retrieved May 11, 2004, from http://www.mitre.org/news/the_edge/march_99/second.html

Martinez, S. G., Talcott, C., Bennett, K. B., Stansifer, C., & Shattuck, L. (2001). Cognitive systems engineering analyses for army tactical operations. In *Proceedings of the Human Factors and Ergonomics Society 45th Annual Meeting* (pp. 523–526). Santa Monica, CA: Human Factors and Ergonomics Society.

Maulsby, D., Greenberg, S., & Mander, R. (1993). Prototyping an intelligent agent through Wizard of Oz. In *Proceedings of the SIGCHI Conference on Human Factors in Computing Systems* (pp. 277–284). New York: ACM Press.

Mayhew, D. J. (1999). *The usability engineering lifecycle: A practitioner's handbook for user interface design.* San Diego, CA: Academic Press.

McDermott, P., Klein, G., Thordsen, M., Ransom, S., & Paley, M. (2000). *Representing the cognitive demands of new systems: A decision-centered design approach* (Rep. No. AFRL-HE-WP-TR-2000-0023). Wright-Patterson Air Force Base, OH: U.S. Air Force Research Laboratory, Human Effectiveness Directorate.

McGraw, K., & Harbison-Briggs, K. (1989). *Knowledge acquisition: Principles and guidelines.* Englewood Cliffs, NJ: Prentice-Hall.

Medical Research Council, Cognition and Brain Sciences Unit. (n.d.). *ICS home page.* Retrieved May 24, 2004, from http://www.mrc-cbu.cam.ac.uk/personal/phil.barnard/ics/

Meister, D. (1989). *Conceptual aspects of human factors.* Baltimore: The Johns Hopkins University Press.

Merisol, M., & Saidane, A. (2000). A tool to support function allocation. In C. Hohnson, P. Palanque, & F. Paterno (Eds.). Paper presented at the 2000 *International Workshop on Safety and Usability Concerns in Aviation.* Retrieved May 24, 2004, from http://www.irit.fr/recherches/liihs/wssuca2000/suca-Merisol.pdf

Merkelbach, E. J. H. M., & Schraagen, J. M. C. (1994). *A framework for the analysis of cognitive tasks* (Rep. No. TNO-TM 1994 B-13). Soesterberg, Netherlands: TNO Human Factors Research Institute.

Militello, L. G., & Hutton, R. J. B. (1998). Applied cognitive task analysis (ACTA): A practitioner's toolkit for understanding cognitive task demands. *Ergonomics, 41*(11), 1618–1641.

Militello, L. G., Hutton, R. J. B., Pliske, R. M., Knight, B. J., & Klein, G. (1997). *Applied cognitive task analysis (ACTA) methodology* (Rep. No. NPRDCTN-98-4). San Diego, CA: Navy Personnel Research and Development Center.

Millen, D. R. (2000). Rapid ethnography: Time deepening strategies for HCI field research. *Proceeding of the Conference on Designing Interactive Systems: Processes, Practices, Methods, and Techniques* (pp. 280–286). New York: ACM Press.

Miller, J. E., Patterson, E. S., & Woods, D. D. (2001). Critiquing as a cognitive task analysis (CTA) methodology. In *Proceedings of the Human Factors and Ergonomics Society 45th Annual Meeting* (pp. 518–522). Santa Monica, CA: Human Factors and Ergonomics Society.

Miller, T. E., Copeland, R. R., Phillips, J. K., & McCloskey, M. J. (1999). *A cognitive approach to developing planning tools to support air campaign planners* (Tech. Rep. No. AFRL-IF-RS-TR-1999-146). Rome, NY: Air Force Research Laboratory, Information Directorate.

Mitchell, C. M. (1999). Model-based design of human interaction with complex systems. In A. P. Sage & W. B. Rouse (Eds.), *Systems engineering & management handbook* (pp. 745–810). New York: Wiley.

Mitchell, D. K. (2000). *Mental workload and ARL workload modeling tools* (Rep. No ARL-TN-161). Aberdeen Proving Ground, MD: Army Research Laboratory.

Naikar, N., Lintern, G., & Sanderson, P. (2002). Cognitive work analysis for air defense applications in Australia. In M. D. McNeese & M. A. Vidulich (Eds.), *Cognitive systems engineering in military aviation environments: Avoiding cogminutia fragmentosa!* (pp. 169–200). Wright-Patterson Air Force Base, OH: Human Systems Information Analysis Center.

Naikar, N., & Sanderson, P. M. (2001). Evaluating design proposals for complex systems with work domain analysis. *Human Factors, 43*(4), 529–42.

Nardi, B. A. (Ed.). (1996). *Context and consciousness: Activity theory and human–computer interaction.* Cambridge, MA: MIT Press.

NASA Ames Research Center. (n.d.-a). *Apex home page.* Retrieved May 24, 2004, from http://www.andrew.cmu.edu/~bj07/apex/index.html

NASA Ames Research Center. (n.d.-b). *MIDAS home page.* Retrieved May 24, 2004, from http://caffeine.arc.nasa.gov/midas/index.html

Neogy, P., Hanson, A. L., Davis, P. R., & Fenstermacher, T. E. (1996). *Hazard and barrier analysis guidance document* (Rep. No. EH-33). Washington, DC: U.S. Department of Energy, Office of Operating Experience Analysis and Feedback.

Nielsen, J. (1993). *Usability engineering.* San Diego: Academic Press.

Nielsen, J., & Mack, R. L. (1994). *Usability inspection methods.* Boston: Wiley.

Nielsen, J., & Phillips, V. L. (1993). Estimating the relative usability of two interfaces: Heuristic, formal, and empirical methods compared. In *Proceedings of the SIGCHI Conference on Human Factors in Computing Systems* (pp. 214–221). New York: ACM Press.

Olson, J. R., & Biolsi, K. J. (1991). Techniques for representing expert knowledge. In K. A. Ericsson & J. Smith (Eds.), *Toward a general theory of expertise* (pp. 240–285). Cambridge, England: Cambridge University Press.

Olson, J. R., & Reuter, H. H. (1987). Extracting expertise from experts: Methods for knowledge acquisition. *Expert Systems, 4,* 152–168.

Patterson, E. S., Woods, D. D., Sarter, N. B., & Watts-Perotti, J. (1999). Patterns in cooperative cognition. In *Proceeding of the Human Factors and Ergonomics Society 43rd Annual Meeting* (pp. 263–266). Santa Monica, CA: Human Factors and Ergonomics Society.

Peterson, B., Stine, J. L., & Darken, R. P. (2000). A process and representation for modeling expert navigators. In *Proceedings of the 9th Conference on Computer Generated Forces and Behavioral Representations* (pp. 459–470). Orlando, FL: Simulation Interoperability Standards Organization.

Pew, R. W., & Mavor, A. S. (Eds.). (1998). *Modeling human and organizational behavior: Applications to military simulations.* Washington, DC: National Academy Press.

Phillips, J., McDermott, P. L., Thordsen, M., McCloskey, M., & Klein, G. (1998). *Cognitive requirements for small unit leaders in military operations in urban terrain* (Research Rep. No. 1728). Arlington, VA: U.S. Army Research Institute for the Behavioral and Social Sciences.

Polk, T. A., & Seifert, C. M. (2002). *Cognitive modeling.* Cambridge, MA: MIT Press.

Polson, P., Rieman, J., & Wharton, C. (1992). *Usability inspection methods: Rationale and examples* (Rep. No. 92-0). Denver: University of Colorado, Institute of Cognitive Science.

Potter, S. S., Elm, W. C., Gualtieri, J. W., Easter, J. R., & Roth, E. M. (2002). Using intermediate design artifacts to bridge the gap between cognitive analysis and cognitive systems engineering. In M. D. McNeese & M. A. Vidulich (Eds.), *Cognitive systems engineering in military aviation environments: Avoiding cogminutia fragmentosa!* (pp. 137–252). Wright-Patterson Air Force Base, OH: Human Systems Information Analysis Center.

Potter, S. S., Elm, W. C., Roth, E. M., & Woods, D. D. (2001). *The development of a computer-aided cognitive systems engineering tool to facilitate the design of advanced decision support systems* (Tech. Rep. No. AFRL-HE-WP-TR-2001-0125). Wright-Patterson Air Force Base, OH: U.S. Air Force Research Laboratory, Human Effectiveness Directorate, Crew System Interface Division.

Potter, S. S., Roth, E. M., Woods, D. D., & Elm, W. C. (2000). Bootstrapping multiple converging cognitive task analysis techniques for system design. In J. M. Schraagen, S. F. Chipman, & V. L. Shalin (Eds.), *Cognitive task analysis* (pp. 317–340). Mahwah, NJ: Lawrence Erlbaum Associates.

Price, H. E. (1985). The allocation of functions in systems. *Human Factors, 27*(1), 33–45.

Randel, J. M., Pugh, L. H., & Wyman, B. G. (1996). *Methods for conducting cognitive task analysis for a decision making task* (Rep. No. TN-96-10). San Diego, CA: Navy Personnel Research and Development Center.

Rasmussen, J., Pejtersen, A. M., & Goodstein, L. P. (1994). *Cognitive system engineering.* Boston: Wiley.

Rasmussen, J., Pejtersen, A. M., & Schmidt, K. (1990). *Taxonomy for cognitive work analysis* (Rep. No. Risø-M-2871). Roskilde, Denmark: Risø National Laboratory.

Raven, M. E., & Flanders, A. (1996). Using contextual inquiry to learn about your audiences. *ACM SIGDOC Asterisk Journal of Computer Documentation, 20*(1), 1–13.

Reason, J. (1990). *Human error.* Cambridge, England: Cambridge University Press.

Redding, R. E. (1992). *A standard procedure for conducting cognitive task analysis.* Mclean, VA: Human Technology. (ERIC Document Reproduction Service No. ED 340 847)

Redding, R. E., Cannon, J. R., Lierman, B. C., Ryder, J. M., Purcell, J. A., & Seamster, T. L. (1991). The analysis of expert performance in the redesign of the en route air traffic control curriculum. In *Proceedings of the Human Factors and Ergonomics Society 35th Annual Meeting* (pp. 1403–1407). Santa Monica, CA: Human Factors and Ergonomics Society.

Reichert, M., & Dadam, P. (1997). A framework for dynamic changes in workflow management systems. In *Proceedings of the 8th International Workshop on Database and Expert Systems Applications* (pp. 42–48). New York: ACM Press.

Rogers, Y. (1997). *A brief introduction to distributed cognition.* Retrieved May, 24, 2004, from the University of Sussex, School of Cognitive and Computing Sciences Web site at http://www.cogs.susx.ac.uk/users/yvonner/papers/dcog/dcog-brief-intro.pdf

Rogers, Y., & Ellis, J. (1994). Distributed cognition: an alternative framework for analysing and explaining collaborative working. *Journal of Information Technology, 9*(2), 119–128.

Rosson, M. B., Carroll, J. M., & Cerra, D. D. (Eds.). (2001). *Usability engineering: Scenario based development of human computer interaction.* New York: Morgan Kaufmann.

Roth, E. M., Malsh, N., & Multer, J. (2001). *Understanding how train dispatchers manage and control Trains: Results of a cognitive task analysis* (Rep. No. DOT/FRA/ORD-01/02). Washington, DC: Federal Railroad Administration, Office of Research and Development.

Roth, E., & O'Hara, J. (2002). *Integrating digital and conventional human-system interfaces: Lessons learned from a control room modernization program* (Rep. No. NUREG/CR-6749). Washington, DC: U.S. Nuclear Regulatory Commission, Office of Nuclear Regulatory Research.

Roth, E. M., & Patterson, E. S. (2005). Using observational study as a tool for discovery: Uncovering cognitive and collaborative demands and adaptive strategies. In H. Montgomery, R. Lipshitz, & B. Brehmer (Eds.), *How professionals make decisions* (pp. 379–392). Mahwah, NJ: Lawrence Erlbaum Associates.

Roth, E. M., Patterson, E. S., & Mumaw, R. J. (2001). Cognitive Systems Engineering: Issues in user-centered system design. In J. J. Marciniak (Ed.), *Encyclopedia of software engineering* (2nd ed., pp. 163–179). New York: Wiley.

Rubin, K. S., Jones, P. M., & Mitchell, C. M. (1988). OFMspert: Inference of operator intentions in supervisory control using a blackboard architecture. *IEEE Transactions on Systems, Man, and Cybernetics, 18*(4), 618–637.

Ryder, J., Weiland, M., Szczepkowski, M., & Zachary, W. (1998). Cognitive systems engineering of a new telephone operator workstation using COGNET. *International Journal of Industrial Ergonomics, 22*(6), 417–429.

Salas, E., & Klein, G. (Eds.). (2001). *Linking expertise and naturalistic decision making.* Mahwah, NJ: Lawrence Erlbaum Associates.

Salvucci, D. D. (2001). Modeling driver distraction from cognitive tasks. In W. D. Gray & C. D. Schunn (Eds.), *Proceedings of the 24th Annual Conference of the Cognitive Science Society* (pp. 792–797). Mahwah, NJ: Lawrence Erlbaum Associates.

Sanderson, P. M., & Fisher, C. (1994). Exploratory sequential data analysis: foundations. *Human–Computer Interaction, 9*(3), 251–317.

Sanderson, P. M., Scott, J. J., Johnston, T., Mainzer, J., Watanabe, L. M. , & James, J. M. (1994). MacSHAPA and the enterprise of exploratory sequential data analysis (ESDA). *International Journal of Human–Computer Studies, 41*(5), 633–681.

Santoro, T. P., Kieras, D. E., & Campbell, G. E. (2000). GOMS modeling application to watchstation design using the GLEAN tool. In *Proceedings of Interservice/Industry Training, Simulation, and Education Conference* (CD-ROM). Arlington, VA: National Training Systems Association.

Schenkman, B. N. (2002). Perceived similarities and preferences for consumer electronic products. *Personal and Ubiquitous Computing, 6*(2).

Schollen, M. J., & Gray, W. D. (2000). Argus prime: Modeling emergent microstrategies in a complex, simulated task environment. In N. Taatgen & J. Aasman (Eds.), *Proceedings of the Third International Conference on Cognitive Modeling* (pp. 260–270). Veenendal, Netherlands: Universal Press.

Schraagen, J. M., Chipman, S. F., & Shalin, V. L. (Eds.). (2000). *Cognitive task analysis.* Mahwah, NJ: Lawrence Erlbaum Associates.

Schreiber, G., Wielinga, B., & Breuker, J. (Eds.). (1993). *KADS: A principled approach to knowledge-based system development.* London: Academic Press.

Schvaneveldt, R. W., Durso, F. T., Goldsmith, T. E., Breen, T. J., Cooke, N. M., Tucker, R. G., et al. (1985). Measuring the structure of expertise. *International Journal of Man–Machine Studies, 23,* 699–728.

Schweigert, W. A. (1994). *Research methods and statistics for psychology.* Pacific Grove, CA: Brooks/Cole.

Scott, A. A., Clayton, J. E. & Gibson, E. L. (1991). *A practical guide to knowledge acquisition.* Reading, MA: Addison-Wesley.

Seamster, T. L., Redding, R. E., & Kaempf, G. L. (1997). *Applied cognitive task analysis in aviation.* London: Ashgate.

Seamster, T. L., Redding, R. E., & Kaempf, G. L. (2000). A skill-based cognitive task analysis framework. In J. M. Schraagen, S. F. Chipman, & V. L. Shalin (Eds.), *Cognitive task analysis* (pp. 135–146). Mahwah, NJ: Lawrence Erlbaum Associates.

See, J. E., & Vidulich, M. A. (1997a). *Computer modeling of operator mental workload during target acquisition: An assessment of predictive validity* (Rep. No. AL/CF-TR-1997-0018). Wright-Patterson Air Force Base, OH: U.S. Air Force Armstrong Laboratory.

See, J. E., & Vidulich, M. A. (1997b). *Operator workload in the F-15E: A comparison of TAWL and Micro Saint computer simulation* (Rep. No. AL/CF-TR-1997-0017). Wright-Patterson Air Force Base, OH: U.S. Air Force Armstrong Laboratory.

Shadbolt, N., & Burton, M. (1990). Knowledge elicitation. In J. R. Wilson & E. N. Corlett (Eds.), *Evaluation of human work: A practical ergonomics methodology* (pp. 321–345). London: Taylor & Francis.

Shepherd, A. (2000). *Hierarchical task analysis.* New York: Taylor & Francis.

Shuler, D., & Namioka, A. (Eds.). (1993). *Participatory design: Principles and practices.* Hillsdale, NJ: Lawrence Erlbaum Associates.

Shute, V., Sugrue, B., & Willis, R. E. (1997, March). *Automating cognitive task analysis.* Paper presented at the Cognitive Technologies for Knowledge Assessment Symposium, Chicago.

Shute, V. J., & Torreano, L. A. (2002). Formative evaluation of an automated knowledge elicitation and organization tool. In T. Murray, S. Blessing, & S. Ainsworth (Eds.), *Authoring tools for advanced technology learning environments: Toward cost-effective adaptive, interactive, and intelligent educational software* (pp. 149–180). Norwell, MA: Kluwer Academic Publishers.

Shute, V. J., Torreano, L. A., & Willis, R. E. (2000). DNA: Providing the blueprint for instruction. In J. M. Schraagen, S. F. Chipman, & V. L. Shalin (Eds.), *Cognitive task analysis* (pp. 71–86). Mahwah, NJ: Lawrence Erlbaum Associates.

Sinclair, M. A. (1990). Subjective assessment. In J. R. Wilson & E. N. Corlett (Eds.), *Evaluation of human work: A practical ergonomics methodology* (pp. 58–88). London: Taylor & Francis.

Smith, B. R. & Tyler, S. W. (1997, March). *The design and application of MIDAS: A constructive simulation for human-system analysis.* Paper presented at the Second Simulation Technology and Training Conference, Canberra, Australia.

Stammers, R. B. (1986). Psychological aspects of simulator design and use. *Advances in Nuclear Science and Technology, 17,* 117–139.

Stammers, R. B., Carey, M. S., & Astley, J. A. (1990). Task analysis. In J. R. Wilson & E. N. Corlett (Eds.), *Evaluation of human work: A practical ergonomics methodology* (pp. 134–160). New York: Taylor & Francis.

Thompson, M., & Wishbow, N. (1992). Improving software and documentation quality through rapid prototyping. In *Proceedings of the 10th Annual International Conference on Systems Documentation* (pp. 191–199). New York: ACM Press.

Thordsen, M. (1991). A comparison of two tools for cognitive task analysis: Concept mapping and the critical decision method. In *Proceedings of the Human Factors and Ergonomics Society 35th Annual Meeting* (pp. 283–285). Santa Monica, CA: Human Factors and Ergonomics Society.

Tijerinia, L., Kiger, S., Wierwille, W., Rockwell, T., Kantowitz, B., Bittner, A., et al. (1996). *Heavy vehicle driver workload assessment task 1: Task analysis data and protocols review* (Rep. No. DOT HS 808 467). Washington, DC: U.S. Department of Transportation, National Highway Traffic Safety Administration.

Trost, W. A., & Nertney, R. J. (1985). *Barrier analysis* (Rep. No. US DOE-76-45/29, SSDC-29). Washington, DC: U.S. Department of Energy.

Vesely, W. E., Goldberg, F. F., Roberts, N. H., & Haasl, D. F. (1981). *Fault tree handbook* (Rep. No. NUREG-0492). Washington, DC: U.S. Nuclear Regulatory Commission, Systems and Reliability Research Office.

Vicente, K. J. (1999). *Cognitive work analysis: Toward safe, productive, and healthy computer-based work.* Mahwah, NJ: Lawrence Erlbaum Associates.

Vicente, K. J. (2000). HCI in the global knowledge-based economy: Designing to support worker adaptation. *ACM Transactions on Computer–Human Interaction, 7*(2), 263–280.

Voigt, J. R. (2000). *Evaluating a computerized aid for conducting a cognitive task analysis.* Unpublished master of science thesis, University of Central Florida, Orlando. (The Defense Technical Information Center [DTIC] Scientific and Technical Information Network [STINET] Service 20000412 035).

Weiland, M., Campbell, G., Zachary, W., & Cannon-Bowers, J. A. (1998). Applications of cognitive models in a Combat Information Center. In *Proceedings of the 1998 Command and Control Research and Technology Symposium* (pp. 770–782). Washington, DC: DoD Command and Control Research Program.

Weiss, S., & Kulikowski, C. (1984). *A practical guide to designing expert systems.* Totowa, NJ: Rowan & Allenheld.

Welbank, M. (1990). An overview of knowledge acquisition methods. *Interacting with Computers, 2,* 83–91.

Whalley, S. P. (1988). Minimising the cause of human error. In G. P. Liberton (Ed.), *10th advances in reliability technology symposium* (pp. 114–128). London: Elsevier.

Wharton, C., Bradford, J., Jeffries, R., & Franzke, M. (1992). Applying cognitive walkthroughs to more complex user interfaces: Experiences, issues, and recommendations. In *Proceedings of the SIGCHI Conference on Human Factors in Computing Systems* (pp. 381–388). New York: ACM Press.

Whitaker-Sheppard, D., & Wendel, T. (1996). *Analysis of the causes of chemical spills from marine transportation or related facilities* (Rep. No. CG-D-08-96). Groton, CT: U.S. Coast Guard Research and Development Center.

Williams, K. E. (2000). An automated aid for modeling human–computer interaction. In J. M. Schraagen, S. F. Chipman, & V. L. Shalin (Eds.), *Cognitive task analysis* (pp. 165–180). Mahwah, NJ: Lawrence Erlbaum Associates.

Wilson, B., Jonassen, D., & Cole, P. (1993). Cognitive approaches to instructional design. In G. M. Piskurich (Ed.), *The ASTD handbook of instructional technology* (pp. 21.1–21.22). New York: McGraw-Hill.

Wilson, B. G., & Madsen-Myers, K. (1999). Situated cognition in theoretical and practical context. In D. Jonassen & S. Lands (Eds.), *Theoretical foundations of learning environments* (pp. 57–88). Mahwah, NJ: Lawrence Erlbaum Associates.

Wilson, M. D., Barnard, P. J., Green, T. R. G., & MacLean, A. (1987). Knowledge-based task analysis for human-computer systems. In G. C. van der Veer, T. R. G. Green, J. M. Hoc, & D.

M. Murray (Eds.), *Working with computers: Theory versus outcome* (pp. 47–88). London: Academic Press.

Wood, L. E. (1997). Semi-structured interviewing for user-centered design. *Interactions, 4*(2), 48–61.

Wood, L. E., & Ford, J. M. (1993). Structuring interviews with experts during knowledge elicitation. *International Journal of Intelligent Systems: Special Issue on Knowledge Acquisition, 8*, 71–90.

Woods, D. D. (1993). Process-tracing methods for the study of cognition outside of the experimental psychology laboratory. In G. Klein, J. Orasanu, R. Calderwood, & C. E. Zsambok (Eds.), *Decision making in action: Models and methods* (pp. 228–251). Norwood, NJ: Ablex.

Wright, G., & Bolger, F. (Eds.). (1992). *Expertise and decision support.* New York: Plenum.

Wright, P., Fields, B., & Harrison, M. (2000). Analysing human-computer interaction as distributed cognition: the resources model. *Human–Computer Interaction, 15*(1), 1–42.

Zacharias, G., Illgen, C., Asdigha, M., Yara, J., & Hudlicka, E. (1996). *VIEW: Visualization and interactive elicitation workstation: A tool for representing the commander's mental model of the battlefield* (Rep. No. A854223). Cambridge, MA: Charles River Analytics.

Zacharias, G. L., Miao, A. X., Illgen, C., & Yara, J. M. (1995). *SAMPLE: Situation awareness model for pilot-in-the-loop evaluation* (Rep. No. R95192). Cambridge, MA: Charles River Analytics.

Zachary, W. W., Ryder, J. M., & Hicinbothom, J. H. (1998). Cognitive task analysis and modeling of decision making in complex environments. In J. A. Cannon-Bowers & E. Salas (Eds.), *Decision making under stress: Implications for training and simulation* (pp. 315–344). Washington, DC: American Psychological Association.

Zachary, W. W., Ryder, J. M., & Hicinbothom, J. H. (2000). Building cognitive task analyses and models of a decision-making team in a complex real-time environment. In J. M. Schraagen, S. F. Chipman, & V. L. Shalin (Eds.), *Cognitive task analysis* (pp. 365–384). Mahwah, NJ: Lawrence Erlbaum Associates.

Zachary, W. W., Ryder, J. M., & Purcell, J. A. (1990). *A computer based tool to support mental modeling for human–computer interface design* (Rep. No. 900831-8908). Fort Washington, PA: CHI Systems.

Zalosh, R., Beller, D., & Till, R. (1996). *Comparative analysis of the reliability of carbon dioxide fire suppression systems as required by 46 CFR, SOLAS II-2, and NFPA 12.* (Rep. No. CG-D-29-96). Groton, CT: U.S. Coast Guard Research and Development Center.

Zsambok, C. E., & Klein, G. (Eds.). (1996). *Naturalistic decision making.* Mahwah, NJ: Lawrence Erlbaum Associates.

Part II

MAKING SENSE OF THINGS

Chapter 4

Criminal Investigative Decision Making: Context and Process

Laurence Alison
The Centre for Critical Incident Research, University of Liverpool, England

Emma Barrett
University of Birmingham, England

Jonathan Crego
National Centre for Applied Learning Technologies, England

The Centre for Critical Incident Research (CCIR), located at the School of Psychology, University of Liverpool, UK, focuses on understanding the process of critical-incident decision management in law enforcement, through collaboration among academics and practitioners. Underpinning this work are principles based on *pragmatic* and *naturalistic* approaches, where researchers consider specific problems faced by practitioners (e.g., Cannon-Bowers & Salas, 1998; Fishman, 1999, 2003, 2004).

In this chapter, we describe how we are using *electronic focus groups* as well as *experimental and observational methods* to increase our understanding of critical-incident management in criminal investigations. One goal is to appreciate the cognitive, social and emotional processes underpinning investigative work (for more details, see Alison, 2005, for a review). Another is to achieve a better understanding of the contexts in which such work is carried out, and how this impacts on the investigative process. In particular, we are exploring how detectives perceive and negotiate more complex and unusual investigative problem situations in which even the most experienced officer may find their expertise stretched to its limit.

The electronic focus group research, conducted with senior law enforcement personnel, has enabled us to compare and contrast different types of police investigation and inquiry. These sessions have helped us to describe (through the practitioners' own words) what it is about critical investigative management that makes such incidents so difficult to coordinate.

Preliminary findings indicate that "organizational structure" has a profound impact on incident management and that, as several previous studies have indicated, effective management of these complex inquiries means that officers that have to be skilled in democratic/participative or "transformational" behaviors (Bruns & Shuman, 1988; Dobby, Anscombe, & Tuffin, 2004), aware of productivity *and* socioemotional factors (e.g., Swanson & Territo, 1982) and able to adapt their leadership style to suit the needs of the situation (e.g., Kuykendall & Usinger, 1982). Thus, working within the constraints and complexities of a very particular, highly stressful organizational climate requires a participative, empathic, and adaptive behavioral style, as well as a variety of conceptual, decision-making skills.

Complementing this work are more traditional studies that focus specifically on the processes of decision making within a more tightly controlled research framework. In combination, the case-based focus group explorations and systematic studies of investigative decision making are helping us to identify the landscape of critical investigative decision management, an area that has, hitherto, received relatively little attention.

ELECTRONIC FOCUS GROUPS:
THE 10KV METHODOLOGY

We have developed an *electronic focus group* method, known as "10,000 Volts"[1] (or 10KV) to collect case-based material on critical incidents (defined as, "any incident where the effectiveness of the police response is likely to have a significant impact on the confidence of the victim, the family and/or the community"; Metropolitan Police Service, 2002). Managing these events requires: (a) effective multiagency coordination, (b) appropriate and accurate assessment of complex and ambiguous information, as well as (c) emotional resolve and (d) effective leadership. Our work on over 40 critical-incident debrief enquiries (including a cabinet ministers' working group, the Metropolitan Police Force's response to the 2004 tsunami, contingency planning for the Athens Olympics, hostage negotiations in Iraq, and the 2005 terrorist bombings in London) has repeatedly identified these as core features of such incidents (Crego & Alison, 2004).

10KV is based, in part, on approaches to learning within the aviation industry. The aviation system works on two levels. First, there is a non-attributable reporting system for pilots and other aviation professionals that enables the safe and blameless reporting of occurrences that endan-

---

[1]This term is based on the observation that critical cases cause massive "jolts" to UK policing practice (the experience of managing such cases has also been likened to receiving a 10,000-volt electric shock).

gered air safety and, perhaps more significantly, any incident that had the potential to endanger air safety. The second level is a repository of essential knowledge, a collection of data from air accident investigations including, where possible, actual recordings from the black boxes, news, and other footage that provides a rich context and enhanced meaning to all professionals who access it. Such a concept applied to a law enforcement context has the opportunity to provide professionals with a rich resource of knowledge.

Thus far, each of our sessions has involved up to 30 participants who play (or have played) important roles within the particular inquiry or inquiries under consideration. On arrival at a debriefing center, participants and debrief facilitators discuss and agree on the purpose and expected outcome of each session. The procedure involves seven stages:

1. *Generation of unstructured accounts:* Each participant is given a laptop computer and asked to reflect on their experiences of managing a specific critical incident, and to identify the issues that they believe contributed significantly to the outcome. The laptops are connected together with collaborative software such that as each individual logs this information, it is simultaneously but anonymously distributed to all participants. We have also conducted sessions where participants can comment on statements and add suggestions, questions, and further information to the already logged comments, thereby facilitating more detailed exposition of particular points. Essentially, this stage of 10KV constitutes an electronic focus group.[2]

2. *Theme building:* Participants are split into syndicate groups to review the collective data held on the system and to generate a list of themes that they collectively feel summarize the large corpus of free narrative statements collected during State 1.

3. *Plenary:* The generated themes are discussed and agreed on.

4. *First sort:* The group organizes all the items generated during State 1 into agreed-on themes.

5. *Summation:* These themes are synthesized into key statements.

6. *Prioritization:* Participants are asked to score the issues against different criteria (e.g., "ease of implementation" or "impact on the inquiry").

7. *Feedback and development:* General comments are sought from participants in terms of developments, feedback, and recommendations.

The anonymity afforded by 10KV's use of a nonattributable reporting system eases the main causes of situational communication apprehension

---

[2]In conventional sessions, there are up to 30 participants. However, in a more recent development of an online facility we can coordinate up to 250 participants contributing remotely. This also enables participants to engage internationally.

(CA), namely subordinate status, conspicuousness, degree of attention, and fear of retribution from others (see, e.g., Clapper & Massey, 1996; Easton, Easton, & Belch, 2003). Anonymity has also been found to reduce inhibitions in electronic focus groups similar to that created by 10KV (see, e.g., DeSanctis & Gallupe, 1987). As a result, the 10KV approach has several distinct advantages as a tool for conducting research with law enforcement personnel, among whom issues of rank and the potential repercussions of admitting to errors or uncertainty could otherwise inhibit frank and candid discussion: (a) participants are able to openly express their views without fear of consequences (see, e.g., Clapper & Massey, 1996; Crego & Alison, 2004); (b) controversial, sensitive, or ethical issues are more easily explored (see, e.g., Clapper & Massey, 1996); (c) individual participation rates tend to be higher (see, e.g., Easton et al., 2003; Nunamaker, Briggs, Mittleman, Vogel, & Balthazard, 1996/1997), and so the "quiet voices" in the group are also heard; (d) participation in general also tends to be more evenly distributed (e.g., see Easton et al., 2003; Nunamaker et al., 1996/1997; Parent, Gallupe, Salisbury, & Handelman, 2000), because the capacity for simultaneous participation reduces individual dominance by making it difficult for any participant to prevent others taking part (Montoya-Weiss, Massey, & Clapper, 1998); (e) ideas are "weighted on their merits rather than on their source" (Nunamaker et al., 1996/1997, p. 178), because the effects of interpersonal differences are lessened (see, e.g., Montoya-Weiss et al., 1998).

The 10KV approach has several further advantages, which it shares with traditional focus group methods:

1. A substantial amount of data can be collected in a relatively short space of time (e.g., see Krueger, 1994; Robinson, 1999; Robson, 2002; Sullivan, 2001). For example, in the first 10KV session (run by Crego, November 2002), 28 senior police officers generated more than 250 statements in just over 20 minutes and developed 15 key themes by the end of the session (Crego & Alison, 2004). To date, researchers have rarely been able to capture in-depth views of individuals at this level of seniority within the police service; as their time is extremely precious, they are reluctant to engage with demanding and resource intensive methods such as individual interviews. The efficiency of 10KV makes it a highly appealing alternative.

2. The experience is empowering, stimulating, and dynamic. This interaction can motivate participants (e.g., see Bryman, 2004; Howitt & Cramer, 2005) and can facilitate the discussion of taboo topics and the participation of more inhibited individuals (e.g., see Kitzinger, 1994; Robinson, 1999; Sullivan, 2001).

3. The issues of the greatest consequence and significance to participants can surface (Bryman, 2004); that is, the group dynamic can help create a focus on the most important topics (Robinson, 1999). Furthermore,

the method encourages the presentation of these issues in the participants' terms: "Group work ensures that priority is given to the respondents' hierarchy of importance, their language and concepts, their frameworks for understanding the world" (Kitzinger, 1994, p. 108).

4. Linked to Item 3, in drawing out these key issues, 10KV assists the development of pertinent research questions for deeper exploration. In other words, it is a valuable starting point for further validation studies, which may include, for example, the use of questionnaires, interviews, and simulations to examine in-depth factors identified by 10KV participants as highly significant.

5. Participants can probe each other's reasons for holding certain views, thereby enabling the researcher to develop an understanding about why people feel the way they do (Bryman, 2004).

6. Participants may become aware of, and agree with, perspectives that they may not otherwise have thought of (Bryman, 2004). Thus, as Kitzinger (1994, p. 112) argues "being with other people who share similar experiences encourages participants to express, clarify or even to develop particular perspectives."

7. The material generated facilitates the extraction of general, abstract themes in law enforcement, while preserving the rich detail that embellishes and provides further subtle shades of those themes.

Ultimately, the success of a focus group is "dependent on the creation of a noninhibiting, synergistic environment in which group members feel comfortable sharing ideas" (Montoya-Weiss et al.,1998, p. 714). The anonymity inherent to and the interaction generated by the 10KV method, creates just such an environment. The process is flexible, dynamic, resource friendly, cost-effective, and encourages candid and open views of participants. We have found it particularly suited to research with law enforcement personnel, among whom issues of rank, and the fear of consequences of admitting to error and uncertainty are especially prominent, and time is precious. Weaknesses of the method include its limited transferability, precision, and predictive utility, and the requirement for further validation.

WHAT HAS 10KV REVEALED?

The Impact of the Organization

Our preliminary findings contend that organizational structure is a powerful and relatively neglected situational factor in research on critical-incident management. Multiagency decision making in such incidents has profound consequences and has been the focus of much work, spanning multiple disciplines and out of which a range of normative, descriptive, and prescriptive theories have emerged. In the experience of conducting our debriefs, neither traditional decision theory (TDT) (Bernoulli, 1738; Savage, 1954), Naturalistic Decision Making (NDM) (Klein, Orasanu, Calderwood, & Zsambok, 1993), nor any other theory in the decision-making literature offers an adequate explanation of how critical decisions are influenced by organizational context. A consistent reference made by participants in our sessions is that even where an individual has access to relevant information, and has the capacity to draw accurate inferences from that information, he or she may not commit to a decision because of real or perceived organizational constraints. For example, in shoot-to-kill decisions, an officer may procrastinate or fail to commit to this decision (despite overwhelming risk) for fear of reprisals and complaints as well as potential ensuing media attention and a lengthy public inquiry. Existing theories of decision making fail to depict these relationships and the very profound impact of perceptions of multiagency support and structure.

In 2004, the UK Policing Skills and Standards Organization accepted that in multiagency working there were many issues that needed to be addressed. These included a lack of information sharing, different agencies working to different standards, and a lack of responsibility and accountability of senior managers for the actions of their staff. Many of these issues have been described as "barriers" to successful multiagency working. Decision theory has overlooked the possibility that organizational structures and climates play an important role in decision making. This shortcoming has arisen both from omission of the influence of contextual factors, for example, in TDT, and from an insufficient view of context as the immediate field setting only, such as in NDM. Rather than constituting the context, the immediate field setting is simply the arena in which a range of competing forces are played out. These forces, and thus decision-making processes, can be completely understood only by recognizing that context is an expansive entity, incorporating practitioners, organizations, and wider social and political systems (Anderson, 2003). Differences in organizational structures and climates result in differences in the way practitioners perceive both situations and potential actions. This, in turn, generates conflict in the multiagency settings—a shared mental model cannot be agreed on, and no response option presents that is acceptable to all organizations.

Desirable Leadership Qualities

Our work has revealed that senior officers must manage not only the incident itself, be it the detection of a crime or resolution of a hostage crisis, but also a team operating within a particular culture and requiring knowledge of the perceptions of the local community, the general public, and the media (see Ogan & Alison, 2005; West & Alison, 2005). In particular, the issues that appear to seem the most difficult and also have the greatest impact are ones that may appear largely beyond the control of the critical-incident manager (media, family liaison, politics). Consistent themes have emerged and are summarized in Table 4.1 as the key components of effective leadership in such incidents. They fall under two broad categories: *interpersonal* and *conceptual skills.*

Leadership and management emerged consistently as important issues that impact on most other issues. However, officers highlighted gaps in Police Service knowledge, both in identifying what constitutes leadership and management, and in determining how the Service should train and select leaders. As one officer commented:

> Leadership—we have spoken a lot about leadership but with little explicit acknowledgement of what it actually is and how to bring it about. How do we encourage our current leaders to develop our leadership and how do we develop our future leaders? I think it is more than just training and it also involves mentoring, rewards, and personal development. (participant in murder inquiry debrief, 2004)

A similar confusion is often apparent in the psychological literature. Zaleznik (1977), for example, argued that leadership research had been almost exclusively about studying managers and supervision and not leadership. Yukl (1994) explains that "managers are oriented toward stability and leaders are oriented toward innovation; managers get people to do things more efficiently, whereas leaders get people to agree about what things should be done" (p. 4). In a policing context, both qualities are important, and, we argue, are the central underlying themes of conceptual and interpersonal skills.

Most contemporary definitions agree that leadership cannot be understood independently of followership. Forsyth (1999), for example, views leadership as a form of power *with* people rather than *over* them. Thus, leadership involves a reciprocal relationship between the leader and the led, in which leaders and members work together to increase their joint rewards. The leader increases the motivation and satisfaction of members by uniting them and changing their beliefs, values, and needs. Leadership involves legitimate influence over others to cooperate, rather than power or force. Within this perspective, one who influences others can only be said to be a leader if those whom he or she leads share the same goals and allow that person to lead them.

The extent to which the central member of a group actually influences the other group members defines the level and form of the leadership that

TABLE 4.1
Leadership Attributes, Divided Into Interpersonal or Conceptual Skills
(Definitional Terms Have Been Developed From Delegates' Comments)

Conceptual Skills

Hypothesis generation

1. Turn initial confusion, complexity, volume of information, and ambiguity into a working model of what is going on.
2. Adapt existing models for the interpretation of the current case.
3. Adapt the model in response to incoming demands.
4. Ensure that the team all have an accurate view of the model (i.e., a "shared mental model").

Open-mindedness

1. Weigh up costs and benefits of competing demands (from staff, in terms of information, resources, management levels).
2. Ensure transparency of decision making.
3. Ensure decisions are made in a positive way.

Prioritization of information and decisions.

1. Recognize *restrictions in current state of systems for collection of intelligence.*
2. Ensure volume of intelligence coming in does not outweigh resources for handling that information.
3. Ensure evaluation of risks is transparent and measured.

Interpersonal Skills

Role selection

1. Harness specialists and effectively delegate work.
2. Foster commitment and investment in the team.
3. Recognize and respond to the need for time off and support.
4. Know what support is available and who can give it.

Hardiness

1. Demonstrate stamina, grit, and determination.
2. Be prepared to make hard decisions, commit to them, and revise them in the light of new information.
3. Be a visible and accessible leader.

Humanity

1. Reassure the public of the service's ability to provide a normal response to issues outside of the immediate inquiry.
2. Share information with the community in order to facilitate their help.
3. Prevent paralysis of community through regular updates.
4. Ensure that one stays "in touch" with the community perception of the service and of the inquiry.
5. Positive reinforcement of all staff at all levels and appreciation of the demands on them ("walking in their shoes")

Impression management

1. Balance the need to encourage a positive external perception of the team with the demands of the enquiry.
2. Ensure the service is fairly represented by the media. Do not underestimate the power of the media or its potential impact on the inquiry.
3. Be aware of the impact of the investigation on the local community and responsive to their interpretation of the service.

---

she or he possesses. Gibb (1969) states that the leader "will be 'accepted' by his . . . group if his behavior helps them to achieve their goals" (p. 341), highlighting that individuals in charge need to be credible. This perspective is similar to that of Lord and Maher's implicit leadership theory (1991) in which they define leadership as the process of being perceived by others as a leader. Thus the theory seeks to define the process in terms of followers' readiness to accept and label a person as a leader. Chemers (2000) presents a similar argument in which influence is dependent on credibility. According to Chemers, in order to influence, leaders must, "first establish the legitimacy of their authority by appearing competent" (p. 40). Chemers goes further in suggesting that one of the key features of effective leadership is effective image management.

In the 10KV debrief, many references were made to the external perceptions of those outside of the inquiry:

> This has been a high profile inquiry with immense pressure placed on certain individuals—yet these people still found time to provide positive feedback and support to others members of staff.

> The ability to co-ordinate a professional credible auditable response to the hundreds of telephone calls received from members of the public who were supporting our request for information.

Thus, credibility in police enquiries is crucial both internally, as a quality of the Senior Investigating Office (SIO) and others who lead the investigative team, and also externally, to the investigative team itself (and indeed to the Police Service as a whole) in, as one participant put it, "achieving [a] successful result under media scrutiny."

As well as managing external and internal perceptions of credibility, police leaders must also drive the inquiry forward. This requires the SIO and the investigative team making sense of a mass of incoming information in an inquiry and developing hypotheses about what has occurred (in the case of a reactive investigation) or what might be about to occur (in the case of a proactive inquiry). In the 10KV debriefs, police officers argued that a central feature of inquiries that have been deemed successful is the rapid management of an initial flood of information, so that the interpretation of this information can assist in streamlining lines of inquiry. One officer described the situation in one particular operation thus: "Although fast moving [the] situation op got slicker and better and coordinated rapidly day by day" (participant in murder inquiry debrief, 2004).

Police investigative leaders not only must develop hypotheses and achieve effective situation assessment (which relates not only to the investigation itself, but also to the organizational and public context in which that investigation occurs), but must also ensure that this understanding is shared among the team. Important to note, the team must also be prepared to accept the potentially serious consequences of actions taken on the basis of this model. The need to ensure that team members are fully committed to the inquiry is exemplified in this participant's comment:

> Teamwork and a collective desire to get the job up. Support for colleagues to avoid any sense of struggle. Maintaining a positive and optimistic manner with staff. Promoting a "can-do" culture. Greater collective understanding as an organization of not just responses to a major incident but how the organization such as ours works—as an organization we have grown. (participant in abduction inquiry debrief, 2003)

Of course, inaction can also prove risky, so any decision (or nondecision) in a critical incident involves a level of associated risk. Successful charismatic leaders have been found to be effective in acting as role models, to have a sense of the collective interests of the group and to be self-confident in their decision (Chemers, 2000). However, such confidence rides heavily on the accurate assessment of the situation and on the extent to which all members of the team are working together on the same view of the problem.

Achieving an understanding of how investigating officers leading a criminal inquiry develop situation awareness cannot be tackled easily through the electronic focus group method described previously. We have, therefore, turned to more traditional methods of data gathering and analysis to explore the psychological mechanisms underpinning the understanding and use of information in criminal investigations. Some of this work is outlined in the following section.

EXPLORING HOW DETECTIVES MAKE SENSE OF INVESTIGATIVE INFORMATION

It takes a talented and robust individual to meet the challenges inherent in a complex, chaotic, high-profile serious crime investigation, and it is acknowledged that "there are very few experienced senior investigators within the police service who could easily manage challenges at this level" (Metropolitan Police Service, 2002, p. 14). There is little understanding of the psychological mechanisms underlying criminal investigation, despite a recognition that the ability to investigate crimes is an essential skill of the "effective detective" (Smith & Flanagan, 2000). Achieving an understanding of these mechanisms and the influences on them would be a valuable step in many areas of policing, from developing effective training to designing efficient decision support systems.

Our conception of investigative situation assessment draws heavily on work on recognition-primed decision making (Klein, 1999), problem representation (Mumford, Reiter-Palmon, & Redmond, 1994), and situation awareness (Endsley, 1995). We suggest that effective investigators must be able to recognize the key cues that signal that a decision problem exists and what the nature of this problem might be. In any given scenario, these cues are likely to be ambiguous and may be quite subtle. The ability to recognize them is likely to be heavily dependent on experience. But as well as this recognitional process, detectives must also "read between the lines" to achieve an understanding of the circumstances of a crime and the characteristics of the offender, and use this understanding to develop lines of inquiry, to gather evidence, and, eventually, to apprehend the offender. The construction of mental representations of investigative decision problems is thus viewed as crucial to the investigative decision-making process (Barrett, 2002).

The construction of investigative mental models involves both recognitional and reasoning processes. Recognition allows the investigator to attend selectively to particularly important pieces of information, whereas reasoning processes, which include the use of heuristics such as story generation (Alison & Barrett, 2004; Pennington & Hastie, 1992; Robinson & Hawpe, 1986), and metacognitive reasoning processes such as critiquing and correcting (Cohen, Freeman, & Thompson, 1998), are used to achieve a plausible and coherent understanding of the investigative problem situation.

Some support for this model comes from a recent ethnographic study of homicide investigations in an English police force, in which the author argued that, "the account of the crime that the police assemble, which is the principal output of the investigative process, is a narrative event" (Innes, 2003, p. 682). Innes noted that such narrative elaboration has two important consequences. First, data that do not fit easily into the chosen narrative framework can be omitted from the account. Second, Innes found that detectives tended to generate inferences that resolve apparent inconsistencies in the absence of solid evidence (Innes, 2002, 2003). Though both strategies ensure that the investigators' view remains coherent, a danger is that officers may disregard crucial evidence that would point toward an alternative (and possibly the correct) version of events. Thus, whereas the construction of a narrative mental representation may be a natural and effective heuristic for achieving an understanding of ambiguous perceived data (Robinson & Hawpe, 1986), the generation of a single, narrative explanation for information in the early stages of an investigation may in fact be detrimental to the outcome of the investigation, because it may result in detectives ignoring alternative explanations for investigative data, or misinterpreting information. An important issue, therefore, is how officers reach an early understanding of an investigation, and whether and how they consider alternative explanations (Ormerod, Barrett, & Taylor, 2005).

Our current research priorities are to understand the cues used by experienced investigators when solving a crime, and to explore the way that these cues are combined with existing knowledge to achieve an understanding of the circumstances of the crime and the characteristics of the offender. This is not an issue that lends itself easily to any focus group method, and so we have turned to observational study and more structured experiments, to explore investigative "sense making" (Klein, Phillips, Rall, & Paluso, chap. 6, this volume; Ormerod, Barrett, & Taylor, 2005).

In 2002, we had the opportunity to study a simulated investigation carried out by detective officers in a UK police force (Barrett, 2002). These officers were relatively new to detective work, although each had at least 2 years' police service and were trained in the legal and procedural basics of investigating serious crimes. During the investigation, officers tended to develop theories about the characteristics of the offenders based on the offenders' reported actions, and when trying to make sense of the investigative data tended to construct narrative explanations to account for the events that led up to and followed the robbery. For example, officers drew together information relating to the victims' profession (jewelers) and the fact that the previous day the victims had brought jewelry home from their shop and constructed a causal scenario in which the offenders found out that the victims were bringing the valuables home that evening and planned the robbery accordingly. This led them to infer that the offenders had "inside knowledge," and thus the officers narrowed the parameters of their inquiry to focus on those who knew about the victims' plans to move the valuables. This process of "reconstructing" past events and using these reconstructions, first, to draw inferences about the goals, characteristics, and actions of actors in the crime, and second to direct investigative action, was a striking feature of the officers' deliberations throughout the simulation.

In a second study (Ormerod et al., 2005), we explored some of the issues surrounding inference generation and person perception in more controlled circumstances. UK police officers with a mean 9.5 years in police service, including experience of experience investigating serious crimes ($M =$ 3.6, SD = 3.1 years) were asked to interpret a series of vignettes describing potential crimes and suggest appropriate lines of inquiry in each case. As with our first study, information concerning the behavior of the actors in the vignettes (offenders, victims, and witnesses) was interpreted as diagnostic of actors' characteristics, leading participants to form impressions of the motivations, characteristics, and goals of these actors, and shaping proposed lines of inquiry. Our results raise important questions about decision making by those experienced in investigating crimes: To what extent does an investigator rely on inferences about the characteristics of offenders (or, for that matter, victims, witnesses, and others involved in the crime) generated from their actions to direct his or her inquiry? How accurate are these inferences? Are there particular actions of offenders that are seen as particularly diagnostic of their characteristics, and to what extent does investigator perception match the reality? Further studies are planned to explore in more detail what actions in particular are considered most important, what interpretations might be attached to these key actions, and the effects of investigator experience and expertise on these interpretations.

These results strengthen our conviction that narrative elaboration, the tendency to develop storylike explanations to account for perceived data, is an important part of investigative decision making. A considerable literature suggests that story generation can be viewed as a cognitive heuristic employed by individuals when trying to come to terms with complex and ambiguous information, particularly when such information relates to social situations (Read, 1987; Read & Miller, 1995; Schank & Abelson, 1995). Pennington and Hastie's classic studies of jury decision making (e.g., Pennington & Hastie, 1992) provide compelling evidence for the use of story construction in a forensic context (see also Wagenaar, Van Koppen, & Crombag, 1993; Wiener, Richmond, Seib, Rauch, & Hackney 2002). It is reasonable, therefore, to hypothesise that investigators achieve an understanding of the information available to them in a complex criminal inquiry by constructing stories: causal scenarios in which sequences of actions are explained by the inference of cause–effect relationships between those actions, and that serve to explain actors' goals and the execution of their plans. Story generation is a good example of an adaptive heuristic (Gigerenzer & Todd, 1999): Cognitive limitations mean that it is easier to deal with complex information that is summarized, or "chunked," into a coherent story. Story construction is also adapted to the nature of the decision environment: Events tend to follow one another in an understandable sequence, outcomes have causes, and, in a general sense, people often do predictable things in well-defined circumstances.

We might expect that narrative elaboration by investigators as they try to make sense of a criminal inquiry will be guided by schemata or prototype mental models (Schank & Ableson, 1995). In the vignette study, the amount of agreement between investigators in their interpretation of these actions varied according to the different scenarios, depending on the degree to which the elements of the vignette mapped onto legal definitions of particular crimes, suggesting that the degree to which detectives engage in narrative elaboration depends on the correspondence between the actions as described and those actions that would define a particular crime in law. Though detectives' schemata are undoubtedly influenced by previous experience, it appears that the law also provides an important structure for detectives engaging in investigative sensemaking.

CONCLUSIONS

Understanding investigative expertise is a critical issue for the UK police service. Detectives work in a rapidly changing and increasingly difficult landscape. They must contend with high levels of both serious and volume crime while taking on new challenges in dealing with skilled criminals carrying out complex crimes such as high-tech crimes and terrorism, and all under increasing public and political scrutiny. Psychological research that illuminates the social, cognitive, and emotional underpinnings of police decision making could make an important contribution to ongoing efforts to improve the training of police skills, yet until recently such research was almost entirely lacking (Barrett, 2005).

Our work has begun to identify the landscape of policing and the focus group research has indicated the range of individuals and groups that officers must deal with in running major inquiries. Particularly difficult are issues associated with managing the incident with respect to the media, the local community, as well as governmental level concerns. This requires that such officers are effective leaders—in this context a term that appears to be defined by distinctive cognitive and interpersonal skills. Subsequent work has begun to focus more directly on the cognitive effort involved in developing situational models of inquiries (case based and experimental). Currently, the definition and nature of expertise in law enforcement is poorly understood and ill defined. Our hope is that both the case-based focus group work and the more traditional experimental methods will help map out the landscape and context of policing as well as the range and effectiveness of the requisite cognitive and interpersonal skills that appear to be emerging as important in managing major incidents.

ACKNOWLEDGMENTS

Support for the preparation of this chapter was provided by Economic and Social Research Council Grant PTA-030-2002-00482 awarded to the second author.

REFERENCES

Alison, L. J. (2005). *The forensic psychologist's casebook: Psychological profiling and criminal investigation.* Devon, England: Willan.

Alison, L. J., & Barrett, E. C. (2004). The interpretation and utilisation of offender profiles: A critical review of "traditional" approaches to profiling. In J. Adler (Ed.), *Forensic psychology: Concepts, debates and practice* (pp. 58–77). Devon, England: Willan.

Anderson, C. J. (2003). The psychology of doing nothing: Forms of decision avoidance result from reason and emotion. *Psychological Bulletin, 129*(1), 139–167.

Barrett, E. C. (2002). *Towards a theory of investigative situation assessment: An examination of the psychological mechanisms underlying the construction of situation models in criminal investigations.* Unpublished masters thesis, University of Liverpool, England.

Barrett, E. C. (2005). Psychological research and police investigations: Does the research meet the needs? In L. J. Alison (Ed.), *The forensic psychologist's casebook: Psychological profiling and criminal investigation* (pp. 47–67). Devon, England: Willan

Bernoulli, D. (1738). Specimen theoriae novae de mensura sortis [Exposition of a new theory of the measurement of risk]. *Commentarii Academiae Scientrum Imperialis Petropolitanae, 5,* 175–192.

Bruns, G. H., & Shuman, I. G. (1988). Police managers' perception of organizational leadership styles. *Public Personnel Management, 17*(2), 145–157.

Bryman, A. (2004). Focus groups. In A. Bryman (Ed.), *Social research methods* (2nd ed., pp. 345–362). Oxford, England: Oxford University Press.

Cannon-Bowers, J. A., & Salas, E. (1998). *Making decisions under stress: Implications for individual and team training.* Washington, DC: American Psychological Association.

Chemers, M. M. (2000). Leadership research and theory: A functional integration. *Group Dynamics: Theory, Research, and Practice, 4,* 27–43.

Clapper, D. L., & Massey, A. P. (1996). Electronic focus groups: A framework for exploration. *Information & Management, 30,* 43–50.

Cohen, M. S., Freeman, J. T., & Thompson, B. (1998). Critical thinking skills in tactical decision making: A model and a training strategy. In J. A. Cannon-Bowers & E. Salas (Eds.), *Making decisions under stress: Implications for individual and team training* (pp. 155–189). Washington, DC: American Psychological Association.

Crego, J., & Alison, L. J. (2004). Control and legacy as functions of perceived criticality in major incidents. *Journal of Investigative Psychology and Offender Profiling, 1,* 207–225.

DeSanctis, G., & Gallupe, R. B. (1987). A foundation for the study of group decision support systems. *Management Science, 33*(5), 589–609.

Dobby, J., Anscombe, J., & Tuffin, R. (2004). Police leadership: Expectations and impact. Home Office Report 20/04. Retrieved September 10, 2005, from http://www.mcb.org.uk/mcbdirect/library/uploads/rdsolr2004.pdf

Easton, G., Easton, A., & Belch, M. (2003). An experimental investigation of electronic focus groups. *Information & Management, 40,* 717–727.

Endsley, M. R. (1995). Toward a theory of situation awareness in dynamic systems. *Human Factors, 37,* 32–64

Fishman, D. B. (1999). *The case for pragmatic psychology.* New York: New York University Press.

Fishman, D. B. (2003). Background on the "Psycholegal Lexis Proposal": Exploring the potential of a systematic case study database in forensic psychology. *Psychology, Public Policy & Law 9,* 267–274.

Fishman, D. B. (2004). Pragmatic psychology and the law: Introduction to Part II. *Psychology, Public Policy & Law, 10,* 3–4.

Forsyth, D. R. (1999). *Group dynamics.* (3rd ed.). Belmont, CA: Wadsworth.

Gibb, C. A. (Ed.). (1969). *Leadership.* New York: Penguin.

Gigerenzer, G., & Todd, P. M. (1999). Fast and frugal heuristics: The adaptive toolbox. In G. Gigerenzer, P. M. Todd, & The ABC Research Group (Eds.), *Simple heuristics that make us smart* (pp. 3–34). New York: Oxford University Press.

Howitt, D., & Cramer, D. (2005). *Introduction to research methods in psychology.* Harlow, England: Pearson Education.

Innes, M. (2002). The "process structures" of police homicide investigations. *British Journal of Criminology, 42,* 669–688.

Innes, M. (2003). *Investigating homicide: Detective work and the police response to criminal homicide.* Oxford, England: Oxford University Press.

Kitzinger, J. (1994). The methodology of focus groups: The importance of interaction between research participants. *Sociology of Health & Illness, 16*(1), 103–121.

Klein, G. (1999). *Sources of power: How people make decisions.* Cambridge, MA: MIT Press.

Klein, G. A., Orasanu, J., Calderwood, R., & Zsambok, C. E. (Eds.). (1993). *Decision making in action: Models and methods.* Norwood, CT: Ablex.

Krueger, R. A. (1994). *Focus groups: A practical guide for applied research* (2nd ed.). Thousand Oaks, CA: Sage.

Kuykendall, J., & Usinger, P. (1982). The leadership styles of police managers. *Journal of Criminal Justice, 10,* 311–321.

Lord, R. G., & Maher, K. J. (1991). *Leadership and information processing: Linking perception and performance.* Boston: Unwin Hyman.

Metropolitan Police Service. (2002). *The Damilola Taylor murder investigation review.* London: Metropolitan Police Service. Retrieved September 10, 2005, from http://image.guardian.co.uk/sys-files/Guardian/documents/2002/12/09/damilola.pdf

Montoya-Weiss, M. M., Massey, A. P., & Clapper, D. L. (1998). On-line focus groups: conceptual issues and a research tool. *European Journal of Marketing, 32*(7/8), 713–723.

Mumford, M. D., Reiter-Palmon, R., & Redmond, M. R. (1994). Problem construction and cognition: Applying problem representations in ill-defined domains. In M. A. Runco (Ed.), *Problem finding, problem solving and creativity* (pp. 3–39). Norwood, NJ: Ablex.

Nunamaker, J. F., Jr., Briggs, R. O., Mittleman, D. D., Vogel, D. R., & Balthazard, P. A. (1996/1997). Lessons from a dozen years of group support systems research: A discussion of lab and field findings. *Journal of Management Information Systems, 13*(3), 163–208.

Ogan, J., & Alison, L. J. (2005). Jack the Ripper and the Whitechapel Murders: A very Victorian critical incident. In L. J. Alison (Ed), *The forensic psychologist's casebook: Psychological profiling and criminal investigation* (pp. 23–46). Devon, England: Willan.

Ormerod, T. C., Barrett, E. C., & Taylor, P. J. (2005, June). *Investigative sense-making in criminal contexts.* Paper presented at the Seventh International NDM Conference, Amsterdam.

Parent, M., Gallupe, R. B., Salisbury, W. D., & Handelman, J. M. (2000). Knowledge creation in focus groups: Can group technologies help? *Information & Management, 38*(1), 47–58.

Pennington, N., & Hastie, R. (2000). Reasoning in explanation-based decision making. *Cognition, 49,* 122–163.

Read, S. J. (1987). Constructing causal scenarios: A knowledge structure approach to causal reasoning. *Journal of Personality and Social Psychology, 52,* 288–302.

Read, S. J., & Miller, L. C. (1995). Stories are fundamental to meaning and memory: For social creatures, could it be otherwise? In R. S. Wyer (Ed.), *Advances in social cognition: Vol. 8. Knowledge and memory: The real story* (pp. 139–152). Hillside, NJ: Lawrence Erlbaum Associates.

Robinson, J. A., & Hawpe, L. (1986). Narrative thinking as a heuristic process. In T. R. Sarbin (Ed.), *Narrative psychology: The storied nature of human conduct* (pp. 111–125). New York: Praeger.

Robinson, N. (1999). The use of focus group methodology—with selected examples from sexual health research. *Journal of Advanced Nursing, 29*(4), 905–913.

Robson, C. (2002). *Real world research: A resource for social scientists and practitioner-researchers* (2nd ed., pp. 284–289). Oxford, England: Blackwell.

Savage, L. J. (1954). *The foundations of statistics.* New York: Wiley.

Schank, R. C., & Abelson, R. P. (1995). Knowledge and memory: The real story. In R. S. Wyer (Ed.), *Advances in Social Cognition: Vol. 8. Knowledge and memory: The real story* (pp. 1–85). Hillside, NJ: Lawrence Erlbaum Associates.

Smith, N., & Flanagan, C. (2000). *The effective detective (Home Office Policing and Reducing Crime Unit: Police Research Series Paper 122.* London: Home Office Research, Development and Statistics Directorate.

Sullivan, T. J. (2001). *Methods of social research.* Orlando, FL: Harcourt College.

Swanson, C. R., & Territo, L. (1982). Police leadership and interpersonal communication styles. In J. R. Greene (Ed.), *Managing police work: Issues and analysis* (pp. 123–139). Beverly Hills, CA: Sage.

Wagenaar, W. A., van Koppen, P. J., & Crombag, H. F. M. (1993). *Anchored narratives: The psychology of criminal evidence.* New York, NY: Harvester Wheatsheaf.

West, A., & Alison, L. (2005). Conclusions: Personal reflections on the last decade. In L. J. Alison (Ed.), *The forensic psychologist's casebook: Psychological profiling and criminal investigation.* Devon, England: Willan.

Wiener, R. L., Richmond, T. L., Seib, H. M., Rauch, S. M., & Hackney, A. A. (2002). The psychology of telling murder stories: Do we think in scripts, exemplars or prototypes? *Behavioral Sciences and the Law, 20,* 119–139.

Yukl, G. (1994). *Leadership in organizations* (3rd ed.). Englewood Cliffs, NJ: Prentice-Hall.

Zaleznik, A. (1977). Managers and leaders: Are they different? *Harvard Business Review, 55,* 67–78.

Chapter 5

What's Burning? The RAWFS Heuristic on the Fire Ground

Raanan Lipshitz
University of Haifa, Israel

Mary Omodei
La Trobe University, Melbourne, Australia

Jim McClellan
La Trobe University, Melbourne, Australia

Alexander Wearing
University of Melbourne, Australia

Experts in any domain find themselves drawn out of comfortable decision-making contexts when the problem at hand is rife with uncertainty. The identification of uncertainty as a constitutive element of decision making was a central contribution of classical decision theory to our understanding of the difficulties decision makers have to overcome. The theory recommends that decision makers first reduce uncertainty through information search and then include uncertainty that cannot be reduced this way as a factor consideration in the selection of their preferred alternative. Lipshitz (1997) dubbed this advice as the RQP heuristic: *Reduce* uncertainty by collecting relevant information' *quantify* the uncertainty that cannot be reduced this way, for example, by expressing it as probability estimates; and *plug* the result into a formula that incorporates uncertainty as a factor in the selection of the preferred alternative.

Owing to extensive research inspired by Kahneman and Tversky's work (Kahneman & Tversky, 2000) on heuristics and biases, it is now well established that people do not—and perhaps cannot—follow the RQP heuristic without assistance, but such assistance is rarely available, or feasible, in real-world situations. The question that presents itself then is, What other methods do people use to cope with uncertainty?

To answer this question, Lipshitz and Strauss (1997) examined a sample of retrospective case reports of real-world decision making under uncertainty. They asked: How do decision makers conceptualize their uncertainty (defined as a sense of doubt that blocks or delays action)? How do they cope with this uncertainty? Are there systematic relationships between different conceptualizations of uncertainty and different methods of coping? Analysis of 102 cases revealed that decision makers encounter three basic types of uncertainty: inadequate understanding, incomplete information, and undifferentiated alternatives. To cope with these uncertainties, decision makers use five strategies of coping:

- Reduction (trying to reduce uncertainty, e.g., through information search).
- Assumption-based reasoning (relying on knowledge and imagination to fill gaps in, or make sense of, factual information).
- Weighing pros and cons of rival options (as in a rough approximation of the subjective expected utility model).
- Suppressing uncertainty (e.g., by ignoring it or by "taking a risk").
- Forestalling (preparing a course of action to counter potential negative contingencies, e.g., building reserves or preparing a worst-case option).

Inadequate understanding was primarily managed by reduction; incomplete information was primarily managed by assumption-based reasoning; and conflict among alternatives was primarily managed by weighing pros and cons. Based on these results and findings from previous studies of naturalistic decision making, Lipshitz and Strauss (1997) hypothesized the RAWFS heuristic (Reduction, Assumption-based reasoning, Weighing pros and cons, Forestalling, and Suppression), which summarizes how decision makers cope with uncertainty without resorting to calculation or the RQP heuristic.

According to the RAWFS heuristic, decision making begins with an attempt to understand, recognize, or make sense of the situation. If this attempt is successful, decision makers initiate a process of serial option evaluation, which they complement, if time permits, by mentally simulating the selected option. When sensemaking fails, decision makers experience the dissonance of inadequate understanding, to which they respond by using reduction or by forestalling. If additional information is not available, decision makers experience a feeling that they lack information, to which they respond by assumption-based reasoning or by forestalling. If decision makers generate two or more good enough options, they experience conflict, to which they respond by weighing pros and cons or by forestalling. Finally, if decision makers either fail to identify a single good enough option, or fail to differentiate among several good enough options, they resort to suppression, forestalling, or the generation of a new alternative.

The RAWFS heuristic is consistent with other findings regarding how decision makers cope with uncertainty in naturalistic settings. Cohen, Freeman, and Wolf (1996) and Klein (1998) noted that experienced decision makers overcome uncertainty owing to missing, unreliable, or ambiguous information by constructing plausible stories (a form of assumption-based reasoning) to make sense of unfolding events and their future implications. Allaire and Firsirotu (1989) and Shapira (1995) described various tactics that decision makers use to cope with uncertainty in organizational settings (i.e., forestalling).

The convergence of findings notwithstanding, Lipshitz and Strauss's (1997) study had three methodological limitations: (a) the use of written self-reports, a form of data that can be questioned on a number of grounds (Ericsson & Simon, 1984); (b) the use of an unspecified sample of decision problems, which may mask potential differences between coping with uncertainty by decision makers in different domains, and (c) the high level of abstraction in which the RAWFS heuristic is framed, which hinders implementation of the model by decision makers in specific domains (Smith, 1997).

The present study was designed to remedy these limitations by testing the RAWFS heuristic in one specific domain—incident management by firefighting unit commanders. In addition, we used a better data collection methodology—cued recall using records of the incidents captured by head-mounted video cameras (Omodei, Wearing, & McLennan, 1997). A second objective of the study was to obtain a deeper understanding of the nature of uncertainty with which firefighters have to cope and the strategies of coping that they employ.

METHOD

Domain and Participants

Our study participants were five male Senior Station Officers (SSO) with the Melbourne Fire and Emergency Service Board. The SSOs' experience as firefighters ranged between 7 and 19 years. SSOs are in charge of general-purpose fire trucks ("pumpers"), and are dispatched to the incident scene (e.g., car or structure fire) by a centralized emergency communication center. The goal is to have an appliance on-scene within 7 minutes of the original alarm call.

The task of a firefighting unit can be divided into three phases: arriving at the scene of incident, handling the incident, and mopping up operations.

The overall procedure that regulates the SSO's attention and actions is dubbed RECEF: rescue (people trapped in the emergency), exposure (prevent fire from expanding beyond its location), containment (contain the fire within its location), extinguish (the fire after it has been contained), and fire duty (do the paperwork reporting the incident). Accordingly, SSOs focus their attention during incident management on: (a) checking for people trapped in the location of fire or accident, (b) preventing fire from spreading outside the burning structure or structures, (c) containing the fire within the burning structure or structures, (d) putting the fire out, and (e) investigating the causes of fire and filling incident report forms.

This sequence is flexible. Depending on the situation at hand, SSOs may note a fact that is relevant to a later stage, engage concurrently in two stages, or skip a stage that turns out to be irrelevant. SSOs' specific responsibilities during the handling of the incident are to determine priorities, direct the deployment of resources, delegate responsibility for different sectors of the incident or for different aspects of managing the incident (e.g., civilian evacuation and liaison with other emergency services), and coordinate operations.

The standard operating procedures established in firefighting organizations, within which SSOs operate, are highly specified and routinized. The organization is hierarchical and position holders relate to one another on the basis of formal authority (Bigley & Roberts, 2001, p. 1282). Specific jobs or roles within the system are specialized and guided by standard operating procedures. Decision making is thus primarily rule based (March, 1994), recognition primed (Klein, 1993), or matching (Lipshitz, 1994). Consequently, firefighters' uncertainty falls into two decision categories: (a) deciding which procedure applies to the given case, but then deciding how to adjust the standard procedure to the situation at hand, and (b) deciding how to cope with uncertainty that is not covered by available procedures.

The highly proceduralized task environment coupled with the high level of training of individual firefighters frees the SSO from micromanaging their actions. As one SSO commented: "They are still in there. I am quite confident about what they are doing. They are trained so that I don't need to be at their backs all the time. Give them a job and it's done. Walk away, they do the job, and I don't have to worry about them." Bigley and Roberts (2001) made a similar observation:

> When a commander believes subordinates possess sufficient experience, training, and resourcefulness to adapt to local conditions, he or she typically leave the task partially unstructured. . . . In other words, supervisors provide subordinates with a degree of latitude to improvise—that is, to activate and coordinate their routines and to apply novel tactics to unexpected problems. (p. 1280)

Procedure

SSOs were equipped with s helmet-mounted microphone and a video camera system, which recorded visual and radio and voice information from the moment of leaving the station to the completion of emergency operations. As soon as practicable following an incident, the SSOs were interviewed using a cued-recall methodology. The SSO reviewed the video and audio records of the incident and reported his associated thinking processes in response to the following instructions:

> We are going to watch replay footage of the incident taken from your helmet camera. As you watch, I want you to take yourself back to being the incident commander. I want you to recall as much as you can of what was going on in your mind when you were managing the incident. I want you to speak these recollections out loud—just begin talking and I will pause the tape so you have plenty of time to recall as much as you can. Your recollections will be recorded into a VHS copy of the original footage of the incident as you saw it, and all the radio and voice communications. We will then have your recollections of the things that were going on in your mind that "drove" your decisions, actions, and communications.

Following these instructions, the helmet camera tape was started for 10 seconds after which the interviewer asked the SSO, "Now, as you look at this first picture of the turnout—what do you recall thinking as your appliance left the station?" As the interview went on, the interviewer encouraged the SSO to recall as much as he could, occasionally using nondirective probes and reminding the SSO to recall what "went in his mind." The 10 cued-recall tapes were transcribed for purpose of analysis. The transcripts were divided into frames, each containing a single communication act from a source to a target. Frames taken during the incident included communications from the SSO to firefighters, the station, or civilians and vice-versa. Frames taken during the cued-recall interview included interchanges between the SSO and the interviewer.

RESULTS

Ten incidents commanded by the five SSOs were analyzed. Although varying in difficulty, they all hinged on the experience of the commanding SSO. Each incident contained between 3 and 21 instances of coping with uncertainty, producing a total of 150 uncertainty-coping tactic pairs. Three incidents involved car fires, four involved large structure fires, and one incident each involved a small structure fire, a gas leak in an apartment, and a smoke detector false alarm. Five of the 10 incidents were classified by the SSO in charge as routine, 1 was classified as moderately complex, and 4 were classified as complex.

Analysis proceeded in two stages, frame-by-frame coding and script analysis. Frame-by-frame coding included the identification of uncertainty and coping tactics that appeared in each frame and mapping them into a set of codes initially based on the classification proposed by Lipshitz and Strauss (1997), and further elaborated as the coding process progressed. The coding was performed by the second author assisted by the first author. The Kappa coefficient between the independent coding by the two authors of the types of uncertainty and coping strategies were $\kappa = .52$ and $\kappa = .58$, respectively. The cases of disagreement were discussed and all were resolved by the two authors.

To provide a better sense of how SSOs coped with uncertainty in the dynamic and evolving incident situations, the quantitative analyses of the frequencies and joint frequencies of types of uncertainty and coping tactics were augmented by constructing short scripts and flowcharts that detailed chronologically each incident's main events and the nature of uncertainty and coping tactics used by the SSO.

Three phases of the overall fire unit's task determined the SSOs' foci of attention and their concerns. During the first stage, *en route to the incident* (from leaving the station to arriving on the scene of the incident), SSOs were concerned with the location of the incident and their whereabouts relative to that location, and with the nature of the incident (e.g., size of fire) and how to best deal with it. During the second phase, *deploying and handling the incident* (from arriving at the scene until the problem is solved or close to be solved), the SSOs' attention was focused on the uncertainties posed by the problem at hand. During the third, *mopping-up* phase, their attention shifted to the causes of the problem and to the mopping-up routine operations (e.g., preparing reports and briefing the police). Owing to these distinct differences between the three phases, the analyses of types of uncertainty and coping strategies were structured accordingly.

SSOs' Uncertainty

The frame-by-frame coding of the incident plus the cued-recall transcripts showed that lack of understanding and inadequate information could not be reliably distinguished in this data set. In addition, consistent with Klein's findings regarding firefighters' decision making (Klein, 1993), there was only one instance of uncertainty owing to undifferentiated alternatives. Thus, virtually all uncertainties identified in the data were classified as inadequate understanding. Further analysis of this category produced three types of inadequate understanding: inadequate understanding regarding

TABLE 5.1
Distribution of Types of Uncertainty Within Incident Phases

Uncertainty	En Route to the Incident	Handling the Incident	Mopping Up	Σ
Regarding the situation	16	75	17	108
Regarding action	2	22	2	26
Regarding the causes		4	12	16
of the incident	—			.11
Σ	18	101	31	150

Note. The proportions refer to the phase totals.

the present situation (e.g., "We couldn't understand about where the floor was, and what was underneath it."), inadequate understanding regarding the required action (e.g., "How do we get into it?"), and inadequate understanding regarding the causes of the incident (e.g., "I am trying get a picture of what exactly has happened.").

Table 5.1 presents the distribution of the different types of uncertainty in the three phases of incident management. Examination of the row margins reveals that SSOs are mostly concerned about the nature of the situation (.72), followed by which action is appropriate (.11) and the causes of the incident (.11). Although this order is retained in the first two phases of the incident management (en route to the incident and handling the incident), the magnitude of the three types of uncertainty conditional on the incident's phase varies significantly from their base probabilities. Thus, while en route to the incident, SSOs experienced more uncertainty regarding the situation, specifically concerning their own location and what to expect once they arrive at the site of the incident (.88 compared to .72 base rate). They experienced less uncertainty regarding which action to take (.11 compared to .17) and were not concerned at all about the causes of the incident (.0 compared to .11). Although most of the uncertainty regarding the nature of the situation occurs in the second, action phase, SSOs experienced less uncertainty regarding the situation in this phase relative to the first phase, mostly because of the increased uncertainty regarding which action to take. Finally, the mopping-up phase is distinguished by heightened uncertainty regarding the causes of the incident to which SSOs at last feel free to turn their attention (.39 compared to a base rate of .11). Thus, task structure clearly shapes the nature of uncertainty that SSOs experience.

SSOs' Coping Tactics

Frame-by-frame coding of the interview transcripts yielded nine coping tactics that could be mapped into three of the five coping strategies comprising the RAWFS heuristic: reduction, assumption-based reasoning, and forestalling. Thus, SSOs did not use either Weighing pros and cons (designed for coping with undifferentiated alternatives) or suppression (which they could apparently do without working on problems within their area of expertise.) *Reduction* included four specific coping tactics:

- *Delaying action,* a form of passive information search (e.g., "We were virtually waiting to see what the feedback from that crew was in regard to what our next plan of attack would be").
- *Prioritizing* so as to focus attention on higher priority objectives (e.g., "The main thing was opening up the rear doors and also having a fan ready, which would make the job much easier").
- *Relying on SOPs* (e.g., "I was looking in there. Just having a quick look, which is just procedure. We all do that").
- *Active information search.*

This latter tactic involved reducing uncertainty by collecting information from crew members and other actors on the scene, taking a tour of the premises, and constantly looking for relevant clues using all one's senses, vision, hearing, smelling, and sensitivity to temperature differences (e.g., "At this stage, I'm back a long way so I can see everything"; "I got them up to have a look to see if it wasn't burning on the roof again"). Information from crew members is useful for reducing uncertainty caused by the fact that the SSO typically cannot observe what is going on all by himself. Other actors can provide important background information such as the layout of the building or the presence of people inside it. (SSOs often plan en route to the incident which information they will seek to obtain this way.) Another important mode of information search is scanning, constantly monitoring the situation to detect details that disconfirm expectations, indicate potentially dangerous developments, or are otherwise significant for preempting the "unexpected" (Weick & Sutcliffe, 2001): "You have got to have your ears open all the time. Like he is saying something. Someone else is saying something. Something is saying something on the radio. You have got to listen and take in what you have to take in."

Assumption-based reasoning also included four tactics:

- *Planning:* Falling back on a plan as guide to resolving uncertainty about the situation ("We always ask questions so we can get a plan in our minds"; "At the moment the plan is that the boys will be at the back. On the ninth floor they will be getting BA [breathing apparatus] for a start. . . . I only worked here for a month or so, but I know what they do").

- *Mental rehearsal:* Imagining potential situations and courses of action prior to the selection of a particular course of action (e.g., "I was thinking that it is LPG and my first thought was I hope it's not burning at the back where the tank is").
- *Mental simulation:* Playing out the implementation of a course of action prior to actual implementation (e.g., "I'm formulating now that we're going have to, we're breaking in, and when we get in we're likely to have to, have a patient and drag her out or do first aid").
- *Conjecturing:* Using assumptions in creating a situation awareness (e.g., "it was reasonable at that stage to assume there was a separation between the two floors").
- *Forestalling:* Finally, SSOs used the strategy of forestalling (e.g., "I'm trying to get as many flares and torches as I can because it's so dark and hazardous . . . when the men go in, make sure there is proper lighting and they can see where they are going"). This strategy was not divided further into subcategories.

In conclusion, the findings replicated those of Lipshitz and Strauss (1997) except for the absence of weighing pros and cons and suppression, and with the addition of several specific reduction and assumption-based reasoning tactics (e.g., prioritizing and conjecturing, respectively).

Table 5.2 presents the distribution of coping strategies within the three incident phases. Reduction was by far the dominant strategy (70%) followed by assumption-based reasoning (21%) and forestalling (.09). Examination of the table reveals that task structure affects the distribution of coping strategies as well as the distribution of types of uncertainty.

Two interesting variations from the base rate pattern of coping strategies occurred en route to the incident. The probability of reduction decreased from .70 to .50 and the probability of assumption-based reasoning increased

TABLE 5.2
Distribution of Coping Strategies Within Incident Phases

Uncertainty	En Route to the Incident	Handling the Incident	Mopping Up	Σ
Reduction	9	72	24	105
	.50	.71	.77	.70
Assumption-based reasoning				32
	8	19	5	
	.44	.19	.16	.21
Forestalling	1	10	2	13
	.00	.10	.07	.09
Σ	18	101	31	150

Note. The proportions pertain to phase totals.

from .21 to .44. The former probability decreases because the firefighters assume that the dispatcher will pass on to them all relevant information available to him so there is no need for them to address questions to him. (The .50 probability represents queries regarding the "pumpers'" location addressed to passers-by or consultation of the city's map.) The increase in assumption-based reasoning represents attempts to construct a mental model of the incident, based on the clues received from the dispatcher, in anticipation of the expected action so as to "hit the ground running."

Table 5.3 presents the joint frequencies of types of uncertainty and strategies of coping. Owing to our inability to differentiate between inadequate understanding of the situation and lack of information, and the absence of uncertainty owing to undifferentiated alternatives, we could not test for presence of the pattern observed by Lipshitz and Strauss (1997). Nevertheless, the marginal probabilities of the coping tactics and the conditional probabilities relevant to uncertainty regarding the situation (which constitute 72% of the observed instances of uncertainty) are consistent with the RAWFS heuristic inasmuch as SSOs relied mostly on reduction tactics, which according to the heuristic are used first.

The low frequencies of uncertainty regarding action and incident causes do not permit unambiguous interpretation of the results regarding these uncertainties. The relatively greater use of assumption-based reasoning to cope with the former, and the absence of forestalling in case of the latter, are consistent, however, with the fact that several tactics of assumption-based reasoning (anticipating action, mental simulation, and, to a certain extent, mental rehearsal) specifically pertain to action. In addition, resolving the uncertainty regarding the causes of fire does not require the SSO to take subsequent action.

Figure 5.1 presents a representative flow-chart summary of one of the ten cases analyzed in the study. On the way to the high-rise-building fire, the SSO was concerned with the presence of people in the vicinity of the fire,

TABLE 5.3
Joint Distribution of Types of Uncertainty and Coping Strategies

Coping ⇒ Uncertainty ⇓	Reduction	Assumption-Based Reasoning	Forestalling	Σ
Regarding the situation	82	18	8	108
	.76	.17	.06	.72
Regarding action	11	10	5	26
	.42	.38	.19	.17
Regarding causes	12	4	—	16
	.75	.25		.11
Σ	105	32	13	150
	.71	.20	.09	

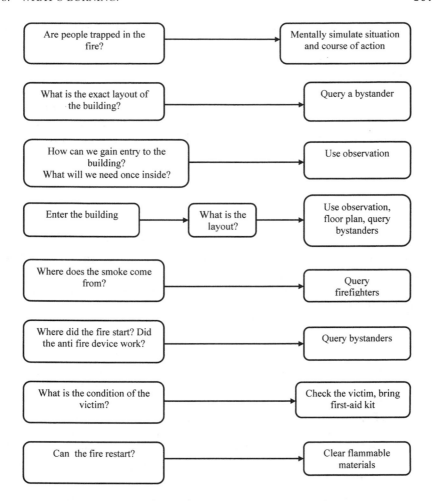

FIG. 5.1. Incident flow Diagram. Phase 1: En route to the Incident.

and mentally simulated various potential courses of action to deal with the obviously nontrivial incident. On arriving at the scene, he questioned bystanders about the exact layout of the building. His next two uncertainties concerned action: how to get into the building, and what would be needed once inside. He coped with the first uncertainty by taking a tour around the building and with the second uncertainty by mentally simulating various possibilities. Having entered the building he was again uncertain about his exact location. He coped with this uncertainty by looking around, consulting the floor plan, and questioning neighbors and residents. Next he worried about the exact location of the fire and about the possibility of people being trapped in the fire. To cope with these uncertainties, he collected information from firefighters who were already inside the building. As the fire got under control, the SSO sent a firefighter to collect information about the source of the fire, and began to wonder about the condition of the person who had been rescued from the fire, and about possible rekindling of the fire. He coped with both issues by a mixture of information search and forestalling tactics (i.e., preparing the first-aid kit and clearing all flammable material). The implication of this and the other nine flow-chart analyses that were constructed in a likewise fashion are discussed in the next section.

DISCUSSION

This study was designed to test the RAWFS heuristic in a specific task environment and with a different methodology than the methodology of its construction, and to obtain an in-depth understanding of SSOs task uncertainties and coping strategies. The following discussion focuses, accordingly, on these issues.

Three findings regarding SSOs' coping with uncertainty are salient from the vantage point of the RAWFS heuristic. The first finding is the absence of uncertainty owing to conflicted or undifferentiated alternatives and of two strategies included in RAWFS: coping, weighing pros and cons, and suppression. The absence of conflicted alternatives is accounted for by the fact that firefighters (similar to most decision makers' naturalistic and organizational settings) employ matching-mode (rule- or recognition-based) decision making (Lipshitz, 1994). The use of the matching also explains the absence of weighing pros and cons, because decision makers using this mode do not compare among alternatives. The absence of suppression is accounted for by the fact that SSOs are highly experienced, and so manage to overcome uncertainty and take action without recourse to this strategy, except for possible rare occasions, which were not represented in the 10 incidents sampled in the present study.

The second salient finding is the influence of the SSOs' task structure on the nature of their uncertainties and coping strategies. This influence can account for the absence of uncertainty owing to conflicted alternatives and the strategy of weighing pros and cons: A task system that relies heavily on routines and is conducted by experienced personnel is more conducive to rule-based than to choice-based decision making. Furthermore, taking into account that SSOs operate in a semantically rich environment, we note that as their focus of attention moves from getting to the scene of the incident, to handling the incident, and to mopping-up operations, their uncertainties change from "Are we on the right track?" and "What lies ahead?" to "Where is the site of fire?" and "How to put the fire out while ensuring the safety of victims, bystanders, and firefighters?" finally to "Is the fire safely out?" and "What caused it in the first place?"

Consistent with the RAWFS heuristic, the findings show that SSOs prefer to use reduction tactics (notably information search), and switch to assumption-based reasoning or forestalling when such tactics are impractical or fail. In particular, SSOs use mental rehearsal (a form of assumption-based reasoning) when no source of information other than their past experience is available to them en route to the incident for preparing them for what lies ahead. This pattern was also reported to Bigley and Roberts (2001) and by one of their informants:

> I start building a picture as soon as I am dispatched. I start thinking of the neighborhood, what's in that neighborhood, what kind of houses we have— big houses, little houses. If it's an industrial area, what kind of chemicals they have. I certainly start the size-up [situation assessment] of it as soon as I read the dispatch. We do the size up right away. (p. 1291)

Once they arrive at the scene of the incident, SSOs begin an intensive information search from whatever source that is available and continue to do so throughout the second phase of handling the incident. Thus, three hallmarks characterize SSOs' coping with uncertainty during incident management:

- Heavy reliance on standard operating procedures, notably the RECEF procedure, which provides them with a basic template to guide their actions.

- Reliance on other people as sources of information, be they subordinate firefighters, people from other engaged units, or bystanders, because SSOs can neither comprehend nor handle incidents by themselves. The upshot of this necessity is intense communication between SSOs and whoever is engaged in the action, including the dispatcher, whom SSOs take care to keep constantly informed of the state of the incident.

- Constant alertness: "You can't have tunnel vision, you have to look and listen to get all the facts that you can. In a split second you need to make a decision where you are going and what you are doing." Such mindful or heedful attention is particularly helpful for noting seemingly insignificant divergences from the normal and other cues that are critical for understanding the situation and for spotting potential problems before they develop into costly and intractable predicaments (Weick & Sutcliffe, 2001).

Regarding the RAWFS heuristic, the present findings help to refine the model and demonstrate its merits and its limitations. Considering the former, the present findings add one tactic of reduction (prioritizing) and one tactic of assumption-based reasoning (mental rehearsal) to the tactics identified by Lipshitz and Strauss (1997). More important, the findings suggest a simplified version of the RAWFS heuristic, one that removes the distinction between inadequate understanding of the situation and lack of information as follows: Decision makers cope with uncertainty by first using reduction tactics. If these fail or prove impractical, decision makers use assumption-based reasoning or forestalling. If decision makers encounter uncertainty owing to conflicted alternatives, they follow reduction tactics that fail to resolve their uncertainty by using weighing the pros and cons of the rival alternatives. Suppression tactics are employed as a last resort.

The RAWFS heuristic might be regarded as a general template or ideal type from which specific models can be crafted to specific domains or task constraints. The heuristic is a toolkit consisting of five basic strategies and a larger number of specific tactics that decision makers can adapt to their specific situations in order to cope with doubts that block or delay their actions. In addition, the heuristic can also be used as a general template for researchers who wish to understand how decision makers cope with uncertainty in specific naturalistic settings.

REFERENCES

Allaire, Y., & Firsirotu, M. E. (1989). Coping with strategic uncertainty. *Sloan Management Review, 30,* 7–16.

Bigley, G. A., & Roberts, K. H. (2001). The incident command system: High reliability organizing for complex and volatile task environments. *Academy of Management Journal, 44,* 1281–1299.

Cohen, M. S., Freeman, J. T., & Wolf, S. (1996). Meta-recognition in time stressed decision making: Recognizing, critiquing and correcting. *Human Factors, 38,* 206–219.

Ericsson, K. A., & Simon, H. A. (1984). *Protocol analysis.* Cambridge, MA: MIT Press.

Kahneman, D., & Tversky, A. (Eds.). (2000). *Frames, values, and choices.* New York: Cambridge University Press.

Klein, G. A. (1993). Recognition-primed decision (RPD) model of rapid decision making. In G. A. Klein, J. Orasanu, R. Calderwood, & C. Zsambok (Eds.), *Decision making in action: Models and methods* (pp. 138–147). Norwood, NJ: Ablex.

Klein, G. (1998). *Sources of power: How people make decisions.* Cambridge, MA: MIT Press.

Lipshitz, R. (1994). Decision making in three modes. *Journal for the Theory of Social Behavior, 24,* 47–66.

Lipshitz, R. (1997). Coping with uncertainty: Beyond the reduce, quantify and plug heuristic. In R. Flin, E. Salas, M. Strub, & L. Martin (Eds.), *Decision making under stress: Emerging themes and applications* (pp. 149–160). Aldershot, England: Ashgate.

Lipshitz, R., & Strauss, O. (1997). Coping with uncertainty: A naturalistic decision making analysis. *Organizational Behavior and Human Decision Processes, 69,* 149–163.

March, J. G. (1994). *A primer on decision making: How decisions happen.* New York: Free Press.

Omodei, M., Wearing, A., & McLennan, J. (1997). Head-mounted video recording: A methodology for studying naturalistic decision making. In R. Flin, E. Salas, M. Strub, & L. Martin

(Eds.), *Decision making under stress: Emerging themes and applications* (pp. 161–169). Aldershot, England: Ashgate.

Shapira, Z. (1995). *Risk taking: A managerial perspective.* New York: Russell Sage Foundation.

Smith, G. F. (1997). Managerial problem solving: A problem-centered approach. In G. Klein & C. Zsambok (Eds.), *Naturalistic decision making* (pp. 371–382), Mahwah, NJ: Lawrence Erlbaum Associates.

Weick, K. E., & Sutcliffe, K. M. (2001). *Managing the unexpected: Assuring high reliability in an age of complexity.* San Francisco: Jossey-Bass.

Chapter 6

A Data–Frame Theory
of Sensemaking

Gary Klein
Jennifer K. Phillips
Erica L. Rall
Deborah A. Peluso
Klein Associates Inc., Fairborn, OH

Throughout the history of psychology, from its earliest philosophical phase through to modern times, it has always been taken as an empirical fact that humans have a faculty to understand. This has been referred to as judgment, apprehension, apperception, and other processes (cf. Woodworth, 1938) in which meanings are ascribed to things that are perceived or known. Typically, both theory and research focused on how people understand individual simple stimuli (e.g., individual letters, words, colored geometric forms, etc.). During the era of "verbal behavior" and the rise of psycholinguistics from the ashes of behaviorism, a process called "comprehension" became the subject of much inquiry. Again, both theory and research focused on how people understand simple stimuli such as words, sentences, or text passages (cf. Clark & Clark, 1977).

During the first decade of information-processing psychology, comprehension was regarded, again in both theory and research, as: "essentially a process of compiling a program written in a natural language into a mental language" (Miller & Johnson-Laird, 1976, p. 631; see also Anderson, 1990, chap. 12). Analyses of semantics and syntax of utterances, combined with information about context, were said to produce an understanding or "meaning representation" as some sort of endpoint in a sequence of mental events (Smith & Spoehr, 1974).

All of these musings, theories, and research studies addressed psychological phenomena from a microcognitive perspective. That is, the goal has been to understand sequences of mental events down to a level of what are believed to be atomic processes and content, such as long-term memory in support of recognition of individual stimuli.

In the context of complex sociotechnical systems, the notion of "comprehension" takes on an entirely new scope, a macrocognitive scope (Klein, Ross et al., 2003). Practitioners in real-world domains must confront and understand complex, dynamic, evolving situations that are rich with various meanings. Though practitioners spend considerable time perceiving stimuli, the stimuli tend to be very complex (e.g., the weather forecaster looks at a satellite image loop) and come in many forms, not just one (e.g., tabular weather data, forecast text messages, computer model outputs, graphics of all sorts, etc.). Furthermore, there is nothing like an "endpoint" since the understanding of dynamic events requires that the understanding itself be dynamic, not a stored, frozen "meaning."

Understanding might piggyback on microcognitive processes and events. The scientific investigation of comprehension at a microcognitive level is certainly possible, and is of value to scientific psychology. However, it must be complemented by an understanding of comprehension at a macrocognitive scale, especially if applied science is to lead to effective change at that scale. Knowing from microcognitive research that there is a 10-millisecond delay in recognizing a target if on the previous trial a false target has appeared at the same location will not take you very far in the design of complex cognitive systems. The domain practitioner's higher purpose is not to perceive stimuli, but is, simply put, *to make sense of things.*

Weick (1995) introduced the process of sensemaking as one of the central cognitive functions that people must carry out in all kinds of natural settings. Sensemaking is initiated when a person (or organization) realizes the inadequacy of the current understanding of events. Sensemaking is often a response to a surprise—a failure of expectations. The surprise can surface contradictory beliefs and/or distrust of messages and other data. Sensemaking enables people to integrate what is known and what is conjectured, to connect what is observed with what is inferred, to explain and to diagnose, to guide actions before routines emerge for performing tasks, and to enrich existing routines. Sensemaking lets people see problems, as when a weather forecaster determines the problem of the day that will need continual monitoring or a business executive diagnoses the major challenges facing a department.

In this chapter, we attempt to explain the nature of sensemaking activities. We define sensemaking as the deliberate effort to understand events. It is typically triggered by unexpected changes or other surprises that make us doubt our prior understanding. We also describe how these sensemaking activities can result in a faulty account of events. Our description incorporates several different aspects of sensemaking:

- The initial account people generate to explain events.
- The elaboration of that account.
- The questioning of that account in response to inconsistent data.
- Fixation on the initial account.
- Discovering inadequacies in the initial account.
- Comparison of alternative accounts.
- Reframing the initial account and replacing it with another.
- The deliberate construction of an account when none is automatically recognized.

Each of these aspects of sensemaking has its own dynamics, which we explore and attempt to describe in the main sections of this chapter. We begin with two examples of sensemaking, to illustrate the nature of sensemaking activities and highlight the differences between such activities and the comprehension of simple stimuli.

Example 1: The Ominous Airplanes.   Major A. S. discussed an incident that occurred soon after 9/11 in which he was able to determine the nature of overflight activity around nuclear power plants and weapons facilities. This incident occurred while he was an analyst. He noticed that there had been increased reports in counterintelligence outlets of overflight incidents around nuclear power plants and weapons facilities. At that time, all nuclear power plants and weapons facilities were "temporary restricted flight" zones. So this meant there were suddenly a number of reports of small, low-flying planes around these facilities. At face value it appeared that this constituted a terrorist threat—that "bad guys" had suddenly increased their surveillance activities. There had not been any reports of this activity prior to 9/11 (but there had been no temporary flight restrictions before 9/11 either).

Major A. S. obtained access to the Al Qaeda tactics manual, which instructed Al Qaeda members not to bring attention to themselves. This piece of information helped him to begin to form the hypothesis that these incidents were bogus—"It was a gut feeling, it just didn't sit right. If I was a terrorist I wouldn't be doing this."

He recalled thinking to himself, "If I was trying to do surveillance how would I do it?" From the Al Qaeda manual, he knew they wouldn't break the rules, which to him meant that they wouldn't break any of the flight rules. He asked himself, "If I'm a terrorist doing surveillance on a potential target, how do I act?" He couldn't put together a sensible story that had a terrorist doing anything as blatant as overflights in an air traffic restricted area.

He thought about who *might* do that, and kept coming back to the overflights as some sort of mistake or blunder. That suggested student pilots to him because "basically, they are idiots."

He was an experienced pilot. He knew that during training, it was absolutely standard for pilots to be instructed that if they got lost, the first thing they should look for were nuclear power plants. He told us that "an entire generation of pilots" had been given this specific instruction when learning to fly. Because they are so easily sighted, and are easily recognized landmarks, nuclear power plants are very useful for getting one's bearings. He also knew that during pilot training the visual flight rules would instruct students to fly east to west and low—about 1,500 feet. Basically students would fly low patterns, from east to west, from airport to airport.

It took Major A. S. about 3 weeks to do his assessment. He found all relevant message traffic by searching databases for about 3 days. He picked the three geographic areas with the highest number of reports and focused on those. He developed overlays to show where airports were located and the different flight routes between them. In all three cases, the "temporary restricted flight" zones (and the nuclear power plants) happened to fall along a vector with an airport on either end. This added support to his hypothesis that the overflights were student pilots, lost and using the nuclear power plants to reorient, just as they had been told to do.

He also checked to see if any of the pilots of the flights that had been cited over nuclear plants or weapons facilities were interviewed by the FBI. In the message traffic, he discovered that about 10% to 15% of these pilots had been detained, but none had panned out as being "nefarious pilots."

With this information, Major A. S. settled on an answer to his question about who would break the rules: student pilots. The students were probably following visual flight rules, not any sort of flight plan. That is, they were flying by looking out the window and navigating.

This instance of sensemaking was triggered by the detection of an anomaly. But we engage in sensemaking even without surprises, simply to extend our grasp of what is going on.

Example 2: The Reconnaissance Team.   During a Marine Corps exercise, a reconnaissance team leader and his team were positioned overlooking a vast area of desert. The fire team leader, a young sergeant, viewed the desert terrain carefully and observed an enemy tank move along a trail and then take cover. He sent this situation report to headquarters. However, a brigadier general, experienced in desert-mechanized operations, had arranged to go into the field as an observer. He also spotted the enemy tank. But he knew that tanks tend not to operate alone. Therefore, based on the position of that one tank, he focused on likely overwatch positions and found another tank. Based on the section's position and his understanding of the terrain, he looked at likely positions for another section and found a well-camouflaged second section. He repeated this process to locate the remaining elements of a tank company that was well-camouflaged and blocking a key choke point in the desert. The size and position of the force suggested that there might be other higher and supporting elements in the area, and so he again looked at likely positions for command and logistics elements. He soon spotted an otherwise superbly camouflaged logistics command post. In short, the brigadier general was able to see and understand and make more sense of the situation than the sergeant. He had much more experience, and he was able to develop a fuller picture rather than record discrete events that he noticed.

As these examples show, sensemaking goes well beyond the comprehension of stimuli. The examples also show that sensemaking can be used for different functions. In Example 1, the purpose of sensemaking was *problem detection*—to determine if the pattern was worth worrying about and monitoring more closely. Weather forecasters are often caught up in the need to identify the problem of the day, the potential storm or disturbance that they will need to track. Example 2 shows the importance of sensemaking for *connecting the dots* and making discoveries. A third function of sensemaking is *forming explanations*, as when a physician diagnoses an illness or a mechanic figures out how a device works. A fourth function is *anticipatory thinking*, to prevent potential accidents. Fifth, people such as pilots engage in sensemaking to *project future states* in order to prepare for them. Sixth, we use sensemaking to *find the levers*—to figure out how to think and act in a situation, as when a project team tries to decide what type of projector to purchase and realizes that the decision is basically a trade-off between cost, size, and functionality. Seventh, sensemaking lets us *see relationships*, as when we use a map to understand where we are located. Eighth, sensemaking enables *problem identification*, as when a physics student tries to find a means of depicting the variables in a homework problem in order to find a solution strategy. Research into sensemaking should reflect these and other functions in order to provide a more comprehensive account.

Our sensemaking research has covered a variety of domains, which was useful for considering different sensemaking functions. Much of the research was supported by the Army Research Institute (Klein, Phillips, Battaglia, Wiggins, & Ross, 2002) and consisted primarily of studies of information operations specialists reviewing a series of messages from a peacekeeping mission. A second research activity was a review of navigation incidents in which people got lost and had to recover their bearings (Klein, Phillips, et al., 2003). A third research project was designed to train sensemaking skills involved in conducting an infantry ambush (Phillips,

Baxter, & Harris, 2003). In addition, we reviewed previous critical-incident studies of neonatal intensive care unit nurses (Crandall & Getchell-Reiter, 1993), firefighters (Klein, Calderwood, & Clinton-Cirocco, 1986), weather forecasters (Pliske, Crandall, & Klein, in press), and Navy AEGIS commanders (Kaempf, Klein, Thordsen, & Wolf, 1996).

THE DATA–FRAME THEORY OF SENSEMAKING

The data–frame theory postulates that elements are explained when they are fitted into a structure that links them to other elements. We use the term *frame* to denote an explanatory structure that defines entities by describing their relationship to other entities. A frame can take the form of a *story*, explaining the chronology of events and the causal relationships between them; a *map*, explaining where we are by showing distances and directions to various landmarks and showing routes to destinations; a *script*, explaining our role or job as complementary to the roles or jobs of others; or a *plan* for describing a sequence of intended actions. Thus, a frame is a structure for accounting for the data and guiding the search for more data. It reflects a person's compiled experiences.

People explore their environment by attending to a small portion of the available information. The data identify the relevant frame, and the frame determines which data are noticed. *Neither of these comes first.* The data elicit and help to construct the frame; the frame defines, connects, and filters the data. "The frame may be in error, but until feedback or some other form of information makes the error evident, the frame is the foundation for understanding the situation and for deciding what to do about it" (Beach, 1997, p. 24).

Sensemaking begins when someone experiences a surprise or perceives an inadequacy in the existing frame and the existing perception of relevant data. The active exploration proceeds in both directions, to improve or replace the frame and to obtain more relevant data. The active exploration of an environment, conducted for a purpose, reminds us that sensemaking is an active process and not the passive receipt and combination of messages.

The data–frame relationship is analogous to a hidden-figures task. People have trouble identifying the hidden figure until it is pointed out to them, and then they can't *not* see it. Once the frame becomes clear, so do the data.[1]

There is an ample literature on frames (Goffman, 1974; Minsky, 1975; Rudolph, 2003; Smith, Giffin, Rockwell, & Thomas, 1986) and similar concepts such as scripts (Schank & Abelson, 1977) and schemata (Barlett,

---

[1]We are indebted to Erik Hollnagel for pointing out this perceptual analogy of the data–frame theory, illustrating how top-down and bottom-up processing occur in parallel.

1932; Neisser, 1976; Piaget, 1952, 1954). Rudolph (2003) asserts that the way people make sense of situations is shaped by cognitive frames that are internal images of external reality. For Minsky, a frame is a type of data structure in which individual cases are defined in terms of a set of features on each of a number of feature parameters, which are used to organize a listing of case attributes and provide a common representational format for all cases of the same type. For Goffman, a frame is a culturally relative system of rules, principles, and so on, that are used to organize society and guide individual behavior. For Schank and Abelson, a script is a regularly occurring sequence of events or activities that can be formulated as a template for structuring a description of particular instances of events or activities of the same kind. For Piaget, a schema is a mental representation of the persistent features or attributes of objects. For Barlett, a schema is a mental representation of the structure of event descriptions (e.g., stories), usually taking the form of regularly occurring and culture-specific sequences of dramatic subevents. For Neisser, a schema is a situation- or domain-specific cognitive structure that directs external information search, guides attention management, organizes information in memory and directs its retrieval, and becomes more differentiated as a function of experience.

Our account draws on the work of these researchers. Rather than trying to untangle the heritage and distinctions between the descriptions of frames, scripts, and schemata, we have chosen to use the term *frame* as a synthesis of these concepts.

Sensemaking is a process of framing and reframing, of fitting data into a frame that helps us filter and interpret the data while testing and improving the frame and cyclically moving forward to further adapt the frame. The purpose of a frame is to define the elements of the situation, describe the significance of these elements, describe their relationship to each other, filter out irrelevant messages, and highlight relevant messages. Frames can organize relationships that are spatial (maps), causal (stories and scenarios), temporal (stories and scenarios), or functional (scripts).

What is not a frame? A frame is not a collection of inferences drawn from the data, although it can include inferences. Paraphrasing Neisser (1976, p. 54), we would say that a frame is that portion of the perceptual cycle that is internal to the perceiver, modifiable by experience, and specific to what is being perceived.

The frame "accepts" information as it becomes available and the frame is changed by that information; it is used to "direct" exploratory activities that make more data available, by which it is further modified.

Our description of sensemaking has some points of overlap with Endsley's (1995) concept of *situation awareness* as a state of knowledge. However, the concept of sensemaking is different from situation awareness in several ways. One difference is that sensemaking is more than accurate retrieval of relevant information and inferences. Sensemaking is directed at performing functions such as the ones described earlier: problem detection, problem identification, anticipatory thinking, forming explanations, seeing relationships, as well as projecting the future (which is Level 3 of Endsley's model).

Another difference is that we are interested in sensemaking as a process, not a state of knowledge. Endsley has referred to situation *assessment* as the process used to achieve the state of situation awareness. Our approach differs from Endsley's in that she is describing how people notice and make inferences about data, whereas we assert that people use their frames to define what counts as data in the first place. We view sensemaking as a process of constructing data as well as meaning.

The data–frame theory of sensemaking consists of nine assertions, which are listed in Table 6.1. The subsequent sections discuss each assertion.

1. Sensemaking Is the Process of Fitting Data Into a Frame and Fitting a Frame Around the Data. We distinguish between two entities: data and frames. Data are the interpreted signals of events; frames are the explanatory structures that account for the data. People react to data elements by trying to find or construct a story, script, a map, or some other type of structure to account for the data. At the same time, their repertoire of frames—explanatory structures—affects which data elements they consider and how they will interpret these data. We see sensemaking as the effort to balance these two entities—data and frames. If people notice data that do not fit into the frames they've been using, the surprise will often initiate sensemaking to modify the frame or replace it with a better one. Another reaction would be to use the frame to search for new data or to reclassify existing data, which in turn could result in a discovery of a better frame. Weick (1995) stated

TABLE 6.1
Assertions of the Data–Frame Theory of Sensemaking

1. Sensemaking is the process of fitting data into a frame and fitting a frame around the data.
2. Therefore, the "data" are inferred, using the frame, rather than being perceptual primitives.
3. The frame is inferred from a few key anchors.
4. The inferences used in sensemaking rely on abductive reasoning as well as logical deduction.
5. Sensemaking usually ceases when the data and frame are brought into congruence.
6. Experts reason the same way as novices, but have a richer repertoire of frames.
7. Sensemaking is used to achieve a functional understanding—what to do in a situation—as well as an abstract understanding.
8. People primarily rely on just-in-time mental models.
9. Sensemaking takes different forms, each with its own dynamics.

that "Sensemaking is about the enlargement of small cues. It is a search for contexts within which small details fit together and make sense . . . It is a continuous alternation between particulars [data] and explanations [frames]" (p. 133).

2. The "Data" Are Inferred, Using the Frame, Rather Than Being Perceptual Primitives. Data elements are not perfect representations of the world but are constructed—the way we construct memories rather than remembering all of the events that took place. Different people viewing the same events can perceive and recall different things depending on their goals and experiences. (See Medin, Lynch, Coley, & Atran, 1997, and Wisniewski & Medin, 1994, for discussions of the construction of cues and categories.) A fire-ground commander, an arson investigator, and an insurance agent will all be aware of different cues and cue patterns in viewing the same house on fire. The commander will look at the smoke to determine how intense the fire is and how much of a risk it might pose to crew members ordered to cut a hole in the roof to ventilate the blaze. The arson investigator will look at the color of the smoke for signs that the fire was accelerated by gasoline or some other means. The arson investigator will track how the fire is spreading in order to infer where it started, whereas the commander will be more interested in gauging where it is going to spread in the future. And the insurance investigator will monitor the extent of damage—by fire, by smoke, by the water used to fight the fire—in order to judge what portions of the house can be recovered.

Because the little things we call "data" are actually abstractions from the environment, they can be distortions of reality. Feltovich, Spiro, and Coulson (1997) described the ways we simplify the world in trying to make sense of it:

- We define continuous processes as discrete steps.
- We treat dynamic processes as static.
- We treat simultaneous processes as sequential.
- We treat complex systems as simple and direct causal mechanisms.
- We separate processes that interact.
- We treat conditional relationships as universals.
- We treat heterogeneous components as homogeneous.
- We treat irregular cases as regular ones.
- We treat nonlinear functional relationships as linear.
- We attend to surface elements rather than deep ones.
- We converge on single interpretations rather than multiple interpretations.

In order to consciously deliberate about the significance of events, we must abstract those events into propositions that we can use in our reasoning. This process can result in distortions of sensemaking.

The process of abstraction is what allows us to function in an infinitely complex environment. Distortion is a small price to pay for being able to engage in abstract reasoning and deliberate sensemaking. If the cues we abstract are oversimplifications, our explanations will also be oversimplified, which is why Weick sees sensemaking as a continuous cycle of moving toward better particulars and better explanations.

Weick posits an additional source of distortion—enactment. People are not passively receiving information but are actively transforming their worlds. "'Distortion' in other words is essentially inevitable because we have a hand—literally—in what we sense. What the world is without our enacting is never known since we fiddle with that world to understand it" (K. E. Weick, personal communication, 2003).

This perspective is different from an information-processing account in which data are perceived and become the basis for generating inferences. This passive view has been criticized from the time of John Dewey (1910). Our research in fields such as information operations, command and control, and intensive care unit nursing has not found neat packets of data presented to decision makers. Instead, expertise was needed to select and define cues.

For example, consider a driver who wants to avoid collisions. One obvious datum is the distance to the car ahead. However, this may not be such a useful cue. After all, that distance is minimal when a driver is engaged in parallel parking, but the car is usually moving too slowly to run much risk. For an experienced driver, spotting a dog running loose can be a more relevant cue than the "safe" current distance to the car ahead.

The identification of data elements depends on background experience and on a repertoire of frames. Our disagreement with an information-processing account of sensemaking is that we do not believe that the inputs are clear-cut cues. We see people as active interpreters and constructors of cues and active judges of what is relevant and deserving of attention. Sensemakers are doing more than registering cues and deriving inferences based on these cues.

3. The Frame Is Inferred From a Few Key Anchors.

When we encounter a new situation or a surprising turn of events, the initial one or two key data elements we experience sometimes serve as anchors for creating an understanding.[2] These anchors elicit the initial frame, and we use that frame to

---

[2]Feltovich, Johnson, Moller, and Swanson (1984) use the term "anchorage points" to refer to the elements used for subsequent elaboration of a frame.

search for more data elements.[3] Klein and Crandall (1995) presented a description of mental simulation that uses a mechanical-assembly metaphor to illustrate how people form explanations and project into the future. In their account, people assemble frames from a limited set of causal factors, gauge the plausibility of the assembly, and then "run" the assembly to imagine how events might have been caused, or to imagine how events might play out.

Following the research of Klein and Crandall (1995) on mental simulation, we further assert that people rely on *at most* three or four anchors in deriving a frame. Recently, Debra Jones (personal communication, May 2003) found that the selection of frames by weather forecasters attempting to understand severe weather conditions tended to be based on only two to three anchors.

Example 3 shows the process of using anchors to select frames. The decision maker treated each data element as an anchor and used it to frame an explanation for an aviation accident. Each successive data element suggested a new frame, and the person moved from one frame to the next without missing a beat.

Example 3: The Investigation of a Helicopter Accident. An accident happened during an Army training exercise. Two helicopters collided. Everyone in one helicopter died and everyone in the other helicopter survived. Our informant, Captain B., was on the battalion staff at the time.

Immediately after the accident, Captain B. suspected that because this was a night mission there could have been some complications due to flying with night-vision goggles that led one helicopter to drift into the other.

Then Captain B. found out that weather had been bad during the exercise, and he thought that was probably the cause of the accident; perhaps they had flown into some clouds at night.

Then Captain B. learned that there was a sling on one of the crashed helicopters, and that this aircraft had been in the rear of the formation. He also found out that an alternate route had been used, and that weather wasn't a factor because they were flying below the clouds when the accident happened. So Captain B. believed that the last helicopter couldn't slow down properly because of the sling. The weight of the sling would make it harder to stop to avoid running into another aircraft. He also briefly suspected that pilot experience was a contributing factor, because they should have understood the risks better and kept better distance between aircraft, but he dismissed this idea because he found out that although the lead pilot hadn't flown much recently, the copilot was very experienced. But Cap-

---

[3]The strategy of using anchors in this way is compatible with the anchoring and adjustment heuristic (Tversky & Kahneman, 1974).

tain B. was puzzled about why the sling-loaded helicopter would have been in trail. It should have been in the lead because it was less agile than the others. Captain B. was also puzzled about the route—the entire formation had to make a big U-turn before landing and this might have been a factor too. So this story, though much different than the first ones, still had some gaps.

Finally, Captain B. found out that the group had not rehearsed the alternate route. The initial route was to fly straight in, with the sling-loaded helicopter in the lead. And that worked well because the sling load had to be delivered in the far end of the landing zone. But because of a shift in the wind direction, they had to shift the landing approach to do a U-turn. When they shifted the landing approach, the sling load had to be put in the back of the formation so that the load could be dropped off in the same place. When the lead helicopter came in fast and then went into the U-turn, the next two helicopters diverted because they could not execute the turn safely at those speeds and were afraid to slow down because the sling-loaded helicopter was right behind them. The sling-loaded helicopter continued with the maneuver and collided with the lead helicopter.

At first, Captain B. had a single datum, the fact that the accident took place at night. He used this as an anchor to construct a likely scenario. Then he learned about the bad weather, and used this fact to anchor an alternate and more plausible explanation.

Next he learned about the sling load, and fastened on this as an anchor because sling loads are so dangerous. The weather and nighttime conditions may still have been factors, but they did not anchor the new explanation, which centered around the problem of maneuvering with a sling load. Captain B.'s previous explanations faded away. Even so, Captain B. knew his explanation was incomplete, because a key datum was inconsistent—why was the helicopter with the sling load placed in the back of the formation?

Eventually, he compiled the anchors: helicopter with a sling load, shift in wind direction, shift to a riskier mission formation, unexpected difficulty of executing the U-turn. Now he had the story of the accident. He also had other pieces of information that contributed, such as time pressure that precluded practicing the new formation, and command failure in approving the risky mission.

4. The Inferences Used in Sensemaking Rely on Abductive Reasoning as Well as Logical Deduction. Our research with information operations specialists (Klein, Phillips, et al., 2003) found that they rely on abductive reasoning to a greater extent than following deductive logic. For example, they were more likely to speculate about causes, given effects, than they were to deduce effects from causes. If one event preceded another (simple correlation) they speculated that the first event might have caused the second.

They were actively searching for frames to connect the messages they were given.

Abductive reasoning (e.g., Peirce, 1903) is reasoning to the best explanation. Josephson and Josephson (1994, p. 5) have described a paradigm case for abductive reasoning:

D is a collection of data (facts, observations, and givens).

H explains D (would, if true, explain D).

No other hypothesis can explain D as well as H does.

Therefore, H is probably true.

In other words, if the match between a set of data and a frame is more plausible than the match to any other frame, we accept the first frame as the likely explanation. This is not deductive reasoning, but a form of reasoning that enables us to make sense of uncertain events. It is also a form of reasoning that permits us to generate new hypotheses, based on the frame we adopt.

Weick (1995) and Lundberg (2000) have argued that effective sensemaking relies more heavily on plausibility, pragmatics, coherence, and reasonableness, the criteria employed by abduction, than on accuracy.

Consider this observation from research by Crandall and Getchell-Reiter (1993) with nurses. A neonatal intensive care unit (NICU) nurse did not use the logical form "if a baby has sepsis, then it will show these symptoms," because the symptoms of sepsis are not straightforward and have to be understood in context. The nurse reasoned: "I am seeing these symptoms (which in my experience have been connected with sepsis); therefore perhaps this baby has sepsis." In deduction, we would call this the fallacy of affirming the consequent ("If A, then B; B, therefore A"). However, the nurse was not engaged in deductive reasoning, and her reasoning strategy seems highly appropriate as a way to generate hypotheses as well as to evaluate them.

Josephson and Josephson (1994) compiled a large set of research efforts examining abductive-reasoning processes used in determining best explanations. These research directions can be useful for further investigation of the reasoning strategies found in sensemaking. (Also see the work of Collins, Burstein, & Michalski, 1987, about a theory of human plausible reasoning, and Marek & Truszczynski, 1993, on the nature of a logical theory of educated and justifiable guesses.) Pennington and Hastie (1993) and Klein and Crandall (1995) have discussed the judgment of plausibility in generating mental simulations and stories.

Our observations with the information operations specialists supported the importance of developing accounts of abductive reasoning. People are explanation machines. People will employ whatever tactics are available to help them find connections and identify anchors. The information operations specialists we studied found connections between all sorts of messages, not just the themes we had intended. They seemed unable to stop themselves from seeing linkages and deriving inferences.

Doherty (1993) showed that this tendency to overexplain can lead to pseudocorrelation (finding connections that do not exist in the data). However, we also need to consider the losses that might accrue if people could not use minimal evidence to speculate about relationships. We assert that abductive reasoning predominates during the process of sensemaking.

5. Sensemaking Usually Ceases When the Data and Frame Are Brought Into Congruence.   Our observations of sensemaking suggest that, as a deliberate activity, it continues as long as key data elements remain unexplained or key components of a frame remain ambiguous. Once the relevant data are readily accounted for, and the frame seems reasonably valid and specified, the motivation behind sensemaking is diminished. Thus, sensemaking has a stopping point—it is not an endless effort to grind out more inferences.

We note that sensemaking may continue if the potential benefits of further exploration are sufficiently strong. Example 2 shows how the brigadier general continued to make discoveries by further exploring the scene in front of him.

6. Experts Reason the Same Way as Novices, but Have a Richer Repertoire of Frames.   Klein et al. (2002) reported that expert information operations specialists performed at a much higher level than novices, but both groups employed the same types of logical and abductive inferencing. This finding is in accord with the literature (e.g., Barrows, Feightner, Neufeld, & Norman, 1978; Elstein, Shulman, & Sprafka, 1978; Simon, 1973) reporting that experts and novices showed no differences in their reasoning processes.

Klein et al. (2002) found that expert and novice information operations specialists both tried to infer cause–effect connections when presented with information operations scenarios. They both tried to infer effects, although the experts were more capable of doing this. They both tried to infer causes from effects. They were both aware of multiple causes, and neither were particularly sensitive to instances when an expected effect did not occur. The experts have the benefit of more knowledge and richer mental models. But in looking at these data and in watching the active connection building of the novices, we did not find that the experts were using different sensemaking strategies from the novices (with a few exceptions, noted later). This finding suggests that little is to be gained by trying to teach novices to think like experts. Novices are already trying to make connections. They just have a limited knowledge base from which to work.

The information operations specialists, both experts and novices, were not going about the task in accordance with the precepts of deductive logic. They were not averse to speculating about causes, given an effect. And this is appropriate. The task was not about making logical deductions. It was about making sense of the situation. The participants were in a speculative mode, not a mode of conclusive reasoning. The only exception was that a few of the novices who had just previously worked as intelligence analysts at first seemed reluctant to speculate too much without more data. None of the experts exhibited this scruple.

Although the experts and novices showed the same types of reasoning strategies, the experts had the advantage of a much stronger understanding of the situations. Their mental models were richer in terms of having greater variety, finer differentiation, and more comprehensive coverage of phenomena. Their comments were deeper, more plausible, showed a greater sensitivity to context, and were more insightful.

The mechanism of generating inferences was the same for experts and novices, but the nature of the experts' inferences was much more interesting. For example, one of the three scenarios we used, "Rebuilding the Schools," included the message, "An intel report designates specific patrols that are now being targeted in the eastern Republica Srpska. A group that contains at least three known agitators is targeting the U.S. afternoon Milici patrol for tomorrow and the IPTF patrol for the next day." One of the experts commented that this was very serious, and generated action items to find out why the posture of the agitators had shifted toward violence in this way. He also speculated on ways to manipulate the local chief of police. The second expert saw this as an opportunity to identify and strike at the violent agitators. In contrast, one of the novices commented, "There is a group targeting U.S. afternoon patrols—then move the patrols." The other novice stated, "The targeting of the patrols—what that means is not clear to me. Is there general harassment or actions in the past on what has happened with the patrols?"

Generally, the novices were less certain about the relevance of messages, and were more likely to interpret messages that were noise in the scenario as important signals. Thus, in "Rebuilding the Schools," another message stated that three teenagers were found in a car with contraband cigarettes. One expert commented, "Business as usual." The second expert commented, "Don't care. You could spend a lot of resources on controlled cigarettes—and why? Unless there is a larger issue, let it go." In contrast, the novices became concerned about this transgression. One wondered if the teenagers were part of a general smuggling gang. Another wanted more data about where the cigarettes came from. A third novice wanted to know what type of suspicious behavior had gotten the teenagers pulled over.

Vicente and Wang (1998) and Klein and Peio (1989) showed that experts can take advantage of redundancies and constraints in situations in order to generate expectancies. The frames that experts have learned provide the basis for these redundancies and constraints.

For instance, if you are a platoon leader in an infantry company, you may be assigned the mission of counterreconnaissance. More specifically, you may be asked to set up an ambush. You would have a script—a general frame—available for ambushes. You would also have a lot of ambush-relevant experiences, such as what has happened during training when you triggered the ambush too early, or about the need to maintain communications between the squads so that the lookouts can easily signal to the fire teams. These causal and functional relationship components are synthesized into your understanding of how to conduct an ambush in the current situation. Your frame/script may have gaps—such as what to do about civilians—but you can fill these gaps (Minsky, 1975, called this "slot-filling") by making assumptions and inferences.

Using the lessons from recent military actions, you may not trust civilians as you might have prior to the U.S. invasion of Iraq. You know that civilians can be booby-trapped as suicide bombers or can call in attacks once you let them go. You can apply this knowledge in a variety of frames, such as dealing with ambushes, guarding a compound, or escorting a convoy.

Experts also appear to have more routines—more ways of accomplishing things, which widens the range of frames they can draw on.[4] In addition, experts have a broad set of associations to draw on. The skilled information operations specialists read more into the messages we showed them—they were aware of distinctions and related knowledge; whereas the novices read the messages at a surface level and lacked a knowledge base for enriching the material.

In addition, experts have a better appreciation of the functions they want to accomplish in a situation. This functional orientation guides their exploration. We discuss the idea of a functional understanding in the next section.

7. Sensemaking Is Used to Achieve a Functional Understanding. In many settings, experienced practitioners want a functional understanding as well as an abstract understanding. They want to know what to do in a situation. In some domains, an abstract understanding is sufficient for experts. Scientists are usually content with gaining an abstract understanding of events and domains because they rarely are called on to act on this understanding. In-

---

[4]The concepts of frames and routines are linked, as described by the recognition-primed decision model (Klein, 1998). Once a person adopts a frame for understanding a situation, the frame will frequently include a set of routines for typical ways of taking action.

telligence officers can seek an abstract understanding of an adversary if they cannot anticipate how their findings will be applied during military missions. In other domains, experts need a functional understanding along with an abstract understanding. Weather forecasters can seek a functional understanding of a weather system if they have to issue alerts or recommend emergency evacuations in the face of a threatening storm.

The experienced information operations specialists we studied wanted both a functional and an abstract understanding. They wanted to know what they could do in a situation, not merely what was going on. For example, one of our information operations scenarios included a theme about refugees in Bosnia preparing for winter. Our material included a message that seven Bosnian families, attempting to reinhabit their homes in Breske, had met with minor skirmishes and were heading to a refugee camp until the situation could be resolved. One of the experts commented, "Is this coordinated with the United Nations organization overseeing refugees in the area? If not, it should be." Other experts made similar comments about what should be done here. None of the novices mentioned taking any action at all. For the three scenarios we used, the experts were almost three times as likely to make comments about actions that should be taken, compared to the novices. The novices we studied averaged 1.5 action suggestions per scenario and the experts averaged 4.4 action suggestions per scenario.

Chi, Feltovich, and Glaser (1981) studied the sensemaking of subjects solving physics problems. The experts showed a functional understanding of the problems: They invoked principles that they could use to solve the problem, not just principles about the abstract form of the problem. Chi et al. found that the rules generated by novices did not contain many actions that were explicitly tied to solution procedures. The novices knew what problem cues were relevant, but did not know what to do with this knowledge.

Charness (1979) reported the results of a cognitive task analysis conducted with bridge players. The sensemaking of the skilled bridge players was based on what they could or could not achieve with a hand—the affordances of the hand they were dealt. In contrast, the novices interpreted bridge hands according to more abstract features such as the number of points in the hand.

Sensemaking involves developing an understanding of what can be accomplished and how capabilities can be expanded. Thus, a novice rock climber surveying a cliff for holds will understand the cliff face differently from someone who is much taller, much stronger, and actually has experience in climbing. The experienced climber will be looking in different places for holds and will have a wider repertoire of frames for detecting potential holds—indentations in the rock, protrusions, patterns that permit different wedging techniques, and so forth. The inspection is not about the nature of the rock face but is about what can and cannot be achieved, and what can be used as a hold.

Rudolph (2003) refers to "action-motivated sensemaking." The anesthesiology residents she studied were not simply attempting to arrive at a diagnosis: They were working with a sophisticated simulation and attempting to provide anesthesia in preparation for an appendectomy. They were interacting with the simulated patient, learning from their interventions.

The experts also showed much more anticipatory sensemaking—identifying actions that needed to be taken. For example, in "Rebuilding the Schools," a message stated that the commander was meeting with the mayor of Brcko. An expert commented that there was a need to prepare the commander for this meeting and to alert the commander to focus on what was currently happening in the schools in Brcko. Experts generated more than four times as many anticipatory action items such as this, compared to novices.

Weick (1995) has taken the functional understanding further, arguing that just as belief can lead to action, action can lead to belief. People sometimes manipulate situations as a way of establishing meaning, a sort of a "ready, aim, fire" mentality. By initiating action, people reduce uncertainty and equivocation. "People act in order to think, but in their acting, shape the cues they interpret" (K. E. Weick, personal communication, 2003).

8. People Primarily Rely on Just-in-Time Mental Models. We define mental models as our causal understanding of the way things work. Thus, we have mental models of the way the brakes in our car work, the way our computer works, and the way our project team is supposed to work to prepare a document. Mental models are another form that frames can take, along with stories, scripts, maps, and so on. Mental models can be stories, as when we imagine the sequence of events from stepping on a brake pedal to slowing down the velocity of our car.

We distinguish between comprehensive mental models and just-in-time (JIT) mental models. A comprehensive mental model captures the essential relationships. An automobile mechanic has a comprehensive mental model of the braking system of a car. An information technology specialist has a comprehensive mental model of the operating system of a computer. In contrast, most of us have only incomplete ideas of these systems. We have some knowledge of the components, and of the basic causal relationships, but there are large gaps in our understanding. If we have to do our own troubleshooting, we have to go beyond our limited knowledge, make some inferences, and cobble together a notion of what is going on—a JIT mental model. We occasionally find that even the specialists to whom we turn don't have truly complete mental models, as when a mechanic fails to diagnose and repair an unusual problem or an information technology specialist needs to consult with the manufacturer to figure out why a computer is behaving so strangely.

The concept of JIT is not intended to convey time pressure. It refers to the construction of a mental model at the time it is needed, rather than calling forth comprehensive mental models that already have been developed.

In most of the incidents we examined, the decision makers did not have a full mental model of the situation or the phenomenon they needed to understand. For example, one of the scenarios we used with information operations specialists contained three critical messages that were embedded in a number of other filler messages: The sewage system in a refugee camp was malfunctioning, refugees were moving from this camp to a second camp, and an outbreak of cholera was reported in the second camp. We believed that the model of events would be clear—the refugees were getting sick with cholera in the first camp because of the sewage problem and spreading it to the second camp. However, we discovered that none of the information operations specialists, even the experts, understood how cholera is transmitted. Therefore, none of them automatically connected the dots when they read the different messages. A few of the specialists, primarily the experts, did manage to figure out the connection as we presented more pointed clues. They used what they knew about diseases to speculate about causes, and eventually realized what was triggering the cholera outbreak in the second camp.

We suggest that people primarily rely on JIT mental models—building on the local cause–effect connections they know about, instead of having comprehensive mental models of the workings of an entire system. These JIT mental models are constructions, using fragmentary knowledge from long-term memory to build explanations in a context. Just as our memories are partial constructions, using fragments of recall together with beliefs, rules, and other bases of inference, we are claiming that most of our mental models are constructed as the situation warrants.

Experienced decision makers have learned a great many simple causal connections, "A" leads to "B," along with other relationships. When events occur that roughly correspond to "A" and "B," experienced decision makers can see the connection and reason about the causal relationship. This inference can become an anchor in its own right, if it is sufficiently relevant to the task at hand. And it can lead to a chain of inferences, "A" to "B," "B" to "C," and so on.

We believe that in many domains, people do not have comprehensive mental models but can still perform effectively. The fragmentary mental models of experts are more complete than those of novices.

Feltovich, Johnson, Moller, and Swanson (1984) have shown that there are domains in which experts do have what seem to be fully worked out mental models. Highly experienced pediatric cardiologists have gained a very sophisticated and in-depth understanding of the mechanism for certain types of heart functions. It may be useful to examine the kinds of environments in which people tend to have more complete mental models as opposed to fragmentary causal knowledge. For example, the cardiac system could be considered a closed system, whereby a highly elaborated understanding of the working of this organ can be developed and refined as science progresses. Contrast a surgeon's mental model of a heart with a soldier's development of multiple frames for learning how to conduct peacekeeping operations in Iraq.

We continue to assert that in most cases even experts are assembling mental models on the spot, drawing from a wider repertoire of cause–effect relationships than novices. One implication of this assertion is that we may be better advised to train people by helping them expand their repertoire and depth of causal relationships, rather than trying to teach fully worked out mental models within a domain.

9. Sensemaking Takes Different Forms, Each With Its Own Dynamics. In studying the process of sensemaking in information operations specialists and in other domains, we found different types. If we ignore these types, our account of sensemaking will be too general to be useful. The next section describes the alternative ways that sensemaking can take place.

THE FORMS OF SENSEMAKING

Figure 6.1 shows seven types of sensemaking we have differentiated thus far: mapping data and frame, elaborating a frame, questioning a frame, preserving a frame, comparing frames, re-framing, and constructing or finding a frame. Although all of them are types of sensemaking, they operate in different ways. The activity of preserving a frame is different from constructing a new frame.

Any of the forms shown in Figure 6.1 can be a starting point for sensemaking. Sometimes data will call for elaborations or changes in a frame, and sometimes a frame will drive the search for data and the definition of what counts as data. Sometimes an anomaly will trigger sensemaking, and at other times a person will have to choose between competing frames that both appear plausible. Therefore, there is no starting point in Figure 6.1.

Some incidents we studied consisted of several of the activities shown in Figure 6.1, whereas other incidents demonstrated only a few. Moreover, the order of activities varied—sometimes people constructed a frame before coming to question it, and at other times people simply recognized frames

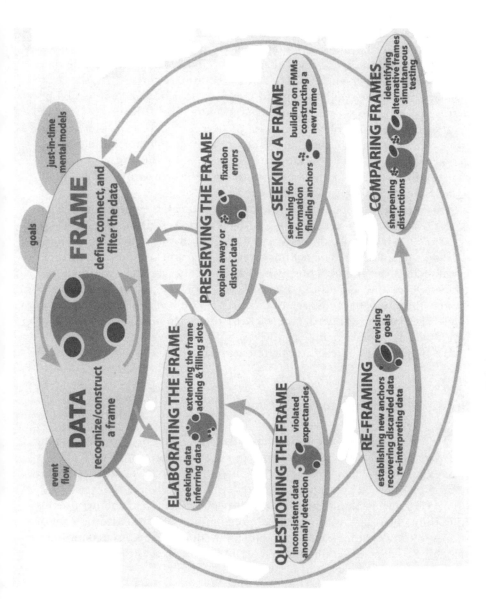

FIG. 6.1.   Sensemaking activities.

133

and never had to do any construction. Sometimes people alter the data to fit the frame, and sometimes they alter the frame.

The cycle of elaborating the frame and preserving it in the face of inconsistent data is akin to Piaget's (1952) function of assimilation. The process of reframing is akin to accommodation.

The top oval in Figure 6.1 is the basic cycle of frames that define data and data that help to elicit frames. In most natural settings, this cycle is ongoing because new events occur and because the actions taken by the person trying to make sense of the events will alter them. Figure 6.1 can be seen as an adaptation of Neisser's (1976) perceptual cycle, in which a person's perceptual exploration and other actions sample and influence the actual world and the information in it, thereby modifying the cognitive map of the world and the understanding of the present environment, which directs further exploration and action.

This section describes each of the seven sensemaking types. Within each of the ovals in Figure 6.1, we show an icon to represent the nature of the sensemaking activity. Actions listed on the left are ways that the frame is driving data collection, and actions listed on the right are ways that the data are affecting the frame. Thus, within each of the activities the data and frame define and update one another. We show reciprocal arrows in the central loop to reflect this relationship. We would have included similar data–frame reciprocal arrows in each of the ovals but decided that this would add too much clutter to the diagram.

We begin with an account of the basic data–frame match.

Sensemaking Attempts to Connect Data and a Frame

The specific frame a person uses depends on the data or information that are available and also on the person's goals, the repertoire of the person's frames, and the person's stance (e.g., current workload, fatigue level, and commitment to an activity).[5]

We view sensemaking as a volitional process, rather than an unconscious one. In many instances the automatic recognition of how to frame a set of events will not require a person to engage in deliberate sensemaking. The matching of data and frame is often achieved preconsciously, through pattern matching and recognition.

We are not dismissing unconscious processes as concomitants of sensemaking. Rather, we are directing our investigation into incidents where deliberate reasoning is employed to achieve some level of under-

---

[5]Weick (1995) has argued that a person's commitment should be counted as an aspect of stance—being committed to an endeavor changes the way events are understood.

standing. Any cognitive activity, no matter how deliberate, will be influenced by unconscious processes. The range between conscious and unconscious processes can be thought of as a continuum, with pattern matching at one end and comparing different frames at the other end. The blending of conscious and automatic processes varies from one end of the continuum to the other.

To illustrate this continuum, consider the recognition-primed decision (RPD) model presented by Klein (1998). Level 1 of the RPD model describes a form of decision making that is based on pattern matching and the recognition of typical situations. The initial frame recognition defines cues (the pertinent data), goals, and expectancies. This level of the RPD model corresponds to the basic data–frame connection shown at the top center of Figure 6.1, and in the icon at the beginning of this section. We do not view this recognitional match as an instance of sensemaking.

Klein also described Level 2 of the RPD model in which the decision maker *does* deliberate about the nature of the situation, sometimes constructing stories to try to account for the observed data. At Level 2, the decision maker is engaging in the different forms of sensemaking. The data–frame theory is an extension of the story-building strategy described in this Level 2.

The initial frame used to explain the data can have important consequences. Thus, a fireground commander has stated that the way an on-scene commander sizes up the situation in the first 5 minutes determines how the fire will be fought for the next 5 hours. Chi et al. (1981), studying the way physics problems were represented, found that novices adopted fairly literal representations, using superficial problem features rather than the pertinent underlying principles at work. In contrast, experts used deeper representations of force, momentum, and energy. Chi et al. argued that a problem was initially tentatively identified through a "shallow" analysis of the problem features (i.e., the initial frame), and then the representation was extended (i.e., elaborated) using knowledge associated with that frame or category.

Elaborating the Frame

As more is learned about the environment, people will extend and elaborate the frame they are using, but will not seek to replace it as long as no surprises or anomalies emerge. They add more details, fill in slots, and so forth.

Example 2 showed how a Marine Corps brigadier general could see much more than a sergeant, using his experience base to direct his exploration. He spotted not just the tank that the sergeant had seen, but also other tanks in the platoon and the command post that the tanks were defending. As the general extended his understanding, he did not replace any of his initial observations or data elements.

Questioning the Frame

Questioning begins when we are surprised—when we have to consider data that are inconsistent with the frame we are using. This is a different activity than elaborating the frame. Lanir (1991) has used the term "fundamental surprise" to describe situations in which we realize we may have to replace a frame on which we had been depending. He contrasted this with situational surprise, in which we merely need to adjust details in the frame we are using.

In this aspect of sensemaking, we may not know if the frame is incorrect, if the situation has changed, or if the inconsistent data are inaccurate. At this point, we just realize that some of the data do not match the frame. Frames provide people with expectations; when the expectations are violated, people may start to question the accuracy of the frame.

Weick (1995) postulated that sensemaking is often initiated by a surprise—the breakdown in expectations as when unexpected events occur or expected events fail to occur. Our research findings support this assertion. Thus, in the navigation incidents we studied, we found that people who got lost might continue for a long time until they experienced a framebreaker, a "moment of untruth," that knocked them out of their existing beliefs and violated their expectancies. However, as shown in Example 2, contrasting the way a Marine brigadier general and a sergeant conducted a reconnaissance mission, there are times when people try to make sense of events because they believe they can deepen their understanding, even without any surprise. Even in the navigation incidents we studied, people were often trying to make sense of events even before they encountered a frame breaker. They actively preserved and elaborated their frames until they realized that their frames were flawed.

Emotional reactions are also important for initiating sensemaking, by generating a feeling of uncertainty or of distress caused by loss of confidence in a frame.

Problem detection (e.g., Klein, Pliske, Crandall, & Woods, in press) is a form of sensemaking in which a person begins to question the frame. However, a person will not start to question a frame simply as a result of receiving data that do not conform to the frame. For example, Feltovich et al. (1984) found that experts in pediatric cardiology had more differentiated frames than novices, letting them be more precise about expectations. Novices were fuzzy about what to expect when they looked at test results, and therefore were less likely to notice when expectancies were violated. As a result, when Feltovich et al. presented a "garden path" scenario (one that suggested an interpretation that turned out to be inaccurate), novices went down the garden path but experts broke free. The experts quickly noticed that the data were not aligning with the initial understanding.

Example 4 describes a fire captain questioning his frame for the progression of a fire involving a laundry chute (taken from Klein, 1998). He realizes that the data do not match his frame when he sees flames rolling along the ceiling of the fourth-floor landing; at this point he recognizes the fire is much more involved than he originally expected, and he is able to quickly adapt his strategies to accommodate the seriousness of the fire.

Example 4: The Laundry Chute Fire. A civilian call came in about 2030 hours that there was a fire in the basement of an apartment complex. Arriving at the scene of the call about 5 minutes later, Captain L. immediately radioed to the dispatcher that the structure was a four-story brick building, "nothing showing," meaning no smoke or flames were apparent. He was familiar with this type of apartment building structure, so he and his driver went around the side of the building to gain access to the basement through one of the side stairwells located to either side of the building. Captain L. saw immediately that the clothes chute was the source of the fire.

Looking up the chute, which ran to the top floor, Captain L. could see nothing but smoke and flames. The duct was constructed of thin metal but with a surrounding wooden skirt throughout its length. Visibility was poor because of the engulfing flames, hampering the initial appraisal of the amount of involvement. Nevertheless, on the assumption that the civilian call came close in time to the start of the fire and that the firefighters' response time was quick, Captain L. instantly assessed the point of attack to be the second-floor clothes chute access point.

Captain L. told the lieutenant in charge of the first arriving crew to take a line into the building. Three crews were sent upstairs, to the first, second, and third floors, and each reported back that the fire had already spread past them.

Captain L. made his way to the front of the building. Unexpectedly, he saw flames rolling along the ceiling of the fourth-floor landing of the glass encased front stairwell external to the building. Immediately recognizing that the fire must be quite involved, Captain L. switched his strategy to protecting the front egress for the now prime objective of search and rescue (S/R) operations. He quickly dispatched his truck crew to do S/R on the fourth floor and then radioed for a "triple two" alarm at this location to get the additional manpower to evacuate the entire building and fight this new front. Seven minutes had now elapsed since the first call.

On arrival, the new units were ordered to protect the front staircase, lay lines to the fourth floor to push the blaze back down the hall, and aid in S/ R of the entire building. Approximately 20 people were eventually evacuated, and total time to containment was about an hour.

Preserving the Frame

We typically preserve a frame by explaining away the data that do not match the frame. Sometimes, we are well-advised to discard unreliable or transient data. But when the inconsistent data are indicators that the explanation may be faulty, it is a mistake to ignore and discard these data. Vaughan (1996) describes how organizations engage in a routinization of deviance, as they explain away anomalies and in time come to see them as familiar and not particularly threatening. Our account of preserving a frame may help to describe how routinization of deviance is sustained.

Feltovich, Coulson, and Spiro (2001) have cataloged a set of "knowledge shields" that cardiologists use to preserve a frame in the face of countervailing evidence. Many of these knowledge shields minimize the importance of the contradictory data. Examples of knowledge shields include arguing from authority, resorting to bad analogies, ignoring secondary effects, arguing from special cases, and arguing that a principle has a restricted applicability.

The result of these knowledge shields is to enable a person to fixate on a frame. We do not want to judge that a person is fixating just because the person merely perseveres in the use of a frame. What, then, are the criteria for a fixation error? De Keyser and Woods (1993) studied fixation errors in the nuclear power plant industry. Their criterion for a fixation error was that the initial account was incorrect plus the decision maker maintained this incorrect explanation in the face of opportunities to revise it.

Rudolph (2003) used a garden path research design to lead participants (medical resident anesthesiologists) into fixation. The garden path paradigm provides salient anchors that suggest an obvious explanation that turns out to be incorrect once subsequent evidence is taken into account. For fixation to occur, the subsequent data are explained away as a by-product of elaborating on a frame. It is not as if people seek to fixate. Rather, people are skilled at forming explanations. And they are skilled at explaining how inconvenient data may have arisen accidentally—using the knowledge shields that Feltovich et al. (2001) described. Rudolph found that some of her participants simply ignored and/or distorted the discrepant data.

The process of preserving an inaccurate frame is well-known in land navigation, where one error leads to another, as waypoints and landmarks are misidentified, and the person "bends the map." In the navigation incidents we studied (Klein, Phillips et al., 2002), we found many examples of fixation—of preserving a frame that should have been discarded. Klein (submitted) has described how people get lost by attaching too much credibility to an inaccurate anchor and then interpreting other data elements to fit the inaccuracy, thereby corrupting the sensemaking process. Thus, an inaccurate anchor that comes early in a sensemaking sequence should have a more serious effect on performance than one that comes later in the sequence.

The phenomenon of fixation can be seen as a simple result of using a frame to direct attention—which is one of the functions of frames. We call this fixation when we are working with a flawed frame, or a frame that is built around flawed data. We call it efficient attention management when our frames are accurate.

Comparing Multiple Frames

We sometimes need to deliberately compare different frames to judge what is going on.

For example, in reviewing incidents from a study of nurses in an NICU (Crandall & Gamblian, 1991), we found several cases where the nurses gathered evidence in support of one frame—that the neonate was making a good recovery—while at the same time elaborated a second, opposing frame—that the neonate was developing sepsis.

Example 5: Comparison of Frames in an NICU.   This baby was my primary; I knew the baby and I knew how she normally acted. Generally she was very alert, was on feedings, and was off IVs. Her lab work on that particular morning looked very good. She was progressing extremely well and hadn't had any of the setbacks that many other preemies have. She typically had numerous apnea episodes and then bradys, but we could easily stimulate her to end these episodes. At 2:30 her mother came in to hold her and I noticed that she wasn't as responsive to her mother as she normally was. She just lay there and half looked at her. When we lifted her arm it fell right back down in the bed and she had no resistance to being handled. This limpness was very unusual for her.

On this day, the monitors were fine, her blood pressure was fine, and she was tolerating feedings all right. There was nothing to suggest that anything was wrong except that I knew the baby and I knew that she wasn't acting normally. At about 3:30 her color started to change. Her skin was not its normal pink color and she had blue rings around her eyes. During the shift she seemed to get progressively grayer. Then at about 4:00, when I was turning her feeding back on, I found that there was a large residual of food in her stomach. I thought maybe it was because her mother had been holding her and the feeding just hadn't settled as well. By 5:00 I had a baby who was gray and had blue rings around her eyes. She was having more and more episodes of apnea and bradys; normally she wouldn't have any bradys when her mom was holding her. Still, her blood pressure hung in there. Her temperature was just a little bit cooler than normal. Her abdomen was a little more distended, up 2 cm from early in the morning, and there was more residual in her stomach. This was a baby who usually had no residual and all of a sudden she had 5 cc to 9 cc. We gave her suppositories thinking maybe she just needed to stool. Although having a stool reduced her girth, she still looked gray and was continuing to have more apnea and bradys. At this point, her blood gas wasn't good so we hooked her back up to the oxygen. On the doctor's orders, we repeated the lab work. The results confirmed that this baby had an infection, but we knew she was in trouble even before we got the lab work back.

In some of our incident analyses, we have found that people were tracking up to three frames simultaneously. We speculate that this may be an upper limit; people may track two to three frames simultaneously, but rarely more than three.

In their research on pediatric cardiologists, Feltovich et al. (1984) found that when the experts broke free of the fixation imposed by a garden path scenario, they would identify one, two, or even three alternative frames. The experts deliberately selected frames that would sharpen the distinctions that were relevant. They identified a cluster of diseases that shared many symptoms in order to make fine-grained diagnoses. Feltovich et al. referred to this strategy as using a "logical competitor set" (LCS) as a means of pinpointing critical details. To Feltovich et al., the LCS is an interconnected memory unit that can be considered a kind of category. Activation of one member of the LCS will activate the full set because these are the similar cardiac conditions that have to be contrasted. Depending on the demands of the task, the decision maker may set out to test all the members of the set simultaneously, or may test only the most likely member. Rudolph (2003) sees this strategy as a means of achieving a differential diagnosis. The anesthesiologists in her study changed the way they sought information—their search strategies became more directed and efficient when they could work from a set of related and competing frames.

We also speculate that if a data element is used as an anchor in one frame, it will be difficult to use it in a second, competing frame. The rival uses of the same anchor may create conceptual strain, and this would limit a person's ability to compare frames. By developing LCSs, experts, such as the pediatric cardiologists studied by Feltovich et al., can handle the strain of using the same anchors in competing frames.

Reframing

In reframing, we are not simply accumulating inconsistencies and contrary evidence. We need the replacement frame to guide the way we search for and define cues, and we need these cues to suggest the replacement frame. Both processes happen simultaneously.

Reframing was illustrated in several of the previous examples. The NICU example (Example 5) shows how the experienced nurse had a "sepsis" frame available and used it to gather evidence until she felt that the risk was sufficiently serious to notify a physician. The helicopter incident (Example 3) showed how Captain B. shifted from one frame to another as he obtained new items of information.

Duncker (1945) introduced the concept that gaining insight into the solution to a problem may require reframing or reformulating the way the problem is understood. Duncker studied the way subjects approached the now-classic "radiation problem" (how to use radiation to destroy a tumor without damaging the healthy tissue surrounding the tumor). As long as the problem was stated in that form, subjects had difficulty finding a solution. Some subjects were able to reframe the problem, from "how to use radiation to treat a tumor without destroying healthy tissue" to "how to minimize the intensity of the radiation except at the site of the tumor." This new frame enabled subjects to generate a solution of aiming weak radiation beams that converged on the tumor.

Weick (1995) has explained that during reframing people consider data elements that had previously been discarded, so that formerly discrepant cues now fit. In this way, sensemaking involves retrospective understanding.

Seeking a Frame

We may deliberately try to find a frame when confronted with data that just do not make sense, or when a frame is questioned and is obviously inadequate. Sometimes, we can replace one frame with another, but at other times we have to try to find or construct a frame. We may look for analogies and also search for more data in order to find anchors that can be used to construct a new frame. The mental simulation account of Klein and Crandall (1995) describes how people can assemble data elements as anchors in the process of building an explanation.

Example 6 summarizes our description of the various aspects of sense-making that are shown in Figure 6.1. It is based on an interview we conducted around a navigation incident (Klein et al., 2002).

Example 6: Flying Blind.   This incident occurred during a solo cross-country flight (a required part of aviation training to become a private pilot) in a Cessna 172. The pilot's plan was for a 45-minute trip. The weather for this journey was somewhat perfect—sunny, warm, few clouds, not much haze. He had several-mile visibility, but not unlimited. He was navigating by landmarks.

The first thing he did was build a flight plan, including: heading, course, planned airspeed, planned altitude, way points in between each leg of the journey, destination (diagram of the airport), and a list of radio frequencies. His flight instructor okayed this flight plan. Then he went through his preflight routine. He got in the airplane, checked the fuel and the ignition (to make sure the engine was running properly), set the altitude (by calibrating the altimeter with the published elevation of the airport where he was located; this is pretty straightforward), and calibrated the directional gyro (DG).

He took off and turned in the direction he needed to go. During the initial several minutes of his flight, he would be flying over somewhat familiar terrain, because he had flown in this general direction several times during his training, including a dual cross-country (a cross-country flight with an instructor) to an airport in the vicinity of his intended course for that day.

About a half hour into the flight, the pilot started getting the feeling that he wasn't where he was supposed to be. Something didn't feel right. But he couldn't figure out where he was on the map—all the little towns looked similar.

What bothered him the most at this point was that his instruments had been telling him he was on course, so how did he get lost? He checked his DG against the compass (while flying as straight and level as he could get), and realized his DG was about 20 to 30 degrees off. That's a very significant inaccuracy. So he stopped trusting his DG, and he had a rough estimate of how far off he was at this point. He knew he'd been going in the right general direction (south), but that he had just drifted east more than he should have.

He decided to keep flying south because he knew he would be crossing the Ohio River. This is a very obvious landmark that he could use as an anchor to discover his true position. Sure enough, the arrangement of the factories on the banks of the river was different from what he was expecting. He abandoned his expectations and tried to match the factory configuration on the river to his map. In this way, he was able to create a new hypothesis about his location.

A particular bend in the river had power plants/factories with large smokestacks. The map he had showed whether a particular vertical obstruction (like a smoke stack or radio antenna) is a single structure or multiple structures bunched together. He noted how these landmarks lined up, trying to establish a pattern to get him to the airport. He noticed a railroad crossing that was crossed by high-tension power lines. He noticed the lines first and thought, "Is this the one crossed by the railroad track that leads right into the airport?" Then he followed it straight to his destination.

In this example, we see several of the sensemaking activities. The pilot started with a good frame, in the form of a map and knowledge of how to use basic navigational equipment. Unknown to him, the equipment was malfunctioning. Nevertheless, he attempted to elaborate the frame as his journey progressed. He encountered data that made him question his frame—question his position on the map. But he explained these data away and preserved the frame. Eventually, he reached a point where the deviation was too great, and where the topology was too discrepant from his expectancies. He used some fragmentary knowledge to devise a strategy for recovering—he knew that he was heading south, and would eventually cross the Ohio River, and he prepared his maps to check his location at that time. The Ohio River was a major landmark, a dominating anchor, and he hoped he could discard all of his confused notions about location and start fresh, using the Ohio River and seeing what map features corresponded to the visual features he would spot. He also had a rough idea of how far he had drifted from his original course, so he could start his search from a likely point. He could not have successfully reoriented earlier because he simply did not have a sufficient set of useful anchors to fix his position.

Example 6 illustrates most of the sensemaking types shown in Figure 6.1. The example shows the initial data–frame match: The pilot started off with a firm belief that he knew where he was. As he proceeded, he elaborated on his frame by incorporating various landmarks. His elaboration also helped him preserve his frame, as he explained away some potential discrepancies. Eventually, he did question his understanding. He did realize that he was lost. He had no alternate frame ready as a replacement, and no easy way to construct a new frame. But he was able to devise a strategy that would let him use a few anchors (the Ohio River and the configuration of factories) to find his position on his map. He used this new frame to locate his destination airport.

All of the activities in this example, and in Figure 6.1, are ways of making sense of events. The nature and the combination of the sensemaking activities are different depending on the demands of the task and the expertise of the sensemaker.

When we think about sensemaking, one source of confusion is to treat different activities as the same. We assert that each of the activities shown in Figure 6.1 has different strategies associated with it, and different obstacles that must be managed. Therefore, we may need to be careful about treating all of these different activities as the same. For instance, the support that we might offer to someone who is tracking two opposing stories or frames might be very different from the support we would give to someone who needs to reach back to prior events to recall which data elements have been explained away and might now be relevant, and these might differ from the help we could offer someone who has become disoriented and is trying to detect one or two anchors from which to derive a new frame.

POTENTIAL APPLICATIONS OF THE DATA–FRAME THEORY

The data–frame theory of sensemaking should have potential to help us design decision support systems (DSSs) and develop training programs. The previous sections have identified a number of ways that sensemaking can break down. The process of abstracting data itself creates distortions. Data can be distorted when people preserve a frame by bending the map. People can fixate on a frame, or they can go to the opposite extreme and keep an open mind, which leads them to miss important connections and hypotheses. In several of our studies, we found that people with a passive attitude of simply receiving messages were likely to miss the important problems that had emerged. Many of our recommendations that follow are aimed at preventing these types of breakdowns.

Decision Support Systems

In designing DSSs, one of the lessons from the literature is that more data does not necessarily lead to more accurate explanations and predictions. Weick (1995) criticized the information-processing metaphor for viewing sensemaking and related problems as settings where people need more information. Weick argued that a central problem requiring sensemaking is that there are too many potential meanings, not too few—equivocality rather than uncertainty. For Weick, resolving equivocality requires values, priorities, and clarity about preferences, rather than more information.

Heuer (1999) made a similar point, arguing that intelligence analysts need help in evaluating and interpreting data, not in acquiring more and more data. The research shows that accuracy increases with data elements up to a point (perhaps 8–10 data elements) and then asymptotes while confidence continues to increase (Oskamp, 1965). Omodei, Wearing, McLennan, Elliott, and Clancy (in press) showed that at some point increasing data may lead to worse decision quality. Lanir (1991) studied cases on intelligence failure, such as the Yom Kippur war, and found that in each case the decision makers already had sufficient data. Their difficulty was in analyzing these data. Gathering additional data would not necessarily have helped them.

On an anecdotal level, Army officers have commented to us during exercises that information technology is slowing the decision cycle instead of speeding it—soldiers are delaying their decisions with each new data element they receive, trying to figure out its significance.

We can offer some guidance for the field of DSSs:

- Attempts to improve judgment and decision quality by increasing the amount of data are unlikely to be as effective as supports to the evaluation of data. Efforts to increase information rate can actually interfere with skilled performance.
- The DSS should not be restricted to logical inferences, because of the importance of abductive inferences.
- The progression of data-information-knowledge-understanding that is shown as a rationale for decision support systems is misleading. It enshrines the information-processing approach to sensemaking, and runs counter to an ecological approach that asserts that the data themselves need to be constructed.
- Data fusion algorithms pose opportunities to reduce information overload, but also pose challenges to sensemaking if the logical bases of the algorithms are underspecified.
- Given the limited number of anchors that are typically used, people may benefit from a DSS that helps them track the available anchors.
- DSSs may be evaluated by applying metrics such as the time needed to catch inconsistencies inserted into scenarios and the time and effort needed to detect faulty data fusion.

Training Programs

The concept of sensemaking also appears to be relevant to the development of training programs. For example, one piece of advice that is often given is that decision makers can reduce fixation errors by avoiding early adoption of a hypothesis. But the data–frame theory regards early commitment to a hypothesis as inevitable and advantageous. Early commitment, the rapid recognition of a frame, permits more efficient information gathering and more specific expectancies that can be violated by anomalies, permitting adjustment and reframing. This claim can be tested by encouraging subjects to delay their interpretation, to see if that improves or hinders performance. It can be falsified by finding cases where the domain practitioner does not enact "early" recognition of a frame.

Feltovich et al. (1984) identified several participants in their study of pediatric cardiology who failed to correctly diagnose the medical problem and who showed a "data-driven dependence" in their diagnosis activity, responding to the most recent strong disease cue in the data. This approach, which is another way of describing continually reframing the situation, is the opposite of fixation.

Rudolph (2003) studied the "diagnostic vagabonding" described by Dörner (1996). She had initially expected that diagnostic vagabonding, with its extensive hypothesis generation, would effectively counter tendencies toward fixation, but her data showed that diagnostic vagabonding was almost as dysfunctional as fixation. She concluded that decision makers need to be sufficiently committed to a frame in order to effectively test it and learn from its inadequacies—something that is missing from the open-minded and open-ended diagnostic vagabonding. Coupled with this commitment to a diagnostic frame is an attitude of exploration for sincerely testing the frame in order to discover when it is inaccurate.

These observations would suggest that efforts to train decision makers to keep an open mind (e.g., Cohen, Adelman, Bresnick, & Freeman, 2003) can be counterproductive.

A second implication of sensemaking is the use of feedback in training. Practice without feedback is not likely to result in effective training. But it is not trivial to provide feedback. Outcome feedback is not as useful as process feedback (Salas, Wilson, Burke, & Bowers, 2002), because knowing that performance was inadequate is not as valuable as diagnosing what needs to be changed. However, neither outcome nor process feedback is straightforward. Trainees have to make sense of the feedback. If stockbrokers perform at or below chance, they can take the good outcomes as evidence of their skill and the bad outcomes as poor luck or anomalies with one segment of the market, or any of the knowledge shields people use. They need to have an accurate frame—describing stock selection as a random walk (Malkiel, 1999) in order to make sense of the feedback they are getting. Feedback does not inevitably lead to better frames. The frames determine the way feedback is understood. Similarly, process feedback is best understood when a person already has a good mental model of how to perform the task. A person with a poor mental model can misinterpret process feedback. Therefore, sensemaking is needed to understand feedback that will improve sensemaking—the same cycle shown in Figure 6.1.

A third issue that is relevant to training concerns the so-called *confirmation bias*. The decision research literature (Mynatt, Doherty, & Tweney, 1977; Wason, 1960) suggests that people are more inclined to look for and take notice of information that confirms a view than information that disconfirms it.

In contrast, we assert that people are using frames, not merely trying to confirm hypotheses. In natural settings, skilled decision makers shift into an active mode of elaborating the competing frame once they detect the possibility that their frame is inaccurate. This tactic is shown in the earlier NICU Example 5 where the nurse tracked two frames. A person uses an initial frame (hypothesis) as a guide in acquiring more information, and, typically, that information will be consistent with the frame. Furthermore, skilled decision makers such as expert forecasters have learned to seek disconfirming evidence where appropriate.

It is not trivial to search for disconfirming information—it may require the activation of a competing frame. Patterson, Woods, Sarter, and Watts-Perotti (1998), studying intelligence analysts who reviewed articles in the open literature, found that if the initial articles were misleading, the rest of the analyses would often be distorted because subsequent searches, and their reviews were conditioned by the initial frame formed from the first articles. The initial anchors affect the frame that is adopted, and that frame guides information seeking. What may look like a confirmation bias may simply be the use of a frame to guide information seeking. One need not think of it as a bias.

Accordingly, we offer the following recommendations:

- We suggest that training programs advocating a delayed commitment to a frame are unrealistic. Instead, training is needed in noticing anomalies and diagnosing them.
- Training may be useful in helping people "get found." For example, helicopter pilots are taught to navigate from one waypoint to another. However, if the helicopters are exposed to antiair attacks, they will typically take violent evasive maneuvers to avoid the threat. These maneuvers not infrequently leave the pilots disoriented. Training might be helpful in developing problem-solving routines for getting found once the pilots are seriously lost.
- Training to expand a repertoire of causal relationships may be more helpful than teaching comprehensive mental models.
- Training in generic sensemaking does not appear to be feasible. We have not seen evidence for a general sensemaking skill. Some of the incidents we have collected do suggest differences in an "adaptive mind-set" of actively looking to make sense of events, as in Example 2. It may be possible to develop and expand this type of attitude.
- Training may be better aimed at key aspects responsible for effective sensemaking, such as increasing the range and richness of frames. For ex-

ample, Phillips et al. (2003) achieved a significant improvement in sensemaking for Marine officers trained with tactical decision games that were carefully designed to improve the richness of mental models and frames.

• Training may be enhanced by verifying that feedback is appropriately presented and understood.

• Training may also be useful for helping people to manage their attention to become less vulnerable to distractions. Dismukes (1998) described how interruptions can result in aviation accidents. Jones and Endsley (1995, 2000) reviewed aviation records and found that many of the errors in Endsley's (1995) Level 1 situation awareness (perception of events) were compounded by ineffective attention management—failure to monitor critical data, and misperceptions and memory loss due to distractions and/or high workload. The management of attention would seem to depend on the way frames are activated and prioritized.

• Training scenarios can be developed for all of the sensemaking activities shown in Figure 6.1: elaborating a frame, questioning a frame, evaluating a frame, comparing alternative frames, reframing a situation, and seeking anchors in order to generate a useful frame.

• Metrics for sensemaking training might include the time and accuracy in detecting anomalies, the degree of concordance with subject-matter expert assessments, and the time and success in recovering from a mistaken interpretation of a situation.

TESTABLE ASPECTS OF THE DATA–FRAME THEORY

Now that we have described the data–frame theory, we want to consider ways of testing it. Based on the implications described previously, and our literature review, we have identified several hypotheses about the data–frame theory:

1. We hypothesize that frames are evoked/constructed using only three or four anchors.

2. We hypothesize that if someone uses a data element as an anchor for one frame, it will be difficult to use that same anchor as a part of a second, competing frame.

3. We hypothesize that the quality of the initial frame considered will be better than chance—that people identify frames using experience, rather than randomly. Based on past research on recognitional decision making, we assert that the greater the experience level, the greater the improvement over chance of the frame initially identified.

4. We hypothesize that introducing a corrupted, inaccurate anchor early in the message stream will have a correspondingly greater negative effect on sensemaking accuracy than introducing it later in the sequence (see Patterson et al., 1998).

5. We hypothesize that increased information and anchors will have a nonlinear relationship to performance, first increasing it, then plateauing, and then in some cases, decreasing. This claim is based on research literature cited previously, and also on the basic data–frame concept and the consequences of adding too many data without a corresponding way to frame them.

6. We hypothesize that experts and novices will show similar reasoning strategies when trying to make inferences from data.

7. We hypothesize that methods designed to prevent premature commitment to a frame (e.g., the recognition/metacognition approach of Cohen, Adelman, Tolcott, Bresnick, & Marvin, 1992) will degrade performance under conditions where active attention management is needed (using frames) and where people have difficulty finding useful frames. The findings reported by Rudolph (2003) suggest that failure to achieve early commitment to a frame can actually promote fixation because commitment to a frame is needed to generate expectancies (and to support the recognition of anomaly) and to conduct effective tests. The data–frame concept is that a frame is needed to efficiently and effectively understand data, and that attempting to review data without introducing frames is unrealistic and unproductive.

8. We hypothesize that people can track up to three frames at a time, but that performance may degrade if more than three frames must be simultaneously considered.

We also speculate that individual differences (e.g., tolerance for ambiguity, need for cognition) should affect sensemaking performance, and that this can be a fruitful line of research. Similarly, cultural differences, such as holistic versus analytical perspectives, should affect sensemaking (Nisbett, 2003). Furthermore, it may be possible to establish priming techniques to elicit frames and to demonstrate this elicitation in the sensemaking and information management activities shown by subjects.

Research into sensemaking can draw from a variety of experimental paradigms. These include the deGroot (1946/1965) method of seeing how people reconstitute a situation once the elements have been erased (also see Endsley, 1995, and Vicente, 1991, for recent applications of this approach), the garden path method of studying how people escape from fixation (e.g., Feltovich et al., 1984; Rudolph, 2003), and analyses of problem representation (e.g., Charness, 1981; Chi et al., 1981; P. J. Smith, McCoy, & Layton, 1997).

SUMMARY

Sensemaking is the deliberate effort to understand events. It serves a variety of functions, such as explaining anomalies, anticipating difficulties, detecting problems, guiding information search, and taking effective action. We have presented a data–frame theory of the process of sensemaking in natural settings.

The theory contains a number of assertions. First, the theory posits that the interaction between the data and the frame is a central feature of sensemaking. The data, along with the goals, expertise, and stance of the sensemaker, combine to generate a relevant frame. The frame subsequently shapes which data from the environment will be recognized as pertinent, how the data will be interpreted, and what role they will play when incorporated into the evolving frame. Our view of the data–frame relationship mirrors Neisser's (1976) cyclical account of perception, in that available information, or data, modifies one's schema, or frame, of the present environment, which in turn directs one's exploration and/or sampling of that environment. The data are used to select and alter the frame. The frame is used to select and configure the data. In this manner, the frame and the data work in concert to generate an explanation. The implication of this continuous, two-way, causal interaction is that both sensemaking and data exploration suffer when the frame is inadequate.

Second, our observations in several domains suggest that people select frames based on a small number of anchors—highly salient data elements. The initial few anchors seem to determine the type of explanatory account that is formed, with no more than three to four anchors active at any point in time.

Third, our research is consistent with prior work (Barrows et al., 1978; Chase & Simon, 1973; Elstein, 1989) showing that expert–novice differences in sensemaking performance are not due to superior reasoning on the part of the expert or mastery of advanced reasoning strategies, but rather to the quality of the frame that is brought to bear. Experts have more factual knowledge about their domain, have built up more experiences, and have more knowledge about cause-and-effect relationships.

Experts are more likely to generate a good explanation of the situation than novices because their frame enables them to select the right data from the environment, interpret them more accurately, and see more pertinent patterns and connections in the data stream.

Fourth, we suggest that people more often construct JIT mental models from available knowledge than draw on comprehensive mental models. We have not found evidence that people often form comprehensive mental models. Instead, people rely on JIT models, constructed from fragments, in a way that is analogous to the construction of memory. In complex and open systems, a comprehensive mental model is unrealistic.

There is some evidence that in domains dealing with closed systems, such as medicine (i.e., the human body can be considered a roughly closed system), an expert can plausibly develop an adequately comprehensive mental model for some medical conditions. However, most people and even most experts rely on fragments of local cause–effect connections, rules of thumb, patterns of cues, and other linkages and relationships between cues and information to guide the sensemaking process (and indeed other high-level cognitive processes).

The concept of JIT mental models is interesting for several reasons. We believe that the fragmentary knowledge (e.g., causal relationships, rules, principles) representing one domain can be applied to a sensemaking activity in a separate domain. If people have worked out complex and comprehensive mental models in a domain, they will have difficulty in generalizing this knowledge to another domain, whereas the generalization of fragmentary knowledge is much easier. Fragmentary knowledge contributes to the frame that is constructed by the sensemaker; fragmentary knowledge therefore helps to guide the selection and interpretation of data. We do not have to limit our study of mental models to the static constructs and beliefs that people hold; we can also study the process of compiling JIT mental models from a person's knowledge base.

Fifth, the data–frame account of sensemaking is different from an information-processing description of generating inferences on data elements. Sensemaking is motivated by the person's goals and by the need to balance the data with the frame—a person experiences confusion in having to consider data that appear relevant and yet are not integrated. Successful sensemaking achieves a mental balance by fitting data into a well-framed relationship with other data. This balance will be temporary because dynamic conditions continually alter the landscape. Nevertheless, the balance, when achieved, is emotionally satisfying in itself. People do not merely churn out inferences. They are actively trying to experience a match, however fleeting, between data and frame.

ACKNOWLEDGMENTS

The research reported in this chapter was supported by a contract with the Army Research Institute for the Behavioral and Social Sciences (Contract 1435-01-01-CT-3116). We would like to thank Mike Drillings, Paul Gade, and Jonathan Kaplan for their encouragement and assistance. We also wish to thank Karl Weick, Arran Caza, Mica Endsley, and Mei-Hua Lin for their helpful suggestions in reviewing our findings and concepts. Karol Ross and Sterling Wiggins made valuable contributions to our work, particular in the early stages. Peter Thunholm provided many useful critiques and ideas of our concepts

REFERENCES

Anderson, J. R. (1990). *The adaptive character of thought.* Mahwah, NJ: Lawrence Erlbaum Associates.

Barlett, F. C. (1932). *Remembering: A study in experimental and social psychology.* Cambridge, England: Cambridge University Press.

Barrows, H. S., Feightner, J. W., Neufeld, V. R., & Norman, G. R. (1978). *Analysis of the clinical methods of medical students and physicians* (Final Report to Ontario Ministry of Health). Hamilton, Ontario, Canada: McMaster University.

Beach, L. R. (1997). *The psychology of decision making—people in organizations.* Thousand Oaks, CA: Sage Publications.

Charness, N. (1979). Components of skill in bridge. *Canadian Journal of Psychology, 33,* 1–16.

Charness, N. (1981). Aging and skilled problem-solving. *Journal of Experimental Psychology: General, 110,* 21–38.

Chase, W. G., & Simon, H. A. (1973). The mind's eye in chess. In W. G. Chase (Ed.), *Visual information processing* (pp. 215–281). New York: Academic Press.

Chi, M. T. H., Feltovich, P. J., & Glaser, R. (1981). Categorization and representation of physics problems by experts and novices. *Cognitive Science, 5,* 121–152.

Clark, H. H., & Clark, E. V. (1977). *Psychology of language.* New York: Harcourt Brace Jovanovich.

Cohen, M. S., Adelman, L., Bresnick, T., & Freeman, M. (2003). *Dialogue as the medium for critical thinking training.* Paper presented at the 6th International Conference on Naturalistic Decision Making, Pensacola Beach, FL.

Cohen, M. S., Adelman, L., Tolcott, M., Bresnick, T., & Marvin, F. (1992). *Recognition and metacognition in commanders situation understanding* (Tech. Rep.). Arlington, VA: CTI.

Collins, A., Burstein, M., & Michalski, R. (1987). *Plausible reasoning in tactical planning* (Annual Interim Report No. BBN 6544). Cambridge, MA: BBN Laboratories.

Crandall, B., & Gamblian, V. (1991). *Guide to early sepsis assessment in the NICU* (Instruction manual prepared for the Ohio Department of Development under the Ohio SBIR Bridge Grant program). Fairborn, OH: Klein Associates.

Crandall, B., & Getchell-Reiter, K. (1993). Critical decision method: A technique for eliciting concrete assessment indicators from the "intuition" of NICU nurses. *Advances in Nursing Sciences, 16*(1), 42–51.

De Keyser, V., & Woods, D. D. (1993). Fixation errors: Failures to revise situation assessment in dynamic and risky systems. In A. G. Colombo & A. Saiz de Bustamente (Eds.), *Advanced systems in reliability modeling.* Norwell, MA: Kluwer Academic.

deGroot, A. D. (1965). *Thought and choice in chess.* The Hague, Netherlands: Mouton. (Original work published 1946)

Dewey, J. (1910). *How we think.* Boston: Heath.

Dismukes, K. (1998). Cockpit interruptions and distractions: Effective management requires a careful balancing act. *ASRS Directline, 10,* 4–9.

Doherty, M. E. (1993). A laboratory scientist's view of naturalistic decision making. In G. A. Klein, J. Orasanu, R. Calderwood & C. E. Zsambok (Eds.), *Decision making in action: Models and methods* (pp. 362–388). Norwood, NJ: Ablex.

Dörner, D. (1996). *The logic of failure.* Reading, MA: Perseus.

Duncker, K. (1945). On problem solving. *Psychological monographs, 5,* Whole No. 270), 1–113.

Elstein, A. S. (1989). On the clinical significance of hindsight bias. *Medical Decision Making, 9,* 70.

Elstein, A. S., Shulman, L. S., & Sprafka, S. A. (1978). *Medical problem solving: An analysis of clinical reasoning.* Cambridge, MA: Harvard University Press.

Endsley, M. R. (1995). Situation awareness and the cognitive management of complex systems. *Human Factors, 37*(1), 85–104.

Feltovich, P. J., Coulson, R. L., & Spiro, R. J. (2001). Learners' (mis)understanding of important and difficult concepts: A challenge to smart machines in education. In K. D. Forbus & P. J. Feltovich (Eds.), *Smart machines in education.* Menlo Park, CA: AAAI/MIT Press.

Feltovich, P. J., Johnson, P. E., Moller, J. H., & Swanson, D. B. (1984). LCS: The role and development of medical knowledge in diagnostic expertise. In W. J. Clancey & E. H. Shortliffe (Eds.), *Readings in medical artificial intelligence: The first decade* (pp. 275–319). Reading, MA: Addison-Wesley.

Feltovich, P. J., Spiro, R. J., & Coulson, R. L. (1997). Issues of expert flexibility in contexts characterized by complexity and change. In P. J. Feltovich, K. M. Ford & R. R. Hoffman (Eds.), *Expertise in context* (pp. 125–146). Menlo Park, CA: AAAI/MIT Press.

Goffman, E. (1974). *Frame analysis: An essay on the organization of experience.* New York: Harper.

Heuer, R. J., Jr. (1999). *Psychology of intelligence analysis.* Washington, DC: Center for the Study of Intelligence, Central Intelligence Agency.

Jones, D. G., & Endsley, M. R. (1995). Investigation of situation awareness errors. In R. S. Jensen & L. A. Rakovan (Eds.), *Proceedings of the Eighth International Symposium on Aviation Psychology* (Vol. 2, pp. 746–751). Columbus: Ohio State University.

Jones, D. G., & Endsley, M. R. (2000). Overcoming representational errors in complex environments. *Human Factors, 42*(3), 367–378.

Josephson, J., & Josephson, S. (Eds.). (1994). *Abductive inference: Computation, technology and philosophy.* New York: Cambridge University Press.

Kaempf, G. L., Klein, G., Thordsen, M. L., & Wolf, S. (1996). Decision making in complex command-and-control environments. *Human Factors, 38,* 220–231.

Klein, G. (1998). *Sources of power: How people make decisions.* Cambridge, MA: MIT Press.

Klein, G. *Corruption and recovery of sensemaking during navigation.* Manuscript submitted for publication.

Klein, G. A., Calderwood, R., & Clinton-Cirocco, A. (1986). Rapid decision making on the fireground. *Proceedings of the Human Factors and Ergonomics Society 30th Annual Meeting, 1,* 576–580.

Klein, G. A., & Crandall, B. W. (1995). The role of mental simulation in naturalistic decision making. In P. Hancock, J. Flach, J. Caird, & K. Vicente (Eds.), *Local applications of the ecological approach to human-machine systems* (Vol. 2, pp. 324–358). Hillsdale, NJ: Lawrence Erlbaum Associates.

Klein, G. A., & Peio, K. J. (1989). The use of a prediction paradigm to evaluate proficient decision making. *American Journal of Psychology, 102*(3), 321–331.

Klein, G., Phillips, J. K., Battaglia, D. A., Wiggins, S. L., & Ross, K. G. (2002). *Focus: A model of sensemaking* (Interim Report–Year 1 prepared under Contract 1435-01-01-CT-31161 [U.S. Department of the Interior] for the U.S. Army Research Institute for the Behavioral and Social Sciences, Alexandria, VA). Fairborn, OH: Klein Associates.

Klein, G., Phillips, J. K., Rall, E. L., Thunholm, P., Battaglia, D. A., & Ross, K. G. (2003). *Focus Year 2 Interim Report* (Interim Report prepared under Contract #1435-01-01-CT-3116 [U.S. Department of the Interior] for the U.S. Army Research Institute for the Behavioral and Social Sciences, Alexandria, VA). Fairborn, OH: Klein Associates.

Klein, G., Pliske, R. M., Crandall, B., & Woods, D. (in press). Problem detection. *Cognition, Technology, and Work.*

Klein, G., Ross, K. G., Moon, B. M., Klein, D. E., Hoffman, R. R., & Hollnagel, E. (2003). Macrocognition. *IEEE Intelligent Systems, 18*(3), 81–85.

Lanir, Z. (1991). The reasonable choice of disaster. In J. Rasmussen (Ed.), *Distributed decision making: Cognitive models for cooperative work* (pp. 215–230). Oxford, England: Wiley.

Lundberg, G. (2000). Made sense and remembered sense: Sensemaking through abduction. *Journal of Economic Psychology, 21,* 691–709.

Malkiel, B. G. (1999). *A random walk down Wall Street.* New York: Norton.

Marek, V. W., & Truszczynski, M. (1993). *Nonmonotonic logic: Context-dependent reasoning.* Berlin: Springer.

Medin, D. L., Lynch, E. B., Coley, J. D., & Atran, S. (1997). Categorization and reasoning among tree experts: Do all roads lead to Rome? *Cognitive Psychology, 32,* 49–96.

Miller, G. A., & Johnson-Laird, P. N. (1976). *Language and perception.* Cambridge, England: Cambridge University Press.

Minsky, M. (1975). A framework for representing knowledge. In P. Winston (Ed.), *The psychology of computer vision* (pp. 211–277). New York: McGraw-Hill.

Mynatt, C. R., Doherty, M. E., & Tweney, R. D. (1977). Confirmation bias in a simulated research environment. *Quarterly Journal of Experimental Psychology, 29,* 85–95.

Neisser, U. (1976). *Cognition and reality: Principles and implications of cognitive psychology.* San Francisco: Freeman.

Nisbett, R. E. (2003). *The geography of thought. How Asians and Westerners think differently . . . and why.* New York: The Free Press.

Omodei, M. M., Wearing, A. J., McLennan, J., Elliott, G. C., & Clancy, J. M. (in press). More is better? Problems of self-regulation in naturalistic decision making settings. In B. Brehmer, R. Lipshitz & H. Montgomery (Eds.), *How professionals make decisions.* Mahwah, NJ: Lawrence Erlbaum Associates.

Oskamp, S. (1965). Overconfidence in case study judgments. *Journal of Consulting Psychology, 29,* 261–265.

Patterson, E. S., Woods, D. D., Sarter, N. B., & Watts-Perotti, J. C. (1998). Patterns in cooperative cognition. In *COOP '98, Third International Conference on the Design of Cooperative Systems.*

Peirce, C. S. (1903). C. S. Peirce's Lowell Lectures of 1903. Eighth Lecture: *Abduction* (MS 475).

Pennington, N., & Hastie, R. (1993). A theory of explanation-based decision making. In G. Klein, J. Orasanu, R. Calderwood, & C. E. Zsambok (Eds.), *Decision making in action: Models and methods* (pp. 188–201). Norwood, NJ: Ablex.

Phillips, J. K., Baxter, H. C., & Harris, D. S. (2003). *Evaluating a scenario-based training approach for Enhancing Situation Awareness Skills* (Tech. Report prepared for ISX Corporation under Contract DASW01-C-0036). Fairborn, OH: Klein Associates.

Piaget, J. (1952). *The origins of intelligence in children.* New York: International Universities Press.

Piaget, J. (1954). *The construction of reality in the child.* New York: Basic Books.

Pliske, R. M., Crandall, B., & Klein, G. (in press). Competence in weather forecasting. In J. Shanteau, P. Johnson, & K. Smith (Eds.), *Psychological exploration of competent decision making.* Cambridge, England: Cambridge University Press.

Rudolph, J. W. (2003). *Into the big muddy and out again.* Unpublished doctoral dissertation, Boston College.

Salas, E., Wilson, K. A., Burke, C. S., & Bowers, C. A. (2002). Myths about crew resource management training. *Ergonomics in Design* (Fall), pp. 20–24.

Schank, R. C., & Abelson, R. P. (1977). *Scripts, plans, goals and understanding: An inquiry into human knowledge structures.* Mahwah, NJ: Lawrence Erlbaum Associates.

Simon, H. A. (1973). The structure of ill-structured problems. *Artificial Intelligence, 4,* 181–201.

Smith, E. E., & Spoehr, K. T. (1974). The perception of printed English: A theoretical perspective. In B. H. Kantowitz. (Ed.), *Human information processing* (pp. 231–275). Hillsdale, NJ: Lawrence Erlbaum Associates.

Smith, P. J., Giffin, W. C., Rockwell, T. H., & Thomas, M. (1986). Modeling fault diagnosis as the activation and use of a frame system. *Human Factors, 28*(6), 703–716.

Smith, P. J., McCoy, E., & Layton, C. (1997). Brittleness in the design of cooperative problem-solving systems: The effects on user performance. *IEEE Transactions on Systems, Man and Cybernetics, 27*, 360–371.

Tversky, A., & Kahneman, D. (1974). Judgment under uncertainty: Heuristics and biases. *Science, 185*, 1124–1131.

Vaughan, D. (1996). *The Challenger launch decision: Risky technology, culture, and deviance at NASA.* Chicago: University of Chicago Press.

Vicente, K. J. (1991). *Supporting knowledge-based behavior through ecological interface design.* Unpublished doctoral dissertation, University of Illinois at Urbana-Champaign.

Vicente, K. J., & Wang, J. H. (1998). An ecological theory of expertise effects in memory recall. *Psychological Review, 106*(1), 33–57.

Wason, P. C. (1960). On the failure to eliminate hypotheses in a conceptual task. *Quarterly Journal of Experimental Psychology, 12*, 129–140.

Weick, K. E. (1995). *Sensemaking in organizations.* Thousand Oaks, CA: Sage.

Wisniewski, E. J., & Medin, D. L. (1994). On the interaction of theory and data. *Cognitive Science, 18*, 221–282.

Woodworth, R. S. (1938). *Experimental psychology.* New York: Holt.

Part III

TOOLS FOR THINKING
OUT OF CONTEXT

Chapter 7

Expert Apprentice Strategies

Laura Militello
Laurie Quill
University of Dayton Research Institute, Dayton, OH

The theme *expertise out of context* calls our attention to studies of situations in which experts find themselves faced with unusual problems for which their prior expertise may or may not prepare them. These sorts of tough and unusual problems represent a fascinating area of study, but this invariably requires the investigator to acquire significant domain knowledge before useful observations can be made or cognitive task analysis procedures conducted. The term *bootstrapping* is generally used to describe the process used by cognitive engineers in learning about the work domain, the system, and the experts' role in the system (Hoffman, 1987; Hoffman, Crandall, & Shadbolt, 1998; Potter, Roth, Woods, & Elm, 2000). More recently, the term *expert apprentice* has been coined to describe the cognitive engineer who becomes expert at bootstrapping (R. R. Hoffman, personal communication, March 23, 2003). The expert apprentice is, by definition, out of context nearly all the time.

For researchers applying cognitive task analysis methods to naturalistic decision-making research, successful data collection depends largely on having a flexible toolkit of knowledge elicitation methods available for bootstrapping. The expert apprentice has experience with a range of methods, and adapts and applies them opportunistically across a range of domains. However, bootstrapping is not simply a matter of methods—which questions are asked, in what order, or whether they are answered as a result of observation, written questionnaires, or live interviews. The stance the investigator adopts as she or he goes about asking questions and conducting

159

research greatly influences the comfort level of the subject-matter experts (SMEs), the type and depth of information elicited, and the insights gleaned from the study. As an expert apprentice, the cognitive engineer shifts into the role of a learner to elicit and explore cognitive data. The learner role allows non-SMEs to explore how the subject-matter expert thinks about the task within the context of a student–teacher relationship.

Strategies for bootstrapping have developed in response to the demands of work, just as the skills of the SME we study have developed in response to the challenges of their work (DiBello, 2001). This chapter examines some of the contextual elements that influence the expert apprentice's choice of methods and strategies for exploring expertise.

This chapter discusses the role of the expert apprentice when bootstrapping strategies are applied as means to investigate complex domains and difficult decisions. For the purpose of illustration, we contrast two approaches. One is a comprehensive approach to bootstrapping in the context of a large sociotechnical system as it unfolds over the course of a multiyear research program. A second case study describes a 6-week project used to inform a single design project, representing a more focused approach to bootstrapping in shorter-term projects focusing on human decision making. Throughout the discussion, we highlight the strategies used by the expert apprentice in the context of different types of projects.

Background

The term *expert apprentice* was suggested by Hoffman (March 23, 2003, personal communication) in the context of organizing the 6th Conference on Naturalistic Decision Making. Hoffman asked participants to reflect on roles they may adopt to elicit knowledge from an experienced worker. During fieldwork studying bus mechanics and how they think about fleet maintenance, DiBello (2001) described her role as "quasi-apprentice." She finds the strategy of asking the experienced maintainer to serve as teacher or master to be very effective because the interviewee is able to take on a familiar role (i.e., training a new employee), and in the role of the apprentice, it is natural for the interviewer to ask lots of questions.

The term *expert apprentice* connotes two important elements. The first, as DiBello describes, is that the investigator takes on the role of learner, and therefore spends some portion of the data collection period learning not just how the domain practitioners do their jobs, but learning *how to do the job* (and how that job relates to other jobs). Several researchers have described this attempt to obtain insight into the first-person perspective of the job; this insight is key to identifying a useful understanding of the cognitive challenges associated with a job (Eggleston, 2003; Hoffman et al., 1998; Morales & Cummings, 2005). The second is that gaining insight into a new domain efficiently and effectively depends not just on the skill of the experts being studied, but also on the skill of the apprentice, in this case the cognitive engineer/bootstrapper.

An expert apprentice possesses the ability to learn a new domain quickly. Although it is tempting to think of the expert apprentice as a true apprentice seeking to achieve some basic level of proficiency in the task of study, this would be an oversimplification. Not only is it impractical in many domains to obtain proficiency without specialized education and training, the expert apprentice retains the perspective and goals of the cognitive engineer throughout. For example, when studying surgeons, the expert apprentice does not attempt to reach the proficiency of a first-year resident. Instead, the goal is to learn enough about how to do the job to gain insight into the cognitive complexities and contextual factors that impact the work of a surgeon. The ways the cognitive engineer chooses to go about this will vary depending on the characteristics of the domain and project constraints. For the sake of illustration, we describe two distinct approaches to bootstrapping, although in reality the expert apprentice moves within and across these approaches, tailoring methods to each project as needed.

The first approach illustrates the type of bootstrapping we have seen evolve when investigators have had an opportunity to work in the same or closely related domains over a period of time. In this context, each project builds upon the last, allowing the cognitive engineer to develop an increasingly comprehensive understanding of the domain, often exploring the work domain at multiple levels. The second approach describes the types of methods that have evolved when investigators have worked across domains, spending limited time within any single domain. Here, the data gathered are by necessity more focused, seeking key cognitive challenges primarily through the eyes of a specific job position or particular team element.

THE COMPREHENSIVE APPROACH

Historically, the cognitive engineer who has found him or herself studying one domain over time in a comprehensive way has been exploring large socio-technical systems such as nuclear power plants (Woods & Roth, 1988), manufacturing systems (DiBello, 2001), aircraft maintenance systems (Quill, Kancler, & Pohle, 1999; Quill, Kancler, Revels, & Batchelor, 2001). In fact, it is likely that the complexity of these large socio-technical systems, paired with the possibility of catastrophic failure, has driven such through examination. In these types of domains, the reliability of the system is highly valued. As a result, many aspects of the work are well documented in the form of technical, procedural, and training manuals. These documents provide an important starting place for the expert apprentice in becoming familiar with key job descriptions, the system, the technology, and the jargon. Another important element of these types of domains is that operations tend to be conducted in a regular, systematic way, and possess built-in redundancy. In this context, bootstrapping strategies such as observation and job shadowing can be conducted without disrupting operations, and can generally be scheduled at the convenience of the investigator.

Examination of relevant documentation and observation of the work domain are often first steps in bootstrapping, and provide the expert apprentice important information about the organizational structure and system objectives (Crandall, Klein, & Hoffman, 2006; Potter et al., 2000; Vicente, 1999). This foundational information is then used to drive future bootstrapping in the form of more focused observations, job shadowing, and interview sessions. Even in this early, exploratory phase, the expert apprentice is reading, watching, and listening for specific types of information. Specifically, the expert apprentice is attempting to build an understanding of the system, understand connections among the components, and identify recurring problems.

Case Study: Aircraft Maintenance

One example of this comprehensive approach to expert apprenticeship can be found in series of studies conducted at the Air Force Research Laboratory in the aircraft maintenance domain. Since 1990, there has been a dramatic increase in the complexity of military aircraft systems. This, combined with pressures to reduce costs, has resulted in a driving need to identify and adopt technologies that will significantly improve the efficiency of the aircraft maintainers. The goal has been (and continues to be) to accomplish these cost savings while sustaining excellent safety records and mission capability reports. The military aircraft maintenance system is highly complex, with many components and a broad range of cognitive challenges. Furthermore, a smoothly running aircraft maintenance system is critical to all air operations. Use of air assets in humanitarian relief efforts, defense operations, and training missions cannot occur without the 24-hour, 7-day-per-week maintenance cycle. The complexity of this domain, combined with the availability of increasingly sophisticated technologies, has called for a more comprehensive approach to bootstrapping and resulted in a series of cognitive analyses in this single domain over an extended period of time (Quill, 2005; Quill & Kancler, 1995; Quill et. al., 1999; Revels, Kancler, Quill, Nemeth, & Donahoo, 2000).

As new technologies have become available to the domain of aircraft maintenance, the research questions and objectives of the various research efforts have evolved. Because of their extended experience in this one domain, the research team has acquired deep knowledge of the maintenance process, as well as an understanding of the historical and current constraints within which aircraft maintainers work. Early bootstrapping efforts focused on obtaining foundational knowledge. Later efforts expanded this to focus on the cognitive elements associated with a specific portion or function of the system while concentrating less on the organizational influences. In this case study, an in-depth understanding of the maintenance system gained over several years of research has provided us with a solid foundation from which to learn more about the specific problem under investigation (Donahoo et al., 2002; Friend & Grinstead, 1992; Kancler et. al., 2005; Revels et. al., 2000). Our primary stance was still one of a learner, but obtaining solid foundational knowledge allowed us to quickly identify connections and discontinuities within the aircraft maintenance system.

A wide range of bootstrapping methods was applied throughout the course of this research. These include standard bootstrapping methods such as shadowing members of aircraft maintenance teams (including mechanics, production superintendents, and wing commanders). In such cases, expert apprentices seek to learn about the cognitive challenges associated with the job by taking on a role very similar to that of an actual apprentice without disrupting the work at hand. An entire day may be spent shadowing one individual as she or he works. This arrangement encourages the domain practitioners to fall into the familiar role of trainer, and the investigator is able to ask questions to direct the conversation toward particular areas of interest. In contrast to a true apprentice, the expert apprentice/ cognitive engineer is not generally guided by a goal to learn to do the job. However, much information about how to do the job is acquired in pursuit of the expert apprentice goal of obtaining a first-person perspective regarding how the system works and what makes the job hard.

Other strategies include interviews and observations with different members of the maintenance team. From these interviews, the cognitive engineer seeks to draw parallels among related jobs and leverage lessons learned for a range of maintenance tasks. For example, many of the issues associated with display of electronic technical manuals also apply to the display of electronic flight manuals. In fact, findings from the study of aircraft maintainers' use of electronic technical manuals have relevance for pilots using electronic flight manuals (Swierenga et al., 2003). Additional bootstrapping activities include participating in regular status briefings and attending seminars about technologies that maintainers use. Furthermore, efforts are made to maintain regular contact with SMEs over time in order to stay abreast of changes within the maintenance industry.

This series of studies of aircraft maintenance and the bootstrapping methods employed are briefly described next.

The Early Years: Shadowing, Functional Task Analysis Interviews, and Simulations. In the early 1990s, research questions focused on where within the maintenance process portable computers would have the most impact. Impact was measured in terms of increased efficiency and reduced likelihood of error. Tools that appeared to hold considerable promise included those that allow the maintainer to carry large amounts of information (i.e., technical data, schematics, etc.) to the flightline, depot, or backshop; those that allow access to information in small spaces; and those that allow data to be recorded in any location on the flightline. However, determining which technologies to use and where they should be inserted presented a challenge. Figure 7.1 depicts a common maintenance situation in which access to technical data and a means of recording measurements or readings might be useful.

In this stage of the research program, job shadowing was used in combination with interviews to assess traditional practices as compared to practices utilizing the new portable computer technology (Quill & Kancler, 1995). Data collection efforts focused on understanding how the job was typically completed and which types of tasks imposed increased workload. From this apprentice stance, investigators generated meaningful recommendations for incorporating portable computers into the maintenance process. A subsequent set of projects aimed at investigating wearable computer technologies included task simulations in which the user was interviewed while performing a simulated maintenance task (Quill, Kancler, Revels, & Masquelier, 1999). This research focused on assessing the suitability of various wearable computer configurations for the task (glasses-mounted vs. head-mounted displays, vest-worn vs. belt-worn central processing units [CPUs]). In this case, bootstrapping was still required, but it

FIG. 7.1.   Aircraft maintainer.

was targeted specifically toward those troubleshooting tasks for which wearable computer technologies might best apply. Learning focused on identifying specific tasks for which the user's hands needed to be free to perform the maintenance and in which extensive mobility was required around the aircraft.

The Later Years: Observations, Interviews, and Simulation.   More recently, research questions have focused on understanding the impact of new technologies on the areas of decision making and collaboration within the maintenance process. To answer these questions, methods shifted from those focusing on behavioral and organizational constraints (i.e., job shadowing, functional task analysis methods) to a combination of observations and cognitive task analysis interviews using both simulated and actual maintenance tasks (Donahoo et. al., 2002; Kancler et. al., 2005; Militello, Quill, Vinson, Stilson, & Gorman, 2003; Quill, 2005).

It is important to note that at this point in the case study, the expert apprentices had up to 15 years of study in this domain. As a result, they had a more comprehensive understanding of the system, were able to integrate knowledge about jobs and tasks, and to incorporate historical knowledge of evolution within the maintenance system.

Based on an understanding of the maintenance process, how flightline maintainers do their jobs, and the interdependencies of various maintenance jobs, investigators targeted elements key to the successful introduction of handheld job aids and collaborative tools. Considerations such as how new devices would impact the larger maintenance process were essential to this research. Mobile computing devices not only allow maintainers to enter data on the flightline rather than in a central location remote from the aircraft, they also permit access to specialized knowledge of remote flightline personnel. For example, one clear benefit of mobile devices on the flightline is that data can be entered as they are gathered, reducing the likelihood that information will be remembered incorrectly or forgotten altogether by the time the maintainer gets back to the office computer. During job shadowing and other observations, investigators observed numerous examples of these errors of commission and omission (Donahoo et. al., 2002). Another benefit of mobile devices is that they afford the flightline technician access to other experts when faced with a challenging troubleshooting problem (Kancler et. al., 2005).

Even in these targeted studies, foundational knowledge of the system obtained during early bootstrapping afforded investigators the ability to anticipate systemic problems. For example, in spite of the fact that real-time data entry has obvious benefits for the individual maintainer, there may be a downside from a team perspective. Mobile devices reduce the number of times per day an individual maintainer is required to visit the office or communicate with others on the flightline. This might eliminate communications that are informal but important for team situation awareness and for mentoring junior-level maintainers. Investigators were thus able to recommend mechanisms for facilitating these important communications, including in-person supervisor sign-offs when the job is complete and on-demand training aids from the schoolhouse (Donahoo et. al., 2002; Gorman, Snead, & Quill, 2003).

This illustrates how domain understanding for an expert apprentice can deepen and become integrated over time. The combination of bootstrapping techniques was helpful in identifying and addressing complex issues surrounding the insertion of the handheld computing technologies into the maintenance process. Recommendations were made concerning which devices to use and how to configure them (Gorman et al., 2003; Kancler et al., 2005), how to display checklist information (Donahoo et. al., 2002), how to display schematics (Quill et. al., 1999), and how to anticipate and prevent potential gaps in communication (Kancler et al., 2005; Quill, 2005). A thorough understanding of particular maintenance activities and the larger maintenance process allowed the research team to make very specific recommendations about implementing a variety of mobile computing devices within the context of cognitively challenging flightline maintenance environments.

In addition to specific research issues identified by the project sponsor, this opportunity to study one domain over time using a range of bootstrapping methods allowed for the identification of widely used, routine patterns of behavior that potentially impact system efficiency and reliability in significant ways. These patterns can be "invisible" to people who work within the system because the routines have been in place for so long and other processes have been created to accommodate them. For example, through the 1990s and early 2000s, initial prioritization of daily maintenance activities had been accomplished early in the shift. Throughout the shift, changes in these priorities were made via radio transmission. After the action was complete, these changes were recorded electronically, but not printed out and disseminated until shift change. Monitoring of activities and progress had been accomplished primarily through visual inspection. Although this process had existed for many years and worked quite well, investigators noted that it could be more efficient if up-to-date information were available electronically throughout the day, reducing reliance on radio transmission, visual inspection, and handwritten notes.

Software to pull data from a range of databases and display up-to-date information in a readable fashion was developed to fill this need. Furthermore, this software was designed to support activity prioritization, status tracking, maintenance scheduling, and improved situational awareness for all aircraft maintenance personnel (Militello et al., 2003). Current research efforts have indicated that many status-tracking and scheduling technologies used by aircraft maintainers can also be used in similar ways by aircraft ground equipment or munitions personnel (Quill, 2005).

This long-term, comprehensive approach to bootstrapping included learning various jobs within the system, staying aware of the technologies involved with the work, and understanding the cognitive complexities of a range of jobs. This extensive bootstrapping effort allowed the research team to make observations targeted to specific research questions and observations about potential system efficiencies noted opportunistically, as well as make recommendations for efficiencies to be gained in complementary settings.

THE FOCUSED APPROACH

We use the term *focused approach* to describe the types of bootstrapping strategies that have developed when cognitive engineers are asked to contribute design concepts (either training design or technology design) working with a short timeline. Historically, these shorter timeline projects have tended to involve systems in which the role of the human is quite prominent. (For examples of studies of systems in which human expertise is prominent see Hoffman, Coffey, & Ford's, 2000, study of weather forecasters or Militello & Lim's, 1995, study of critical-care nursing.) The research project is more likely to be framed around the topic of human decision making than system reliability. In addition, the domains in which these projects take place tend to present more obstacles to observation than the large sociotechnical systems described earlier. Because the object of study is generally a human decision maker, the events to be observed do not lend themselves to job shadowing. For example, though researchers are often permitted to observe military exercises, interrupting the flow of the game with questions could prove dangerous in live-fire exercises. Instead, questions must be deferred until a break in the exercise, or until a debrief session following the exercise. In health care domains, it is often difficult for researchers to obtain permission to observe health care teams because of patient confidentiality and liability issues. Though firefighters are often amenable to allowing researchers to observe operations, the occurrence of fires is quite unpredictable. The dynamic nature and unpredictability of these types of events generally precludes the use of job-shadowing techniques. Observations may be more feasible, but the researcher often must be available for a window of time, in hopes that an interesting and observable event will occur.

Because of the difficulties involved in conducting observations and using job-shadowing techniques, expert apprentices working in these domains have come to rely more heavily on interviewing. This is not to say that observation never occurs, but that given the time pressure and domain constraints, observation techniques are often less feasible in these projects. Because interviews are generally used instead of observation techniques (or to supplement very limited observation opportunities), the expert apprentice is often relying on a set of practitioners' recollections of challenging incidents to obtain a first-person view of the challenges of the domain.

For the expert apprentice working on these short-term projects, interviewing becomes especially important, particularly if observation is not possible. Furthermore, if time is limited for the research, interviewing may be the most efficient way to unpack the key cognitive challenges of the task. Incident-based interview strategies represent one type of technique well suited to the goal of developing descriptive models of human decision making for tasks in which the human role has significant impact on the outcome (Crandall et al., 2006). In this case, a view of the technology, the work, and the organization is provided through the eyes of the human in the context of a challenging incident. A comprehensive understanding of these elements is not sought. Rather, they are considered important only as they impact (either hinder or facilitate) the human's cognitive processes. The interview strategies are deliberately developed to highlight perceptual information such as critical cues and cue clusters, how the practitioner makes sense of these cues and other information in the environment, what sorts of assessments are derived, what decisions are made, and what judgments are rendered, as well as goals that drive behavior at different points in time. Rather than developing a broad, deep, and comprehensive understanding of a single domain over time, expert apprentices who specialize in these methods seek to develop skills at recognizing, unpacking, and describing cognitive aspects of expert performance for a narrower set of job tasks within a single domain, or an understanding of selected aspects of a range of domains.

Expert apprentices working on short-term projects tend to spend a very brief amount of time gathering foundational knowledge as compared to the more comprehensive approach described earlier in which there is time and opportunity to engage in thorough document analysis. The cognitive engineer must be able to gather enough background information to conduct an efficient cognitive task analysis in a matter of days or weeks. Time is spent obtaining and reading materials on the topic including training manuals, procedural manuals, news articles, and books. However, not all work domains are well documented, and documents that do exist are not always easily accessible. The search for background information may extend to unexpected sources at times. For example, Clancy's 1984 novel, *The Hunt for Red October,* was read by investigators who were preparing for a project investigating sonar operators (D. W. Klinger, personal communication, January

12, 2002). Preliminary interviews can be conducted with practitioners or other researchers who are familiar with the domain, to obtain overview information about the organizational structure and the nature of the work. Researchers can find information via Internet searches. Sometimes, researchers can participate in or observe a training course or a portion of the training. In summary, early information can be extremely opportunistic and highly dependent on the creativity of the investigator in seeking and obtaining relevant information.

Also during this early opportunistic phase, the cognitive engineer is likely to be working on tailoring knowledge elicitation strategies based on details learned while collecting background information. If observations will be possible, observation strategies can be discussed with the team. A plan will be generated for deciding how people will participate in the observations, who will fulfill which roles, and how to focus the observations. If interviews will be the sole source of data, a specific interview strategy will be selected. If it seems likely that domain practitioners, especially the experts, will have a rich experience base to draw on, interviewing techniques such as the Critical Decision Method (Hoffman et al., 1998; Klein, Calderwood, & MacGregor, 1989) are often used. If practitioners are in a domain for which there is limited opportunity for real-world experiences (e.g., Army commanders are unlikely to participate in more than five or six real-world conflicts in the course of a career), researchers will consider developing a simulation or scenario around which the interviews can be structured. If, before interviews begin, there is still relatively little known about the domain, the Knowledge Audit (Klein & Militello, 2005; Militello & Hutton, 1998) may be used to survey the cognitive aspects of the task, before more in-depth interviews are conducted.

As the expert apprentice moves from the early opportunistic phase to the more purposeful cognitive task analysis interviews, the stance remains very open. Although a knowledge elicitation strategy has been selected, this activity is goal directed. In pursuing project goals and working within real-world constraints, the application of cognitive task analysis remains somewhat flexible. In our experience, when unexpected observation opportunities arise, we intentionally attempt to work them into the plan. If the selected interview strategy is not yielding the desired results, we adapt or discard it for a more promising technique. If the selected practitioners are not available, we make every effort to find other people with relevant experience.

The term *expert apprentice* may seem less directly relevant here because there is little or no opportunity for the cognitive engineer to act as apprentice by shadowing an expert and asking questions along the way. However, the need to fulfill the role of the learner remains in effect. The aspect of the cognitive engineer's stance that differentiates it as an *expert apprentice stance* is the shift from user-interface designer, research psychologist, or industrial engineer to expert apprentice. It is true that short-term projects often do not allow for the type of job shadowing in which the practitioner is asked to take on the role of master in order to train the investigator in doing the job. Instead, more focused interview techniques are employed to get at a first-person perspective. Incident-based interview techniques encourage the practitioner to recount challenging incidents. As the investigator walks through these incidents with the practitioner, timelines and cognitive probes are used to thoroughly explore the sequence of events and how this sequence is understood and interpreted by the practitioner (Hoffman et al., 1998; Klein et al., 1989). The stance of learner allows the investigator to ask the SME to explain things more fully, and reexplain things not fully understood.

The following case study illustrates the expert apprentice assuming a more focused approach to bootstrapping.

Case Study: B-2 In-Flight Troubleshooting

Snead, Militello, and Ritter (2004) conducted a study of in-flight troubleshooting by B-2 pilots. The study was undertaken in the context of an effort to transition paper checklists into an electronic checklist. Of the checklists to be transitioned, the emergency procedures were those of greatest concern to the Air Force, as they are most likely to be used under time pressure, during nonroutine events, and in conditions of risk. The design of a prototype checklist was already well under way when the research team was asked to conduct a cognitive task analysis focusing on the use of emergency procedure checklists in the pilot decision process. The study was characterized as a means to better understand pilot troubleshooting so that the prototype would be designed to facilitate rather than interfere with or hinder pilot decision making during in-flight troubleshooting incidents.

This case study is offered as a contrast to the aircraft maintenance case study discussed earlier in this chapter. In this case, the project timeline was approximately 6 weeks, access to documentation of the B-2 was quite limited (due in part to classification issues, but also because of a lack of contact within the B-2 community), and observation of any sort did not seem feasible at the outset (due to the lack of an advocate in the B-2 community).

Investigators began by reviewing a paper version of the B-2 checklist to understand how it was organized and what information could be found there. Information was obtained on the Internet regarding the types of missions flown by the B-2, the aircraft capabilities, as well as a photo of the B-2 cockpit. Permission was obtained to visit Whiteman Air Force Base and interview six experienced B-2 pilots. In preparation for the data collection trip, investigators developed an interview guide consisting of Critical Decision Method (Klein et al., 1989) probes aimed at obtaining examples of two types of troubleshooting incidents: (a) those in which the paper checklists had proved helpful and (b) incidents in which problem-solving strategies beyond those found in paper checklists were needed. Observation opportunities were not anticipated during the visit.

In response to the interview probes, interviewees explained that few troubleshooting incidents arise in the B-2 because the aircraft is relatively new and technologically advanced. For those interviewees who did not have firsthand experience troubleshooting in the B-2, interviewers adapted the interview to include troubleshooting incidents in other aircraft (i.e., earlier in the interviewee's career). As these incidents were related, interviewees were able to discuss how the strategies used might be similar or different in the B-2. As investigators continued to ask questions aimed at unpacking the troubleshooting task through the eyes of the pilot, one interviewee offered to provide a tour of the B-2 to aid investigators in envisioning the different information sources and their location in the cockpit.

Investigators took advantage of the opportunity to see the B-2 cockpit firsthand and discuss troubleshooting as it occurs in that setting. During this discussion, the interviewee suggested that a session in the simulator might provide a better understanding of the dynamic nature of the task. Investigators were then provided an opportunity to fly the simulator for a context-based experience of different warnings, cautions, and advisories that occur in flight.

At the end of the 2-day data collection trip, investigators did not have a comprehensive understanding of the B-2 in the Air Force system. They did, however, learn enough about in-flight troubleshooting in the B-2 to inform prototype design. Although this entire study took place in the span of 6 weeks, investigators were able to conduct focused interviews and take part in an opportunistic tour of the work space (i.e., the cockpit and simulator session). In this short time span, and with limited access to documentation, practitioners, and the work setting, investigators were able to map out the decision flow as it occurs during in-flight troubleshooting, highlighting the use of paper checklists and other resources. A discussion of characteristics of the paper checklist that are highly valued by pilots, as well as aspects that are considered to be in need of improvement, was delivered to prototype designers. In addition, leverage points for improving checklist functionality in a digital format were identified and shared with the larger project team.

CAN AN APPRENTICE BE TOO EXPERT?

We are often asked if being an SME in the domain of interest makes the job of the expert apprentice easier. This question was also of interest in the era of first-generation expert systems, in a debate concerning whether it is better for the knowledge elicitor to become an SME or better to train the SME in the ways and means of knowledge elicitation (see, e.g., Westphal & McGraw, 1998).

In fact, hybrid SME/cognitive engineers are generally eagerly sought after to fill out a project team, but rarely available. The difficulty is that becoming an expert in any domain, whether it be piloting or cognitive engineering, generally requires education as well as an extensive experience base. Finding a person with both the educational knowledge and the real-world expertise in two domains (such as piloting and cognitive engineering) is rare, but highly valued.

It is important to mention that we have encountered one important drawback when working with a practitioner who is serving in the role of interviewer: The skills and experience necessitated by one's role as a practitioner limit the ability to take on the role of the apprentice convincingly. The SME's task is to know the subject in question, whereas the apprentice's task is to learn about the domain in question. For a practitioner-interviewer, a somewhat telegraphic discussion can take place in which the interviewee does not describe fully what was seen, heard, smelled, felt, because of a tacit understanding that the practitioner-interviewer already knows about the material being discussed. The interviewer can fail to ask for this sort of detail, either because she or he can imagine what it must have been like and doesn't realize it has not been articulated, or also because it feels socially awkward to ask this practitioner to state what seems patently obvious (to another practitioner). In either case, the resulting information is the investigator's inference about the perceptual information available and not the interviewee's actual experience. The expert apprentice, on the other hand can (and often should) deliberately ask naive questions under the guise of gaining additional detail about the topic at hand. This is not to say that a practitioner cannot gather useful data during the bootstrapping phase, but that appearing genuine in the role of apprentice will likely be a challenge for the practitioner-interviewer.

On the other hand, there is the danger of going into interviews with insufficient background knowledge, particularly if incident-based interview techniques are used. The danger is that the bulk of the interview time can be spent explaining jargon, organizational constraints, equipment, and so forth, so that little time is spent unpacking the cognitive challenges of the job. This can seriously limit the effectiveness of a project with a short timeline and limited access to the domain practitioners. It is important to note that some interview techniques are designed so that they can be used to scaffold interviews in which the interviewer has engaged in limited bootstrapping (i.e, concept mapping, the knowledge audit) (Cooke, 1994; Klein & Militello, 2005; Militello & Hutton, 1998; Novak & Gowin, 1984).

One strategy commonly used to ensure that investigators have access to enough background knowledge is including a practitioner-consultant as an "informant" on the research team. The practitioner-consultant receives explanation of the ways and means, and rationale of cognitive systems engineering and cognitive task analysis. After that, the practitioner-consultant can participate in observations and help interpret events for the rest of the team. The practitioner-consultant can also play a role in prepping interviewers, and can provide a reality check in postinterview data analysis. If the practitioner-consultant participates in the interview sessions, it is important to define roles and take steps to ensure that the interviewee does not feel judged or evaluated by the practitioner-consultant. Having another practitioner in the room can be intimidating to an interviewee, especially if the practitioner-consultant is perceived as having more experience or higher rank than the practitioner being interviewed. In any case, having an SME-consultant in the interview should be handled with some finesse. For example, we often explicitly diminish or downplay the practitioner-consultant's experience in order to focus the conversation on the knowledge of the practitioner being interviewed.

CONCLUSIONS

The goal of this chapter was to discuss the role of the expert apprentice and associated bootstrapping strategies as they are applied to the investigation of complex domains and difficult decisions. The researcher's stance is important in developing skills as an expert apprentice. In the context of conducting cognitive task analyses, it is important that the investigator take on the stance of a learner. This represents a deliberate shift for many investigators, who are often quite accomplished in their own areas of expertise (e.g., training design, system design, user-interface design, experimental design). Additionally, this "learner's attitude" adopted by researchers in these scenarios contrasts with more commonly adopted stances, such as the precision-oriented stance that an experimenter might adopt when conducting controlled laboratory studies, or the more authoritative stance one might adopt when instructing participants to complete a workload assessment inventory.

An equally important issue for the expert apprentice is determining which bootstrapping methods to use and how to adapt them for a specific project. As cognitive engineers, our strategies for eliciting knowledge have developed in response to the demands of work. In the context of large socio-technological systems, we have seen the evolution of more comprehensive, thorough techniques that often include careful study of documentation followed by job-shadowing techniques that mimic, to some extent, the experiences an actual apprentice might have. In-depth, focused interviews are likely to follow examination of foundational information about the work domain, various job positions, technological functions, and work procedures have been thoroughly examined. In contrast, shorter-term projects often focus on the human component of the system from the very beginning, exploring the work domain, procedures, and technological functions only as they impact the individual's (or team's) ability to make key decisions, solve problems, and form judgments. In the context of this more focused approach, acquisition of foundational information is more opportunistic than systematic. The expert apprentice will often move very quickly to focused, in-depth interview techniques.

The term *expert apprentice* appropriately connotes the stance of the cognitive engineer as a learner. Although this appears quite simple on the surface, becoming an expert apprentice requires skill. Successful expert apprentices are able to shift from their own area of expertise into the role of a learner. Furthermore, they have a set of techniques that they can adapt and tailor to put the SME at ease and ensure effective data collection for each project. Constraints that impact the choice of technique include the time available for bootstrapping, access to practitioners, opportunities for job shadowing and observation, as well as the overall project goals. An expert apprentice is able to take these and other project constraints into account to develop a data collection plan—and to improvise, changing the plan as opportunities appear (or disappear) and an understanding of the domain evolves.

The Salient Challenge

There is a salient issue that we have not discussed in this chapter, but have deliberately saved for last. Hoffman raised a series of questions focusing on this issue at the NDM6 conference:

> How is it that some people have managed to become expert apprentices? Is experience at conducting cognitive task analysis all there is to it? How is it that some scientists in the fields of human factors and cognitive engineering, those who are in academia, have trained people to adopt the challenges of cognitive systems engineering, and how are their students trained in conducting cognitive task analysis? As technology advances, as the world undergoes rapid change, as cognitive work (including work using advanced technologies) becomes more the norm across sectors of society, business, government, and the military—how can we deliberately go about training a next generation of "expert apprentices"? How do we prepare people to help design advanced technologies by deliberately placing themselves "out of context" every time they take on a new project?

We invite the community of human factors and cognitive systems engineering specialists, and academia broadly, to seriously consider the implications for training, and perhaps new forms of training programs. Our hope is that the discussion of expert apprenticeship in this chapter will serve as a jumping-off point for those interested in becoming a cognitive engineer as well as those providing training and education for future cognitive engineers.

REFERENCES

Clancy, T. (1984). *The Hunt for Red October.* Annapolis, MD: Naval Institute Press.

Cooke, N. J. (1994). Varieties of knowledge elicitation techniques. *International Journal of Human–Computer Studies, 41,* 801–849.

Crandall, B., Klein, G., & Hoffman, R. (2006). *Minds at work: A practitioner's guide to cognitive task analysis.* Cambridge, MA: MIT Press.

DiBello, L. (2001). Solving the problem of employee resistance to technology by reframing the problem as one of experts and their tools. In E. Salas & G. Klein (Eds.), *Linking expertise and naturalistic decision making* (pp. 71–94). Mahwah, NJ: Lawrence Erlbaum Associates.

Donahoo, C. D., Gorman, M. E., Kancler, D. K., Quill, L. L, Revels, A. R., & Goddard, M. (2002). Point of maintenance usability study final report (Tech. Rep. No. AFRL-HE-WP-TR-2002-0100). Wright-Patterson AFB, OH: AFRL/HESR.

Eggleston, R. G. (2003). Work-centered design: A cognitive engineering approach to system design. In *Proceedings of the Human Factors and Ergonomics Society 47th annual meeting* (pp. 263–267). Santa Monica, CA: Human Factors Society.

Friend, J., & Grinstead, R. (1992). *Comparative evaluation of a monocular head mounted display device versus a flat screen display device in presenting aircraft maintenance technical data.* Unpublished master's thesis, Air Force Institute of Technology, Wright-Patterson, AFB, OH.

Gorman, M., Donahoo, C., Quill, L., Jernigan, J., Goddard, M., & Masquelier, B. (2003). *Point of maintenance ruggedized operational device evaluation and observation test report.* (Tech. Rep. No. AFRL-HE-WP-TR-2002-0251). Wright-Patterson AFB, OH: AFRL/HESR.

Gorman, M., Snead, A., & Quill, L. (2003). Maintenance mentoring system: Nellis Predator study (Report No. UDR-TR-2003-00138). Dayton, OH: University of Dayton Research Institute.

Hoffman, R. R. (1987). The problem of extracting the knowledge of experts from the perspective of experimental psychology. *The AI Magazine, 8,* 53–67.

Hoffman, R. R., Coffey, J. W., & Ford, K. M. (2000). *A case study in the research paradigm of human-centered computing: local expertise in weather forecasting* (Report on the contract, "human-centered system prototype." Washington, DC: National Technology Alliance.

Hoffman, R. R., Crandall, B. & Shadbolt, N. (1998). Use of the critical decision method to elicit expert knowledge: A case study in the methodology of cognitive task analysis. *Human Factors, 40,* 254–276.

Kancler, D. E., Wesler, M., Bachman, S., Curtis C., Stimson D., Gorman, M., et al. (2005). Application of cognitive task analysis in user requirements definition and prototype design. In *Proceedings of the Human Factors and Ergonomics Society 49th Annual Meeting* (pp. 2045–2049). Santa Monica, CA: Human Factors and Ergonomics Society.

Klein, G. A., Calderwood, R., & MacGregor, D. (1989). Critical decision method for eliciting knowledge. *IEEE Transactions on Systems, Man, and Cybernetics, 19,* 462–472.

Klein, G., & Militello, L. G. (2005). The knowledge audit as an approach for cognitive task analysis. In B. Brehemer, R. Lipshitz, & H. Montgomery (Eds.). *How professionals make decisions* (pp. 335–342). Mahwah, NJ: Lawrence Erlbaum Associates.

Militello, L. G., & Hutton, R. J. B. (1998). Applied cognitive task analysis (ACTA): A practitioner's toolkit for understanding cognitive task demands [Special issue]. *Ergonomics: Task Analysis, 41,* 1618–1641.

Militello, L. G., & Lim, L. (1995). Patient assessment skills: Assessing early cues of necrotizing enterocolitis. *The Journal of Perinatal & Neonatal Nursing, 9*(2), 42–52.

Militello, L. G., Quill, L., Vinson, K., Stilson, M., & Gorman, M. (2003). Toward developing situation awareness evaluation strategies for command and control environments. In *Proceedings of the 47th Human Factors and Ergonomics Society Meeting* (pp. 434–438). Santa Monica, CA: HFES.

Morales, D., & Cummings, M. L. (2005). UAVs as tactical wingmen: Control methods and pilots' perceptions. *Unmanned Systems, 23*(1), 25–27. Retrieved September 13, 2005, from http://www.auvsi.org/us/

Novak, J. D., & D. B. Gowin. (1984). *Learning how to learn.* New York: Cambridge University Press.

Potter, S. S, Roth, E. M., Woods, D. D., & Elm W. C (2000). Bootstrapping multiple converging cognitive task analysis techniques for system design. In J. Schraagen, V. Shalin, & S. Chipman (Eds.), *Cognitive task analysis* (pp. 317–340). Mahwah, NJ: Lawrence Erlbaum Associates.

Quill, L. L. (2005, August). *Smart systems for logistics command & control: How can we make logistics command and control technologies smart?* Poster presented at the Integrated Systems Health Management Conference. Wright-Patterson AFB, OH.

Quill, L. L., & Kancler, D. E. (1995). Subjective workload measurement: An aid in evaluating flightline maintenance systems. In *Proceedings of the Human Factors and Ergonomics Society 39th annual meeting* (pp. 1208–1213). Santa Monica, CA: Human Factors Society.

Quill, L. L, Kancler, D. E., & Pohle, P. (1999). *Preliminary recommendations for the electronic display of graphical aircraft maintenance information.* (Tech. Rep. No. AFRL-HE-WP-TR-1999-0196). Wright-Patterson AFB, OH: AFRL/HESR.

Quill, L., Kancler, D. E., Revels, A. R., & Batchelor, C. (2001). Application of information visualization principles at various stages of system development. In *Proceedings of the Human Factors and Ergonomics Society 44th Annual Meeting* (pp. 1713–1717). Santa Monica, CA: Human Factors and Ergonomics Society.

Quill, L. L., Kancler, D. E., Revels, A. R., & Masquelier, B. L. (1999). Synthetic Environments don't have to be digital. In *Proceedings of the Interservice/Industry Training, Simulation, Education Conference* (pp. 1005–1011). Arlington, VA: National Training Systems Association.

Revels, A. R., Kancler, D. E., Quill, L. L., Nemeth, K. N., & Donahoo, C .D. (2000). *Final report: Technician performance with wearable PC/HMD* (Contract No. F41624-98-C-5004). Wright-Patterson AFB, OH: Armstrong Laboratory Logistics Research Division.

Snead, A. E., Militello, L. G., & Ritter, J. A. (2004). Electronic checklists in the context of B-2 pilot decision making. In *Proceedings of the 48th Human Factors and Ergonomics Society Meeting* (pp. 640–644). Santa Monica, CA: HFES.

Swierenga, S., Walker, J., Snead, A., Donahoo, C., Quill, L., & Ritter, J. (2003). *Electronic Flight and Technical Manual Design Guidelines* (Tech. Rep. No. AFRL-HE-WP-TR-2003-0161). Wright-Patterson Air Force Base, OH: AFRL/HEAL.

Vicente, K. J. (1999). *Cognitive work analysis.* Mahwah, NJ: Lawrence Erlbaum Associates.

Westphal, C. R., & McGraw, K. (Eds.). (1998). Special Issue on Knowledge Acquisition, *SIGART Newsletter,* No. 108. New York: Special Interest Group on Artificial Intelligence, Association for Computing Machinery.

Woods, D. D. & Roth, E. M. (1988). Aiding human performance: I. Cognitive analysis. *Le Travail Humain, 51,* 39–64.

Chapter 8

Play a Winning Game:
An Implementation of
Critical Thinking Training

Karel van den Bosch
Marlous M. de Beer
TNO Human Factors, Soesterberg, Netherlands

World events of the last two decades demand a reorientation of military missions. Military deployment has shifted away from preparing for major large-scale conflicts toward coalitional, peacekeeping, and other forms of newer types of operations. Thus, military expertise is challenged by having to work "out of context." Military commanders need the tactical knowledge and skills for successfully preparing, executing, and managing operations in new forms of unpredictable, unstable, and complex conditions (Lussier, 2003).

The Royal Netherlands Army, in particular, has ascertained that their current training programs are unable to bring personnel to the required levels of tactical competence for new contexts of operations. In addition, tactical knowledge is "seeping out" of the organization to such an extent that operational readiness is at risk (Benoist & Soldaat, 2004). For personnel who continue service in the military, there may also be an issue of "shallowness" of the level of tactical knowledge and expertise, relative to the new contexts for military operations. Commanders often have basic knowledge of tactical procedures, but current training programs, designed for an older context and era, provide insufficient opportunities for learning so that personnel can recognize the situational subtleties that make the application of a new tactical procedure appropriate (or inappropriate). This problem has been denoted as "lack of reference," and is a prime example of the challenges created by "expertise out of context."

To bring trainees up to the level of tactical experts, they need:

- Intensive, deliberate, and reflective practice (Ericsson, Krampe, & Tesch-Römer, 1993).
- Active engagement in situation assessment and decision making in representative and relevant cases (Klein, 1998).
- Practice at studying cases from different perspectives, acknowledging the relevance of the right cues and their interdependencies (Cohen, Freeman, & Thompson, 1998).

Clearly, new concepts for training tactics are needed.

Studies of tactical decision making have shown that experts have large collections of schemas, enabling them to recognize a large number of situations as familiar. When faced with a complex and unfamiliar tactical problem, experts collect and critically evaluate the available evidence, seek consistency, and test assumptions underlying an assessment. The nature of this process can be described as experience-based iterative problem solving in which correcting for one problem sometimes leads to identification of another problem (Cohen et al., 1998). Expert decision makers thus try to integrate the results into a comprehensive, plausible, and consistent story that can explain the actual problem situation. Novices, on the other hand, often consider aspects of the situation literally (vs. conceptually), and take cues, factors, or features separately and treat them as being independent.

In order to help trainees to become tactical experts, newer concepts for training are based on the known characteristics of experts. A major change in the view of how decision making should be trained has been brought about by Naturalistic Decision Making (NDM) research (e.g., Klein, 1998; Klein, McCloskey, Pliske, & Schmitt, 1997). Klein and colleagues have shown that experts rely on their accumulated experience by recognizing pattern similarity between actual and stored decision-making situations. This has been formalized in the recognition-primed decision (RPD) model (Klein, 1993). There is evidence that trainees in the military regard the core idea of RPD to be naturally applicable to their tactical planning and results of RPD training suggest several benefits over traditional models of decision making that have been in use by the military. Ross and colleagues used the RPD model to develop the recognition planning model (RPM; Ross, Klein, Thunholm, Schmitt, & Baxter, 2004). Instead of generating and comparing COAs (courses of action) as the MDMP (military decision-making process) requires, RPM stimulates sizing up situations and facilitating replanning as part of the cycle of continuously improving and adjusting a COA. The researchers assessed feasibility and acceptability of this method in a 2-week field study at the Fort Leavenworth Battle Command Battle Laboratory (Ross et al., 2004). An ad hoc Objective Force Unit of Action (UA) staff received 2 days of training in the RPM, using electronic tactical decision games to practice this type of decision making. The next phase of the experiment included 5 days of exercising the RPM by planning and executing different missions. The researchers used observations, questionnaires, and in-depth interviews of key personnel to collect data during the experiment. Findings showed that face validity for the RPM was high, and that participants had little trouble using the RPM.

Trainees not only need to know how and when to make use of stored experiences, they also need to know how to handle novel and ambiguous situations. This includes, for example, how to deal with conflicting and unreliable data, when to abandon an assessment in favor of an alternative one, when to stop thinking and start acting. It is these skills that are addressed in a new training concept, referred to as *critical-thinking training* (Cohen et al., 1998).

Critical thinking can formally be defined as asking and answering questions about alternative possibilities in order to better achieve some objective (Cohen, Adelman, Bresnick, Freeman, & Salas, 2003). The critical-thinking strategy involves a problem-solving approach to new and unfamiliar situations. It is a dynamic and iterative strategy, consisting of a set of methods to build, test, and critique situation assessments. These methods are to some extent generalizable but they can best be taught if grounded in a specific domain and with trainees who already have a basic level of knowledge of that domain. Effective critical-thinking training combines instruction with realistic, scenario-based practice (Cohen et al., 1998). The design of exercise scenarios is very important because these must provide opportunities to practice critical-thinking strategies.

Critical thinking has been successfully used to modify and improve the command planning process (Pascual, Blendell, Molloy, Catchpole, & Henderson, 2001, 2002). In their first study, Pascual et al. (2001) compared the command planning processes undertaken by two constructed Joint Task Force Headquarters teams, using a traditional command estimate process and subsequently, a set of alternative planning processes (including RPM). Data were collected through the use of experiment observation proforma, questionnaires, video analysis, and team debriefs. Many participants felt that RPM more closely matched real-world planning processes of experienced planners and better supported the use of intuitive decision-making styles. However, one often-reported weakness of the RPM approach was the absence of formal checklists and a detailed procedural model of the proposed processes. Pascual et al. (2002) subsequently developed and evaluated a refined version of the RPM. This time participants pointed out that the revised RPM should be used by experienced planners because it fits their intuitive and naturalistic problem solving, whereas the traditional estimate process might still be best for inexperienced planners.

Effects of critical-thinking training have been studied in a series of field experiments (e.g., Cohen & Freeman, 1997; Klein et al., 1997) with encouraging results. In these studies, trainees received critical-thinking training in scenario-based exercises. Performance was compared to control subjects who did not receive training exercises but participated in activities that are not directly related to tactical command, like filling out psychological test forms, or discussing work-related issues. This allows for the possibility that not critical thinking, but mere participation in scenario-based exercises, accounts for the effects.

In the present chapter, we report on two training studies in which we administered tactical exercises to two groups of officers. One group received critical thinking training whereas the other group received standard training. Performance of both groups was compared during and after training. In the first study, we trained the individual military decision maker using a simple training environment. Commanders make the decisions, but tactical command is typically performed as a team. We therefore carried out a second study investigating the effects of critical-thinking training for teams.

The next section describes how the critical-thinking training was developed. After that the training studies and the results are described. Next an account of our experiences in putting critical thinking into practice is presented.

DEVELOPING THE CRITICAL-THINKING TRAINING

In scenario-based training, trainees prepare, execute, and evaluate exercises that are situations representative of the real world (albeit simplified). It provides trainees with the opportunity to gain experience under controlled and safe conditions (Farmer, Van Rooij, Riemersma, Jorna, & Moraal, 1999). By participating in training scenarios, trainees may gain knowledge about typical problems and their solutions, thereby increasing their experiential knowledge of situation–response relationships. For critical-thinking training, practice involves the following.

Building a Story. A story is a comprehensive assessment of the situation, in which all the existing evidence is incorporated and explained, and assumptions about uncertain aspects of the situation are made explicit. Past, present, and future are addressed in the story. The purpose of story building is to keep trainees from assessing situations solely on the basis of individual or isolated events. Instead, trainees are taught how they can integrate the available information into its context, which may include elements such as the history of events leading to the current situation, the presumed goals and capacities of the enemy, the opportunities of the enemy, and so on.

Testing a Story.   Testing a story aims at identifying incomplete and conflicting information. Trainees have to correct these problems by collecting more data, retrieving knowledge from memory, making assumptions about the missing pieces of the story, or by resolving conflicts in the argumentation.

Evaluating a Story.   After a story is constructed, it should be evaluated for its plausibility. The decision maker has to take a step back, identify critical assumptions that remain hidden, and play the devil's advocate by attempting to falsify these assumptions, that is, by explaining how an assumption can be false and building an alternative story.

Time Management.   Critical thinking is not always the appropriate decision strategy. Decision makers have to evaluate the time available and the consequences of their actions. In stressful situations such as those often encountered by military commanders, there may be little time to spare. The decision maker often has to act immediately unless the risk of a delay is acceptable, the cost of an error is high, and the situation is nonroutine or problematic (Cohen et al., 1998). Critical-thinking training focuses on the way trainees apply these criteria.

Intentional Introduction of Ambiguous, Incomplete, and Inconsistent Information.   This enables students to produce different explanations for events, recognize critical assumptions of situation assessments, critique and adjust assumptions and explanations, and mentally simulate outcomes of possible decisions.

The following general guidelines were presented to subject-matter experts (SMEs) who were asked to develop scenarios for our critical-thinking training. An abridged example based on these guidelines is added in *italics:*

1. Select a target story representative of the tactical issue associated with one of the learning objectives. *Domain: Ground-to-air defense. Learning objective: Trainee needs to develop a defense plan using personnel, weapon and sensor systems to defend assets (cities, airports, C2 stations, water supplies, etc.). In setting the priorities, trainee must take into account the political situation, the recent military developments, logistic constraints, and so on. Target story is the detachment of a Reaction Force (RF) in a NATO country that is in conflict with a neighbor country. Due to an arrival delay of the commander, the battle captain of the Advanced Party is now in charge of formulating the defense plan.*

2. Use your experience or imagination to develop a basic scenario for this target story.

Basic scenario: Water shortage has caused a conflict between NATO country (Greenland) and its neighbor country (Redland). Greenland has the only natural water reservoirs in the region at its disposal. However, Greenland has supplied less water than agreed upon to Redland. Redland has announced plans to take over the reservoir area in order to guarantee their water supply at all times. NATO has reacted by deploying an RF. Basic scenario is that, despite their statements, Redland will not capture the water reservoirs by force (insufficient military power, escalation of conflict), nor will they attack the reservoirs by poisoning them (lose-lose situation). Redland is unlikely to attack the city, because of the presence of a large population that has the same religion as the principal religion of Redland. Focus of defense should therefore be the airports, the C2 structures, and own-troops areas.

3. Modify basic scenario in such a fashion that it allows for alternative interpretations (e.g., by making information incomplete or unreliable, by introducing information that is inconsistent with other information sources, or by making crucial information inaccessible to the decision maker).

Alternative interpretations: During the scenario new information becomes available (some as a result of actions or requests of the trainee, some inserted by the scenario leader), such as escalating political statements of Redland; ostentatious statements of military support by Orangeland (a military powerful neighbor of Redland); intelligence information indicating that Redland might have new tactical ballistic missiles capable of precision bombardments (striking certain areas of a city only), and so on.

4. Use the target and alternative stories to imagine what actions you would predict trainees would take to verify or retrieve information. Determine whether the scenario leader should release this information or not (or might again provide incomplete or unreliable information).

STUDY 1

The first study was conducted in the domain of air defense by the Royal Netherlands Air Force, in particular the Tactical Command Station (TCS) of a ground-to-air defense battalion. In an office room, trainee-officers played air-defense scenarios under supervision of a scenario leader, who was a domain expert. The study was conducted in a small, quiet room, using paper and pencil, with the trainee sitting at one desk and the supervisor at another.

The trainee played the role of battle captain; the scenario leader played all other functions (lower and higher control), and introduced the scripted events in the scenario (e.g., battle damage reports, information about enemy movements, identified radar tracks). Prior to each training scenario, the trainee was provided with a short description of the political, military, and civil background situation.

Method

A training posttest design was used, as depicted in Table 8.1.

Sixteen officers took part in the study. All of them were operators of a ground-to-air defense battalion, with different backgrounds and (tactical) ex-

TABLE 8.1
Experimental Design

Condition	Instruction	Training	Test
Critical Thinking group (N = 8)	Instruction and demos in critical thinking	Scenarios 1–6, with support in critical thinking; process and outcome feedback	Scenarios 7–8; without support; no feedback
Control group (N = 8)	No specific instruction	Scenarios 1–6, no support; outcome feedback only	Scenarios 7–8; without support; no feedback

perience levels. Age of the participants ranged from 26 to 40 years, with an average of 32. Participants were evaluated according to their tactical education and experience, and matched pairs of trainees were randomly assigned to the conditions. Experience level was equally spread over conditions.

The critical-thinking group received a critical-thinking tutorial, followed by a demonstration in which two scenario leaders (one of them played the role of trainee) showed how critical thinking should be used in the scenarios. Trainees of the control group were instructed to run the scenarios as a normal command post exercise.

Two sets each consisting of three scenarios were used. Two scenario leaders were available. Order of scenario sets and assignment of sets to scenario leaders were counterbalanced. While performing the scenarios, all trainees were asked to think aloud in order to give the scenario leader access to the assumptions and reasoning underlying the assessments and decisions. At predetermined moments, the scenario leader "froze" the scenario for interventions. After each scenario, the scenario leader filled in an evaluation form.

For the critical-thinking group, critical-thinking supporting schemes were available during training. As an example, Figure 8.1 shows a scheme that covers the part of the critical-thinking cycle in which the story is tested.

At scenario freezes, and after completing the scenario, the scenario leader provided support and feedback on the critical-thinking process (e.g., by asking "Which alternative explanations are possible?" or "How can you verify that assumption?"). For the control group, trainees received outcome feedback only (e.g., "That was a good decision," or "You should have issued that request earlier").

For the posttraining test, two test scenarios and two scenario leaders were available. Order of scenario and assignment of scenario to scenario leader were counterbalanced. All trainees were asked to think aloud. No support or feedback was given.

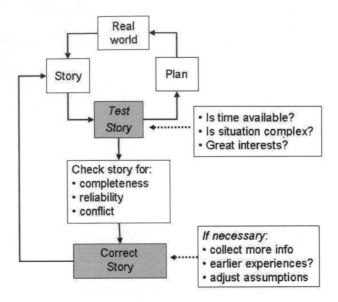

FIG. 8.1.   Supporting scheme for story testing.

To investigate the effects of critical-thinking training, both outcome and process measures were used. Outcome measures assess the quality of the end result (what is actually achieved?); process measures refer to the quality of the strategies, steps or procedures used to accomplish the task. The scenario developer specified the required outcome (e.g., order, information request, plan) for each scenario event. Outcome measures were two types: *result* (the assessed quality and timeliness of the plan, communication, actions) and *contingency plans* (assessed degree of anticipation to alternative courses of events in plan and the quality of precautionary measures).

Process measures refer directly to the critical-thinking skills. Process measures were grouped into the following two types: *information processing* (selecting relevant information, story building, identifying of incomplete or conflicting information) and *argumentation* (the explaining for missing or conflicting evidence, criticizing assumptions, coming up with alternative explanations).

Trainee performance on all four measures was assessed by the instructor. A 10-point scale was used. A verbal description was used for each scale point, ranging from "very poor" for score 1, to "excellent" for score 10. Prior to the experiment, instructors had used the results of pilot subjects (using samples of the same scenarios) to come to a common understanding of assigning scores.

Results

Figure 8.2 shows median scores on the test scenarios. Kruskal–Wallis tests showed significant differences for *contingency plans* only [$H(1) = 3.91$, $p <$ 0.05]. Scores on the variables *information processing, argumentation,* and *result* showed a similar pattern, but the differences between groups were not statistically significant [$H(1) = 1.62$, $p = 0.21$; $H(1) = 2.08$, $p = 0.15$; and $H(1) = 1.23$, $p = 0.27$, respectively].

Discussion

The study offered the battalion commander insight into the tactical competencies of his personnel. This critical-thinking training requires making processes, knowledge, and reasoning more explicit by means of thinking aloud. These features enabled the scenario leaders (squadron leaders in daily life) to see lacunae in tactical competencies of the participating officers that had remained concealed during the exercises that constitute normal training. Furthermore, comparing the average scores on the dependent measures suggested better performance of the critical-thinking group, although the difference reached statistical significance for the contingency measure only.

However, it is necessary to point out that the present study has a number of methodological limitations as well as some limitations in scope. We begin

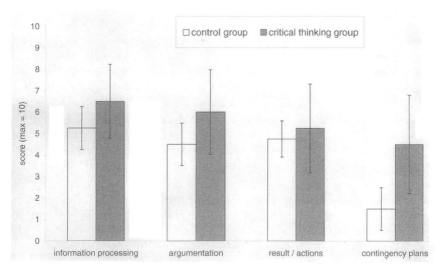

FIG. 8.2. Median scores on the test scenarios.

by addressing three methodological limitations. First, the scoring of trainee performance was conducted by the scenario leaders, who had knowledge of the study's goal and design. Practical limitations forced us to do it this way, but it is not ideal as it may have introduced a confound. More independent assessments are needed. A second problem in making comparisons between groups are the differences in time-on-task. Although both groups received the same number of scenarios and an equivalent number of training sessions, the length of a session was generally slightly longer for the critical-thinking group than for the control group (approximately 30–40 minutes for the critical-thinking group; 20–25 minutes for the control group per training session). It is possible that these time-on-task differences account (partly) for the effect. Third, it may be argued that subjects of the critical-thinking group knew that they were getting special training, whereas subjects of the control group did not. In other words, any differences might be the result of a Hawthorne effect. In order to truly rule out the possibility of a Hawthorne effect, a "yoked" control condition is necessary (a group receiving a new form of training that is unrelated both to critical thinking and to current training practice). However, there are arguments making the Hawthorne effect as the sole factor explaining between-subjects differences in this study unlikely. Although the control group's training conformed to everyday training (accent on military tactical issues; little or no guidance or feedback on task processes), the form in which they received training was new to them (one-to-one with an instructor; unusual setting: classroom instead of training in operational task environment; presence of scientists; required to talk aloud while carrying out the task, etc.). Thus, subjects of the control group had to deal with another form of training too. "Newness of training" is, therefore, in our opinion not likely to be the sole explaining factor.

The scope of the present study does not fully cover the potentially wide range of critical-thinking training applications. First, the effect of training was studied in a simple task environment. This proved to be suitable for practicing critical thinking, possibly because the absence of any kind of distracting events (incoming messages, alarms on displays, etc.) kept students focused on the key issues: situation assessment and decision making. Eventually, however, critical-thinking skills need to be applied in the real world. For reasons of transfer, it is necessary to investigate whether critical-thinking skills can be successfully trained and applied in higher fidelity task environments. Another aspect of the present study is that it focuses on the individual commander, whereas tactical command is typically performed as a team. Critical-thinking training for teams may be especially helpful in promoting shared mental models (Stout, Cannon-Bowers, & Salas, 1996), and in preventing "group think" by making team members more critical of implicit or explicit assumptions. In another study, we addressed the effects of critical-thinking training for teams in a high-fidelity tactical simulator.

STUDY 2

This study investigated the effects of critical thinking applied to the training of command teams operating in more realistic task environments. It was conducted in the domains of "anti-air warfare" (AAW) and "anti-surface warfare" (ASuW) at the Operational School of the Royal Netherlands Navy. Eight officers and eight petty officers participated in the study. Participants were anti-air-, surface-, or subsurface-warfare instructors at the Tactical Education Department. Age of the officers ranged from 29 to 32 years, with an average of 31. Average on-board experience for this group was 5 years. Age of the petty officers ranged from 29 to 46 years, with an average of 39, and an average on-board experience of 14 years. Participants were grouped according to expertise in either air defense (AAW) or (sub)surface defense (ASW/ASuW) teams. Teams were composed of an officer and a petty officer. Trainees played single-ship/single-threat scenarios in a high-fidelity tactical simulator. This is shown in Figure 8.3. Scenarios were developed by two instructors at the Operational School.

Method

A training posttest design was used. The supervising project officer arranged the eight participating teams according to their tactical education and operational experience, and assigned teams randomly to either the

FIG. 8.3. Research setting.

critical-thinking training group or the control group. The supervising project officer also selected two instructors for the study. They were randomly assigned to train either the critical-thinking teams or the control teams.

Prior to the experiment, instructors who were assigned to train the critical-thinking training teams were extensively briefed on the critical-thinking training method, as well as on how to support trainees in the application of critical-thinking processes.

Instructors assigned to the control team were not informed about the concepts of critical thinking. They were told to support the teams as they would normally during training. Instructors trained one team at a time. The briefing, training, and testing required 4 days per team.

The first day of the study was used for briefing and instruction of the teams. The experimenter and the assigned instructor briefed the critical-thinking team on the principles of critical thinking and showed them how to apply these principles in paper-based demonstration scenarios. The control group instructor briefed his team on the itinerary of the coming days, and discussed a paper-based demonstration scenario with them.

On the second day, teams received two interactive role-playing scenarios in a staff room under supervision of their instructor. On the third day, teams received two scenarios in the tactical simulator. A scenario run took approximately 2 hours. The instructor interventions were the same as in Study 1.

In training the critical-thinking group, the instructor encouraged his team to explicitly execute all critical-thinking components and he provided extensive guidance and feedback during and after the scenarios. For the control group, the instructor supported the control group teams as in normal training, which means they received domain-specific outcome-related feedback, and general encouragement.

On the fourth and final day, teams were tested on two test scenarios in the simulator. Their instructors were not present during the test. The supervising project officer assigned two subject-matter experts to carry out performance evaluations. These evaluators were not informed about the goals, nature, or design of the study. Assignment of evaluators to teams was counterbalanced (each evaluator rated two critical-thinking teams and two control teams). They assessed performance of trainees individually, as well as that of the team. Evaluators received the scenarios on paper. Markers in the scenario description prompted the evaluators to score trainee and team performance at that particular moment, on specified performance criteria. The same outcome and process measures as in the previous study were used. In addition, performance with respect to *time management* and *team behavior* (communication, supportive behavior, coordination, leadership) was scored. Because the evaluators were used to the NATO 4-point performance-rating scale, it was decided to use this scale in this study. The verbal descriptions for the four scale points were, respectively: 1 = Unsatisfactory, 2 = Marginal, 3 = Satisfactory, and 4 = Excellent.

Prior to the experiment, the experimenter briefed the evaluators about the scoring procedure and how to use the scale. The results of pilot subjects were used to arrive at a common interpretation of performance measurement.

Results

Data on individual as well as on team performance were collected during training and test. For reasons of brevity, performance data of the teams on the two test scenarios are reported only. Figure 8.4 shows the results on the test scenarios.

Kruskal–Wallis tests showed statistically significant differences for *argumentation* [$H(1) = 7.5$, p. < 05], *time management* [$H(1) = 11.4$, p < .05], *contingency plans* [H1) = 5.6, p < .05], and *team skills* [$H(1) = 8.7$, p < .05]. Performance on *information processing* and *actions* did not significantly differ between groups.

DISCUSSION

Critical-thinking training produced positive effects on the *process* of team tactical command (i.e., better argumentation, more effective time management, and better team skills) as well as on the *outcomes* (i.e., more and better contingency plans). Furthermore, we observed that instructions to think

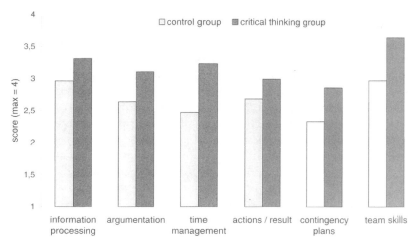

FIG. 8.4.   Median results on the test scenarios.

critically induced team members to clarify their assumptions and perspectives on the situation to one another. *For example: Officer: "Considering all information so far, we can be pretty sure that our position is known to the Huang-Fens* [enemy ships]. *What would then be the reason for the Zulfiquar* [enemy ship] *to move around in the open and almost act like a sitting duck? Could it be that our position is yet unknown?" Petty officer:"I suppose they do know where we are, and that the Zulfiquar is acting as a distractor."* This type of interaction helps in making assumptions and reasoning explicit and is considered to be important for developing shared mental models and for coordinating team actions (Stout et al., 1996). Apparently, introducing the devil's advocate procedure in training brings about changes in task performance and team behavior that are known to be important for good team performance. This is a significant observation, because in the Navy, the allocation of tasks between the officer and the petty officer is strictly defined, organized hierarchically according to rank. Officers tend not to involve petty officers in tactical issues, and it is certainly highly unusual for petty officers to comment on officers' assessments or decisions. This is clearly a waste of expertise, as petty officers in the Netherlands normally have substantial operational experience. In the present study, the control group teams showed the normal rank-based behavior: petty officers carrying out the tasks assigned to them and officers making the decisions. Teams in the critical-thinking group on the other hand, showed a different behavior pattern. Officers in this group tended to actively involve petty officers in tactical assessment, for example, by inviting them to play the role of devil's advocate. And the petty officers appeared not to be inhibited to ask officers to clarify their assessments, to make critical comments, and to suggest alternative possibilities.

In contrast with the first study, training was provided in a dynamic and interactive task environment. We observed that in this type of setting, training sometimes became too hectic to fully achieve the objective of this type of training: reflective and critical task performance. This suggests that critical thinking may need to be practiced first in more simple learning environments (e.g., by preparatory paper-and-pencil and role-playing scenarios) before introducing this type of training in simulator exercises. Disruptive effects of distracting events on critical thinking during simulator training may be overcome by introducing time-outs during the exercise, offering students the opportunity to perform the required behavior.

Again, it is necessary to point out that due to organizational and logistic constraints, the present study does not fulfill all methodological requirements. First, there is the issue of the training of instructors. It is, of course, necessary that the instructors responsible for delivering the critical-thinking training have a thorough understanding of the concept, rationale, and procedures of critical thinking. We therefore invested quite some effort to prepare the instructors for their task. Overall the instructors did a fair job. However, understanding a concept, and implementing this concept into training are not entirely the same. "Old" habits of instruction occasionally popped up, as, for example, pointing out the significance of a particular cue instead of letting students discover that for themselves. It is possible that if instructors were trained longer, effects of critical thinking would have been more outspoken. A second issue is the fact that we had only two evaluators available. Both evaluators rated critical thinking teams and control teams, but a team's performance was never rated by both evaluators. Therefore, no interrater reliability data can be computed. Although we invested efforts to make evaluators arrive at conjoint performance evaluations, common interpretation does not necessarily mean that their ratings were reliable or valid.

The results of the present and earlier studies (Cohen & Freeman, 1997; Cohen, Freeman & Thompson, 1997; Cohen et al., 1998; Freeman & Cohen, 1996) warrant further research. First of all, hitherto evidence on the effects of critical thinking is explorative. We need experimentally controlled studies that can unequivocally demonstrate whether critical thinking indeed has beneficial effects, and if so, identify the mechanism that explain such effects. The military, however, does not want to wait for such studies to be conducted. Their hands-on experiences with critical-thinking training and the results obtained so far convinced them that this type of training is a valuable solution to their current and urgent need: adequate tactical training. The urge to make use of critical thinking in real training programs requires answers to questions such as: "How can we integrate critical-thinking training into an existing curriculum?", "How should critical thinking be introduced to students?", and "What are students' and teachers' experiences in using this method?". The Royal Netherlands Navy asked us to help them with the implementation of critical thinking in a module of their tactical training program. In the next section, we give an account of our experiences.

PUTTING CRITICAL-THINKING TRAINING INTO PRACTICE

Our training studies have shown positive results for critical-thinking training. These training studies were specifically designed to study the effects of our training manipulation. They were mini training courses, conducted under controlled conditions. This is different from any normal training program. In a regular training program, instructors usually have the freedom to treat every student differently, according to the individual student's needs. In our training experiment, instructors had to treat all students similarly and according to specific experimental protocols. Another difference is the short duration of a training study as compared to a regular training program. A short training intervention may not provide the opportunity for trainees to really master critical-thinking skills. In this section, we report findings and observations of putting critical-thinking training into practice.

Recently, the Operational School of the Royal Netherlands Navy revised its training program for CIC (Command Information Centre) commanders. Two separate curricula, one for the air warfare commander and another for the surface and subsurface commanders, were merged into one. This reorganization offered an opportunity to identify shortcomings of the existing programs, and to bring about improvements in training concepts, methods, and materials for the new training program.

A major criticism of the old training programs was that a wide gap exists between theoretical lessons and practical exercises in the tactical trainers. Theoretical tactical lessons emphasized learning tactical procedures and capacities of sensor and weapon systems. The relevance of the materials and exercises to tactical situation assessment and the implications for decision making often remained implicit. When, later in the training course, trainees were required to bring this knowledge into use during exercises in the tactical simulators, they often lacked the skills to do so.

It was concluded that theoretical lessons should be redesigned in such a fashion that (a) students can develop a satisfactory repertoire of tactical patterns, and (b) there is sufficient opportunity to practice situation assessment and decision-making skills. This should better prepare students for training exercises on the tactical simulator, and for the on-board exercises.

The Approach

To achieve the objectives, we decided to embed critical thinking into Tactical Decision Games (TDGs) exercises. TDGs are a paper-and-pencil training technique that can be used to present tactical problems to trainees. They generally consist of a written description of a tactical problem accompanied by a schematic map. TDGs can be administered individually or to groups. They can be static, requiring trainees to develop a detailed and founded plan, but they can also be dynamic through the use of role players, who introduce events to which trainees must respond. TDG exercises have been used successfully to acquire contextualized tactical knowledge and understanding (Crichton, Flin, & Rattray, 2000; Klein, 1998).

The Training

The training block "Theory of Anti-Surface Warfare (ASuW)" was selected for the revised training program. A series of four TDG exercises (of different tactical complexity) was developed by an instructor at the Operational School, using the scenario guidelines set out earlier in this chapter. Exercises consisted of a problem and mission description, accompanied by a tactical map. See Table 8.2 for an example TDG, and Figure 8.5 for the accompanying map.

Prior to the training sessions, instructors were instructed extensively on the concept and principles of critical thinking. Observation protocols and performance measures were designed to support instructors in their tasks.

For the trainees, we developed an instruction book on critical thinking within the context of surface warfare, including self-study questions and exercises. In a 2-hour classroom instruction on critical thinking, we familiarized students with TDGs and explained what would be expected from them in the TDG sessions. TDGs were administered to groups of four students. By turns (so that each of the students got the chance to consider from a distance and gain insight into the group's performance), one of them was assigned the role of observer using a scoring form to evaluate his group on the following dimensions: information selection and acquisition, argumentation and reasoning, planning and contingency planning.

In addition, an experimenter-observer also evaluated the group's performance.

TABLE 8.2
A TDG Exercise

Mission: The mission of the TaskGroup (TG) 540.01 is the safe arrival of Her Majesty *Rotterdam* in the harbor of Bluton.

History and setting: On land, at sea and in the air there have been hostilities and incidents between Amberland and Blueland/NATO, yet there are still merchants and civil aircrafts in the area.

Task: You perform the role of ASuW Commander. Develop your plan and at least one contingency plan.

Tactical issues (for instructors): When students develop (contingency) plans, instructors observe whether the following tactical issues are taken into account: the unknown contacts in the area (where are the Kimons [hostile ships]?); the interception of an enemy Atlantique radar-signal (are we identified/classified?); the possible intentions of the Huang Fens (hostile ships) and the Zulfiquar (enemy ship, equipped with ballistic weapons only); the status of Greenland (on our side?); can we expect a coordinated attack?; is an enemy suicide mission likely? Do students consider the option to split up, is the option to change course and/or speed considered?

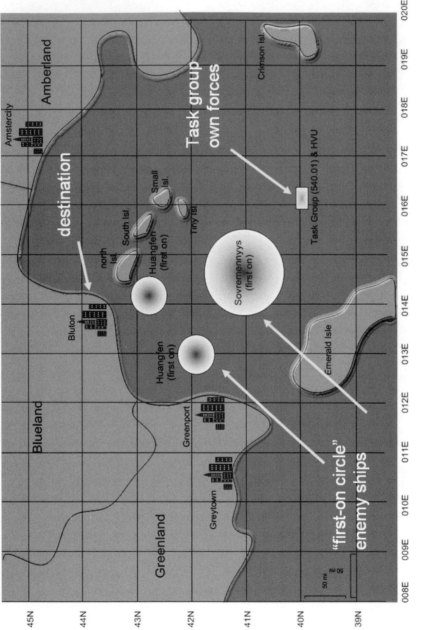

FIG. 8.5.   The map that accompanied the scenario laid out in Table 8.2.

Students were asked to clarify their assessments in their discussions, thus giving observers and the instructor access to the assumptions and reasoning underlying their decisions. In order to enhance critical-thinking processes, the instructor requested the students to execute specific critical-thinking tasks, such as "Now try to finalize your initial assessment into a story," or "Now test your story on conflicting, unreliable or incomplete information," or "Identify a critical assumption in your story and apply the advocate-of-the-devil technique." After completion, each group presented their assessments, plans, and contingency plans to the other groups. Tactical key decisions were discussed collectively.

Findings

The majority of students were enthusiastic and motivated to cooperate. They felt that critical thinking helped them to systematically assess a situation, integrate different observations into a coherent story, identify uncertainties and justify assumptions, and come up with (contingency) plans. They appreciated the exercises as a suitable method for consolidating and applying their tactical knowledge, and for practicing their skills in tactical assessment and decision making.

Although the majority of students were distinctly positive, there were also some individuals who failed to appreciate the purpose of the critical-thinking concept. It appeared that some of these students lacked the domain knowledge required to conduct critical thinking as intended. For instance, they were unable to identify a critical assumption in their assessment, or were unable to judge the tactical relevance of ambiguous information. As a result, trainees applied the critical-thinking method in an obligatory fashion, more like a checklist to be completed, rather than as an approach to reflect on the quality of tactical assessments. During after-action reviews they were reluctant to elaborate on alternative assessments, because they considered them to be "too unlikely."

Instructors were of the opinion that the present concept of training helps trainees to make the necessary shift from passive classroom learning to (inter)active scenario-based training. They felt that the required elaboration on the tactical issues presented in the TDGs helps students to develop tactical schemes, and that critical thinking helps shape the necessary strategic skills.

The importance of adequate training of instructors can not be overstated, because they have to be able to simultaneously teach, guide, monitor, and assess critical-thinking processes. We observed that sometimes instructors fell back on traditional teaching techniques and presented their solutions to students too quickly.

The introduction of the critical-thinking TDG as an alternative concept to training formed a good start toward meeting the objectives of the Operational School, and received support from those involved—students, teachers, and management.

GENERAL DISCUSSION

The international political developments of the last decade incontrovertibly show that military operations are becoming more likely to be carried out under dynamic, complex, and uncertain conditions. However, current training programs have difficulty in bringing military commanders up to the required level of tactical competence associated with these types of operations. Commanders do master the basic tactical procedures, but are often unable to acknowledge the relevance of contextual information and take situational subtleties into account when assessing a situation and making decisions. Less experienced commanders are inclined to focus on isolated cues and tend to take them at face value. Furthermore, they are often not explicitly aware of the assumptions they maintain, hence are less critical about them, and are more likely to "jump to conclusions."

Critical-thinking training has been claimed to be successful. It specifically addresses the two key aspects of expertise: availability of elaborated mental schematas and the ability to build, test and critique situation assessments for novel and ambiguous situations. The benefit of critical thinking is that it enables commanders to sort out what is really important, to address conflicts in available information, and to anticipate and prepare for possible deviations from the expected course of events. *For example, the bearing of a track homing on the own-ship suggests enemy threat, whereas its low speed indicates a routine patrol. In this case the commander has to recognize this information as conflicting, and further investigate.*

The present study indicates that training commanders in critical thinking is indeed beneficial. The two studies into the effects of critical-thinking training show that it not only improves the *process* of tactical decision making, but also has a positive effect on the outcome. Furthermore, experiences and observations during the implementation of critical thinking in an actual training course of the Netherlands Navy show that all those involved consider the method helpful.

The present studies yielded results and observations that are helpful as a first step toward setting up critical-thinking training. In order to start critical-thinking exercises, participants must have sufficient domain knowledge and should master the elementary tactical procedures. Apparently, it is best to start critical-thinking training in a simple learning environment, with sufficient time and support to practice all components of the strategy. When students have sufficiently integrated critical thinking into their tactical planning and command, then exercises in more dynamic and interactive learning environments are in order. Critical thinking is also possible in the training of teams. We observed that students explicated assumptions and argumentations, which is known to be important for developing shared situational awareness among team members (Stout et al., 1996).

The modest successes reported here form only the first steps in the introduction of critical-thinking training in military training programs. In order to expand our knowledge of the effects of critical-thinking training and to survey its opportunities and restrictions, it is important not only to establish the short-term effects of training, but also to investigate how acquired knowledge and skills transfer to novel situations and tasks. In addition, it is of interest to observe how critical-thinking skills are retained during periods of little or no practice. The long-term effects of critical-thinking training are not yet determined. We hope to be able to conduct further evaluations in the near future.

REFERENCES

Benoist, L. J. T., & Soldaat, P. B. (2004). Opleiding en Training van commandanten in de KL [Education and training of commanders in the Royal Netherlands Army]. *Militaire Spectator, 3*, 145–153.

Cohen, M S., Adelman, L., Bresnick, T.A., Freeman, J., & Salas, E. (2003, May). Dialogue as medium (and message) for training critical thinking. In *Proceedings of the 6th NDM Conference*, Pensacola Beach, FL.

Cohen, M. S., & Freeman, J. T. (1997). Improving critical thinking. In: R. Flin, E. Salas, M. Strub, & L. Martin (Eds.), *Decision making under stress: Emerging themes and applications* (pp. 161–169). Brookfield, VT: Ashgate.

Cohen, M. S., Freeman, J. T., & Thompson, B. B. (1997). Training the naturalistic decision maker. In C. Zsambok, & G. Klein (Eds.), *Naturalistic decision making* (pp. 257–268). Mahwah, NJ: Lawrence Erlbaum Associates.

Cohen, M. S., Freeman, J. T., & Thompson, B. B. (1998). Critical thinking skills in tactical decision making: a model and a training strategy. In J. A. Cannon-Bowers & E. Salas (Eds.), *Making decision under stress: Implications for individual and team training* (pp. 155–190). Washington, DC: American Psychological Association.

Crichton, M., Flin, R., & Rattray, W. A. R. (2000). Training decision makers: tactical decision games. *Journal of Contingencies and Crisis Management, 8*, 208–217.

Ericsson, K. A., Krampe, R. T., & Tesch-Römer, C. (1993). The role of deliberate practice in the acquisition of expert performance. *Psychological Review, 100*, 363–406.

Farmer, E. W., van Rooij, J. C. G. M., Riemersma, J. B. J., Jorna, P. G. A. M., & Moraal, J. (1999). *Handbook of simulator-based training.* Aldershot, England: Ashgate.

Freeman, J. T., & Cohen, M. S. (1996, June). Training for complex decision making: A test of instruction-based training on the recognition/metacognition model. In *Proceedings of the Third Symposium on Command and Control Research and Decision Aids*. Monterey, CA.

Klein, G. (1993). A recognition-primed decision (RPD) model of rapid decision making. In: G. Klein, J Orasanu, R. Calderwood, & C. E. Zsambok (Eds.), *Decision making in action: models and methods* (pp.138–147). Norwood, NJ: Ablex.

Klein, G. (1998). *Sources of power: How people make decisions.* Cambridge, MA: MIT Press.

Klein, G., McCloskey, M., Pliske, R., & Schmitt, J. (1997). Decision skills training. In *Proceedings of the Human Factors Society 41st Annual Meeting* (pp. 182–185). Santa Monica, CA: Human Factors Society.

Lussier, J. (2003). Intelligent tutoring systems for command thinking skills. *ARI (Army Research Institute) Newsletter, 13,* 7–9.

Pascual, R. G., Blendell, C., Molloy, J. J., Catchpole, L. J., & Henderson, S. M. (2001). *An investigation of alternative command planning processes* (Report No. DERA/KIS/SEB/WP010248). Fort Halstead, Kent, England: DERA.

Pascual, R. G., Blendell, C., Molloy, J. J., Catchpole, L. J., & Henderson, S. M. (2002). *A second investigation of alternative command planning processes* (Report No. DERA/KIS/SEB/WP0102613/1.0). Fort Halstead, Kent England: DERA.

Ross, K. G., Klein, G. A., Thunholm, P., Schmitt, J. F., & Baxter, H. C. (2004, July–August). The recognition-primed decision model. *Military Review,* pp. 6–10.

Stout, R. J., Cannon-Bowers, J. A., & Salas, E. (1996). The role of shared mental models in developing team situational awareness: implications for training. *Training Research Journal, 2,* 85–116.

Chapter 9

Measurement of Initiative in High-Speed Tactical Decision Making

Herbert N. Maier
TactiCog, Salisbury, NC

Katherine H. Taber
Hefner VA Medical Center, Salisbury, NC

Most situations calling for expertise require a progressive sequence of decision–action sets, each with a dependence on the outcome of the previous
ones. We are interested in tasks where the expert is in competition. An example domain that might come to mind would be tactical military action
such as air warfare. A competitor adds a high level of complexity by pur-
posefully interfering with the subject's efforts. The dynamic that emerges
from this iterative interaction increases in dimensionality, and appears to
take on a life of its own. The dynamic of oppositional expertise is one thing
that makes sporting events intense, and compelling to the spectator. In
high-stakes occupations, such as hostage negotiations, 911 dispatching, or
combat, managing this oppositional dynamic is often a life-or-death issue.
In striving for dominance, each competitor wants to stabilize the situation
in his own favor (Battig, 1972). Seizing the initiative and restricting the adversary's options reduces one's own uncertainty and complexity (Schmitt &
Klein, 1996).

Researchers in the areas of human factors and cognitive engineering often find themselves having to work out of context, studying domains with
which they are not familiar. The challenges of applied research are well-
known, including such factors as access to domain practitioners, the constraints of having to conduct research in field settings, and so on. An alternative approach is to attempt to bring some of the complexity of the world
into the lab, by studying representative tasks, or stripped-down ones. We
found ourselves at something of an advantage when it comes to the tough

199

choices for lab versus field studies, because in our case we were not entirely "out of context." The domain was one with which we were already familiar. But we still had to face the challenge of "bringing the world into the lab."

Research into oppositional expertise must re-create challenges with sufficient dimensionality and representativeness to stimulate these idiosyncrasies of adaptation and resource recruitment in a measurable way, and the measurement strategy must allow for the evaluation of individual differences. The study reported in this chapter is the first step at incorporating an "oppositional will" into research on practice schedules (Brady, 1998; Magill & Hall, 1990). Random practice is an excellent procedure for presenting uncertain input in a way that the varieties of input are still limited enough to be clearly tracked. Substituting an interactive partner for the researcher-scripted input string in traditional practice schedule research produces a "system." This system is much more realistic than any script, because the dyad is freed to find its own dynamic, without losing measurability. The goal of our research was to develop and demonstrate methods for measuring the shifting distribution of cognitive resources between incoming loads, outgoing decisions, and their interaction's impact on the situation. We report on a method-development study of decision making on the part of practitioners of the martial art, Wing Chun.

This chapter is organized as follows. First, we use the problem of finding the right domain and right method in order to frame the theoretical notions and perspective that we brought to bear in forming the study. Next, we present a brief description of the observed activity and the nature of the interpersonal dynamic expected. We then discuss the method and measures, followed by a discussion of the results and then a general discussion, in which we return to our theoretical notions.

THEORETICAL AND METHODOLOGICAL CONSIDERATIONS

The resource distribution activity involved in adversarial performance cannot be observed if the participant is simply responding to input, with no influence on the input stream. The opportunity to influence the future of the situation is a necessary challenge to total allocation for the incoming problem, because a well-chosen influence might reduce the difficulty of the next cycle. Management of cognitive resources can be put on a budgetary or economic basis by referring to an older term for situational awareness, residual attention (O'Hare, 1997). Whether by superiority in knowledge, knowledge organization, heuristics, or even actual increased neurological capacity, the struggle in competition often comes down to putting the opponent into working memory deficit while maintaining a surplus oneself.

Significant characteristics of this activity are: uncertainty of next input, recognition/planning/anticipation under extreme time pressure, high variance of situational intensity, and multiple effects of one's own actions. These multiple effects include: change of situation to new conditions with new alternatives, and the experience of reducing one's own workload by taxing the opposition's resources.

Dimensionality of the testing situation must be sufficient to induce idiosyncratic approaches. The activity must be small enough to map thoroughly, but large enough to offer alternatives to induce choice, which creates uncertainty. Visible physical action eliminating self-reporting increases confidence in the data; it also makes cycle times short enough to preclude reflection and second-guessing.

In an example of a dimensionally simple situation, a pilot must hold the aircraft stable through the seemingly random individual buffets of air turbulence. This can be represented by a random practice model such as that used by Wulf (1991). Children practiced throwing beanbags at a basket at varying distances, either repetitiously (block practice) or randomly. Hitting the basket at various distances would represent keeping the plane stable through the varying gusts, which are not intentionally directed. Managing this single-dimensional variance over a sequence of throws is unsatisfactory as an analogue for the multidimensional challenges encountered in expert decision making. The uncertainty of a progression of multidimensional states comes from innate situational factors, other players, and feedback from the actual (as opposed to anticipated) effect of one's own actions. No matter how many similar situations an expert has successfully handled in the past, the current one is different from the previous situations in some way. Studies such as that of Jodlowski, Doane, and Brou (2003), in which participants had to recognize that an instrument on an aircraft was malfunctioning and work around it, showed how an unanticipated (nonordinary) situation could reduce an expert's performance to novice levels. This relates to Schmitt and Klein's (1996) point, in that the expert's overlay of pattern does not include this condition. Such work shows important gaps in training, but is still single-dimensional, and actually quite static. A higher-dimensional model is needed to approach challenges that are more complex.

Decision-making research in chess or in small tactical exercises in the military often has participants verbally report their thought processes. This is possible due to the time available for each decision in the design, and necessary due to the broad range of options available at each decision point. A problem with this approach is that any form of self-reporting inevitably introduces an additional task with all its support functions including editing, thus possibly altering the system being studied.

The activity observed in the present study—martial arts competition—has a limited set of actions and options, therefore a limited set of available decisions. The activity occurs at a high rate of speed, limiting reflective thought and precluding verbalization. The limited set of actions is physical, therefore directly observable. This compact system gives a unique opportunity to observe high-speed, iterative decision making.

In searching for relevant prior literature, it appeared that the theoretical notions we have discussed have not been directly researched. Several fields of literature touched on aspects, but no theory, set of dimensions, or measures was found. Our theoretical description is compatible with basic cognitive theory as used in educational psychology, motor learning, knowledge organization, and other fields. Thus, it may turn out that generalization to activities other than the martial arts is possible.

METHODS

Observed Activity and Analysis Method

The martial arts are nearly unmined by any sort of cognitive research. We looked specifically at decision making on the part of practitioners of the martial art, Wing Chun. In working at a closer range than better-known arts such as Karate or Tae Kwon Do, Wing Chun relies upon overtly stated principles of physical and cognitive organization to push the opponent to overload through complexity and time pressure. The close range is qualitatively unique in that two or more objectives are usually accomplished at once, versus one objective in more familiar martial arts. Wing Chun practice is a high-risk, high-stakes activity involving cognitive and movement complexity. It continually requires judgment and both tactical and strategic decision making. The practitioner must constantly update a mental model, with a different option set for each situation, and replan the initiative to be taken. As is shown herein, the activity requires continuous problem recognition and decision making, maintenance of situation awareness, continual and rapid replanning and option generation, and continuous management of attention and uncertainty. Thus, this activity is macrocognitive as defined by Klein et al. (2003).

At the time of this research, the first author had had 17 years of experience at Wing Chun practice, and at the teaching of Wing Chun practice. Given also the training the first author had received in cognitive science, it was easy to see how aspects of the Wing Chun practice drills resonate with ideas in published research on situational awareness (e.g., Endsley & Bolstad, 1994; Endsley & Smith, 1996; Roscoe, 1993, 1997), cognitive load theory (e.g., Sweller, Van Merrienboer, & Paas, 1998), practice schedules (e.g., Li & Wright, 2000), and knowledge organization (e.g., Schvaneveldt, 1990; Wyman & Randel, 1998).

The focus of this research is on using a network structure to organize decision-making options for training and application under pressures of time, risk, and uncertainty. The small network observed in this study is a subset of a network structure inherent in the larger system from which this set of drills was drawn. Six basic Wing Chun drills instantiate the network in this study. Following is a description, first of the drills, then of the network structure.

For each of the two participants in a Wing Chun drill, the hands alternate action, one hand blocking and the other hand punching. Partners in a dyad engage such that each partner's block connects with the other partner's punch. The six drills are easily distinguished in video and photographic records. Figure 9.1 shows that Drills 1 and 2 are mirror images of each other, as are Drills 3 and 4, and Drills 5 and 6, expressing both left- and right-handed versions of each movement. In Drills 1 and 2, palm blocks the elbow of the punch; in Drills 3 and 4, palm blocks the fist of the punch; and in Drills 5 and 6, wrist blocks wrist of the punch (see also http://www.tacticog.com/research/drillnetwork1.htm).

Each drill alone could continue indefinitely with partners alternating roles of punching and blocking, as shown in Panel a of Figure 9.2. Prolonged repetition of any one drill is referred to as block practice (Magill & Hall, 1990). Additional knowledge is required to execute the transitions linking these isolated drills into a network of continuous action (Panel b in Figure 9.2). At each cycle through a drill, each partner has several choices: Continue in the same drill or transition to one of the other drills accessible from that point in the network. Both partners are encouraged to transition drills as often as possible, forcing the other partner to respond effectively to the transition while also planning his or her own transitions. Thus, the network structure generates simultaneous foreground, background, and secondary tasks for both partners. Because neither partner knows what the other partner is going to do, the networked drill structure is an empirical example of a random practice schedule, which results in higher levels of mental workload than block practice. There is a substantial literature (e.g.,

FIG. 9.1. In order, views of Drills 1, 2, 3, 4, 5, and 6.

Brady, 1998) on the benefits of random practice to long-term retention and to transfer tasks. These benefits are credited to elevated levels of cognitive load, presumably including recognition and response time-sharing with planning and execution.

Panel c of Figure 9.2 shows a traditional generic form of a drill network. This form is clearly expandable to any number of nodes, pruned to fit the tactical and logistical constraints of any specific set of drills such as the present one. Training flexibility of tactical options and fast recognition-based choicing is fundamental to responsiveness and initiative. Using networks as a foundation of training communicates a high density of information in an integrated fashion. The essence of decision making is selection of an alternative, under constraints, as an act of initiative.

The structure of the partner interaction enforces macrocognitive activity from the interaction of time pressure and the requirement to divide attention between primary tasks of vigilance and response to transitions from the partner, and planning one's own transitions, as well as relevant support functions. A response must be completed in less than half a cycle of the activity (approximately 0.45 second), or it will fail utterly. This urgency would tend to preempt resources that might be invested in building long-term advantage. Planning done as a background task pays off in pushing the urgency of responding onto the partner. This imposition can reduce the partner's investment in planning, and if he is close to his maximum capacity, push him into dysfunction, concretely modeling the aphorism that "the best defense is a good offense."

The transitions provide visible evidence of decision making. We expected that partner pairs would display measurably different levels of ex-

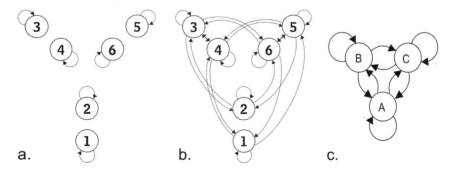

FIG. 9.2. a) Six drills are learned in isolation. b) Transitions linking the drills contain technical knowledge not found in the drills themselves, including tactical or logistical conditions prohibiting some connections. c) One traditional representation of a generic network that is expanded and pruned to suit specific situations.

pertise as indicated by frequency of transitions between drills and overall utilization of the full range of drills from which to choose. Because individuals follow different paths of experience in their learning, we also expected individualization in patterns of activity.

Participants

Participants were 12 men, age 18 to 24, possessing a general familiarity with and a strong interest in Wing Chun. The all-male sample was a consequence of the preponderance of men in the martial arts. The fact that they were all undergraduate college students probably impacted their level of seriousness and focus. Participants had two to six semesters of moderately supervised practice with the activity we were to observe. Their semiformal background in this skill set consisted of 2 to 4 hours of instruction each semester from the first author, and irregular weekly practice under the faculty adviser of the club, who is also a student of the first author. These far-from-ideal training conditions would be expected to mitigate effect size of training, providing a challenging test for the proposed method.

Procedures

Each participant was partnered with 10 other members of the sample for 2-minute periods. Activity was recorded in digital video for later data extraction. Prior to extracting data from the videos, two pairs of participants were selected as approximating the upper and lower quartiles in experience. We designate the more experienced participants as "Pair 1," and the less experienced participants as "Pair 2." Individuals in each pair were labeled L or R based on left/right position in the camera's view, and thus the four participants are referred to as 1L, 1R, 2L, and 2R. In Pair-1, 1L had six semesters' prior involvement; 1R had four. In Pair-2, 2L had four semesters' involvement; 2R had two.

Video was transferred to a personal computer, and observed at slow motion using video-editing software ("Premiere" version 6.0, Adobe Systems Inc., 345 Park Ave., San Jose, CA 95110-2704). Shot at approximately 30 frames per second, each 2-minute record consisted of approximately 3,600 frames.

From the video, an event was tallied each time a participant blocked a punch. Some of these events continued activity in the same drill (see Figure 9.1, Panel c); some transitioned (via a link) to a different drill (node), requiring the partner to recognize and follow the change. An illustration of this recording process is shown in Table 9.1. Events were categorized as reflecting:

C ("continue")—the expected block for the ongoing drill.

I ("initiate")—a block that transitions to a different drill.

A ("accommodate")—the expected block for drill partner has just transitioned to.

O ("overload")—a participant nonresponding to an action (freezing) or giving one or more inappropriate responses (flailing).

C, I, and A are thus directly derived by comparing the drill number at each event with the drill number of the previous event. O ("overload") is an exception to this, as the dimension represented by drill numbers has no place for any form of success/failure of execution of a particular instance of drill. The O category anchored a scale of cognitive load on the presumption that an excessive cognitive load, shown by confusion, is higher than loads that are managed successfully.

In this study, only one rater was used. There are several reasons why this is satisfactory. The first is that we conducted a visual comparison of photos of the three drills, showing how clearly distinct they are. It is unlikely that an experienced observer would mistake one drill for the other. The second is that in a comparison of the results shown as event series in Figures 9.2 and 9.3, it would take a huge disparity between raters to make the results from Pairs 1 and 2 resemble each other. Likewise, in Table 9.4, common levels of rater disparity would not eliminate contrasts of two-to-one or six-to-one.

RESULTS AND DISCUSSION

The drill data are presented in summary form, and through Markov Matrix form. Drill data are then interpreted through the CIAO scale, first in summary form, then through a modified time-series process.

TABLE 9.1
Illustration of Recording Process

	1L			1R	
Event #	*Drill*	*CIAO*	*Event #*	*Drill*	*CIAO*
5	1	C	6	1	C
7	1	C	8	6	I
9	6	A	10	6	C

Note. In this illustrative example, Drill 1 continues until Participant 1R transitions to Drill 6 in Event 8.

Drill Data

In a 2-minute period, Pair 1 completed 264 cycles (132 per person, cycle time mean: 0.91 second, mode: 0.93, frequency of 1.10 Hz), and Pair 2 completed 270 cycles (135 per person, cycle time mean: 0.88 second, mode: 0.80, mean frequency of 1.12 Hz), so their speeds were fairly equivalent. Table 9.2 compares the number of cycles each pair spent in each drill. Drills 1, 2, 3 and 4 were performed most frequently by both pairs. Pair 2 did not perform Drill 6 at all.

Markov Transition Matrices, shown in Table 9.3, tally changes in the state of a system by comparing consecutive pairs of states, in this case, which drill is being performed. The matrix row represents the drill performed first; the matrix column represents the drill performed second. Bolded cells represent events in which a drill follows itself ("continue" on the CIAO scale). This count is subtracted from the total counts for each column and row to calculate the number of transitions to a new drill. Italic cells represent transitions that are discouraged by logistic and/or tactical concerns (see nonlinked nodes in Panel b of Figure 9.2).

Note that in Table 9.3 there is considerable activity in grayed-out, prohibited cells. Patterns of performance errors like these are valuable to the instructor/trainer, as they point to misconceptions that must be addressed. These procedural errors are comparable to the points of procedural failure Voskoglou (1995a, 1995b) showed in students' solving of math problems. Even so, as transitions were performed and therefore workload borne, these were counted for our analysis of amounts of activity. This incorrect learning is credited to inadequate supervision for this group's needs. The patterns of these errors would help an instructor focus future training time on specific errors, improving practice, in the same way that Voskoglou noted or that Britton and Tidwell (1991) noted in redesigning reading material. Additional discussion of the use of Markov processes is presented by Maier (2001).

The great majority of activity is in the diagonal line of cells representing sequences in which a drill followed itself (bold numbers), indicating that both pairs were much more likely to continue doing the same

TABLE 9.2
Total Counts of Events Counted in Each Drill,
as Designated in Panel b of Figure 9.1

| | Drill ID | | | | | | |
	1	2	3	1	5	6	Totals
Pair 1	71	54	49	71	7	12	264
Pair 2	41	41	76	99	13	0	270

TABLE 9.3
Markov Transition Matrices for Pairs 1 and 2

Pair 1	Drill ID						Totals	
	1	*2*	*3*	*4*	*5*	*6*	*Drills*	*Transitions*
1	**60**	*6*	*0*	0	0	*6*	72	12
2	*0*	**48**	0	*0*	6	0	54	6
3	*4*	0	**41**	3	*1*	0	49	8
4	7	*0*	4	**59**	0	*0*	70	11
5	*0*	0	2	5	**0**	0	7	5
6	0	*0*	2	*4*	0	**6**	12	8
Drill Totals	71	54	49	71	7	12	264	—
Total Transitions	11	6	8	12	7	6	—	50

Pair 2	Drill ID						Totals	
	1	*2*	*3*	*4*	*5*	*6*	*Drills*	*Transitions*
1	**34**	*6*	*0*	1	0	*0*	41	7
2	5	**33**	2	*0*	*1*	0	41	8
3	*1*	0	**67**	7	*1*	0	76	9
4	1	2	5	**90**	1	*0*	99	9
5	*0*	0	2	1	**10**	0	13	3
6	0	*0*	0	*0*	0	**0**	0	0
Drill Totals	41	41	76	99	13	0	270	—
Total Transitions	7	8	9	9	3	0	—	36

drill rather than transition to a new one. Pair 1 shifts more activity from the diagonal outward into transitions, as expected based on their greater experience. But on the other hand Pair 1 and Pair 2 explored the space of possibilities to about the same extent. Pair 1 also shows a balanced level of activity between Drills 1-2 and 3-4, with a deficiency in Drills 5-6, whereas Pair 2 shows a nearly linear gradation in activity decreasing from 3-4 to 1-2 to 5, with a total lack of activity in 6, also in keeping with their lesser experience.

It can be seen that each pair favored particular trajectories through the network. Pair 1 shows an interesting pattern involving Drill 5. Seven transitions enter Drill 5, but none continue there; all seven immediately pass on to either Drill 3 or 4—a disturbing lack of variability, given the instructions to vary as much as possible. This pattern is discussed further, later. Pair 1 also shows very low activity in Drill 6, which is equally split between continuing in 6 and transitioning to 3-4. Pair 1 presents distinct clusters of activity indicating a strong habit of circulating in one direction only, that being from 1-2 to 5-6 to 3-4 and back to 1-2. Pair 2 presents an oscillation from 1-2 through 3-4 to 5 and back through 3-4 to 1-2. This lack of activity between 1-2 and 5 means that both are treated as dead-ends. Each pair's unique deficiency in certain regions of the network offers a kind of predictability and vulnerability to an opponent—a lack of preparedness to cope with the unpracticed transitions.

Traditional summary statistics of event counts (number of cycles spent in each drill) actually offered very little to define or compare specific performances, as literally thousands of different sample sequences could generate exactly the same total counts. Markov matrices improved on this time-blindness by allowing the inspection of two-event sequences. In focusing specifically on transitions, this procedure was the first breakthrough in showing quantitatively what Panels b and c in Figure 9.2 show in principle. However, as Voskoglou (1994, 1995a, 1995b) found in using Markov processes to observe mathematics problem solving, the transitions were still total counts, failing to differentiate large numbers of possible samples. Markov processes, like summary statistics, are probabilistic; they cannot show the evolving structure of an iterative system. Only a time-series perspective looks at sufficiently long sequences to show this.

Performance time of individual events was filtered out, in order to focus on ordinality and sequencing. So, an actual timeline was reduced to an event line, illustrated in Figure 9.3. Note that differences between Pair 1 and Pair 2 are great enough to be apparent even to superficial appraisal of their event-lines.

The greater overall richness in activity of Pair 1 is even clearer in an event series than it was in the Markov matrices. Using Drill 1 as a baseline, Pair 1 cycles through baseline five times between Events 1 and 100, whereas Pair 2 makes the circuit only once. This is more than twice the difference in the pairs' total transitions (50:36, see Table 9.3). This contrast is repeated again between Events 100 and 235, where Pair 1 cycles widely through the network to return to baseline five times for Pair 2's once. This cyclicity and 5:1 shortening of wavelength adds another dimension that might be termed "ordinal complexity" to the contrast.

The pattern concerning Drill 5 in Pair 1, which we noted earlier in discussing Markov matrices, is quite visible in Pair 1's event series as tiny spikes up to five at Event numbers 48, 118, 141, 184, 204, 236, and 260. Additionally, the graph shows these events as being the consistent harbinger of nearly stereotypical sequences, suggesting very high predictability of the next two to three transitions. They also exhibit an increasing trend, smoothly replacing the decreasing trend of spikes into Drill 6, which are extremely similar as signposts to even more stereotyped sequences through Drill 3 or 4, back to 1. Additional discussion of the use of events series is presented by Maier (2004).

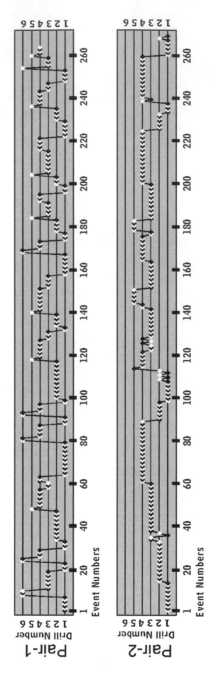

FIG. 9.3. Event series for Pairs 1 and 2. The drill performed is graphed against event number described in Table 9.1. Black dots indicate left partner; white dots indicate right partner.

210

ANALYSIS USING THE CIAO SCALE

Though events in terms of drill numbers give insight into the relative skill levels and possible training deficiencies of the two pairs, it cannot give insight into the personal dynamic within each pair. The CIAO scale was developed specifically to address this emergent aspect of the practice. In terms of data analysis, this scale further partitions the residual variance.

Data are presented in Table 9.4. Pair 1 "initiated" more (51 times vs. 33 times) and "overloaded" less (1 vs. 9) than Pair 2. Thus, the CIAO scale also captures some aspects of the differences in expertise between these pairs. It is seen here that Pair 2 (initiate ~2:1) is better matched than Pair 1 (initiate ~4:1). Within Pair 1 most of the transitions were "initiated" by 1-R, whereas the single "overload" was evidenced by 1-L, combining to indicate a strong general dominance. In Pair 2, 2-R "initiated" more and "overloaded" less, supporting the suggestion that these measures give insight into the workings of relative dominance of two partners. Thus, CIAO differentiates individual partners within the pairs on dominance.

As with drill data, the clear contrasts in CIAO summary statistics are enriched by viewing the order and clustering in an event series, illustrated in Figure 9.4, graphing the CIAO performance interpretation scale for "initiative" against event number. Note the dramatic differences between Pair 1 and Pair 2 in frequency of both "initiations" and "overloads." The difference in frequency and pattern of initiations and overloads becomes much easier to appreciate.

The distinctive sequences of CIAO values reveal much about interactions between partners in each pair. The most fundamental sequence, C-I-A-C, in which one partner initiates, the other accommodates, and the pair then settle into a different drill, is seen in both pairs. Also evident in both pairs is a C-I-A-I-A-C sequence, in which the same partner initiates twice, whereas the other partner accommodates. This sequence is quite frequent in Pair 1, where CIAO summary statistics (Table 9.4) show Partner 1-R initiated four times as often as 1-L. The ability to view longer strings of activity is particularly valuable. The basic sequence of C-I-A-C forms a recognizable

TABLE 9.4
Summary Scores of Participants After Interpretation of Event Data
Through the CIAO Scale

Pair	Person	Continue	Initiate	Accommodate	Overload
1	L	89	11	31	1
	R	81	40	11	0
2	L	105	13	11	6
	R	105	20	7	3

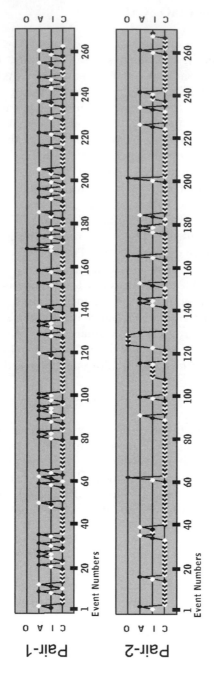

FIG. 9.4. CIAO series for Pairs 1 and 2. Black dots indicate left partner; white dots indicate right partner.

212

spike. The double spike of the C-I-A-I-A-C variant shows the greater initiative of Partner 1-R.

A more demanding sequence in which "initiate" immediately follows "initiate," such as C-I-I-A-C, is observed in Pair 1 only when 1-L responds to a 2-5 initiation with an immediate 5-4 or 5-3 initiation, generating two consecutive initiations. All of 1-L's initiations occur as immediate or one-off responses to an initiation from 1-R. This can be an effective form of counterfighting, but under the instructions given in this study, it is interpreted as low initiative. This low score in general initiative on the part of Partner 1-L makes it difficult to assess the limits of Partner 1-R. On the other hand, 1-R never once directly counterinitiates against 1-L, even in the most stereotyped situation. This suggests a person who is playing his own game, not really interacting with his partner. The fact that 1-L maintained a lower overload count than either member of Pair 2, despite a much higher ordinal complexity of his challenge, speaks well of his level of coping; he successfully accommodated a much heavier onslaught than either 2-L or 2-R received. Ostensibly, this could be because he found a maintainable equilibrium between the loads of "initiate" and "accommodate."

Though Pair 2 shows lower levels of initiation overall, their "initiation" events are more clustered, for example, at Events 238 to 240 (three consecutive initiations) and Events 109 to 114 (seven consecutive initiations). A third such episode at Event 123 caused a discrepancy between Pair 2's total initiations in Table 9.3 (36) and in Table 9.4 (33) by generating a cascade of overloads beginning at Event 124, seen in Figure9.3. This moment of total disarray for both partners includes very "game" but ineffective attempts to recover by initiating a different drill. It is considered consistent with the general pattern because the first overload, beginning the cascade, was in response to an initiation. Under definitions generating the CIAO scale, this is in an "accommodate" situation, when time pressure is maximized. Additional discussion of the CIAO scale is presented by Maier (2004).

It would be useful to a trainer to know such facts as:

1. The large number of events remaining in "continue" (Pair 1: 170/264 64%, Pair 2: 210/270 77%) show a huge amount of room remaining available for further growth.
2. In Pair 2, all of the overloads involved transitions between Drill 3 and Drill 4, recommending a focus for further practice.
3. Both pairs utilized less of the available range than desired.
4. And thus, both pairs were quite limited in their sequencing.
5. The high number of prohibited transitions indicate significant misunderstanding of the system. All of these facts concern insufficient knowledge and practice of transitions or links.

DISCUSSION AND CONCLUSIONS

This research model differs crucially from those in which a participant simply responds to a researcher-designed sequence of inputs. The fact that each participant in a pair repeatedly changes the future course of the activity makes those changes (transitions) central to any discussion of the results. Transitions are the core of the study. No run is random, as each event is an individual's reaction to the current situation and overall history of the run; but no two runs are identical. This is what brings the activity closer to real-world activity. It also means that analysis techniques must be adapted to fit the structure.

Key Results

Results here distinguish between dyads by overall transition counts and preferences for certain patterns of responding, and between individuals by willingness to initiate transitions and success in responding to the partner's initiative. Occasional overload events give evidence that each pair is maintaining a high but manageable level of cognitive load. Changing the course of the activity, as expressed by initiating a transition, is decision-based action, which is possible only if cognitive resources are available above minimal coping. Thus, these preliminary results support the hypothesis that the studied activity provides a useful model for analysis of high-speed tactical decision making, that is, oppositional expertise.

Both the Markov process and event-series perspectives clearly showed for both pairs that the great majority of drill cycles were a continuation of the preceding drill rather than a transition to a new drill. There was room for a substantial increase in initiative and for concentrating several dozen "decision incidents" into each minute of valuable practice time. The novel coding scale developed for this study shows promise for revealing important dimensions of macrocognition, especially management of attentional resources and response to cognitive overload. It provides a way to analyze an emergent property of each individual within the interpersonal dynamic of the partner pair.

Method Generalizability?

Through references to research in conventional piloting, combat flight, dispatching, practice schedules, and learning in mathematics (tracked with Markov Processes) and history (tracked with conceptual networks), we have shown connections to fields well beyond the martial arts, where this model originated.

Hunt and Joslyn (chap. 2, this volume) discuss limited attentional resources supporting multiple processing streams that must not only all be finished on time, but be selectively integrated into choosing an action, followed by integrating the action's results into the next action. Cyclic scanning of information streams, whether they are mirrors in a car, instruments in an airplane, or hands of a training partner, all seem suitable content to a network model. All time-constrained activities require judicious dynamic allocation of attention across a world of multiple dimensions. As cyclicity is fundamental to attentional scanning, it is also fundamental to action. Whether the action is a keystroke, dispatching a police car, or quarterly maintenance on a ship's equipment, actions recur over some time period. They must be linked to themselves and to the other actions forming their contexts. Checklists can share steps—they are simply single chains in an encompassing network. The ubiquitous use of Markov Processes in many fields to track actions within a possibility space supports this. Voskoglou (1994, 1995a, 1995b) applied Markov Chains to locating precise points of procedural failure in mathematics students. Like other activities, martial arts practice shows diverse events. Some of these events recur more often than others, and some work very well or very poorly. Composition of these productions, basic to cognitive theory (e.g., Gagne, Yekovich, & Yekovich, 1993), generates condition-based sequences. Steps at which two or more sequences cross become nexuses in a network.

The network model that is this study's focus can be found, albeit in a different form, in a traditional martial art training activity, but no pertinent literature was found to provide a proven methodology. Commonly used research designs do not incorporate the iteration of response to response. There is not yet an abundance of researchable models in which the participant's actions actually influence the future course of the activity. This is a direction in which research needs to become enabled, and it is hoped that this study contributes. We have demonstrated the value of a method for studying interactive tactical decision at high speeds, and without the possible confounds of either technological systems sandwiched between the humans, or any form of reporting by the participants required. We have also proposed and demonstrated an interpretive scale that provides insight into a difficult-to-access quality of human interaction that is essential to military training and personnel selection in terms of flexibility and vulnerability, as well as many other occupations in which decisiveness and initiative are essential under high stakes and time pressure.

Models and methods often are developed in some particular venue, by detailed description of a specific activity. Klein's (1997) recognition-primed decision-making model derived from watching fire ground commanders. Hunt and Joslyn (chap. 2, this volume) observed public safety dispatchers, then weather forecasters. During the first stages of any development cycle, the question can be raised, "What has this to do with any other activity?" The preliminary results in this study are insufficient to support general claims. However, the representation and analysis methods we developed might be adaptable in the study of other types of activities, not limited to oppositional expertise.

The Study of Initiative

Webster (Merriam-Webster, 1993) defines initiative as "energy or ability displayed in initiating something." This study advances the importance of decisions and actions as quantifiable expressions of initiative, making aspects of this ephemeral attribute measurable. The fact that, for both pairs in our study, the individual who overloaded most also initiated least, fits the intuition that initiative is cognitively expensive. In addition, the fact that Pair 1, who initiated more, followed fairly stereotyped sequences suggests two dimensions of performance, transition generation and ordinal complexity, between which an individual's cognitive resources are budgeted. This indicates that a dyadic system emerges from the iterative interaction in this activity, and that this system dynamically distributes cognitive load across at least two performance dimensions in each of two participants, not across some single dimension.

Training for Adaptability

The importance of immediate context to each recognition–decision event is revealed by ordinal complexity seen in an event-series view of the iterative interaction between two partners. The human tendency to compose sequences of experiences and then to automate expectations and actions is a two-edged sword. It contributes to overrunning an opponent when one holds the initiative, but gives the initiative-taking opponent reliable opportunities to entrap. Both generating and coping with ordinal complexity should be a goal of training. Endsley and Robertson (2000) could as well be referring to lawyers, doctors, or corporate directors when they wrote, "Less effective pilots appeared to apply the same strategies in all cases rather than matching their strategy to the situation" (p. 352). This agrees with Schmitt and Klein (1996) urging development of skill for improvisation and adaptability, which is exactly what research shows to be the targeted strength of random practice schedules. The predictability of stereotyped sequences in the present study is especially spotlighted in event series. The significance of predictable, stereotypical sequences would depend on the situation. Between team members, it would indicate a cohesiveness of procedure. In combat or other competition, it would be a point of high vulnerability. It is commonly understood that surprising or confusing an opponent is a reliable and desirable way to overcome him. Defining the situation in such a way that his resources are misallocated creates vulnerability.

This line of research revealed some important dimensions, which are traded off against each other differently by different people. Identified while still being learned, these errors can be targeted for additional training. Presentation of skilled activity as a network with multiple links departing each node gives structure to learning criteria for the alternatives. Multiple links entering each node encourage abstraction/conceptualization in showing that the same or similar situation can appear in multiple contexts. Both of these factors can support the development of what Jodlowski et al. (2003) called adaptive expertise. The weight of the accumulating evidence would encourage testing our theory and approach in other activities where a small network of alternative actions or information streams can be selected from the larger set.

Both routine and adaptive expertise (Jodlowski et al., 2003) get pushed to new capacities by situations in which a dynamic emerges from iterative cycles of initiative interacting with external change. The network structure studied here presents a small closed system with sufficient states to make each cycle uncertain. In this study, routine expertise is tested and extended by increasingly frequent passages through familiar states, and familiar changes; adaptive expertise is tested and extended by increasingly varied sequences of states, in which familiar states look new in a different context.

REFERENCES

Battig, W. F. (1972). Intratask interference as a source of facilitation in transfer and retention. In R. F. Thompson & J. F. Voss (Eds.), *Topics in learning and performance* (pp. 131–159). New York: Academic Press.

Brady, F. (1998). A theoretical and empirical review of the contextual interference effect and the learning of motor skills. *Quest, 50,* 266–293.

Britton, B. K., & Tidwell, P. (1991). Shifting novices' mental representations of texts toward experts': diagnosis and repair of novices' mis- and missing conceptions. In *Proceedings of the Thirteenth Annual Conference of the Cognitive Science Society* (pp. 233–238). Mahwah, NJ: Lawrence Erlbaum Associates.

Endsley, M. R., & Bolstad, C. A. (1994). Individual differences in pilot situation awareness. *International Journal of Aviation Psychology, 4,* 241–264.

Endsley, M. R., & Robertson, M. M. (2000). Training for situation awareness in individuals and teams. In M. R. Endsley & D. J. Garland (Eds.), *Situation awareness analysis and measurement* (pp. 349–365). Mahwah, NJ: Lawrence Erlbaum Associates.

Endsley, M. R., & Smith, R. P. (1996). Attention distribution and decision making in tactical air combat. *Human Factors, 38,* 232–249.

Gagne, E. D., Yekovich, C. W., & Yekovich, F. R. (1993). *The cognitive psychology of school learning* (2nd ed.). New York: Addison Wesley Longman.

Hunt, E., & Joslyn, S. (2003, May 15–17). *The dynamics of the relation between applied and basic research.* Paper presented at the Naturalistic Decision Making 6 conference, Pensacola Beach, FL.

Jodlowski, M. T., Doane, S. M., & Brou, R. J. (2003). Adaptive expertise during simulated flight. In *Proceedings of the Human Factors and Ergonomics Society 47th Annual Meeting* (pp. 2028–2032). Santa Monica, CA: Human Factors and Ergonomics Society.

Klein, G. (1997). Making decisions in natural environments (Special Report 31). Arlington, VA: Research and Advanced Concepts Office, U.S. Army Research Institute for the Behavioral and Social Sciences.

Klein, G., Ross, K. G., Moon, B. M., Klein, D. E., Hoffman, R. R., & Hollnagel, E. (2003, May/June). Macrocognition. *IEEE Intelligent Systems*, pp. 81–85.

Li, Y., & Wright, D. L. (2000). An assessment of the attention demands during random- and blocked practice schedules. *The Quarterly Journal of Experimental Psychology, 53a,* 591–606.

Magill, R. A., & Hall, K. G. (1990). A review of the contextual interference effect in motor skill acquisition. *Human Movement Science, 9,* 241–289.

Maier, H. N. (2001, March 21). *Structuring random practice through Markov processes.* Paper presented at the Educational Research Exchange, Texas A&M University, College Station. Retrieved November 1, 2005, from http://www.tacticog.com/research/ere01_markovpaper.htm

Maier, H. N. (2004). *Measurement of cognitive load in a traditional martial arts training model.* Texas A&M University, College Station. Retrieved November 1, 2005, from http://www.tacticog.com/research/dissertation_access.htm

Merriam-Webster. (1993). *Webster's new encyclopedic dictionary.* New York: Black Dog & Levanthal.

O'Hare, D. (1997). Cognitive ability determinants of elite pilot performance. *Human Factors, 39,* 540–552.

Roscoe, S. N. (1993). An aid in the selection process—WOMBAT. *Civil Aviation Training, 4,* 48–51.

Roscoe, S. N. (1997). *Predicting human performance.* Quebec: Helio Press.

Schmitt, J. F. M., & Klein, G. A. (1996, August). Fighting in the fog: Dealing with battlefield uncertainty. *Marine Corps Gazette, 80*(8), pp. 62–69.

Schvaneveldt, R. W. (Ed.). (1990). *Pathfinder associative networks: Studies in knowledge organization.* Norwood, NJ: Ablex.

Sweller, J., van Merrienboer, J. J. G., & Paas, F. G. W. C. (1998). Cognitive architecture and instructional design. *Educational Psychology Review, 10,* 251–296.

Voskoglou, M. G. (1994). An application of Markov chain to the process of modeling. *International Journal of Mathematical Education, Science and Technology, 25,* 475–480.

Voskoglou, M. G. (1995a). Measuring mathematical model building abilities. *International Journal of Mathematical Education, Science and Technology, 26,* 29–35.

Voskoglou, M. G. (1995b). Use of absorbing Markov chains to describe the process of mathematical modeling: A classroom experiment. *International Journal of Mathematical Education, Science and Technology, 26,* 759–763.

Wulf, G. (1991). The effect of type of practice on motor learning in children. *Applied Cognitive Psychology, 5,* 123–134.

Wyman, B. G., & Randel, J. M. (1998). The relation of knowledge organization to performance of a complex cognitive task. *Applied Cognitive Psychology, 12,* 251–264.

Chapter 10

Dialogue as Medium (and Message) for Training Critical Thinking

Marvin S. Cohen
Cognitive Technologies, Inc., Arlington, VA

Leonard Adelman
George Mason University

Terry Bresnick
F. Freeman Marvin
Innovative Decisions, Inc., Vienna, VA

Eduardo Salas
University of Central Florida

Sharon L. Riedel
Army Research Institute, Arlington, VA

CRITICAL THINKING

An NDM Approach to Individual and Joint Reasoning

Naturalistic Decision Making (NDM) has correctly emphasized the importance of nondeliberative, recognition-based decision making by experienced individuals in relatively unstructured and time-constrained situations (Klein, 1993; Orasanu & Connolly, 1993). This is, of course, not the whole story. Attention has also been paid to mental simulation (Klein, 1998); to metacognitive processes that critique, evaluate, and improve recognitional products when more time is available (Cohen, Freeman, & Wolf, 1996); and to collective processes that produce team adaptation (Serfaty, Entin, & Johnston, 1998), implicit coordination (Orasanu, 1993), shared mental models (Cannon-Bowers, Salas, & Converse, 1993), and the "team mind" (Klein, 1998). Still, it is probably fair to say that NDM has not yet evolved its own coherent approach to the more deliberative reasoning that sometimes

219

occurs in non-routine real-world situations—often referred to as *critical thinking* and including the kinds of processes to which logic, probability theory, and decision theory are sometimes applied. Similarly, it is probably fair to say that NDM needs to look more deeply into how effective teams manage diverging perspectives and opinions, and lay the groundwork for coordination despite *differences* in their mental models.

The present chapter is meant to stimulate ideas about what a naturalistic approach might look like. In that approach, we argue, deliberative reasoning and disagreement are aspects of the same phenomenon. We propose a theory in which critical thinking is understood as dialogue with oneself or others, in which coordinated roles for questioning and defending mental models account for the structure associated with different types of reasoning, in which persuasion is the counterpart of drawing an inferential conclusion, and in which normative evaluation in terms of reliability emerges as a natural perspective on the process as a whole. Mental models must ultimately originate in recognitional processes, but they can be evaluated, compared, and modified during critical dialogue, which aims at more complete and accurate judgment. In the team context, critical dialogue presupposes mutual knowledge but also seeks to expand the sphere in which mutual knowledge holds. The theory, therefore, helps bridge gaps between intuition and deliberation, between individual and team cognition, between thought and communication, and between description and evaluation.

This first section motivates a theory of critical thinking as dialogue, while the second section outlines the theory's basic elements and introduces an example. The next three sections discuss its major components: mental models, processes of challenging and defending, and monitoring for reliability. The final two sections briefly summarize preliminary training results and conclusions.

Limitations of the Internalist Paradigm

Definitions of critical thinking tend to have a common theme. Siegel (1997) says that "the beliefs and actions of the critical thinker, at least ideally, are *justified* by reasons for them which she has properly evaluated" (p. 14, italics in original). According to this and many other definitions of both critical thinking and rationality (Cohen, Salas, & Riedel, 2002; Johnson, 1996, ch. 12), critical thinking is the deliberate application of normative criteria *directly* to fully articulated conclusions and reasons. The purpose is to use "reason" to free individuals from habitual and conventional thought, which have no rational standing. The means of doing so are *individualist* and *internalist* because they assume that everything relevant to the rational justification of a belief or decision is found by inspecting the contents of one's own conscious mind (Feldman & Conee, 2000; Plantinga, 1993a, pp. 3–29). In this chapter, we motivate, describe, and illustrate an approach to critical thinking that applies pragmatic *externalist* criteria to what is, in some important senses, a *collective* process.

Some persons know more than others, but according to the internalist paradigm everyone has equal access to their own mental life and to the light of reason. It would be unfair, then, to fault the rationality of a decision that fails only because of substantive ignorance or violation of non-logical criteria. By the same token, rationality gets no credit for good intuitive judgment honed by practice in a domain. Expert physicians who cannot explain diagnoses that save patient's lives (Patel, Arocha, & Kaufman, 1999, p. 82) are not good critical thinkers. Unable to trace a logical path to their decision, it is as if they got it right by accident (Klein, 2000). Sosa (1991) dubbed this the "intellectualist model of justification" (p. 195); Evans and Over (1996) call it "impersonal rationality"; Hammond (1996) refers to it as the "coherence theory of truth." Less often noticed is the prominence it gives to universalism and egalitarianism over effectiveness. Appropriate reasons and evaluative criteria must be experienced as such by all. Critical thinking is an individual intellectual effort, but fairness places everyone at the same starting line (Plantinga, 1993a, pp. 3–29; Klein, 2000).

This paradigm disregards he importance of recognitional skills in proficient problem solving (Chase & Simon, 1973) and decision making (Klein, 1993). For both experts and non-experts, most of the processes that generate judgments and decisions are tacit. "Reasons" for decisions, if verbalizable at all, are implicit in experiences extending over long periods of time, and sometimes over multiple generations. The habit of asking for reasons, if applied "systematically and habitually" as critical-thinking proponents sccm to urge (e.g., Siegel, 1997, p. 16), is either endless or circular, and must come to a stop at acceptable assumptions rather than rock-solid foundations (Day, 1989). There never will be an exhaustive set of criteria for directly certifying rational judgments and decisions about the world. Instead, critical thinkers need a method for determining *which* conclusions to critically examine on any particular occasion and *which* reasons, if any, to try to articulate and defend (Evans & Over, 1996).

The internalist climate does not encourage prescriptive research about time-sensitive cognition. Textbooks still regard critical-thinking as a form of inner intellectual purity, and tend to emphasize rules from formal or informal logic, probability, or decision theory. The internalist paradigm prompts questions about the usefulness of training individuals or teams to think critically in practical domains like the Army tactical battlefield: Will assembling a "valid" chain of reasons take too much time or weaken the will to fight? Will implicit egalitarianism supplant experience and proven proficiency, stifle dissent and innovation, or disrupt leadership and coordination? An alternative approach to critical thinking, from the outside in, provides more encouraging answers.

An External Perspective

Our objective was to place critical thinking in a more realistic context for individual practitioners and teams. To accomplish this, we sought a conceptualization that would:

- Capture the idea of thinking about thinking without demanding that all (or even most) decision making be deliberative, or that all (or even most) reasons be made explicit.
- Be adaptable in time-constrained situations (A case can be made that all thinking, even in science, involves tradeoffs with time and other resources).
- Take account of both opportunities and obstacles due to social and organizational relationships.
- Be easy to teach, practice, and evaluate in real-world contexts.
- Take account of and enhance the effectiveness of strategies already used successfully by proficient decision makers, such as recognition (Klein, 1993), story building (Cohen et al., 1996; Pennington & Hastie, 1988), metacognition (Cohen et al., 1996; Nelson, 1992), and consultation or constructive debate with others (Van Eemeren, Grootendorst, Blair, & Willard, 1987).

These objectives commit us to a *naturalized epistemology* (Kornblith, 1994) that focuses evaluation on external facts about the world rather than "infallible" logical and introspective truth (e.g., Lipshitz & Cohen, 2005; Papineau, 2003; Sosa & Kim, 2000). This should not be confused with directly claiming, trivially, that the beliefs to be justified are factual. The pragmatic approach focuses evaluation on other facts, about the *processes* or *mechanisms* (e.g., perception, memory, recognition, problem-solving, reasoning, testimony, and even habit and custom) that contributed to a conclusion instead of the conclusion itself. Individual beliefs or decisions receive their warrant indirectly based on the more general reliability of belief-generating and belief-testing methods. Thus, an individual belief or decision is warranted if it was produced (or is now sustained) by recognizable methods that yield a sufficiently high proportion of confirmed beliefs or successful decisions under relevant conditions, including available time and the costs of errors (Goldman, 1992, ch 6; Nozick, 1981; Plantinga, 1993b; Rescher, 1977b). This turns justification into an empirical question that can be asked and answered by outside observers and studied scientifically (Gigerenzer, 2000; Hammond & Stewart, 2001).

Critical thinking has a different look and status in the externalist perspective. It is:

• *Context-sensitive:* The appropriate depth and structure of argument will vary with the economics of the situation (i.e., the costs and potential benefits of avoiding errors in that context). Justification of judgments and decisions cannot be a matter of fixed formal criteria, because there is no *a priori* enforceable end to the demand for reasons.

• *Collective:* Nothing restricts critical thinking to an individual consciousness. Social practices that actively seek information, expose views to outside challenge, draw on past cases, or appeal to cultural and historical paradigms may increase the reliability of results (Goldman, 1992, ch 10). Judgments and decisions may be based on knowledge embedded in long-term cultural practices or artifacts, and distributed among group members (Hutchins, 1995).

• *Intuitive:* Because tacit processes are inherently non-verbalizable and because implicit reasons for a judgment can never be exhaustively spelled out, there will always be a major residual dependence on the reliability of relatively automatic perceptual and inferential processes, or *recognition*.

• *Feasible:* Feasibility is a prerequisite for effectiveness, so it is built into externalist criteria. They will often favor strategies that are closely related to the way people already think over formally rigorous methods that are "fair" but virtually impossible to implement (Lipshitz & Cohen, 2005).

• *Pluralistic:* No single method monopolizes reliability. Critical thinking itself is not necessary for rationality, because recognitional processes or customary practices may more reliably achieve individual or group goals in familiar situations or when time is limited (Cohen et al., 1996). Critical thinking itself encompasses a range of strategies that probe more or less deeply depending on the task and context.

The Disagreement Heuristic

Critical-thinking texts have surprisingly little to say about disagreements *among* critical thinkers. They focus instead on finding faults in arguments by authors who are not present to respond (e.g., Govier, 1987; Johnson, 1996). Discussions of both formal and informal logic (Johnson & Blair, 1994) imply that if everyone reasoned correctly from shared premises, they would arrive at consistent conclusions. Because everyone has equal access to proper reasoning, the responsibility of the individual critical thinker extends no further than stating her premises and getting her own reasoning right. Once *my* arguments satisfy appropriate criteria, any disagreement must be due to *your* laziness or stupidity (or evil intent). To some extent, the pervasive interest in systematic mistakes, fallacies (Hansen & Pinto, 1995), reasoning errors (e.g., Wason & Johnson-Laird, 1972), and decision biases (e.g., Kahneman, Slovic, & Tversky, 1982), reinforces this condescending attitude toward those with differing views (an attitude that may be reflected in our current political discourse). It gives people a sense of intellectual rectitude that ill prepares them for constructive intellectual exchanges with others. Under the illusion that a single line of reasoning can be validated in isolation, there is scant motivation for seriously considering opposing points of view and little chance of intellectual synergy.

Unfortunately, the same attitude is sometimes adopted (for different reasons) by naturalistic researchers. The reasons include exaggerated emphasis on the role of pattern recognition and fast intuitive judgment in expertise (e.g., Gladwell, 2005), and on consensus and homogeneous knowledge in team research (see Cooke, Salas, Kiekel, & Bell, 2004, for the latter point). As Shanteau (1998) points out, consensus among experts in the same domain is not only expected, it is frequently regarded as part of the definition of expertise. Experts can have no use for exploration and evaluation of alternative viewpoints on the topics they have "mastered." This view assumes away the possibility of novel and complex situations that no learned pattern fits perfectly (Cohen et al., 1996) or problems that are at the cutting edge of experts' own knowledge. Yet such situations are not uncommon in complex and dynamic domains, including science, some areas of medicine, and military tactics (Shanteau, 1992, 1998; Wineburg, 1998). Shanteau's (1992) research suggests that expert skills include the ability to identify exceptional situations or novel problems where previous methods do not apply, and to use appropriate strategies such as collaboration with other experts.

Suppose we take seriously the fallibility of *both* intuitive expert judgments *and* rational arguments about the physical and social world: No matter how expert or cogent they respectively are, their conclusions are incompletely grounded by evidence, experience, or any other reliable method, especially in unusual situations. We can no longer assume that if one argument or expert judgment by itself appears acceptable, then conflicting arguments or judgments can be rejected sight unseen. Viewed in isolation from one another, opposing arguments or expert judgments may each appear sufficient to establish their respective conclusions. Flaws such as overlooked evidence, hidden assumptions, or limitations in an expert's assumptions or range of experience may not surface except in the light shed by a *competing* argument or judgment—if we are lucky enough to have alternative perspectives available.

There is ample empirical evidence (e.g., Amason & Schweiger, 1997; Gigone & Hastie, 1997; Jehn, 1997) that groups make better decisions in complex, nonroutine tasks when there are disagreements among members. Empirically successful interventions in group decision making have generally worked by increasing either the likelihood of divergence among viewpoints or the awareness of preexisting inconsistency in opinions (Tasa & Whyte, 2005). As by-products, intragroup disagreement facilitates a more thorough understanding of the problem (Schweiger, Sandburg, & Ragan, 1986) and an expanded body of shared knowledge (Amason, 1996). The *disagreement heuristic*, then, is to focus individual and collective attention on issues where views diverge. It works because both arguments and intuitions are fallible, because intrapersonal and interpersonal resources are limited, and, we hypothesize, because disagreement is more likely to surface on issues that are not yet reliably grounded enough in the relevant context.

The benefits of disagreement depend on the availability of others who have dissenting views. Groups can compensate at least partly for the absence of spontaneous dissent by *stimulating* it. The quality of group decisions is typically improved when some members are assigned the role of *devil's advocate* or *dialectical opponent* (Katzenstein, 1996; Schweiger et al., 1986; Schwenk, 1989; Schwenk & Cosier, 1980). Henry (1995) found that asking team members to critique one another led to a significantly higher probability that they would surface and use important information possessed uniquely by particular members. Individual critical thinkers can compensate by *simulating* dissent, that is, imaginatively constructing alternative perspectives from which to critique and improve their own solutions. On many occasions, the most reliable way for both teams and individual thinkers to organize, understand, and evaluate positions on complex or novel problems, when the stakes are high and time is available, is to conduct a dialogue with competent opponents—*actual* if possible, *stimulated or simulated* if not.

CRITICAL THINKING AS DIALOGUE

Walton (1998) defined a dialogue as any *characteristic type of multiperson exchange that depends on mutual expectations about purposes, roles, and constraints* (Walton & Krabbe, 1995, p. 66). The theory of dialogue blends descriptive and normative concerns (Van Eemeren & Grootendorst, 1992; Walton, 1998). Researchers start in bottom-up fashion with real-world conversations and the settings in which they occur, identify recurrent recognizable *types* and the *purposes* they serve, and build idealized models of the procedures that dialogues use to reliably achieve their purposes. Using that approach, Walton identified six dialogue types: persuasion, negotiation, deliberation, inquiry, information seeking, and even quarreling.

Dialogue rules have normative force for participants who mutually recognize one another's desire to cooperate to achieve the relevant dialogue goal

(Grice, 1989). Dialogue theory directly maps descriptive analyses of actual exchanges onto prescriptive process constraints to identify where they diverge and why. Real conversations can be compared to ideal models to evaluate the quality of the real-world conversations, to improve the models, or both. Dialogue theory thus provides an empirical basis for prescriptive evaluation of actual conversations. (For discussion of how normative and empirical concerns interact, see Cohen, 1993a, and Lipshitz & Cohen, 2005.)

Two related models—Walton's (1998) *persuasion dialogue* and Van Eemeren & Grootendorst's (1992) *critical discussion*—provide a starting point for our concept of critical thinking as critical dialogue. In each case, argument is examined as it actually occurs among participants with a difference of opinion and a shared task: not as logical relations among premises and conclusions but as a dynamic exchange of reasons for and against a conclusion (Hamblin, 1970; Rescher, 1977a; Van Eemeren & Grootendorst, 1983; Van Eemeren, Grootendorst, Jackson, & Jacobs, 1993). The result is an idealized description of the thinking *process* rather than of the thinking *product*, a schematization that specifies the purpose, the roles played, rules for each player, and criteria for determining the outcome.

Elements of the Theory

Critical thinking is critical dialogue—whether by a single individual, a co-located group, or team members distributed in time and place. Critical dialogue is a process of challenging and defending alternative possibilities to resolve a disagreement, with the long-term purpose of improving performance in a task. This definition implies the three layers shown in Figure 10.1. Alternative possibilities correspond to *mental models* of the situation or plan (Johnson-Laird, 1983). In practical contexts, these often take the form of stories that weave together evidence, hypotheses, actions, and desired or undesired outcomes (Pennington & Hastie, 1993a). Challenges are *questions* about mental models intended to multiply the number of possible scenarios (hence, increase doubt) by pointing out gaps in evidence, uncover-

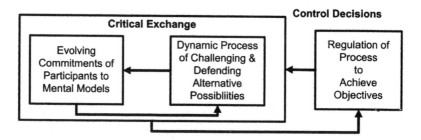

FIG. 10.1.   Overview of the theory of critical thinking as dialogue.

ing unreliable assumptions, or constructing conflicting accounts, while defenses are *answers* intended to eliminate those possibilities (hence, reduce doubt). Finally, the pragmatic purpose of the exchange implies the need for implicit or explicit *monitoring* to ensure its cost-effectiveness by selecting reliable methods and implementing them appropriately.

Implicit in the same definition is the orchestration of three *roles* or processes, corresponding roughly to pronouns *I, you,* and *they.* The *first-person* process recognizes, constructs, and reconstructs mental models of *my* situation understanding or plans. A *second-person* process takes a step back and iteratively critiques *your* hypotheses and plans, requiring a defense from the first-person role. A *third-person* role monitors and regulates what *they* are up to with respect to a larger task purpose.

These elements of critical dialogue combine to form a prototype rather than necessary and sufficient conditions (Lakoff, 1987): To the extent that an individual or collective process resembles critical dialogue in these respects, it is not only thinking but *critical thinking.* In particular, the first-person role requires the defense and repair of mental models; the second-person role requires the generation of competing mental models through the vivid enactment of alternative perspectives. The third-person role requires the use of strategies that reliably serve the relevant practical goals.

The Bridgeton Training Example

The following three sections elaborate on each component of critical dialogue. In doing so, they draw on the example in Figure 10.2 and Figure 10.3. Figure 10.2 is the map for a tactical decision game (TDG) called *Bridgeton Crossing,* from the *Marine Corps Gazette* (Klein, McCloskey, Thordsen, Klinger, & Schmitt, 1998). We used tactical decision games, including this one, for critical-dialogue training of active-duty Army officers. The task is to take the part of the decision maker in the scenario (an officer at a specified echelon) and quickly come up with a written plan in the form of a brief operations order and scheme of maneuver for subordinate units (Schmidt, 1994). In pretests and posttests, officers first worked individually for 15 minutes, then worked as a group for 30 minutes. Written products were collected from both phases.

In *Bridgeton Crossing,* the company's mission is to advance north without getting decisively engaged and to secure a safe river crossing for the division. Bridgeton, the only known crossing site, was the original objective, but the town was seized the previous night by an enemy mechanized force. The latest orders from division were to scout for other fordable locations. Since then, however, the situation has changed: (a) Only a few enemy scout vehicles (rather than a full mechanized force) have been observed this morning

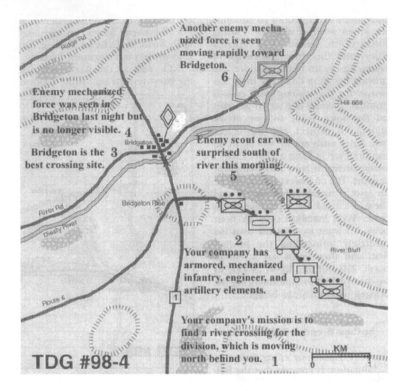

FIG. 10.2.   Map for tactical decision game *Bridgeton Crossing* taken with permission from the *Marine Corps Gazette* (April 1998). Annotations (not present in training) are numbered in the order in which the corresponding information is obtained in the scenario.

in Bridgeton by 2nd Platoon; (b) an enemy scout car was flushed by 1st Platoon on the south side of the river; (c) a different enemy mech force is reported by 2nd platoon to be moving quickly down Route 6 toward Bridgeton. The company commander must quickly decide what to do.

Figure 10.3 is an edited and compressed version of the conversation among a group of officers during a post-training practice, who play the roles of (a) a *proponent*, (b) an *opponent*, and (c) a dialogue *monitor*, respectively.

MENTAL MODELS IN CRITICAL DIALOGUE

Mental models represent what critical dialogue is about. Each model is a possible understanding of the situation, including observations, interpretations, actions, and predicted future events. A claim is accepted if it is true in

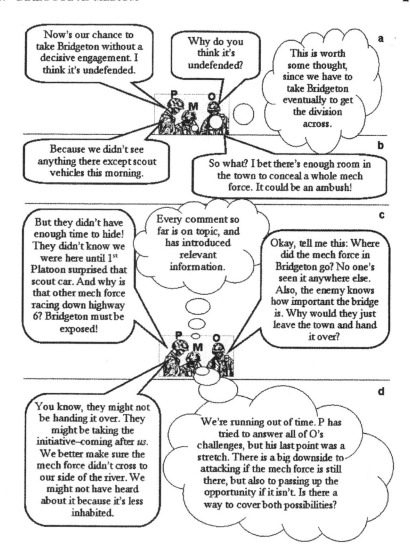

FIG. 10.3. An illustrative critical dialogue for the tactical decision game in Figure 10.2. P = Proponent. O = Opponent. M = Monitor.

all the models under consideration, rejected if true in none, and *uncertain*, or subject to doubt, when it is true in some models and false in others. The *probability* of the claim can be represented by the proportion of mental models in which it is true (Johnson-Laird, Legrenzi, Girotto, Legrenzi, & Caverni, 1999) or the relative frequency of scenarios in which it occurs (Gigerenzer, 2000, ch 7). On the basis of these simple principles, different critical dialogue roles can be distinguished by their different attitudes toward the uncertainty of a claim. The opponent tries to increase uncertainty by introducing models in which the claim is false or by rejecting those in which it is true; the proponent tries to decrease uncertainty by rejecting models in which the claim is false or introducing models in which it is true. We can categorize dialogue moves at a more detailed level in terms of their preconditions and intended effects on participants' mental models, and track progress toward consensus as the number of models under consideration alternately increases and decreases under the influence of the participants' opposing strategies.

Mental model has sometimes been dismissed as merely another general term for knowledge representation (Rickheit & Sichelschmidt, 1999). While this may be true in some applications, our interest in the mental model concept is focused on distinctive features that expose deep connections between *reasoning* and *discourse* (Johnson-Laird, 1983; Oakhill & Garnham, 1996; Rickheit & Habel, 1999), as well as between intuitive processing and reasoning (Evans & Over, 1996; Legrenzi, Girotto, & Johnson-Laird, 1993). The functional and structural features that distinguish mental models from other proposed knowledge structures precisely match the requirements of critical dialogue:

(1) Mental models of the situation are created by relatively automatic processes, which continue to modify the model as new information about the situation arrives via perception, discourse comprehension, inference, memory, or imagination (Johnson-Laird, 1983; Kessler, Duwe, & Strohner, 1999). They account for the coherence of extended discourse by providing an enduring, evolving, and integrated representation of the situation being discussed (Anderson, Howe, & Tolmie, 1996; Oakhill, 1996). (Cognitive linguists such as Langacker (2000) are studying how syntactic devices function as instructions for the addition, deletion, or modification of model elements. By contrast, conventional linguistics posits a redundant, syntactically structured level of "sentence meaning" which cannot be integrated with perceptual, linguistic, or inferential inputs.) Linguistic support for mental models as semantic representations includes reference to objects understood only from the context rather than from surface constructions(Garnham, 1996, 2001). Experimental support includes prediction of memory centered on objects rather than surface descriptions, such as failure to recollect descriptions of items that are no longer part of the situation, and false recall of descriptions that apply to objects that were never explicitly described that way (Stevenson, 1996). Because mental models figure in both non-deliberative and deliberative processing, they predict reasoning errors caused by recognitional priming of irrelevant information (Evans & Over, 1996; Legrenzi et al., 1993), and by explicit representation

of only the most salient possibilities consistent with a discourse (Johnson-Laird, 2003; Legrenzi et al., 1993).

(2) Once models are created, recognitional processes are applied to them to identify emergent relationships that were not explicit in any of the inputs. Mental models thus account for novel conclusions that people draw from a combination of information sources and background knowledge without explicit rules of inference, based on relational isomorphism with the represented situations (Rickheit& Sichelschmidt, 1999; Schnotz & Preuss, 1999). Mental models also account for errors that result from flawed understanding of the relational structure of the domain (Anderson et al., 1996). Mental models extend the isomorphism principle to logical and probabilistic relationships, allowing inferential conclusions to be recognitionally "read off" the set of mental models that results from combining all the premises (Johnson-Laird, 1983; Johnson-Laird & Byrne, 1991). Probability judgments based on the relative number of scenarios in which an event occurs will be dependent on the readiness with which scenarios come to mind, and can be affected by salience, recency, prototypicality, and other factors that may not be representative of the relevant population, especially in novel situations (Kahneman, Slovic, & Tversky, 1982).

(3) Once a candidate conclusion has been recognized, it can be verified by searching for mental models in which the claim is false (precisely the opponent's role in critical thinking). The common semantic foundation (models as possible states of affairs) and the common verification process (searching for counterexamples) provide an integrated account of deductive, assumption-based, and probabilistic inference (Tabossi, Bell, & Johnson-Laird, 1999). Differences are accounted for by different constraints on verification—viz., whether a counterexample is fatal, can be explained away, or quantified as risk, respectively—rather than by discrete normative rule systems. The failure to generate and consider relevant alternative possibilities is a frequent cause of errors in reasoning and decision making (Legrenzi & Girotto, 1996; Evans, 1996; Byrne, Espino, & Santamaria, 2000). For example, people interpret the probability of the evidence conditional on their initial hypothesis as support for that hypothesis; often, however, they do not even try to determine the probability of observing the same evidence in scenarios that embed alternative hypotheses (Legrenzi et al., 1993; Doherty, Mynatt, Tweney, & Schiavo, 1979). Whether it is worthwhile trying to construct alternative scenarios, of course, is a function of the pragmatic context, which is seldom captured in laboratory studies of decision biases. Mental models also support cognitive collaboration among individuals with divergent views by allowing each one to represent others' mental models of the same objects (Fauconnier, 1994; Johnson-Laird, 1983, pp. 430–438).

Example

Figure 10.4, which illustrates some of the above functions, is a hypothesis about the opponent's mental model at the end of the exchange in Figure 10.3b. Different mental spaces or partitions (Fauconnier, 1994) correspond to time (the situation last night, the situation now, and the near future) and to alternative possibilities (corresponding to divergent views by the proponent and opponent about the present and future). Object identity as well as time can be tracked across these spaces, representing changes in relationships among the same objects (i.e., the proponent's belief that the enemy unit in Bridgeton last night has left) as well as different views about the same object at the same time (i.e., that the enemy unit is or is not in Bridgeton now). These features illustrate coherence of discourse rooted in a grounded representation of the situation under discussion and conversational interaction guided by representations of the other's beliefs about the same objects.

A participant is committed to a claim if it is true in each situation (mental model) that he or she is actively considering. According to Figure 10.4, the opponent believes there is agreement that an enemy mechanized unit was in Bridgeton last night. In addition, according to the opponent, the

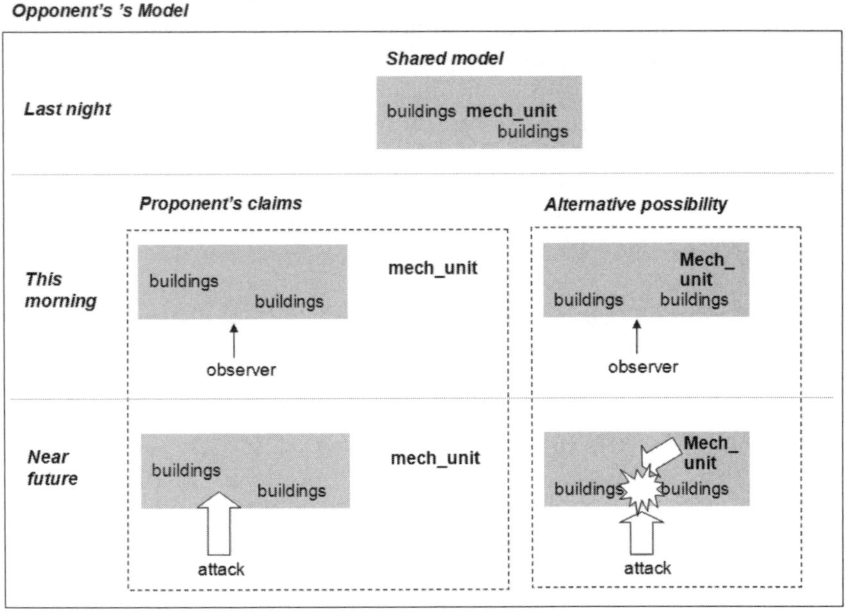

FIG. 10.4. A spatial view of the opponent's situation understanding at phase (b) in Figure 10.3. Time in the scenario is represented from top to bottom.

proponent is committed to the claim that the unit is no longer in Bridgeton and thus also to the claim that his company can attack and take the city without a fight. The opponent's *uncertainty* corresponds to active consideration of both the proponent's model, showing the unit not in Bridgeton, and an alternative model in which the proponent's claim is false.

To create this alternative model, the opponent must show how the unit might still be in Bridgeton despite not being observed. He elaborates the original shared mental model by adding an observer corresponding to friendly forces on a hill south of the river, and by using a schematic spatial representation to mentally test whether buildings might conceal enemy forces from that observer (Figure 10.4, middle right). Because of its consistency with testimonial and observational evidence, and its satisfaction of constraints associated with relational isomorphism to a real situation, success in constructing such a model is prima facie evidence for the *possibility* that the enemy is concealed in Bridgeton (Johnson-Laird, Girotto, & Legrenzi, 2004). Moreover, in the course of his attempt to construct a model of the *proponent's* position (Figure 10.4, lower left), the opponent is able to "read off" conclusions implicit in its relational structure (for example, *If the unit left the town, it must be somewhere else.*) These implications raise questions about the plausibility of the proponent's model and thus supply ammunition for the opponent's subsequent challenges in Figure 10.3c: *If the unit left the town, where did it go? And why have we not observed it somewhere else?* These features illustrate error due to initially overlooking a possibility (i.e., concealment of the unit), the use of relational isomorphism to generate novel insights without inference rules, and assumption-based reasoning (e.g., to accept the proponent's argument, one must assume that the enemy is not concealed).

CHALLENGING AND DEFENDING MENTAL MODELS

Types of Dialogue

A dialogue is a communicative exchange *of a recognizable type* (Walton, 1998) in which two or more individuals cooperate to bring about changes in their cognitive states. Some familiar dialogue types are distinguishable by whether the participants primarily want to change intentions, beliefs, or emotions, whether players in different dialogue roles are expected to experience the same changes, and the balance of cooperation and competition in determining the direction of change. In *inquiry* and *deliberation*, for example, participants cooperate to fill collective gaps in their shared understanding (beliefs) and plans (intentions), respectively. *Information transfer* (with roles of asking and telling) and *action commitment* (with roles of requesting and offering) are cooperative responses to an imbalance across members of the team; some parties are expected to exert an asymmetrical influence on the beliefs or intentions, respectively, of others. By contrast, *negotiation* and *persuasion* highlight competition over whose cognitive state will be most changed. Negotiation is prompted by divergence of values and involves offers and counteroffers in symmetrical efforts by each side to steer decisions about limited resources toward their own goals. Persuasion dialogue (of which critical dialogue is a special case) is prompted by divergence of claims rather than values, and participants use challenges and defenses to steer one another's beliefs toward a favored conclusion. In practical contexts, critical dialogues target decisions as well, since divergent beliefs typically include predictions of action outcomes or inferences about action preconditions. In a *simple* critical or persuasion dialogue, there is an asymmetry between the role of proponent, who has the burden of proof, and the role of opponent or challenger, who only needs to create doubt about the proponent's claim. In a *compound* persuasion or critical dialogue, however, each participant plays both roles, as proponent for his or her favored claim and as opponent for claims by others (Von Eemeren & Grootendorst, 1992). As in sports, a cooperative goal of playing the game coexists with a competitive goal of winning it, and all players reap benefits from skillful effort.

Rules of Argumentation

A move is permissible in critical dialogue if it promotes the participants' shared purpose, convergence of belief on true claims (Walton & Krabbe, 1995). Not surprisingly, these rules systematically enforce the *disagreement heuristic*. For example, a challenge is ruled irrelevant unless it targets claims to which the defender is committed and to which the challenger is not, and which have a bearing on a task decision. When a commitment is challenged, the recipient of the challenge must on the next turn either defend it by providing reasons (i.e., try to remove the opponent's doubt by introducing claims that are inconsistent with the alternative possibilities) or retract it (i.e., accept the alternative possibilities). If reasons are offered in defense of a commitment, the reasons in turn become commitments and are themselves subject to subsequent challenge. If a commitment is retracted, other commitments that depended upon it are in jeopardy (because it eliminated mental models in which they were false). They may be given a new defense, retracted, or retained subject to later challenge. Commitments cannot be defended or retracted, however, *unless* they have been either directly challenged or indirectly placed in jeopardy (the disagreement heuristic again). If a party does not directly or indirectly challenge the other party's assertion at the first opportunity (i.e., the next turn), it is *conceded*. Concessions allow other parties to get on with their case without stopping to defend every step. Later, however, with a larger view of how a step relates to the claim of interest, a challenger may go back and probe key elements of the other party's case more vigorously. Thus, a concession is unlike a commitment in being retractable at any time, even in the absence of a challenge or threat. These rules force participants to set aside areas of agreement and inconsequential disagreement and focus attention and discussion on disagreements that most strongly influence shared task outcomes, in pursuit of both consensus and accuracy.

Knowledge, Risk, and Ignorance

Critical dialogue alternately increases the number of possible scenarios under consideration (through the opponent's challenges and the proponent's retractions) and decreases them (through the proponent's defenses and the opponent's concessions). Nevertheless, both participants tend to increase the total number of claims under discussion, hence, the number of dimensions along which mental models vary (Rescher, 1977a). The aim to persuade means that each has an incentive to introduce factors not considered by the other but which the other is likely to understand, believe, and find convincing as a reason to accept, doubt, or deny a claim. (An exception is that the challenger may occasionally only request reasons for a claim or point out an apparent incoherence in the proponent's case, without adding a substantive new issue herself.) Despite opposing short-term goals, therefore, the parties generally collaborate to increase the fineness of discriminations among possible situations (i.e., the *resolution* of the situation picture) and indirectly, the scope of the knowledge network brought to bear on the disputed claims. Critical dialogue should mitigate the *common knowledge effect*, a phenomenon in which group members tend to focus on information that other group members already have (Stasser, 1999) and make more use of it in decisions (Gigone & Hastie, 1997), even when unshared information is more important. Whether or not consensus is achieved, critical dialogue should increase the participant's understanding of the situation, increase the amount and impact of shared knowledge, and (at the very least), improve their understanding of what the other participants believe or want.

Mental models generalize the notion of a *commitment store* found in current theories of dialogical reasoning (Hamblin, 1970; Rescher, 1977a; Walton & Krabbe, 1995) and as a result, provide a better formulation of rules for permissible dialogue moves (such as *assert, challenge,* and *concede*). Rules are conditional on a participant's commitment store, which is usually envisaged as a single list of propositions that the participant has asserted and not subsequently retracted. If instead we identify it with the *set* of mental models that the participant is currently entertaining, we can conditionalize role-specific dialogue rules and objectives on the possible combinations of propositions entertained by each participant. For example, the opponent in simple critical dialogue needs to demonstrate not the falsity but merely the *possibility* that the proponent's claim is false, i.e., that there is at least one plausible mental model in which it fails. By contrast, the proponent's burden of proof requires her to show the *falsity* of claims whose mere possibility weakens her case. Treating an opponent's challenge or a proponent's retraction as equivalent to a negative assertion (that is, including the *negation* of the challenged or retracted claim in a single list of commitments) is too strong, while dropping both the claim and its negation from the list throws away the fact that the claim has received attention and been challenged (hence, if the proponent wants to retain it or reintroduce it later, she must offer a positive defense). Adding alternative mental models in which the claim is false captures this distinction well.

Model-based commitment stores also clarify the semantics of concessions, as an element in assumption-based reasoning. A claim is conceded, or accepted as a provisional assumption, when it is true in all or most mental models *considered by the participant thus far*, but the participant is unwilling to bet that a more thorough search of relevant cases, or discovery of hitherto neglected key variables, would not yield counterexamples. A commitment on the other hand *is* a bet that exploration of additional variables and scenarios will not overturn the claim.

The generalized notion of commitment store corresponds to an expanded notion of *situation awareness*. It includes not just perceptions, interpretations, and predictions (Endsley, 2000), but also awareness of what questions are currently *relevant* and the present degree and type of *uncertainty* regarding them. Shared awareness of relevant issues and their associated uncertainty is important in implicit coordination skills like *pushing* needed information to team members without being asked (Serfaty et al., 1998). Pushing information depends on shared awareness of relevance and uncertainty, not specific beliefs about the issue in question.

Mental models and dialogue rules help provide a unified framework for thinking strategies that people use in different contexts. For example, in a well-understood domain, challenges may be implicitly restricted to a known set of issues, and each party aims for the other's commitment. The limiting case of this is deductive reasoning, where the relevant knowledge can be encapsulated in a fixed set of premises. Neither party is allowed to add or subtract information during this part of the discussion, exhaustive search for counterexamples is therefore feasible, and the proponent must show that each case in which the claim is false is inconsistent with the premises (Hintikka, 1999). In a less structured or more novel situation, when there is no clear way to limit the processes of proposing and rebutting counterexamples, the parties may aim merely for a *concession* by the other side: a willingness to act on the overall set of assumptions that is currently the most plausible.

In probabilistic reasoning, the participants aim to converge on a *probability*—effectively, a willingness to bet at certain odds—rather than on a conclusion. Counterexamples (i.e., scenarios in which the claim is false) are not fatal as in deductive reasoning, and do not need to be explained as in assumption-based reasoning, but help quantify risk (the proportion of scenarios in which the claim is true). New factors, or conditioning variables, are introduced into the discussion until the proponent and opponent are in sufficiently close agreement on how to bet (Shafer & Vovk, 2001). In domains where there is substantial ignorance about what the critical factors are and how they influence the claim, agreement on a particular probability may be arrived at more by concession (i.e., as the most plausible current estimate) than by firm commitment.

Example

Figure 10.5 illustrates the way commitment stores evolve via challenges and defenses. Rows correspond to claims whose truth and falsity differentiate among possible views of the situation. Time in the real world flows from top to bottom, and arrows represent cause–effect relationships among the corresponding states of affairs and events. Time in the dialogue, on the other hand, flows from left to right. Shading indicates the dialogue phases in which a proposition is present in the commitment store of a particular party, either by assertion or concession. Striped shading represents disagreement (i.e., a challenge without a retraction); thus, the claim is present in one commitment store while its negation is present in the alternative.

This diagram makes three features of critical dialogue evident: (a) Mental models are elaborated by introducing new variables (represented by the increased shading further to the right). (b) Shared awareness grows as each participant introduces factors that the other participant concedes as a basis for further reasoning and action (represented by the proportion of non-striped shading). (c) Newly introduced factors are inferentially relevant because of their causal links to disputed claims. Most of the reasoning in this dialogue can be construed as either explanatory (inferring causes from effects) or predictive (inferring effects from causes).

Figure 10.5 also supports a more detailed analysis of the tactics of the two participants. In particular, three kinds of challenges by the opponent stimulate model development by the proponent:

(1) Exposing *gaps* in the proponent's case. The proponent's first turn (Figure 10.5a) leaves him committed to a simple causal scenario: *Bridgeton is undefended & We attack* → *We take Bridgeton without a fight*. The opponent's first challenge (Figure 10.5a) *Why do you think it's undefended?* is a request for reasons that implicitly raises the possibility that *Bridgeton is defended* and that if we attack, *there will be a fight* (based on the Gricean, 1989, assumption that

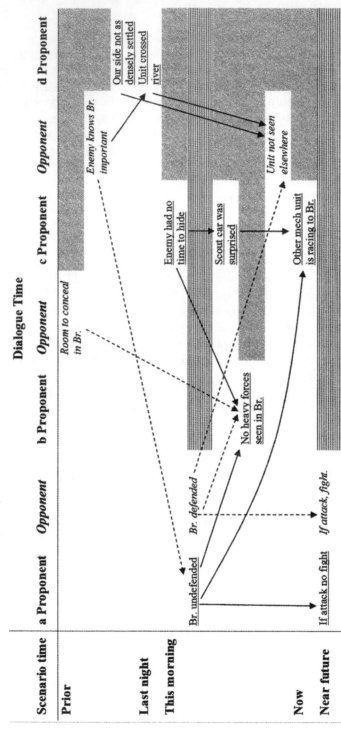

Dialogue Time

FIG. 10.5. Over the course of the dialogue in Fig. 10.2, proponent and opponent flesh out a pair of alternative scenarios or mental models. Stripes indicate disagreement, and shading indicates consensus. Arrows indicate causal relationships. Events introduced by the opponent are in italics: causal relationships used by opponent are represented by dotted-line arrows.

the opponent is cooperatively playing his role in the dialogue). The proponent attempts to eliminate this alternative model by adding *No heavy forces seen in Bridgeton*, an observation whose most natural causal explanation is that *Bridgeton is undefended.*

(2) Exposing and challenging the proponent's *assumptions.* In Figure 10.5b, the opponent concedes *No heavy forces seen* but adds that there was *Room for concealment.* The latter offers an alternative explanation of the former and thus makes it consistent with the possibility that *Bridgeton is defended.* (The proponent's original argument, based on *No heavy forces seen,* implicitly assumed room for concealment was not available.) In his next turn (Figure 10.5c), the proponent uses the same tactic. He concedes *Room for concealment* but denies an assumption (that the enemy had enough time to hide) implicit in its use by the opponent to explain *No heavy forces seen.* The proponent now has a coherent causal sequence: *Bridgeton is undefended & Enemy was not aware of us in time to hide (despite room for concealment)* → *No heavy forces seen in Bridgeton.* Support for *Enemy was not aware of us in time to hide* comes from the fact that it nicely explains the perceptual judgment that *The scout car was surprised.* In case this is still insufficient to eliminate the opponent's alternative model, the proponent adds another perceptual judgment, *Enemy mech unit racing toward Bridgeton,* whose most obvious explanation is that *Bridgeton is undefended* (in combination with the assumption that the enemy has just learned of our presence from the scout car). At this point in the dialogue, the proponent's disputed claim that *Bridgeton is undefended* has been situated within a more detailed, coherent, and observationally grounded causal picture.

(3) Offering evidence that directly *conflicts* with the original claim (rather than demanding or attacking one of the proponent's arguments for the claim). The opponent now goes on the offensive (Figure 10.5c), offering one claim, *Enemy knows importance of Bridgeton,* that predicts *Bridgeton is defended* and another claim, *Enemy unit not seen elsewhere,* that is explained by it. In his final turn (Figure 10.5d), the proponent again concedes these new claims, but exposes hidden assumptions in their inferential use. *Enemy knows importance of Bridgeton* does not predict *Bridgeton is defended* if there is a better way to achieve the same end, for example, to *Protect Bridgeton from our side of river.* He also offers an alternative explanation of *Enemy unit not seen elsewhere* by pointing out that *Our side of river is less densely settled.*

MONITORING CRITICAL DIALOGUE

Decisions about process reliability require a third role in critical dialogue, which we call the *dialogue monitor.* Although this corresponds to a more objective, *third-person* stance, it does not offer certainty or infallibility, but rather a distinctive *perspective*—that of an external, neutral observer (Brandom, 2000; Shafer, 2001). The outer layer of critical dialogue, *control decisions* (Figure 10.1), corresponds to individual or team *metacognition*, or cognition about cognition (Cohen et al., 1996; Hinsz, 2004). Metacognition includes (a) dynamic awareness of dialogue processes in real time and (b) general knowledge about those processes and their outcomes (Nelson, 1992; Metcalf & Shimura, 1994). Accumulated metacognitive knowledge assesses the reliability of judgment and decision processes by the relative frequency of confirmed beliefs and successful outcomes that they produce (Brunswick, 2001; Hammond, 1993; Gigerenzer, 2000), or indirectly, by comparison with expert performance (Orasanu & Connolly, 1993; Cohen et al. 1996; Lipshitz & Cohen, 2005). This perspective is as different from those of the proponent and opponent as they are from one another, although the roles may in practice be combined in a single person or distributed across team members.

Critical dialogue, as a coordinated activity, presupposes shared knowledge of how it is to be carried out. This knowledge can be summarized as a set of state variables and their causal relationships, as shown in Figure 10.6. The transitions in Figure 10.6 can be divided into four functional phases, which need not be entirely discrete in time (Cohen, 2004): (1) Recognition of the initial state involving disagreement about a significant claim, (2) formulation of a cognitive and communicative intent to resolve the disagreement by critical dialogue, (3) exchange of challenges and defenses, and (4) resolution of the disagreement. These phases are further elaborated in Table 10.1.

The dialogue monitor's job includes three regulatory functions—*facilitator, referee,* and *judge*—focused on enforcing constraints associated with different phases (Figure 10.6). The monitor is a *facilitator* in Phases 1 and 2, when a problem is recognized and a dialogue type is selected to resolve it, and in Phases 3e and 4a, when the problem is resolved and the dialogue is concluded. The monitor functions as a *referee* during the exchange of challenges and defenses in Phase 3. Finally, the monitor serves as a *judge* when determining how disagreement is settled in Phase 4.

Facilitator

The *facilitator* is the top-level control of the dialogue process. This function is especially prominent in Phase 1, where the facilitator detects and diagnoses a pragmatically problematic cognitive state, and in Phase 2, where the facilitator assesses such conditions as motive, means, opportunity, and intention to determine what method will be used to resolve the problem, by whom, and at what time and place.

Selection of an appropriate dialogue method requires matching features of the situation to templates for different types of dialogue (Cohen, 2004). Thus, the facilitator begins Phase 1 by recognizing the initial state or prob-

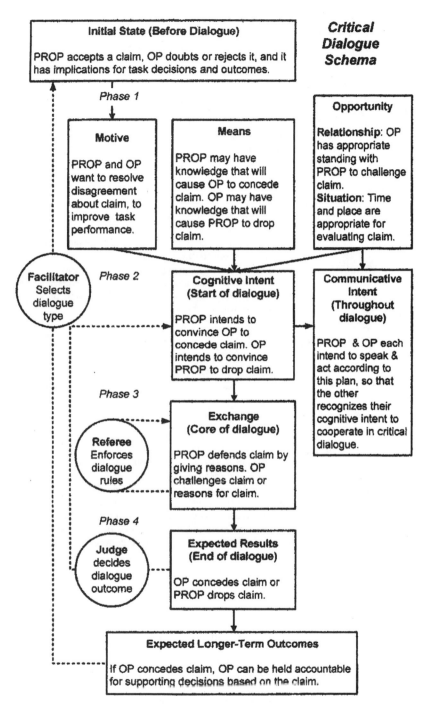

Critical Dialogue Schema

Initial State (Before Dialogue)

PROP accepts a claim, OP doubts or rejects it, and it has implications for task decisions and outcomes.

Phase 1

Motive

PROP and OP want to resolve disagreement about claim, to improve task performance.

Means

PROP may have knowledge that will cause OP to concede claim. OP may have knowledge that will cause PROP to drop claim.

Opportunity

Relationship: OP has appropriate standing with PROP to challenge claim.
Situation: Time and place are appropriate for evaluating claim.

Facilitator
Selects dialogue type

Phase 2

Cognitive Intent (Start of dialogue)

PROP intends to convince OP to concede claim. OP intends to convince PROP to drop claim.

Communicative Intent (Throughout dialogue)

PROP & OP each intend to speak & act according to this plan, so that the other recognizes their cognitive intent to cooperate in critical dialogue.

Phase 3

Referee
Enforces dialogue rules

Exchange (Core of dialogue)

PROP defends claim by giving reasons. OP challenges claim or reasons for claim.

Phase 4

Judge
decides dialogue outcome

Expected Results (End of dialogue)

OP concedes claim or PROP drops claim.

Expected Longer-Term Outcomes

If OP concedes claim, OP can be held accountable for supporting decisions based on the claim.

FIG. 10.6.   A schema for critical dialogue, which functions as a shared plan, an explanatory pattern, and a source of normative constraints.

TABLE 10.1

Phases and Tasks in Critical Discussion

Stage	Tasks
1. Recognizing disagreement	a. Individuals think about problem separately. (Group is more effective after members have thought about issues independently, even if just for a short time.) b. Express own views. c. Learn what others' positions are and why. Ask for clarification if not clear.
2. Prioritize issues for critical discussion	a. Recognize and expand areas of agreement (e.g., quickly settle minor differences and distinctions without a difference). b. Recognize and understand significant disagreements, that is, those that have implications for actions and outcomes. c. Determine what disagreements are important enough to critically discuss. Prioritize them. If there is no disagreement, determine the most important issues for which there is uncertainty. (Look at actual disagreements first, because an uncertainty is more likely to be significant if people have actually adopted different positions on it.) d. For high priority issue(s), quickly: Decide approximately how much time you have. Decide who plays primary roles of proponent and opponent. (If players have competing claims, each plays both roles.) Designate someone to be the monitor. This may be someone with no other role, or it may be the proponent and opponent jointly. If more than three people, assign teams to roles.
3. Challenge-defend	a. Parties take turns and do not interrupt one another. b. On each turn, for each claim that has been challenged, the proponent must defend it with reasons, modify it or other claims to avoid the objection, or retract it. c. On each turn, the opponent must challenge the proponent's position or else give up opposition. A challenge may demand a reason for any claim, present reasons against any claim, question the soundness of any inference from some claims to others, point out an incoherence among claims, or present an alternative coherent viewpoint. d. Monitor makes sure rules are being followed. e. Monitor watches time, keeps discussion going when appropriate, and closes it when costs outweigh benefits.
4. Resolution	a. End the discussion when parties resolve initial disagreement, or Monitor declares time is up. b. Identify recommendation of the group: This may be by concession of one of the parties, or else a decision by the monitor (or in some cases, an external superior). c. Monitor summarizes the strengths and weaknesses of each side, and explains why decision was made.

lem: Is there an imbalance of information, intentions, or affect among individuals within the group; a collective gap with respect to external knowledge, plans, or affect; or a divergence among team members in task-relevant beliefs, values, or emotions? Different types of dialogues correspond to different symptoms. In the case of critical dialogue, the initiating problem is disagreement in beliefs or decisions that are consequential for a task.

Having identified a problem, in Phase 2 the facilitator assesses whether other preconditions for the candidate dialogue type are satisfied. For example, how much does the imbalance, gap, or divergence of views matter for the task at hand? In the case of critical dialogue, are the parties *motivated* to resolve their disagreement or do they have a hidden agenda, such as challenging authority? Do they have the *means* to make a persuasive case, including requisite knowledge skills, and aptitudes? With regards to *opportunity*: Is an immediate decision required? What are the costs of delaying action? How much time will be needed for discussion? Are the time and place appropriate? Are there obstacles to critical discussion such as status differences and personality?

An initial problem state, together with motive, means, and an actual or anticipated opportunity, leads to a *cognitive intention* on the part of one or more of the parties to remove the imbalance, gap, or divergence by initiating an appropriate type of dialogue. This cognitive intention gives rise to a *communicative intention* (Sperber & Wilson, 199; Grice, 1989), to make other individuals aware of the cognitive intention and invite them to share it by taking on complementary dialogue roles. The verbal and nonverbal cues used to initiate critical dialogue may be quite subtle and indirect, because of potential loss of face in the process of challenging and being challenged (Brown & Levinson, 1987; Cohen, 2004). Critical discussion may be invited by remarks that just barely evoke the schema, such as "If we're wrong, we're in trouble" (partially matching the motive component), "Are you sure?" (partially matching the motive and means conditions), or even more indirectly, "We have a few minutes before we have to go" (partially matching the opportunity component). Conversely, an invitation to critical dialogue may be signaled by jumping straight into Phase 3 (e.g., the opponent's question, *Why do you believe that Bridgeton is undefended?*) and may be accepted in the same manner (e.g., the fact that the proponent answers with a reason). The communicative intention is achieved when the appropriate cognitive intent is mutually understood and shared.

Facilitation can proactively test for and *shape* conditions favorable for different types of dialogue. When employed skillfully by team leaders, proactive facilitation is likely to improve mutual trust and coordination among team members (Cohen, 2004). A proactive facilitator learns to detect and even elicit consequential disagreements in social contexts where politeness, personality, custom, or deference to authority may discourage dissent

(Brown et al., 1987). For example, is someone's body language signaling discomfort with a decision? Does someone's background or previously expressed views suggest that she might disagree with a decision? Other proactive skills include mitigation of the intimidating effects of status differences, active creation of times and places where discussion is possible, efforts to motivate individuals to express and defend their views, and techniques for *declining* to engage in dialogue when conditions are not right without discouraging open discussion when appropriate in the future.

The facilitator continues to monitor dialogue conditions during Phase 3, to determine when the exchange of challenges and defenses should conclude. Phase 3 questions are: Is the dialogue still productive? Has the importance of the issues under discussion changed, either based on external information or on results of the dialogue itself? Has the cost of delaying action changed? Does the cost of further delay still outweigh the importance of the issues? Here too the facilitator may be more proactive, as shown in Table 10.2 (among the materials we used in training). Part of the facilitator's skill is stimulating productive discussion when it seems to have petered out. Phase 4 begins when the facilitator declares time up.

According to Paul (1987), Siegel (1997), Missmer (1994), and others, critical thinking is not simply a set of skills, but implies a more enduring disposition (including temperament and motivation) to put the relevant skills to use. This notion is captured in the idea of proactive facilitation of internal dialogue. To be an effective individual critical thinker, a person must internalize and enact that role, including both sensitivity to signs of doubt about her own beliefs and decisions and (we emphasize) recognition of the need, at an appropriate point, to stop questioning and act.

TABLE 10.2
Guidelines for Stimulating Critical Discussion

Guideline	Example
Prioritize use of time	What are the top priority issues? Which need an immediate decision? What do we already agree on, and where do we disagree?
Keep discussion going if someone concedes too soon	We still have some time. Don't give up yet. Work harder to come up with a better defense or a modification that meets the objections. An infallible crystal ball says your position is correct—Explain how this can be so despite the objections. If it is true, how could you show it?
Energize discussion if it gets into a rut or peters out.	Don't repeat the same points. Come up with new ideas. An infallible crystal ball says the position is false despite this evidence—Explain how that could be so. The crystal ball says there are other problems with the position—What are they?
Call foul if a party violates the rules	Aren't you changing the subject? Let's stay on this another moment.

After the dialogue ends, the facilitator's function is to *learn*: to track and record outcomes, including the effects of the resulting cognitive state changes on on-going or future tasks. The facilitator thereby enhances knowledge about the benefits and costs of different dialogue methods under different conditions. A proactive facilitator actively creates opportunities to develop better ways of thinking and deciding.

Referee

Although the facilitator has selected the most reliable method for the purpose at hand, performance can be degraded if participants violate constraints conducive to the shared purpose. It is necessary, then, to supplement the facilitator's role with the function of a *referee*, who calls "fouls" when the contending parties violate norms associated with a particular dialogue type. The referee's task is to keep the exchange of challenges and defenses on track in Phase 3.

Critical-thinking and informal-logic textbooks identify numerous supposed *fallacies* in ordinary reasoning (e.g., Govier, 1997; Johnson & Blair, 1994). These include, for example: attacking a strawman position, arguing with respect to an issue other than the one under discussion, attacking the motives or character of opponents rather than their reasons, assuming what is to be proven, and appealing to authority, pity, passions, threats, or the popularity of a viewpoint to make one's case. More specific errors include inferring causality from association, affirming the consequent (viz., concluding A from B together with *If A then B*), false analogy, overgeneralization, overlooking distinctions, and slippery slope. Some simple examples of fouls that were defined and illustrated for our dialogue training are shown in Table 10.3.

Efforts to provide a unified "theory of fallacies," generally from an internalist point of view, have been unsuccessful (Hamblin, 1970; Woods & Walton, 1989). Fallacies cannot be formally defined as errors in logic, probability, or decision theory. Some fallacies, like begging the question, are logically correct, and others (e.g., affirming the consequent) may be correct if understood within a different inferential framework, such as causal reasoning instead of deductive logic. As Walton (1995) points out, for virtually any so-called fallacy there are contexts in which arguments of the same apparent form make perfect sense; similar points have been made about alleged decision and judgment biases (Gigerenzer & Murray, 1987; Cohen, 1993b). An alternative, quasi-externalist approach is to characterize fallacies as violations of Grice's (1958) principle of cooperation in conversation (or dialogue). This has been pursued by Van Eemeren & Grootendorst (1992) for fallacies of reasoning and by Schwarz (1996) for decision and judgment biases. A problem pointed out by Walton (1995) is that the gran-

TABLE 10.3
Two Rules for Critical Dialogue, With Examples

Rule	Fouls to avoid	Examples of foul
A. Don't suppress disagreement, or prevent each other from defending or challenging positions.	No intimidation by use of authority or expertise	If I want your views, I'll ask for them.
	Don't distort others' views (create a strawman)	So, you cowards just want to cut and run?
	No personal attacks on competence or motives	
	No appeals to sympathy of other party	Give me a break! No one ever accepts my ideas. Just go along with me this one time!
B. Whoever makes a claim has to defend it if asked to do so, or else concede.	Don't ask others to rely on your personal guarantee.	I'm the expert here. I don't have to defend my views.
	Don't declare your conclusion to be obvious.	Everybody knows that . . .
	Don't turn the tables.	Well, I'd like to see you prove that I'm wrong.
	Don't bargain. Settle issues on the merits.	I'll let you have your way on the 1st platoon if you'll accept my suggestion on the tanks.

ularity of Grice's principles is too coarse by themselves to account for standard distinctions among fallacies. Moreover, while principles of cooperation seem to apply directly to some fallacies, such as begging the question, using threats to suppress disagreement, changing the subject, reversing the burden of proof, or attacks on character, there does not seem to be a necessary connection to other errors (e.g., false analogy, slippery slope). Finally, this proposal seems to make social convention the entire basis for argument evaluation to the exclusion of reliably fitting reality.

Cooperation can, however, be combined with externalist reliability to produce a general explanation of so-called fallacies. A unified classification scheme emerges when we identify different levels at which moves by dialogue participants might threaten the reliability of a dialogue process:

(1) Some fallacies involve moves belonging to the *wrong dialogue type,* resulting in an interaction that will not reliably achieve the presumptive shared goal (Walton, 1998). For example, citing one's own authority to cause a cognitive change in another person may be appropriate when giving orders (a special case of requesting action commitment) or imparting

specialized information, but not under conditions appropriate for deliberation, inquiry, negotiation, or persuasion. Appealing to consequences (e.g., threats) or emotions is appropriate in negotiation and quarreling, but generally not in persuasion. Bargaining about the conclusion of a critical discussion (*I'll accept your view on issue A if you'll accept my view on issue B*) may be very cooperative, but it is out of bounds in critical dialogue—because the outcomes of a bargaining process are unreliably linked, if at all, to the truth of what is finally agreed upon.

(2) Another class of fallacies involves importing behavior from a *different dialogue phase*. For example, it is the facilitator's job in Phases 1 and 2 to determine the topics, times, and participants that are most likely to produce a productive critical dialogue. It is fallacious, however, during challenge and defense in Phase 3 for the *opponent* or *proponent* to change the topic, claim that the answer is obvious, cut off the discussion, or personally attack the motives or character of other participants. The dialogue monitor makes such decisions on pragmatic grounds from a perspective that does not favor one side over the other, while the contenders (when acting in those roles) are liable to be influenced by partisan factors not correlated with reliability.

(3) Within the challenge and defend phase, fallacies may be associated with the proponent's refusal to defend arguments or the opponent's use of inappropriate challenges (Walton, 1996a). Walton encapsulates these ideas in the notion of an *argumentation scheme* (1996a): that is, a characteristic way of reasoning that specifies (a) a presumptive conclusion, (b) evidence whose presence is typically sufficient for creating the presumption of truth and shifting the burden of proof to the opponent, and (c) a set of critical questions that an opponent can ask to shift the burden of proof back to the proponent. For example, an argument from analogy may be challenged by pointing out differences between the current problem and the supposed analog, or by bringing forward different analogies that point toward different conclusions; expert testimony may be challenged by asking whether the expert's opinion is clear, if other experts agree, if the claim is really in the expert's area of specialization, and so on; a slippery slope argument can be challenged by asking what process, if any, would lead from a small compromise to an unacceptable extreme; generalization from cases may be challenged by questioning the sample size or by pointing out relevant but overlooked distinctions. Named fallacies like these (false analogy, appeal to expertise, slippery slope, overgeneralization, and neglecting distinctions) cannot be automatically identified based on surface features of an argument. They come into being in a dialogical context when opponents raise questions associated with a particular type of argument which proponents fail to answer (while also refusing to retract the challenged claim). It is plausible to suppose that such maneuvers—to the degree that opponents'

questions are informed and well motivated—reduce the reliability of the dialogue as a generator of accurate beliefs.

The same principle works in reverse to prevent the opponent from using frivolous questions to prolong a critical dialogue indefinitely (Walton, 1996a). Once an argument of a certain type has been made, the proponent is entitled to the associated default conclusion at least as a working assumption. The opponent must either concede it or challenge it by means of specifically appropriate types of questions. For example, the opponent's first challenge merely asked for reasons (*Why do you believe Bridgeton is undefended?*) and was answered by citing visual observations (*We only saw scout vehicles there*). The report of visual observation, however, cannot be challenged merely by asking for reasons (e.g., *Why do you believe you saw only scout vehicles there?*). Challenges to a visual observation would have to introduce appropriate substantive issues (e.g., *Were the viewing conditions good? Has the observation been confirmed by multiple observers?*).

Nonsubstantive challenges are inappropriate against arguments based on the normal operation of sense perception, deeply entrenched common sense, an accepted tenet of the proponent and opponent's shared community, or a belief manifested in the opponent's own statements or actions (Walton, 1996a). If the opponent is unable or unwilling to explain why she finds beliefs such as these unconvincing, the proponent cannot know where to *begin* in providing reasons that the opponent *would* accept. The opponent leaves the proponent no ground to stand on unless she states her reasons for doubt in such cases, to which the proponent can then respond. Legitimate questions target the reliability of the connection between the evidence presented and the truth of the presumptive conclusion. In combination with the evidence, therefore, the answers to critical questions should form a coherent, *linked* argument for the claim (Walton, 1996b).

Questions the referee asks during Phase 3 include: Are the participants interacting within the constraints associated with the current dialogue's purpose, setting, and phase? Does each participant understand and correctly describe the other's positions and arguments? Is the opponent upholding her responsibility to present relevant counterarguments or concede? Is the proponent upholding her responsibility to answer relevant challenges or retract?

Judge

The facilitator regulates the time taken for dialogue, and the referee regulates how the exchange of challenges and defenses is conducted. But there is no guarantee that the available time, as determined by the facilitator, is sufficient for the contending parties themselves to reach agreement. Reliability thus requires the function of a *judge,* authorized to review the outcome of the exchange and if necessary, decide the issue unilaterally. As in other aspects of the monitor's job, the specific method of judging should be reliably associated with true or successful results under the prevailing circumstances.

Time constraints are not the only reason to have a judge. Even if the contenders reach an accord, it may be based on factors that are not reliably correlated with the truth of the views in question. There are two sets of factors that contribute to this result: the *local* nature of argumentation and the *inside* perspective of the contenders. A more holistic and external point of view may enable the judge (or the participants themselves in the role of judge) to reach more reliable conclusions than the participants in the roles of proponent and opponent.

Argumentation stimulates the construction of mental models, but it is flawed as a method for evaluating them (Pennington & Hastie, 1993b). Points won in debate need not correspond to overall plausibility. For example, suppose the proponent offers two independently sufficient reasons, A and B, for the conclusion C. The opponent rebuts A by evidence $R1$ and rebuts B by evidence $R2$. Neither party notices that $R1$ and $R2$ are mutually incompatible. Thus, the opponent appears to have won on points, but the proponent's claim, C, remains plausible because at most only one of the ways it could be true has been refuted. Similarly, the proponent's case may contain an inconsistency that escapes the opponent's notice. For example, as Pearl (1986) points out, an argument from evidence A to conclusion B may be quite plausible (e.g., A almost always has B as an effect); and an argument from B to conclusion C may also be quite plausible (e.g., B is almost always due to C); but the proponent cannot concatenate these two arguments to yield an inference of C from A because, in this example, A and C are competing explanations of B.

Opponent and proponent in critical dialogue are free to challenge the overall coherence of the other's position. An advantage of mental models is that they encourage critical dialogue participants to look at the case as a whole and ask whether the other participant has succeeded in putting together a complete, plausible, and consistent story (cf. Pennington & Hastie, 1988, 1993b). The local character of most argument can distract the proponent and opponent from doing this, however, and the judge may have a better view of what each has accomplished.

The second advantage of the judge is an external point of view. The output of critical dialogue is a cognitive product, including new beliefs or intentions and the reasons for adopting them. One of the unique characteristics of a *cognitive* product is that there may be two complementary approaches to trusting it. *Internal trust* is the likelihood that an assessment is accurate or that a plan will achieve desired outcomes as seen from inside the situation model or plan itself, e.g., the proportion of envisioned scenarios that end in accomplishment of the specific mission. *External trust* is an estimate of the relative frequency of satisfactory outcomes associated with similar decision processes, pursued for a similar amount of time by similarly qualified participants under similar circumstances (Cohen, Parasuraman, Serfaty, & Andes, 1997).

The proponent and opponent are likely to base their estimates of trust only on the internal model space. In doing so, they neglect the past record. For example, inexperienced planners may be confident that all foreseeable contingencies have been accounted for. Experienced implementers, however, who have executed plans developed under similar circumstances, are sure that something unexpected will go wrong (even if they do not know precisely what) unless the plan is vetted far more thoroughly. Even experienced planners are likely to overestimate of the current chances of success because they discount other's experiences and emphasize the unique aspects of the present case (Buehler, Griffin, & Ross, 2002). John Leddo (unpublished experiment) found that people playing the role of implementer were less overconfident than planners. The judge in critical dialogue calibrates the outputs of the proponent and opponent with respect to reality, just as an experienced implementer grounds the ideas of a novice planner. In such circumstances, general strategies suggested by the judge, such as buying extra time and maintaining operational flexibility, may be associated with more reliable achievement of objectives.

Questions the judge might ask during Phase 4 are: Did the participants agree on the issue in question? What is their degree of confidence? To what degree is the outcome of the plan predictable based on their confidence and other indicators? What is the quality of the participants' final mental models? Are they complete, plausible, and consistent? Are any characteristics of the topic, time, place, participants, or process associated with errors in conclusions or confidence? Under these conditions, what is the best procedure for reconciling the opinions of the participants?

Example

What conclusion should be drawn from the dialogue shown in Figure 10.3? Standard critical-thinking and informal logic texts decide by applying internalist criteria to individual arguments (e.g., by assessing the *acceptability* of premises, their individual *relevance*, and their collective *sufficiency* for the conclusion). From this point of view, we might conclude that the proponent comes out ahead, since he has rebutted every one of the opponent's arguments, while the opponent has left some of the proponent's arguments unanswered (those in Phase (c)). Most of our trained groups, however, were not confident enough on this basis to risk a flat out attack against Bridgeton. The reason can be discovered by reading down the rightmost

proponent and opponent columns in Figure 10.5, including all propositions and causal relationships to which that participant has become committed (and not retracted) along the way.

For the opponent, three claims were conceded but not woven into his
scenario's explanatory web: *Scout car was surprised, Enemy mech force racing toward Bridgeton,* and *Enemy was not aware of us in time to hide.* These stand out
like sore thumbs in the alternative scenario and detract from its plausibility.
By the same token, a holistic look at the proponent's story raises interesting
new questions: Is it consistent to say that the enemy did not have time to
hide in Bridgeton but did have time to launch an initiative across the river?
Is it consistent to say that such an initiative across the river is a way of protecting Bridgeton, while at the same time explaining the mech force racing
down Highway 6 as a response to Bridgeton's vulnerability? Would it be safe
to attack Bridgeton with part of our force if there were an enemy force on
our side of the river? A mental model display like Figure 10.5 encourages
recognition of emergent relationships that influence plausibility but remain invisible in the serial examination of arguments (cf. Pennington &
Hastie, 1988).

Under these circumstances, a judge might fairly conclude, an attack on
Bridgeton must include careful planning for contingencies, and if possible,
steps should be taken to avoid decisive engagement if the town is still occu-
pied. After training, several groups decided to buy time by destroying the
mech force on Highway 6 from positions on our side of the river before ap-
proaching the town, and to hedge against the possibility of decisive engagement by having engineer units simultaneously continue the search for alternate crossing sites.

TRAINING CRITICAL DIALOGUE

Our training package, called *Critical Thinking through Dialogue,* begins with
facilitator skills: a discussion of what a critical dialogue is, the conditions under which it may be appropriate, and how it can improve accuracy and consensus. The instructor then explains the four phases of critical dialogue in
terms of characteristic tasks and rules (Table 10.1), provides examples of
how rules tend to be violated (Table 10.2), elaborates the role of the dialogue monitor in facilitating and refereeing (Table 10.3), and provides
guided dialogue practice and feedback in all three roles. An especially important point for military participants is helping them understand that dialogue roles are not associated with rank.

Training took 60 minutes in our initial test, preceded and followed by
tests with tactical decision game scenarios, as described previously. One of
two lead instructors conducted each session, with support from one of two

retired Army officers. Fifty-four active duty Army officers participated, in 20 groups of 2 to 4 officers each. Variation in group size was outside our control, but critical dialogue is designed to work with variable numbers of participants sharing roles in different ways. The following tentative conclusions are based on analysis of seven randomly chosen groups.

Findings

A dialogue is evaluated, from an externalist point of view, by measuring how well the constraints associated with the three monitor functions were satisfied: (1) *Was the correct process chosen?* (2) *Was it carried out properly by opponents and proponents?* (3) *Were conclusions satisfactory?* Dialogue training appeared to have a positive effect on all three of these categories. Significant differences between groups before and after training included the following:

(1) In accord with *facilitator* functions in Phases 1 and 2, groups were better at choosing relevant topics to discuss. They were significantly more likely after training to explicitly note areas of agreement and disagreement, setting aside areas of agreement and focusing discussion on disagreements.

(2) In accord with *referee* functions in Phase 3, proponents and opponents were significantly less likely to interrupt one another after training. Moreover, they were significantly more likely to ask for and give reasons rather than merely assert positions.

(3) Critical dialogue also affected the substance of group decisions, in accordance with *judge* functions in Phase 4. After training, groups generated and accepted significantly more new options than before training. More precisely, when we compared the maps and operations orders that participants produced as individuals with those produced by the group acting as a whole, we found that trained groups were more likely to adopt courses of action for major ground maneuver units that no individual group member had used when working individually. The reasons for this result need further investigation. First, note that there was no explicit requirement or encouragement in training to perform an "option generation" step. Thinking up more courses of action need not always be the best way to improve decisions; it may steal time that could be put to better use improving a single good option in response to specifically identified problems (Simon, 1997). However, new courses of action *should* be generated or recalled by trained groups when previous solutions are no longer salvageable. We do not know whether the novel courses of action adopted by trained groups were generated for the first time in the group or had been silently considered and rejected by individuals working alone. The most likely cause in either case was exposure of team members' individual proposals to serious challenge. In any case, it is clear that these groups did not

rely on either a bargaining strategy or a cherry-picking strategy, which would have merely traded off and combined pieces of their individual proposals.

CONCLUSIONS

Disagreement in a group or team sometimes leads to an exchange of reasons for and against a viewpoint, i.e., argumentation. A consequence of argumentation, when properly conducted, is the surfacing of assumptions about the domain that participants do and do not share, and the refinement and elaboration of knowledge. Even if the disagreement is not resolved, the exchange may increase the base of knowledge brought to bear on a decision, and improve the degree to which knowledge is shared, thus improving coordination in the future. If argumentation does resolve a disagreement, it may correct errors in the beliefs of one or more of the parties, augmenting both the accuracy and the consistency of shared knowledge and increasing the chances of mission success.

The rationale for framing individual critical thinking as a type of dialogue is the functional similarity between persuading another individual to accept or reject a position and determining for oneself whether the position is acceptable or not (Walton & Krabbe, 1995, p. 26). The primary goal of our research was to determine the practical value of critical dialogue training for improving *both* individual critical-thinking *and* cognitive collaboration in real-world contexts. Several substantive hypotheses are implicit in the critical dialogue theory:

(1) Effective critical thinking demands a vivid interplay between real or imagined opposing perspectives on the issue in question (cf. Vocate, 1994). Mental model representations capture different levels and types of uncertainty, permit holistic comparisons, stimulate new challenges and defenses, and show how the process and its results relate to the world.

(2) Theories of reasoning and decision making must draw on principles from both discourse and cognition. Different types of reasoning (e.g., deduction, induction, or assumption-based) and decision making (e.g., optimizing or satisficing) correspond to variations in implicit constraints on how mental models may be challenged and defended by participants playing different roles: for example, how burden of proof is allocated, the ability to introduce new information and to retract or revise claims, handling of inconsistency and counterexamples, the standard of proof, the depth to which underlying assumptions may be challenged, and the stopping rule. Mental models and dialogue rules together afford a general, unified, and naturalistically plausible account of critical thinking—which

contrasts sharply with the conventional picture of a heterogeneous set of discrete reasoning frameworks (e.g., predicate calculus, propositional calculus, and probability) that untutored individuals must somehow chose from.

(3) Critical thinking is regulated by the *reliability* and *cost-effectiveness* with which it produces successful beliefs and actions in the relevant types of situations (Gigerenzer, 2000; Hammond, 1996; Lipshitz & Cohen, 2005; Rescher, 1977b, 2001). There is no *logically* enforceable end to potential challenges and defenses and no *formal* criterion for the completeness or sufficiency of an argument. This is why critical thinking cannot be fully understood or evaluated as a product (i.e., a fixed set of premises that either do or do not entail a conclusion) but only pragmatically in terms of context-sensitive costs and benefits.

The functional analogy between critical thinking and specific types of dialogue may be more than coincidence. Some developmental psychologists (e.g., Rogoff, 1990; Tomasello, 1999; Vygotsky, 1986) propose that thought first develops as internalized speech and that we learn to reflect on and evaluate our own thoughts by responding to the questions and answers of others. Moreover, as noted by Rieke and Sillars (1997), dialogue *continues* to be the natural format of adult thinking: "Research suggests that critical thinking is really a mini-debate that you carry on with yourself. What is often mistaken for private thought is more likely an 'internalized conversation' (Mead [1967]), an 'internal dialogue' (Mukarovsky [1978]), or an 'imagined interaction' (Gotcher and Honeycutt [1989])."

There is an even simpler, practical reason for a dialogue-based theory of critical thinking. Thinking skills are not only shaped by social interaction but continue to be manifested in social contexts (Hutchins, 1995). Much critical thinking takes place in a team or group context, in which dialogue plays a direct role in decision making. Dialogues are the interactions by means of which members of a team pool information and insights to solve a problem, resolve competing goals, build up shared understanding of the situation and task, and over time construct relationships and commitments that improve team cohesiveness and trust as well as coordination (Amason & Schweiger, 1997; Cohen, 2004). Our fourth and final substantive hypothesis is that the fastest road to improving critical thinking in both an individual and a team is training for critical dialogue.

The cognitive and social skills required by reasoning together in conversation have drawn increasing interdisciplinary attention from experimental, developmental, and social psychologists (e.g., Anderson et al., 1996; Clark, 1996, 1992; Koslowski, 1996; Kuhn, 1991; Molder & Potter, 2005). The perspective on critical thinking as critical dialogue may be an opportunity for naturalistic decision-making researchers to add their own distinctive contributions to those efforts, by looking at how critical discussions are carried off in the real world.

ACKNOWLEDGMENTS

This research was sponsored under Contract No. DASW01-02-C-0002 to Cognitive Technologies, Inc., from the Army Research Institute, Fort Leavenworth Field Unit, Leavenworth, Kansas.

REFERENCES

Amason, A. C. (1996). Distinguishing the effects of functional and dysfunctional conflict on strategic decision making: Resolving a paradox for top management teams. *Academy of Management Journal, 39,* 123–148.

Amason, A. C., & Schweiger, D. M. (1997). The effects of conflict on strategic decision making effectiveness and organizational performance. In C. K. W. De Dreu & E. van de Vliert (Eds.), *Using conflict in organizations* (pp. 101–115). London: Sage.

Anderson, T., Howe, C., & Tolmie, A. (1996). Interaction and mental models of physics phenomena: Evidence from dialogues. In J. Oakhill & A. Garnham (Eds.), *Mental models in cognitive science* (pp. 247–274). East Sussex, England: Psychology Press..

Brandom, R. (2000). Knowledge and the social articulation of the space of reasons. In E. Sosa & J. Kim (Eds.), *Epistemology: An anthology* (pp. 424–432). Malden, MA: Blackwell.

Brown, P., & Levinson, S. C. (1987). *Politeness: Some universals in language usage.* Cambridge, England: Cambridge University Press.

Brunswick, E. (2001). Representative design and probabilistic theory in a functional psychology. In K. R. Hammond & T. R. Stewart (Eds.), *The Essential Brunswick* (pp. 135–156). Oxford, England: Oxford University Press.

Buehler, R., Griffin, D., & Ross, M. (2002). Inside the planning fallacy: The causes and consequences of optimistic time predictions. In T. Gilovich, D. Griffin, & D. Kahneman (Eds.), *Heuristics and biases: The psychology of intuitive judgment* (pp. 250–70). Cambridge, England: Cambridge University Press.

Byrne, R. M. J., Espino, O., & Santamari, C. (2000). Counterexample availability. In W. Schaeken, G. De Vooght, A. Vandierendonck, & G. d'Ydewalle (Eds.), *Deductive reasoning and Strategies* (pp. 97–110). Mahwah, NJ: Lawrence Erlbaum Associates.

Cannon-Bowers, J., Salas, E., & Converse, S. (1993). Shared mental models in expert team decision making. In N. Castellan (Ed.), *Individual and group decision making* (pp. 221–246). Hillsdale, NJ: Lawrence Erlbaum Associates.

Chase, W., & Simon, H. A. (1973). Perception in chess. *Cognitive psychology, 4,* 55–81.

Clark, H. (1992). *Arenas of language use.* Chicago, IL: University of Chicago Press.

Clark, H. (1996). *Using language.* Cambridge, England: Cambridge University Press.

Cohen, M. S. (1993a). Three paradigms for viewing decision biases. In G. Klein, J. Orasanu, R. Calderwood, & C. E. Zsambok (Eds.), *Decision making in action: Models and methods.* Norwood, NJ: Ablex

Cohen, M. S. (1993b). The naturalistic basis of decision biases. In G. Klein, J. Orasanu, R. Calderwood, & C. E. Zsambok (Eds.), *Decision making in action: Models and methods* (pp. 51–99). Norwood, NJ: Ablex.

Cohen, M. S. (2004). Leadership as the orchestration and improvisation of dialogue: Cognitive and communicative skills in conversations among leaders and subordinates. In D. V. Day, S. J. Zaccaro, & S. M. Halpin (Eds.), *Leader development for transforming organizations: Growing leaders for tomorrow*. Mahwah, NJ: Lawrence Erlbaum Associates.

Cohen, M. S., Freeman, J. T., & Wolf, S. (1996). Meta-recognition in time stressed decision making: Recognizing, critiquing, and correcting. *Human Factors, 38*(2), 206–219.

Cohen, M. S., Parasuraman, R., Serfaty, D., & Andes, R. (1997). *Trust in decision aids: A model and a training strategy*. Arlington, VA: Cognitive Technologies, Inc.

Cohen, M. S., Salas, E., & Riedel, S. (2002). *Critical thinking: Challenges, possibilities, and purpose*. Arlington, VA: Cognitive Technologies, Inc.

Cooke, N., Salas, E., Cannon-Bowers, J. A., & Stout, R. J. (2000). Measuring team knowledge. *Human Factors 42*(1), 151–173.

Day, T. J. (1989). Circularity, non-linear justification, and holistic coherentism. In J. W. Bender (Ed.), *The current state of the coherence theory*. Dordecht, Holland: Kluwer Academic Publishers.

DeKleer, J. (1986). An assumption-based truth maintenance system. *Artificial Intelligence, 28*, 127–162.

Doherty, M., Mynatt, C. R., Tweney, R., & Schiavo, M. (1979). Pseudodiagnosticity. *Acta Psychologica, 43*, 111–121.

Ehrlich, S. (1996). Applied mental models in human-computer interaction. In J. Oakhill & A. Garnham (Eds.), *Mental models in cognitive science* (pp. 223–245). East Sussex, England: Psychology Press.

Endsley, M. R. (2000). Theoretical underpinnings of situation awareness: A critical review. In M. R Endsley & D. J. Garland (Eds.), *Situation awareness analysis and measurement* (pp. 3–32). Mahwah, NJ: Lawrence Erlbaum Associates.

Endsley, M. R., & Garland, D. J. (Eds.). (2000). *Situation awareness analysis and measurement*. Mahwah, NJ: Lawrence Erlbaum Associates.

Evans, J. S. T. (1996). Afterword: The model theory of reasoning: Current standing and future prospects. In J. Oakhill & A. Garland (Eds.), *Mental models in cognitive science* (pp. 319–27). East Sussex, England: Psychology Press.

Evans, J. S. T., & Over, D. E. (1996). *Rationality and reasoning*. East Sussex, England: Psychology Press.

Fauconnier, G. (1994). *Mental spaces: Aspects of meaning construction in natural language*. Cambridge, England: Cambridge University Press.

Feldman, R., & Conee, E. (2000). Evidentialism. In E. Sosa & J. Kim (Eds.), *Epistemology: An anthology*. Oxford England: Blackwell.

Garnham, A. (1996). The other side of mental models: Theories of language comprehension. In J. Oakhill & A. Garnham (Eds.), *Mental models in cognitive science* (pp. 35–52). East Sussex, England: Psychology Press.

Garnham, A. (2001). *Mental models and the interpretation of anaphora*. East Sussex, England: Psychology Press.

Gentner, D., & Stevens, A. (1983). *Mental models*. Hillsdale, NJ: Lawrence Erlbaum Associates.

Gigerenzer, G. (2000). *Adaptive thinking: Rationality in the real world*. Oxford, England: Oxford University Press.

Gigerenzer, G., & Murray, D. J. (1987). *Cognition as intuitive statistics*. Hillsdale, NJ: Lawrence Erlbaum Associates.

Gigone, D., & Hastie, R. (1997). The impact of information on small group choice. *Journal of Personality and Social Psychology, 72*(1), 132–140.

Gladwell, M. (2005). *Blink: The power of thinking without thinking*. New York: Little, Brown.

Goldman, A. I. (1992). *Liaisons: Philosophy meets the cognitive and social sciences*. Cambridge MA: MIT Press.

Gotcher, J. M., & Honeycutt, J. M. (1989). An analysis of imagined interactions of forensic participants. *The National Forensic Journal, 7,* 1–20.

Govier, T. (1987). *Problems in argument analysis and evaluation.* Dordrecht, Netherlands: Foris Publications.

Govier, T. (1997). *A practical study of argument.* Belmont, CA: Wadsworth.

Grice, P. (1989). *Studies in the way of words.* Cambridge, MA: Harvard University Press.

Hamblin, C. H. (1970). *Fallacies.* Newport News, VA: Vale Press.

Hammond, K. R. (1996). *Human judgment and social policy: Irreducible uncertainty, inevitable error, unavoidable injustice.* Oxford, England: Oxford University Press.

Hammond, K. R., & Stewart, T. R. (Eds.). (2001). *The Essential Brunswick: Beginnings, explications, applications.* Oxford, England: Oxford University Press.

Hansen, H. V., & Pinto, R. C. (1995). *Fallacies: Classical and contemporary readings.* University Park: Pennsylvania State University.

Hinsz, V. B. (2004). Metacognition and mental models in groups: An illustration with metamemory of group recognition memory. In E. Salas & S. M. Fiore (Eds.), *Team cognition: Understanding the factors that drive process and performance* (pp. 33–58). Washington, DC: American Psychological Association.

Hintikka, J. (1999). *Inquiry as inquiry: A logic of scientific discovery.* Dordrecht, Netherlands: Kluwer Academic Publishers.

Hutchins, E. (1995). *Cognition in the wild.* Cambridge, MA: The MIT Press.

Jehn, K. A. (1997). Affective and cognitive conflict in work groups: Increasing performance through value-based intragroup conflict. In C. K. W. De Dreu & E. van de Vliert (Eds.) *Using conflict in organizations* (pp. 87–100). London: Sage.

Johnson, R. H. (1996). *The rise of informal logic: Essays on argumentation, critical thinking, reasoning and politics.* Newport News, VA: Vale Press.

Johnson, R. H., & Blair, J. A. (1994). *Logical self-defense.* New York: McGraw-Hill.

Johnson-Laird, P. N. (1983). *Mental models.* Cambridge, MA: Harvard University Press.

Johnson-Laird, P. N. (2003). Illusions of understanding. In A. J. Sanford (Ed.), *The nature and limits of human understanding: The 2001 Gifford Lectures at the University of Glasgow* (pp. 3–25). London: T & T Clark.

Johnson-Laird, P. N., & Byrne, R. M. (1991). *Deduction.* Hillsdale, NJ: Lawrence Erlbaum Associates.

Johnson-Laird, P. N., Girotto, V., & Legrenzi, P. (2004). Reasoning from inconsistency to consistency. *Psychological Review, 111,* 640–661.

Johnson-Laird, P. N., Legrenzi, P., Girotto, V., Legrenzi, M. S., & Caverni, J.-P. (1999). Naive probability: A mental model theory of extensional reasoning. *Psychological Review, 106,* 62–88.

Johnson-Laird, P., & Shafir, E. (1993). *Reasoning and decision making.* Amsterdam: Elsevier.

Kahneman, D., Slovic, P., & Tversky, A. (1982). *Judgment under uncertainty: Heuristics and biases.* New York: Cambridge University Press.

Kahneman, D., & Tversky, A. (1982). Intuitive prediction: Biases and corrective procedures. In D. Kahneman, P. Slovic, & A. Tversky (Eds.), *Judgment under uncertainty: Heuristics and biases* (pp. 414–421). Cambridge, England: Cambridge University Press.

Katzenstein, G. (1996). The debate on structured debate: Toward a unified theory. *Organizational Behavior and Human Decision Processes, 66*(3), 316–332.

Kessler, K., Duwe, I., & Strohner, H. (1999). Grounding mental models—Subconceptual dynamics in the resolution of linguistic reference in discourse. In G. Rickheit & C. Habel (Eds.), *Mental models in discourse processing and reasoning* (pp. 169–194). Amsterdam: Elsevier.

Klein, G. (1993). A Recognition-Primed Decision (RPD) model of rapid decision making. In G. A. Klein, J. R. Orasanu, R. Calderwood, & C. E. Zsambok (Eds.), *Decision making in action: Models and methods* (pp. 138–147). Norwood, NJ: Ablex.

Klein, G. (1998). *Sources of power.* Cambridge, MA: MIT Press.

Klein, G., McCloskey, M., Thordsen, M., Klinger, D., & Schmitt, J. F. (1998). Bridgeton Crossing: Tactical decision game #98-4. *Marine Corps Gazette*(4), 86.

Klein, P. (2000). A proposed definition of propositional knowledge. In E. Sosa & J. Kim (Eds.), *Epistemology: An anthology* (pp. 60–66). Oxford, England: Blackwell.

Kornblith, H. (1994). *Naturalizing epistemology.* Cambridge, MA: MIT Press.

Koslowski, B. (1996). *Theory and evidence: The development of scientific reasoning.* Cambridge: MIT Press.

Kuhn, D. (1991). *The skills of reasoning.* Cambridge, England: Cambridge University Press.

Lakoff, G. (1987). *Women, fire, and dangerous things: What categories reveal about the human mind.* Chicago, IL: University of Chicago Press.

Langacker, R. W. (2000). *Grammar and conceptualization.* The Hague: Mouton de Gruyter.

Legrenzi, P., & Girotto, V. (1996). Mental models in reasoning and decision-making processes. In J. Oakhill (Ed.), *Mental models in cognitive science* (pp. 95–118). East Sussex, England: Psychology Press.

Lipshitz, R., & Cohen, M. S. (2005). Warrants for prescription: Analytically and empirically based approaches to improving decision making. *Human Factors, 47*(1), 121–130.

McDermott, D., & Doyle, J. (1980). Non-monotonic Logic. *Artificial Intelligence, 13,* 41–72.

Mead, G. H. (1967). *Mind, self, and society: From the standpoint of a social behaviorist.* Chicago, IL: University of Chicago Press.

Metcalfe, J., & Shimamura, A. P. (1994). *Metacognition.* Cambridge, MA: MIT Press.

Missimer, C. (1994). Why two heads are better than one: Philosophical and pedagogical implications of a social view of critical thinking. In K. S. Walters (Ed.), *Re-thinking reason: New perspectives in critical thinking.* Albany: State University of New York Press.

Molder, H. T., & Potter, J. (Eds.). (2005). *Conversation and cognition.* Cambridge, England: Cambridge University Press.

Mukarovsky, J. (1978). *Structure, sign and function: Selected essays.* New Haven, CT: Yale University Press.

Nelson, T. (1992). *Metacognition: Core readings.* Boston: Allyn & Bacon.

Nozick, R. (1981). *Philosophical explanations.* Cambridge MA: Harvard University Press.

Oakhill, J., & Garnham, A. (Eds.). (1996). *Mental models in cognitive science.* East Sussex, England: Psychology Press.

Oakhill, J. (1996). Mental models in children's text comprehension. In J. Oakhill & A. Garnham (Eds.), *Mental models in cognitive science* (pp. 77–94). East Sussex, England: Psychology Press.

Orasanu, J. (1993). Decision making in the cockpit. In E. L. Wiener, R. H., & B. G. Kanki (Eds.), *Cockpit resource management* (pp. 137–172). New York: Academic Press.

Orasanu, J., & Connolly, T. (1993). The reinvention of decision making. In G. Klein, J. Orasanu, R. Calderwood, & C. Zsambok (Eds.), *Decision making in action: Models and methods* (pp. 3–20). Norwood, NJ: Ablex.

Papineau, D. (2003). *The roots of reason: Philosophical essays on rationality, evolution, and probability.* Oxford: Clarendon Press.

Patel, V. L., Arocha, J. F., & Kaufman, D. R. (1999). Expertise and tacit knowledge in medicine. In R. H. Sternberg & J. A. Horvath (Eds.), *Tacit knowledge in professional practice.* Mahwah NJ: Lawrence Erlbaum Associates.

Paul, R. W. (1987). Critical thinking in the strong sense and the role of argumentation in everyday life. In F. H. Eemeren, R. Grootendorst, J. A. Blair, & C. A. Willard (Eds.), *Argumentation: Across the lines of discipline.* Dordrecht, Netherlands: Foris Publications.

Pearl, J. (1989). *Probabilistic reasoning in intelligent systems: Networks of plausible inference.* San Mateo, CA: Morgan Kaufmann.

Pennington, N., & Hastie, R. (1988). Explanation-based decision making: Effects of memory structure on judgment. *Journal of Experimental Psychology: Learning, Memory, & Cognition* *14*(3), 521–533.

Pennington, N., & Hastie, R. (1993a). A theory of explanation-based decision making. In G. Klein, J. Orasanu, R. Calderwood, & C. Zsambok (Eds.), *Decision making in action: Models and methods* (pp. 188–201). Norwood, NJ: Ablex.

Pennington, N., & Hastie, R. (1993b). Reasoning in explanation-based decision making. In P. Johnson-Laird & E. Shafir (Eds.), *Reasoning and decision making* (pp. 123–165). Amsterdam: Elsevier.

Plantinga, A. (1993a). *Warrant: The current debate.* New York: Oxford University Press.

Plantinga, A. (1993b). *Warrant and proper function.* New York: Oxford University Press.

Rescher, N. (1977a). *Dialectics: A controversy-oriented approach to the theory of knowledge.* Albany: State University of New York Press.

Rescher, N. (1977b). *Methodological pragmatism: A systems-theoretic approach to the theory of knowledge.* New York: New York University Press.

Rescher, N. (2001). *Cognitive pragmatism: The theory of knowledge in pragmatic perspective.* Pittsburgh, PA: University of Pittsburgh Press.

Rickheit, G., & Habel, C. (Eds.). (1999). *Mental models in discourse processing and reasoning.* Amsterdam: Elsevier.

Rickheit, G., & Sichelschmidt, L. (1999). Mental models: Some answers, some questions, some suggestions. In G. Rickheit & C. Habel (Eds.) *Mental models in discourse processing and reasoning* (pp. 9–41). Amsterdam: Elsevier.

Rieke, R. D., & Sillars, M. O. (1997). *Argumentation and critical decision making.* New York: Addison-Wesley Educational Publishers.

Rogoff, B. (1990). *Apprenticeship in thinking: Cognitive development in social context.* Oxford, England: Oxford University Press.

Schmitt, J. F. C. (1994). *Marine Corps Gazette's mastering tactics: A tactical decision games workbook.* Quantico, VA: Marine Corps Association.

Schnotz, W., & Preuss, A. (1999). Task-dependent construction of mental models as a basis for conceptual change. In G. Rickheit & C. Habel (Eds.), *Mental models in discourse processing and reasoning* (pp. 131–68). Amsterdam: Elsevier.

Schwarz, N. (1996). *Cognition and communication: Judgmental biases, research methods, and the logic of conversation.* Mahwah, NJ: Lawrence Erlbaum Associates.

Schweiger, D., Sandburg, W., & Ragan, J. W. (1986). Group approaches for improving strategic decision making: a comparative analysis of dialectical inquiry, devil's advocacy and consensus. *Academy of Management Journal, 29,* 51–71.

Schwenk, C. (1989). A meta-analysis on the cooperative effectiveness of devil's advocacy and dialectical inquiry. *Strategic Management Journal, 10,* 303–306.

Schwenk, C., & Cosier, R. (1980). Effects of the expert, devil's advocate, and dialectical inquiry methods on prediction performance. *Organizational Behavior and Human Performance, 26,* 409–424.

Serfaty, D., Entin, E. E., & Johnston, J. (1998). Team adaptation and coordination training. In J. A. Cannon-Bowers & E. Salas (Eds.), *Decision making under stress: Implications for training and simulation* (pp. 221–245). Washington, DC: American Psychological Association.

Shafer, G. (2001). Nature's possibilities and expectations. In V. F. Hendriks, S. A. Pedersen, & K. F. Jorgensen (Eds.), *Probability theory: Philosophy, recent history and relations to science* (pp. 147-166). Dordrecht, Netherlands: Kluwer.

Shafer, G., & Vovk, V. (2001). *Probability and finance.* New York: Wiley.

Shanteau, J. (1992). The psychology of experts: An alternative view. In G. Wright & F. Bolger (Eds.), *Expertise and decision support* (pp. 11–23). New York: Plenum.

Shanteau, J. (1998). *Why do experts disagree with each other?* Paper presented at the fourth conference on Naturalistic Decision Making, Washington, DC.

Siegel, H. (1997). *Rationality redeemed: Further dialogues on an educational ideal.* New York: Routledge.

Simon, H. A. (1997). *Models of bounded rationality: Empirically grounded economic reason.* Cambridge MA: MIT Press.

Sosa, E. (1991). *Knowledge in perspective: Selected essays in epistemology.* New York: Cambridge University Press.

Sosa, E., & Kim, J. (2000). *Epistemology: An anthology.* Malden MA: Blackwell.

Sperber, D., & Wilson, D. A. (1995). *Relevance: Communication and cognition.* Oxford: Blackwell.

Stasser, G. (1999). The uncertain role of unshared information in collective choice. In L. L. Thompson, J. M. Levine, & D. M. Messick (Eds.), *Shared cognition in organizations: The management of knowledge* (pp. 49–69). Mahwah, NJ: Lawrence Erlbaum Associates.

Tabossi, P., Bell, V. A., & Johnson-Laird, P. N. (1999). Mental models in deductive, modal, and probabilistic reasoning. In G. Rickheit & C. Habel (Eds.), *Mental models in discourse processing and reasoning* (pp. 299–332). Amsterdam: Elsevier.

Tasa, K., & Whyte, G. (2005). Collective efficacy and vigilant problem solving in group decision making: A non-linear model. *Organizational behavior and human decision processes,* 96(2), 119–129.

Tomasello, M. (1999). *The cultural origins of human cognition.* Cambridge, MA: Harvard University Press.

Van Eemeren, F. H., & Grootendorst, R. (1983). *Speech acts in argumentative discussions.* Dordrecht, Holland: Foris.

Van Eemeren, F. H., & Grootendorst, R. (1992). *Argumentation, communication, and fallacies: A pragma-dialectical perspective.* Mahwah, NJ: Lawrence Erlbaum Associates.

Van Eemeren, F. H., Grootendorst, R., Blair, J. A., & Willard, C. A. (Eds.). (1987). *Argumentation: Perspectives and approaches.* Dordecht, Netherlands: Foris.

Van Eemeren, F. H., Grootendorst, R., Jackson, S., & Jacobs, S. (1993). *Reconstructing argumentative discourse.* Tuscaloosa: University of Alabama Press.

Vocate, D. R. (1994). *Intrapersonal communication: Different voices, different minds.* Hillsdale, NJ: Lawrence Erlbaum Associates.

Vygotsky, L. (1986). *Thought and language.* Cambridge, MA: MIT Press.

Walton, D. N. (1995). *A pragmatic theory of fallacy.* Tuscaloosa AL: University of Alabama Press.

Walton, D. N. (1996a). *Argumentation schemes for presumptive reasoning.* Mahwah, NJ: Lawrence Erlbaum Associates.

Walton, D. N. (1996b). *Argument structure: A pragmatic theory.* Toronto: University of Toronto Press.

Walton, D. N. (1998). *The new dialectic: Conversational contexts of argument.* Toronto: University of Toronto Press.

Walton, D. N., & Krabbe, E. C. W. (1995). *Commitment in dialogue: Basic concepts of interpersonal reasoning.* Albany: State University of New York Press.

Wason, P., & Johnson-Laird, P. N. (1972). *Psychology of reasoning: Structure and content.* Cambridge, MA: Harvard University Press.

Wineburg, S. (1998). Reading Abraham Lincoln: An expert/expert study in the interpretation of historical texts. *Cognitive Science,* 22(3), 319–346.

Woods, J., & Walton, D. N. (1989). *Fallacies: Selected papers 1972–1982.* Dordecht, Holland: Foris.

Part IV

COPING WITH UNCERTAINTY
IN A CHANGING WORKPLACE

Chapter 11

Dealing With Probabilities: On Improving Inferences With Bayesian Boxes

Kevin Burns
The MITRE Corporation, Bedford, MA

Uncertainty is everywhere in real-world decision making. In this chapter, I review the results of laboratory experiments on how people deal with uncertainty and I present a support system called *Bayesian Boxes* that can help to improve inferences under uncertainty.

This work is motivated by a case study of human error in situation awareness (Burns, 2005a), which showed that a decision maker's actions could be explained by a bounded-Bayesian model. In other words, human errors are not necessarily the result of non-Bayesian biases, but rather can be attributed to Bayesian reasoning in the context of limited data and limited resources for getting more data. This insight is consistent with a growing body of research that uses Bayesian modeling to explain cognition (Knill & Richards, 1996), and yet it is also contrary to another body of research that suggests people are biased (i.e., non-Bayesian) in probabilistic reasoning (Kahneman, Slovic, & Tversky, 1982). This raises the basic question of exactly when people are Bayesian or not, and why. It also raises the applied question of how inferences can be improved or de-biased. This chapter addresses both of these questions.

To set the stage, it is useful to distinguish between *ecological reasoning* under uncertainty and *mathematical reasoning* under uncertainty. The former is concerned with reasoning in which uncertainty is dealt with qualitatively or implicitly, whereas the latter is concerned with reasoning in which uncertainty is expressed quantitatively as probability. In general, previous research (cited earlier) suggests that people are fairly good at ecological reasoning under uncertainty but not so good at mathematical reasoning under uncertainty—which is not surprising because the natural world rarely expresses itself explicitly as probabilities. Thus, consistent with the title of this volume, mathematical reasoning under uncertainty can be seen as a problem of *expertise out of context.*

In the field of Naturalistic Decision Making (NDM), one might ask why we should care about human performance in mathematical reasoning. That is, if people are good at dealing with uncertainty in ecological situations, then why should NDM care about bounds or biases in reasoning about probabilistic information? The reason is that probabilistic reasoning is inherent to job tasks in many domains of real-world expertise. One example would be probability-of-precipitation estimates by weather forecasters (see Hoffman, Trafton, & Roebber, 2005). Beyond this single example, in our modern world people are increasingly being challenged to deal with probabilities in military, medical, and other high-stakes domains of concern to NDM. In fact, it is precisely because of the high stakes involved that uncertainties are often expressed as numerical probabilities in these domains, so that formal methods might be employed by people and systems to manage these uncertainties. The problem then is that people must deal with numerical probabilities that systems provide in the form of statistical reports (e.g., in medical diagnosis), sensor reliabilities (e.g., in military intelligence), and so forth. The bigger problem is that humans are often relied on to understand and compensate for the limits of these systems (statistics, sensors, etc.), which ironically exacerbates the quandary of *expertise out of context.*

Also in the field of NDM, one might ask why we should expect to gain any useful insights into naturalistic performance via laboratory experiments like those employed here. The reason is that some of these laboratory experiments present participants with problems of probabilistic reasoning in the dynamic context of game tasks where feedback and noise simulate the conditions of real-world decision making. These sorts of experiments are clearly not NDM, but they pose cognitive challenges that are more like NDM than the hypothetical questions often used in research on judgment and decision making (JDM). Thus experiments with game tasks can help to extend and apply the work of JDM to the field of NDM.

Here the basic research challenge is to better understand how people deal with probabilities in uncertain and dynamic situations. Then armed with this better understanding, the applied research challenge is to engineer a support system that can leverage human powers of ecological reasoning, and thereby help to solve a problem of *expertise out of context.*

In the following, I outline the basic problem in a section titled "What Is The Correct Answer?" This section provides a short tutorial on the mathematics of Bayesian inference. I then highlight the research challenge in a section titled "Why Do People Get It Wrong?" This section reviews the results of laboratory experiments that shed light on when people are Bayesian or not, and why. Finally, I address the applied challenge in a section titled "What Can Be Done About It?" This section presents the design of a support system called *Bayesian Boxes,* which has been shown to improve human performance in Bayesian inference.

WHAT IS THE CORRECT ANSWER?

Imagine that you are a military commander with limited information about a possible target, called a track. Your job is to identify the track as Friend or Foe, so you can choose a course of action. You have information from two independent sources, for example, an intelligence source and a surveillance source. The first source reports "Foe." This source has a success rate of 0.80 in correctly reporting Friend or Foe, so based on this source the chances are 80% that the track is a Foe. The second source reports "Foe." This source has a success rate of 0.67 in correctly reporting Friend or Foe, so based on this source the chances are 67% that the track is a Foe. With the information presented, what do you think are the chances that the track is a Foe?

When presented with this question, many people say the chances are somewhere between 67% and 80%. Other people use only the most (or least) reliable source and say the chances are 80% (or 67%). Still other people multiply the two probabilities and say the chances are around 50%. The correct answer is developed next in a short tutorial on the mathematics of Bayesian inference. Then, in the following section, insights into cognitive biases are developed by reviewing a series of experiments involving static questions like the aforementioned as well as dynamic challenges in a game task.

Bayes Rule

In the previous example, two probabilities ($P_1 = 80\%$ and $P_2 = 67\%$) are given about one track. The problem is to combine these two probabilities P_1 and P_2 to obtain a third probability P_3 about the track. The mathematical procedure for combining P_1 with P_2 to compute P_3 is known as Bayes Rule, which in simplified form can be expressed as follows (a more comprehensive form is derived later):

$$P_3 = (P_1 * P_2) / \{(P_1 * P_2) + [(1 - P_1) * (1 - P_2)]\} \tag{1}$$

Substituting the values $P_1 = 80\%$ and $P_2 = 67\%$ into this equation yields $P_3 = 89\%$. So, the correct answer is that there is an 89% chance that the track is a

Foe. Notice that P_3 is higher (not lower) than P_1 and P_2. Now, to gain more insight into Bayes Rule, consider the following question:

> Imagine that you hold a poker hand of 4 Kings and 1 Queen. You shuffle these five cards, then close your eyes and select one of the cards at random. My eyes are open so I can see the card you selected but your eyes remain closed. After you select a card, I roll a fair die and say either "King" or "Queen" according to the following rule: If the die shows a number 1 through 4, I tell the truth about your selected card; if the die shows the number 5 or 6, I tell a lie about your selected card. After rolling the die, I look at the number on the die and say, "King." Based on the number of Kings and Queens, the chances are $\frac{4}{5} = 80\%$ that your selected card is a King. Based on my report from the die, the chances are $\frac{4}{6} = 67\%$ that your selected card is a King. What do you think are the chances that your selected card is a King?

This problem is mathematically (but not psychologically) equivalent to the Friend/Foe problem because you are given two independent probabilities (80% and 67%) about one thing (the selected card). Again the answer is 89% and again most people get it wrong. Following, I analyze the King/Queen problem because it serves to illustrate an important feature of Bayes Rule, namely that the probabilities are *conditional*. Later I return to the Friend/Foe problem to illustrate another feature of Bayes Rule, namely that the procedure is *sequential*.

Conditional Probabilities

The poker hand contains 4 Kings and 1 Queen. So, using K to denote King, you start with the *prior* knowledge that $P(K) = 80\%$. The problem statement then provides additional information about the *likelihood* that your card is a King, $P(k|K) = 67\%$, where k means "I report King" and K means "the card is a King." $P(k|K)$ is a conditional probability (of k given K) because my report is conditional (dependent) on the card I see.

At this point you know $P(K) = 80\%$ and $P(k|K) = 67\%$ but what you want to know is $P(K|k)$. That is, you want to know the conditional probability that the selected card is a King (K) given that I report it is a King (k). You can get what you want from what you know via a fundamental property of probabilities, namely:

$$P(K,k) = P(k) * P(K|k) = P(K) * P(k|K) \tag{2}$$

This equation simply says that there are two equivalent ways to compute the joint probability $P(K,k)$ of having selected a King (K) and having received a report (k) that the selected card is a King. The desired probability $P(K|k)$ can now be obtained by rearranging this equation as follows:

$$P(K|k) = P(K) * P(k|K) / P(k) \tag{3}$$

where $P(k)$ is defined by another fundamental property of probabilities, namely:

$$P(k) = P(K) * P(k|K) + P(\sim K) * P(k|\sim K) \tag{4}$$

Here, the notation $\sim K$ means not-K (i.e., Queen). Equation 4 simply says that the probability of k is computed as the sum of probabilities for all the ways that k can occur, that is, the probability of k and K [first term, because $P(k,K) = P(K) * P(k|K)$] plus the probability of k and $\sim K$ [second term, because $P(k,\sim K) = P(\sim K) * P(k|\sim K)$]. Substituting Equation 4 into Equation 3 yields the following formula for Bayes Rule:

$$P(K|k) = P(K) * P(k|K) / [P(K) * P(k|K) + P(\sim K) * P(k|\sim K)] \tag{5}$$

In the King/Queen problem (and the Friend/Foe problem) there are two factors that simplify this equation. First, the card must be a King or not-King, so $P(\sim K) = 1 - P(K)$. Second, the report is either right (truth) or wrong (lie), so $P(k|\sim K) = P(\sim k|K) = 1 - P(k|K)$. Thus, Equation 5 can be re-written as follows:

$$P(K|k) = P(K) * P(k|K) / \{P(K) * P(k|K) \\ + [1 - P(K)] * [1 - P(k|K)]\} \tag{6}$$

This is Bayes Rule, expressed in a form that explicitly shows the conditional probabilities for the King/Queen problem. The simpler form shown in Equation 1 comes from substituting $P_1 = P(K)$, $P_2 = P(k|K)$, and $P_3 = P(K|k)$.

At first glance Equation 6 may appear to be a rather complex and nonintuitive formula. However, as seen in Equation 1, there are really just two terms, one given by $P_1 * P_2$ and the other given by $(1 - P_1) * (1 - P_2)$. Furthermore, each of these terms as well as Bayes Rule itself can be expressed in a graphic format that is actually quite intuitive (which is illustrated later in this chapter). For now, I focus on the algebraic equation and its practical implications.

The mathematical advantage of Bayes Rule is that it allows one to compute $P(K|k)$ if one knows $P(K)$ and $P(k|K)$, that is, if one has some prior knowledge $P(K)$ and if one knows the conditional likelihood $P(k|K)$. Thus, in Bayesian terminology, $P(K)$ is called the *prior* probability, $P(k|K)$ is called the *likelihood* function, and the result $P(K|k)$ is called the *posterior* probability.

The practical advantage of Bayes Rule is that one often knows $P(k|K)$ because the world works from *cause* (K) to *effect* (k), yet after the fact, one needs $P(K|k)$ to infer the likely cause (K or ~K) of an observed effect (k). Bayes Rule provides the mathematical machinery for making this inference.

In fact, the process of Bayesian inference can be seen as the essence of any diagnosis to achieve *situation awareness.* In less formal terms, situation awareness has been defined (Endsley, 1988) as the "perception of the elements in the environment . . . and projection of their status in the near future" (p. 97). In more formal terms (Burns, 2005a) I showed that posterior probabilities computed from Bayes Rule can be used to explain a person's perception of a current situation. That is, given a frame of discernment (i.e., set of hypotheses), the Bayesian posterior can be taken as a model of the percept (also see Knill & Richards, 1996). I also showed how prior-weighted likelihoods can be used to make causal projections of future situations and thereby model a person's expectations. As such, a Bayesian framework is useful for understanding how people achieve *situation awareness,* even if the decision makers who achieve this situation awareness do not realize that they are in fact performing Bayesian inferences (or some approximation thereto).

Sequential Procedure

The Friend/Foe problem is slightly different from the King/Queen problem because both P_1 and P_2 are conditional probabilities (likelihoods). That is, using F to denote "Foe" and using f_n to denote "source n reports Foe" (where $n = 1$ or 2), the two likelihoods $P_1 = P(f_1|F)$ and $P_2 = P(f_2|F)$ are given by the success rate of source 1 and source 2 in correctly reporting Foe. The Friend/Foe problem is also more complex than the King/Queen problem because Bayes Rule must be applied twice. Starting with a noninformative prior $P_0 = P(F) = 50\%$ and using the likelihood $P_1 = P(f_1|F) = 80\%$, the first application of Bayes Rule yields the following result:

$$P(F|f_1) = P(F) * P(f_1|F) / \{P(F) * P(f_1|F) + [1 - P(F)] * [1 - P(f_1|F)]\} = 80\% \qquad (7)$$

Notice that the posterior $P(F|f_1) = 80\%$ has the same value as the likelihood $P(f_1|F) = 80\%$, because the prior $P(F) = 50\%$ is noninformative. The posterior $P(F|f_1) = 80\%$ then becomes the prior for a second application of Bayes Rule. Because the second source is independent of the first source, as stated in the Friend/Foe problem, the likelihood can be written as $P(f_2|F,f_1) = P(f_2|F) = 67\%$. Then, using $P(F|f_1) = 80\%$ as the prior and $P(f_2|F,f_1) = 67\%$ as the likelihood, the second application of Bayes Rule yields the new posterior $P(F|f_2,f_1)$ = 89%. (The same result would be obtained if the likelihoods were used in reverse order, that is, first 67% then 80%.) Additional data $P(f_3|F,f_2,f_1)$ from another source can then be used in the same way, to update $P(F|f_2,f_1)$ and compute $P(F|f_3,f_2,f_1)$, and so on.

WHY DO PEOPLE GET IT WRONG?

Preliminary experiments on the Friend/Foe problem showed that few people reported a posterior that was even in the ballpark of the Bayesian answer (89%). Instead, most people reported a posterior that was *lower, not higher,* than 80%. Many people even reported values lower than 67%. This led me to perform a series of experiments, starting with the King/Queen problem, to gain further insight into cognitive strategies for combining probabilities. In the following, I begin with some historical background and then I review my experimental design and results.

In a pioneering experiment on Bayesian inference, Edwards and Phillips (1964; also see Edwards & von Winterfeldt, 2000) used a simulated radar mission to test human performance in Bayesian inference. Their finding was that people are generally conservative in estimating posterior probabilities (Edwards, 1982); that is, humans fail to extract as much certainty as they can from the data that they get. However, the radar mission was so complex that only a few people could be tested and the findings were difficult to generalize. This led to further experiments on simpler problems, such as a bag and poker chip task, which involved sampling poker chips of two different colors (e.g., red and blue) from a bag that concealed the colors of the chips (Phillips & Edwards, 1966). This experiment tested participants on multistage (sequential) Bayesian updates in repeated sampling of chips from the bag with replacement of chips to the bag. Subsequent experiments by other researchers have focused on even simpler problems in verbal formats, such as the two-color cab problem (Tversky & Kahneman, 1982). In this problem, participants are given statements about the base rate (prior) for each cab color and the likelihood that an accident involved a cab of a given color. They must then perform a one-shot Bayesian update in order to estimate the posterior chances that a given color cab was involved in the accident.

The poker chip task illustrates three limitations of most experiments on Bayesian inference (also see McKenzie, 1994, for a review). First, the task tests only one likelihood, which is given by each bag's mix of red and blue chips, because the chips are sampled with replacement (i.e., the mix is fixed). Second, although the likelihood is presumably paired with a different prior for each sequential update in repeated sampling, it is not clear that people treat their internally generated prior (which is the posterior

from the previous update) the same as they would treat an externally specified prior. Third, the domain (bag) is sampled and judgments are made
without feedback. This does not reflect the context of naturalistic environments where decisions have consequences that give decision makers some
measure of their performance. Similar limitations arise in research on the
cab problem.

My experiments include hypothetical questions, but also a game task
that differs in two important ways from the cab problem and the poker chip
task. First, each participant was tested on a wide range of prior/likelihood
pairs. Second, priors and likelihoods were explicitly provided to participants (not inferred by participants) and experienced directly by participants in a temporal environment—via a deck of cards (which is depleted
with time) and via reports from "spies" (whose actual reliabilities match the
reported reliabilities). This game task, called Spy TRACS, was somewhat
similar to (but simpler than) the radar task of Edwards and Phillips (1964)
in that it simulated (to a limited extent) the uncertain and temporal conditions of NDM (Zsambok & Klein, 1997).

I performed a series of three experiments referred to as *Card Quiz* (Experiment 1), *Bird Quiz* (Experiment 2), and *Spy TRACS* (Experiment 3).
The experiments were administered via a written playbook that contained
the questions for Card Quiz/Bird Quiz and the instructions for Spy TRACS,
which was played on a personal computer with a mouse interface. Participants were 45 volunteers, mostly professional scientists or engineers with
undergraduate or graduate degrees. Each participant took all three experiments in the same order (Experiment 1, 2, 3), in about 30 to 45 minutes.
They then took a short break, which was followed by an additional experiment (discussed later in this chapter as Experiment 4) along with a questionnaire and interviews.

Experiment 1: Card Quiz

The Card Quiz comprised four questions referred to as Q1, Q2, Q3, and
Q4. Q1 was simply a screening question to establish that a person could
mentally convert the fraction $\frac{4}{5}$ to the probability 80%. Q2, Q3, and Q4
were variations of the King/Queen problem with different prior/likeli-
hood values, as shown in Table 11.1. Q2 was the same as the King/Queen
problem presented earlier; that is, prior = 80% and likelihood = 67%.

The participants' responses were classified as one of four Types: (A) Anchoring, (B) Bayesian, (C) Compromising, or (O) Other. Table 11.2 and
Figure 11.1 define and display the Types for Q2, and the Types are similar
for Q3 and Q4. Note that any response in the ballpark of the Bayesian response is classified as Type B. For example, even though the Bayesian posterior for Q2 is 89%, any response > 80% is classified as Bayesian. Note also

TABLE 11.1
Prior and Likelihood Values in Card Quiz

Question	Prior	Likelihood	Bayesian Posterior
Q2	80%	67%	89%
Q3	80%	33%	66%
Q4	80%	83%	95%

TABLE 11.2
Types of Possible Responses for Q2

Type	Description
A	Anchoring: Posterior equal to 67% or 80%
B	Bayesian: Posterior greater than 80%
C	Compromising: Posterior between 67% and 80%
O	Other: Posterior less than 67%

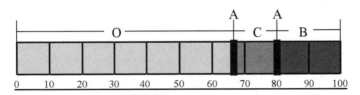

FIG. 11.1.   Ranges of values for Q2 response types.

that Type A includes Anchoring at either the prior or the likelihood. Finally, note that the Bayesian posterior for Q3 lies between the prior and likelihood. Thus, for Q3 there is a Type B/C that does not distinguish between Type B and Type C.

Figure 11.2 shows the experimental results for Q2, Q3, and Q4. Although all four Types of response are found, Type A is the dominant mode for each question. For Q2 and Q4, about 50% of participants are Type A, about 30% are Type O, about 10% are Type B, and less than 10% are Type C. These results are significantly different from chance, $c^2(3,45) = 44$, $p < 0.001$. The results are similar for Q3 (but Type B/C does not distinguish between Type B and Type C).

Further breakdown of these results showed that most (80%–90%) of the Type A participants were anchored to the prior rather the likelihood. This is interesting because it is contrary to the well-known *base rate neglect* effect found in the cab problem (Tversky & Kahneman, 1982) where people ignore (or discount) the prior in favor of the likelihood. This base rate neglect presumably arises because the likelihood has a more causal basis (see the earlier

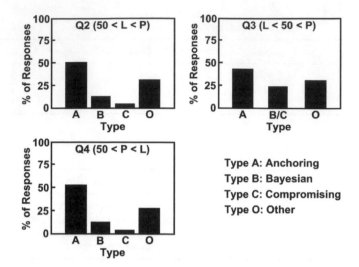

FIG. 11.2.   Results of Card Quiz (Experiment 1). P refers to the prior and L refers to the likelihood.

discussion of cause-and-effect under Bayes Rule) than the prior; that is, the base rate (prior) simply reflects a population density (see Birnbaum & Mellers, 1983). The questions in the Card Quiz were designed to make the prior and likelihood equally causal, because both are governed by random processes, and the finding is that people favor (not discount) the prior because they judge it to be more relevant than the likelihood for various other reasons (also see Koehler, 1996). For example, in postquiz questioning, some participants said that they favored the prior because it was a "physical" probability whereas the likelihood was only a "verbal" probability.

Experiment 2: Bird Quiz

The Bird Quiz had only one question, which was as follows:

> Imagine that you are eating dinner with a friend. Out the window is a bird feeder that attracts only Red birds and Blue birds. One bird is at the feeder and the bird looks Red to you. You are pretty sure the bird is Red, but you are not certain because it is getting dark outside. You ask your friend, "Is that a Red bird or a Blue bird?" He replies, "It looks Red to me." After hearing this from your friend, are you less sure or more sure (or unchanged) that the bird is Red?

Notice that this question is qualitatively the same as Q2 (and Q4) of the Card Quiz. That is, assuming that you and your friend see bird colors better than chance, and assuming that your friend usually does not lie to you, then both the prior and the likelihood are > 50% Red. Thus, responses for the Bird Quiz can be compared to responses for the Card Quiz (Q2 and Q4) using the same Types defined earlier. For the Bird Quiz, responses are classified as Type A, Type B, and Type C/O (which does not distinguish between types C and O).

Figure 11.3 shows the results for the Bird Quiz. Here, about 90% of participants are Type B, about 10% are Type A, and no participants are Type C/O. These results are significantly different from those of the Card Quiz, $\chi^2(2,45) = 712$, $p < .001$. Thus, unlike the Card Quiz (Figure 11.2) where most people are *not* Bayesian (in quantitative reasoning), the Bird Quiz (Figure 11.3) shows that most people *are* Bayesian (in qualitative reasoning). This result is remarkable because, unlike the Card Quiz, there is actually a normative basis for Anchoring in the Bird Quiz; that is, if a person believes that his friend's report is not independent because the limits of human eyes are a common cause of errors in both his and his friend's vision. Nevertheless, the experimental finding is that most people are *not* Anchoring in the Bird Quiz and they *are* Anchoring in the Card Quiz.

The results of the Bird Quiz are encouraging because they suggest that Bayesian inference is *not* counterintuitive; that is, *Bayesian inference can actually be "intuitive" when the problem is framed in a more natural way that allows people to reason with familiar heuristics* (also see Burns, 2005a). As such, I consider the non-Bayesian biases observed in the Card Quiz (Experiment 1) to be a leverage point for the engineering of a system that supports mathematical reasoning about probability by tapping into ecological reasoning about uncertainty. I present such a system, called *Bayesian Boxes,* in a later section of this chapter. But first I must present the results from Experiment 3, which showed yet a different result from Experiments 1 and 2.

Experiment 3: Spy TRACS

Experiment 3 was performed with Spy TRACS, which is a solitaire game played with a special deck of two-sided cards. This game is somewhat like

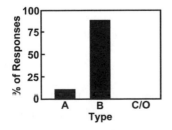

FIG. 11.3.   Results of Bird Quiz (Experiment 2).

the bag and poker chip task discussed earlier, in that the front of each card is one of two colors (red or blue). But Spy TRACS is unlike the poker chip task in several other important respects. One difference is that the backs of the cards in the deck, which are like the shapes of the chips in the bag, are not all the same. In TRACS there is one of three different shapes shown on the back of each card—triangle, circle, or square—and the red:blue odds are different for each shape. The red:blue odds, which are given by the proportion of each shape/color (back/front) card type in the whole deck, are 3:1 for triangles, 1:1 for circles, and 1:3 for squares. This structure of the TRACS deck has experimental advantages for testing a wide range of prior-likelihood pairs in one-shot Bayesian updates. It also has practical advantages in simulating (albeit simply) the clue–truth structure of real-word diagnoses, where truths (like the color on the front of a card) must be diagnosed from clues (like the shape on the back of the card) given probabilistic information in the form of priors and likelihoods that relate the clues to truths. In this way, the problem in Spy TRACS is similar to a radar mission where the front of the card (like ground truth) must be diagnosed from the back of the card (like a track) given a prior (from intelligence) and likelihood (from surveillance).

Three other differences between the poker chip task and Spy TRACS arise in the way the game is played. First, Spy TRACS presents a wide range of priors as the deck changes with time, because the game involves sampling without replacement as cards are dealt from the deck. Second, Spy TRACS is played on a computer where the player is given data for the prior and likelihood both numerically and graphically, and the player also reports his or her posterior in this dual format. The prior is given by the remaining number of each shape/color card type in the deck (i.e., red triangles, blue triangles, etc.) as the deck is depleted in play, and the likelihood comes from a simulated "spy" that reports the color of a given shape with a specified reliability. Finally, Spy TRACS involves making Bayesian inferences in a game where judgments lead to choices and those choices have consequences on a game score. Players see the outcome of their judgments (right or wrong) and choices (good or bad) so they get some feedback on how well they are doing, much like they would in the real world.

Details of the TRACS cards and Spy TRACS rules are provided elsewhere (Burns, 2001, 2004; 2005b). Here I focus on how Experiment 3 (Spy TRACS) relates to Experiment 2 (Bird Quiz) and Experiment 1 (Card Quiz).

Spy TRACS is different from the Bird Quiz and the Card Quiz in that it immerses players in a *synthetic task environment* (Gray, 2002). Thus, although the card game is far simpler than a war game or real-life challenge in NDM, it is also far different than the static questions used in Experiments 1 and 2, which are similar to the tasks employed in previous research on Bayesian inference in JDM. Also, because the prior changes with time as cards are dealt and removed from play, each game of Spy TRACS can test a wide range of prior/likelihood pairs rather than just a few points as tested in the Card Quiz. For comparison these test points can be grouped into three sets S2, S3, and S4 where the qualitative relation between prior and likelihood is the same as Q2, Q3, and Q4 (respectively) of the Card Quiz. And that is what is done here, in order to compare the results of Spy TRACS to the results of the Card Quiz.

Figure 11.4 shows the experimental results for S2/S3/S4 of Spy TRACS. Comparing this to Figure 11.2 (Q2/Q3/Q4 of the Card Quiz) shows an interesting difference. That is, in Figure 11.4 the dominant mode of player response is Compromising, whereas in Figure 2 the dominant mode of player response is Anchoring. In both cases, only a small fraction (10%–20%) of players are even qualitatively Bayesian, which is remarkably different from the Bird Quiz (Figure 11.3) where most of the same people (~90%) were qualitatively Bayesian. These results for Spy TRACS are significantly different from those of the Card Quiz and the Bird Quiz, $\chi^2(3,45) = 118$, $p < .001$ and $\chi^2(2,45) = 231$, $p < .001$, respectively.

Why were people Bayesian on the Bird Quiz but not Bayesian in Spy TRACS or the Card Quiz? Based on postgame interviews and questionnaires, it appears that most people simply did not know how to put the two numbers (prior and likelihood) together for Spy TRACS and the Card Quiz. In the Card Quiz, which presented the numbers in paragraph form, people usually ignored one source (the likelihood) because they thought it was less relevant than the other source (the prior) to which they were An-

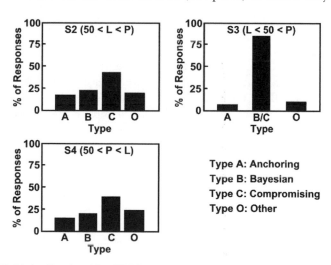

FIG. 11.4. Results of Spy TRACS (Experiment 3). P refers to the prior and L refers to the likelihood.

chored. In Spy TRACS, which presented the data in graphical form, people usually considered both sources equally but the result was a Compromising posterior that was still not in the ballpark of the Bayesian posterior (except for S3 where the Compromising and Bayesian responses are, by definition, in the same ballpark).

Yet in the Bird Quiz, where the prior and likelihood were presented in paragraph form but in qualitative (non-numerical) terms, most people were intuitively Bayesian. Based on interviews and questionnaires, this success on the Bird Quiz appears to be because the problem was framed in a way that allowed people to "see" the problem/solution as one of accumulating a "line" of support for each hypothesis, and in fact some people used these words in their interviews. Motivated by this finding, I developed a support system called *Bayesian Boxes* that frames the numerical problem in a way that helps people see the Bayesian answer as well as the Bayesian basis that underlies this answer.

WHAT CAN BE DONE ABOUT IT?

Bayesian Boxes is a support system designed to provide a graphic representation of the basic computation (Bayes Rule) needed to make probabilistic diagnoses. The system works by mapping the Bayesian-mathematical problem to a colored-ecological display (see Vicente & Rasmussen, 1992). The Bayesian computation involves two products, $P * L$ and $(1 - P) * (1 - L)$, where P refers to the prior and L refers to the likelihood. The graphic representation shows these products as boxes (colored areas) that represent the combined evidence (area) in support of each hypothesis (color). This is illustrated in Figure 11.5.

Bayesian Boxes has been implemented in a prototype system that works like a colored calculator. Figure 11.5 is a screen shot of the dynamic display using dark gray and light gray as the two colors. To use the calculator, a person simply moves the cursor (bold black "+") until P and L represent the prior and likelihood (inputs). The Bayesian posterior B is computed and illustrated at the top (output). For example, Figure 11.5 shows that a prior of 80% dark (20% light) and a likelihood of 67% dark (33% light) combine to yield a posterior of 89% dark (11% light). Thus, using dark to denote "Foe" and light to denote "Friend," Figure 11.5 shows how Bayesian Boxes represents the solution to the introductory Friend/Foe problem.

Besides this result, Bayesian Boxes also shows the underlying reason. That is, Figure 11.5 shows a dark box with area = $P * L = 80\% * 67\%$ and a light box with area = $(1 - P) * (1 - L) = 20\% * 33\%$. These two areas represent the two basic products that appear in Bayes Rule. The visual display of these boxes promotes an intuitive feel for why 89% is the correct answer;

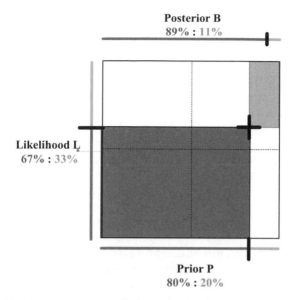

FIG. 11.5. A support system called Bayesian Boxes that provides a graphic display of Bayes Rule. A user sets P and L as inputs and the system gives B as output. The system displays the inputs and result as lines. The underlying reason for the result is displayed as boxes.

that is, the answer B = 89% dark is simply the fraction of the colored area (dark + light) that is dark. The white areas do not count.

As described previously, the support system does two things: it calculates *what* (the Bayesian posterior) and it illustrates *why* (the Bayesian principle). For example, at the top of Figure 11.5 a user can *see* that the *line* for the Bayesian posterior (B), which is an *update* of the prior (P), is a greater percentage dark than the prior shown at the bottom. This helps people overcome the non-Bayesian biases (Anchoring or Compromising or Other) that were observed in the Card Quiz and Spy TRACS. A user can also see the underlying principle, that is, as products (boxes) of probabilities (lines) where the ratio of boxed areas determines the Bayesian posterior. This is similar to the *frequency format* that other researchers have found to improve human performance in Bayesian inference (see Gigerenzer & Hoffrage, 1995).

An additional experiment, Experiment 4, was performed with Spy TRACS to test the effectiveness of Bayesian Boxes. In this experiment, participants of Experiments 1, 2, and 3 were given a short break, after which they were given a one-page tutorial that explained the notion of Bayesian inference without any equations but with a working version of Bayesian Boxes on a personal computer. The tutorial guided the participant through several sample calculations so they could get a feel for how the colored calculator worked on the kind of problem faced in Spy TRACS. After this tutorial, which lasted about 5 to 10 minutes, the colored calculator was taken away. The participant then played two more games of Spy TRACS without Bayesian Boxes. The *after* results (Experiment 4) were compared to the *before* results (Experiment 3) as an empirical measure of whether exposure to the colored calculator improved people's intuitive understanding of Bayesian inference.

The experimental results are shown in Figure 11.6. Comparing these results to those of Figure 11.4, which were obtained *before* exposure to Bayesian Boxes, the percentage of Type B responses *after* exposure to Bayesian Boxes has approximately doubled and the dominant mode of human response has changed from Compromising to Bayesian. The difference (before vs. after) is significant, $\chi^2(3,45) = 58$, $p < .001$.

The purpose of this experiment was to measure whether exposure to Bayesian Boxes in the short tutorial had any effect on people's intuitive understanding of Bayesian inference. The results suggest it did. In addition to these results, a postgame questionnaire asked participants if they thought that Bayesian Boxes was a "lot," "some," or "no" help to them in combining probabilities. More than half of the participants said it was a "lot" of help and all but a few of the remaining participants said it was "some" help. Many participants even reported that they tried to mentally visualize Bayesian Boxes as they played Spy TRACS.

Based on these tests and polls, I believe that Bayesian Boxes has the potential for improving probabilistic reasoning in military, medical, and other domains of NDM. Thus, further efforts are aimed at extending the conceptual design and developing a practical system that can be used in the face of

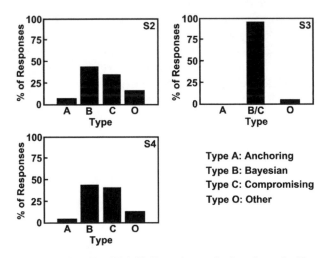

FIG. 11.6.   Results of Spy TRACS (Experiment 4) after players had been exposed to Bayesian Boxes.

real-world complexities (Burns, 2006). These complexities include additional hypotheses (i.e., when there are more than just two possibilities), conditional dependencies (i.e., when the sources are not independent), and more formidable likelihood functions (i.e., when a wrong report can be either a *false negative* P(~k|K) or a *false positive* P(k|~K) and P(~k|K) ≠ P(k|~K).

In addressing these complexities, the focus of the system design continues to be on displaying both computational results (*what*) along with conceptual reasons (*why*), because this dual display is the whole point of Bayesian Boxes. Whereas visual displays of results (*what*) are commonplace, visual displays of reasons (*why*) are less common but still important if designers expect users to reconcile algorithmic results with intuitive beliefs. In fact, I suspect that an intuitive illustration of reasons (*why*) is the missing link in many of the support systems that people "refuse to use."

CONCLUSION

In this chapter, I reviewed the results of laboratory experiments on human performance in Bayesian inference. I also outlined a support system called *Bayesian Boxes* that was shown to improve human performance in a game task of Bayesian inference. The experiments help to relate research in JDM to problems in NDM, and the support system holds promise for practical applications.

The focus of my experiments and design efforts is on probabilistic reasoning, because there is a growing need for people to deal with numerical probabilities in the high-stakes and time-pressed situations of NDM. In these cases, it is often difficult and dangerous to collect data—and time windows are critically limited. When faced with these constraints it is vital for decision makers to extract all of the certainty that they can from the data that they get, so that they can take proper and timely actions. This presents a problem of *expertise out of context,* because human competence in ecological reasoning is not readily transferred to human performance in mathematical reasoning. My approach with Bayesian Boxes was to develop an ecological interface that could help put human expertise back in context—and thereby improve inferences in dealing with probabilities.

REFERENCES

Birnbaum, M. H., & Mellers, B. A. (1983). Bayesian inference: Combining base rates with opinions of sources who vary in credibility. *Journal of Personality and Social Psychology, 45,* 792–804.

Burns, K. (2001). TRACS: A tool for research on adaptive cognitive strategies. Retrieved October 24, 2005, from http://www.tracsgame.com

Burns, K. (2004). Making TRACS: The diagrammatic design of a double-sided deck. *Diagrammatic Representation and Inference, Proceedings of the 3rd International Conference on Diagrams* (pp. 341–343). Cambridge, England.

Burns, K. (2005a). Mental models and normal errors. In H. Montgomery, R. Lipshits, & B. Brehmer (Eds.), *How professionals make decisions* (pp. 15–28). Mahwah, NJ: Lawrence Erlbaum Associates.

Burns, K. (2005b). On TRACS: Dealing with a deck of double-sided cards. In *Proceedings of the IEEE Symposium on Computational Intelligence in Games.* Retrieved April 4 from http://csapps.essex.ac.uk/cig/2005/papers/p1024.pdf

Burns, K. (2006). Bayesian inference in disputed authorship: A case study of cognitive errors and a new system for decision support. *Information Sciences, 176,* 1570–1589.

Edwards, W. (1982). Conservatism in human information processing. In D. Kahneman, P. Slovic, & A. Tversky (Eds.), *Judgment under uncertainty: Heuristics and biases* (pp. 359–369). New York: Cambridge University Press.

Edwards, W., & Phillips, L. D. (1964). Man as transducer for probabilities in Bayesian command and control systems. In M. W. Shelly & G. L. Bryan (Eds.), *Human judgments and optimality* (pp. 360–401). New York: Wiley.

Edwards, W., & von Winterfeldt, D. (2000). On cognitive illusions and their implications. In T. Connolly, H. R. Arkes, & K. R. Hammond (Eds.), *Judgment and decision making: An interdisciplinary reader* (pp. 592–620). New York: Cambridge University Press.

Endsley, M. (1988). Design and evaluation for situation awareness enhancement. In *Proceedings of the 32nd Annual Meeting of the Human Factors Society* (pp. 97–101). Santa Monica, CA: Human Factors and Ergonomics Society.

Gigerenzer, G., & Hoffrage, U. (1995). How to improve Bayesian reasoning without instruction: Frequency formats. *Psychological Review, 102,* 684–704.

Gray, W. (2002). Simulated task environments: The role of high-fidelity simulations, scaled worlds, synthetic environments, and micro worlds in basic and applied research. *Cognitive Science Quarterly, 2,* 205–227.

Hoffman, R. R., Trafton, G., & Roebber, P. (2005). *Minding the weather: How expert forecasters think.* Cambridge, MA: MIT Press.

Kahneman, D., Slovic, P., & Tversky, A. (Eds.). (1982). *Judgment under uncertainty: Heuristics and biases.* New York: Cambridge University Press.

Knill, D. C., & Richards, W. (Eds.). (1996). *Perception as Bayesian inference.* Cambridge, England: Cambridge University Press.

Koehler, J. (1996). The base rate fallacy reconsidered: Descriptive, normative, and methodological challenges. *Behavioral and Brain Sciences, 19,* 1–53.

McKenzie, C. R. (1994). The accuracy of intuitive judgment strategies: Covariation assessment and Bayesian inference. *Cognitive Psychology, 26,* 209–239.

Phillips, L. D., & Edwards, W. (1966). Conservatism in a simple probability inference task. *Journal of Experimental Psychology, 72,* 346–357.

Tversky, A., & Kahneman, D. (1982). Evidential impact of base rates. In D. Kahneman, P. Slovic, & A. Tversky (Eds.), *Judgment under uncertainty: Heuristics and biases* (pp. 153–160). New York: Cambridge University Press.

Vicente, K. J., & Rasmussen, J. (1992). Ecological interface design: Theoretical foundations. *IEEE Transactions on Systems, Man, and Cybernetics, SMC-22,* 589–606.

Zsambok, C. E., & Klein, G. (1997). *Naturalistic decision making.* Mahwah, NJ: Lawrence Erlbaum Associates.

Chapter 12

What Makes Intelligence Analysis Difficult?: A Cognitive Task Analysis

Susan G. Hutchins
Naval Postgraduate School, Monterey, CA

Peter L. Pirolli
Stuart K. Card
Palo Alto Research Center, Palo Alto, CA

In this chapter, we describe research involving a Cognitive Task Analysis (CTA) with intelligence analysts, focusing on the strategies used by experts who are confronted with tough scenarios and unusual tasks. We present what we have learned regarding how experienced practitioners deal with the extremely challenging task of intelligence analysis by summarizing a set of 10 CTA interviews conducted with intelligence analysts to identify leverage points for the development of new technologies.

The challenges facing practitioners in the modern world where expertise gets "stretched" by dynamics and uncertainty, a theme for this volume, characterize many of the problems that are now experienced by intelligence analysts. Part of the effort reported in this chapter is aimed at building up an empirical psychological science of analyst knowledge, reasoning, performance, and learning. We expect this will provide a scientific basis for design insights for new aiding technologies. In addition, this psychological research should yield task scenarios and benchmark tasks that can be used in controlled experimental studies and evaluation of emerging analyst technologies.

INTELLIGENCE ANALYSIS

The hallmark of intelligence analysis is the ability to sort through enormous volumes of data and combine seemingly unrelated events, to construct an accurate interpretation of a situation, and make predictions about complex, dynamic events. These volumes of data typically represent an extensive and far-ranging collection of sources, and are represented in many different formats (e.g., written and oral reports, photographs, satellite images, maps, tables of numeric data, to name a few). As part of this process, the intelligence analyst must assess the relevance, reliability, and significance of these disparate pieces of information. Intelligence analysis also involves performing complex reasoning processes to determine "the best explanation for uncertain, contradictory and incomplete data" (Patterson, Roth, & Woods, 2001, p. 225).

The nature of the data, the complex judgments and reasoning required, and a sociotechnical environment that is characterized by high workload, time pressure, and high stakes, all combine to create an extremely challenging problem for the intelligence analyst. High levels of uncertainty are associated with the data, when "deception is the rule." Because the validity of the data is always subject to question, this impacts the cognitive strategies used by analysts (Johnston, 2004). Moreover, the complex problems to be analyzed entail complex reasoning that mixes deduction, induction (where the conclusion, though supported by the premises, does not follow from them necessarily), and also abductive reasoning, which is when one attempts to determine the best or most plausible explanation for a given set of evidence (Josephson & Josephson, 1994; see also chap. 6, this volume). Finally, high stakes are associated with the pressure not to miss anything and to provide timely, actionable analysis. Potentially high consequences for failure—where analysis products have a significant impact on policy— also contribute to make the task challenging as decision makers, senior policymakers, and military leaders use the products of analysis to make high-stakes decisions involving national security.

A number of reports have emerged that provide normative or prescriptive views on intelligence analysis (e.g., Clark, 2004; Heuer, 1999). There have been very few that provide empirical, descriptive studies of intelligence analysis. One exception is Johnston's (2005) study of the intelligence analysts' community where he characterizes analysts and the cultural environment in which they work as part of his effort to identify and describe conditions and variables that negatively affect intelligence analysis. Cultural conditions include a risk-adverse environment with a conservative management that emphasizes avoiding error as opposed to imagining surprises. This organizational context also includes an extreme pressure to avoid failure, which produces a criticism taboo, such that analysts are reluctant to criticize fellow analysts. These organizational aspects of intelligence analysis create a climate where analysts are susceptible to cognitive biases, such as confirmation bias. Due to the high consequences of failure, and the concomitant pressure to avoid failure, Johnston reports a long-standing bureaucratic resistance to implementing a program for improving analytical performance as that would imply the current system is less than optimal. Criteria used for deciding on promotions, that is, the quantity of reports produced counts more than the quality of reports, result in analysts placing greater emphasis on reporting than on conducting in-depth analysis.

Regarding the process of analysis, Johnston (2005) characterizes analysts as using personal, idiosyncratic approaches where some 160 analytical methods are employed. Johnston reports on the lack of a standardized analytic doctrine, such that no guidelines exist to indicate which approach is the most effective method for a particular case. The lack of a body of research along these lines leads to an inherent inability to determine performance requirements, and individual performance metrics for the evaluation and development of new analytic methodologies. Johnston advocates development of formalized methods and processes because there is currently no formal system for measuring and tracking the validity or reliability of analytic methods due to the idiosyncratic tradecraft approach.

It is likely that there are many previous studies of intelligence analysis, which we might look back on and say involved some form of CTA, that will never become part of the public literature because of the classified nature of the work involved. Despite the spottiness of open literature, what does exist reveals that intelligence analysis is a widely variegated task domain. There is no such thing as "the" intelligence analyst. This means that it is important to be careful in making generalizations from any circumscribed types of intelligence tasks or types of analysts. It is equally important not to be daunted by the vastness of the domain, and to start the investigative venture somewhere.

Intelligence analysis is commonly described as a highly iterative cycle involving requirements (problem) specification, collection, analysis, production, dissemination, use, and feedback. It is an event-driven, dynamic process that involves viewing the information from different perspectives in order to examine competing hypotheses and develop an understanding of a complex issue. The critical role of the human is to add "value" to original data by integrating disparate information and providing an interpretation (Krizan, 1999). This integration and interpretation entails difficult, complex judgments to make sense of the information obtained. According to a number of discussions of the work of intelligence analysts (e.g., Millward, 1993; Moore, 2003) reasoning involves the analysis of collected and created evidence, then the sorting of significant from insignificant evidence, followed by deduction, induction, and abduction.

Warning-oriented intelligence includes supporting the need for senior policymakers to avoid getting surprised (Bodnar, 2003). Analysts need to "provide detailed enough judgments—with supporting reporting—so that both the warfighter and the policymaker can anticipate the actions of potential adversaries and take timely action to support US interests" (Bodnar,

2003, p. 6). For example, the intelligence analyst needs to make predictions regarding what the adversary has the capability to do and how likely it is that he or she will act. These predictions need to include what actions can be taken to change, or respond to these actions, and the probable consequences of those actions (Bodnar, 2003).

Table 12.1 presents an analysis of problem types that Krizan derives from Jones (1995) and course work at the Joint Military Intelligence College. A range of problem types, from simplistic to indeterminate, are explicated by characterizing each level of the problem along several dimensions, such as type of analytic task, analytic method, output, and probability of error.

Figure 12.1 presents another way of characterizing the domain of intelligence analysis developed by Cooper (2005). Along one axis there are various types of intelligence, along a second are different accounts (topics), and along a third axis are different types of products. The different types of intelligence (or "sources") are functionally organized into: *human source intelligence* (HUMINT), which includes field agents, informants, and observers (attaches); *imagery intelligence* (IMINT), which includes photo, electrooptical, infrared, radar, and multispectral imagery from sources such as satellites; *signals intelligence* (SIGINT), which includes communications, electronic, and telemetry; *measurement and signatures intelligence* (MASINT), which includes acoustic and radiation signals; *open-source intelligence* (OSINT), which includes public documents, newspapers, journals, books, television, radio, and the World Wide Web; and *all-source intelligence*, which involves all of the aforementioned.

Domains (or topics) may address terrorism, military, politics, science and technology, or economics. Product types range from those that are close to the raw data, through those that involve increasing amounts of analysis that may eventually lead to national-level estimates and assessments. As in any hierarchically organized information system, this means that information is filtered and recoded as the analysis process progresses from lower to higher levels.

Techniques to Enhance Processing of Intelligence Data

Recent world events have focused attention on some of the inherent challenges involved in performing intelligence analysis (viz., National Commission On Terrorist Attacks, 2004). As a result, a great deal of research is being conducted to develop new training, tools, and techniques that will enhance the processing of intelligence data. As one example, support and training in the organizing and piecing together aspects of intelligence analysis and decision making has been identified as an area that is greatly in need of more basic and applied research. One current research thread that seeks to address this need is the Novel Information from Massive Data

TABLE 12.1
Intelligence Analysis Problem Types

Characteristics	Problem Types				
	Simplistic	Deterministic	Moderately Random	Severely Random	Indeterminate
What is the question?	Obtain information	How much? How many?	Identify and rank all outcomes	Identify outcomes in unbounded situation	Predict future events/situations
Role of facts	Highest	High	Moderate	Low	Lowest
Role of judgment	Lowest	Low	Moderate	High	Highest
Analytical task	Find information	Find/create formula	Generate all outcomes	Define potential outcomes	Define futures factors
Analytical method	Search sources	Match data to formula	Decision theory; utility analysis	Role playing and gaming	Analyze models and scenarios
Analytical instrument	Matching	Mathematical formula	Influence diagram, utility, probability	Subjective evaluation of outcomes	Use of experts
Analytic output	Fact	Specific value or number	Weighted alternative outcomes	Plausible outcomes	Elaboration on expected future
Probability of error	Lowest	Very low	Dependent on data quality	High to very high	Highest
Follow-up task	None	None	Monitor for change	Repeated testing to determine true state	Exhaustive learning

Note. From Krizan (1999), based on course materials of the Joint Military College, 1991, by Thomas H. Murray, Sequoia Associates Inc.

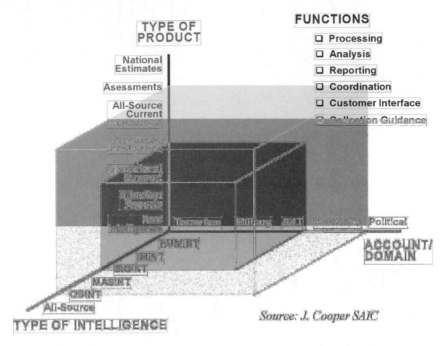

FIG. 12.1. Types of intelligence, domains, functions, and products. From
Cooper (2005). Reprinted by permission.

(NIMD) Program where the goal is to develop an "information manager"
to assist analysts in dealing with the high volumes and disparate types of
data that inundate intelligence analysts (Advanced Research and Development Activity [ARDA], 2002). The NIMD research program seeks to develop techniques that structure data repositories to help highlight and interpret novel contents and techniques that can accurately model and draw
inferences about rare events and sequences of events (widely and sparsely
distributed over time).

Connable (2001) asserts that the intelligence process would be well
served by enhancing the ability to leverage open sources, particularly because open sources provide the intelligence community with between 40%
and 80% of its usable data (Joint Military Intelligence Training Center,
1996). As an example, one of our study participants worked on a strategic
analysis assignment regarding the question of whether President Estrada,
of the Philippines, was going to remain in power or be removed from office.
For the analysis, 80% of the information the analyst needed was found in
open-source material. Information-foraging theory (Pirolli & Card, 1998,
1999) is being applied in this research on tasks that involve information-
intensive work where the approach is to analyze the tasks as an attempt by

the user to maximize information gained per unit time. A computational model of the intelligence analysis process will be developed as a result of this CTA research and used to support tool prototyping and testing.

The goals for the research described in this chapter are threefold. One purpose of this first CTA phase was to yield "broad brushstroke" models of analyst knowledge and reasoning. A second purpose of this research was to identify leverage points where technical innovations may have the chance to yield dramatic improvements in intelligence analysis. A third purpose of the CTA phase was to guide the development of benchmark tasks, scenarios, resources, corpora, evaluation methods, and criteria to shape the iterative design of new analyst technologies. A CTA procedure is typically used to identify the decision requirements, and the knowledge and processing strategies used for proficient task performance. The following section presents a brief description of CTA and describes specific techniques that are representative of CTA methods.

COGNITIVE TASK ANALYSIS

CTA refers to a group of methods that are extensively used in cognitive systems engineering and naturalistic decision-making applications. Klein's (2001) definition of CTA is "methods for capturing expertise and making it accessible for training and system design" (p. 173). Klein delineated the following five steps: (a) identifying sources of expertise, (b) assaying the knowledge, (c) extracting the knowledge, (d) codifying the knowledge, and (e) applying the knowledge. System design goals supported by CTA include human–computer interaction design, developing training, tests, models to serve as a foundation for developing an expert system, and analysis of a team's activities to support allocation of responsibilities to individual humans and cooperating computer systems. (An extensive treatment of CTA methodology appears in Crandall, Klein, & Hoffman, 2006.)

Different CTA methods are used for different goals (Hoffman, Shadbolt, Burton, & Klein, 1995). Our goals for conducting a CTA were twofold. Our first goal was to capture data that will provide input to support development of a computational model of the intelligence analyst's processes and analytic strategies. Our second goal was to identify leverage points to inform the development of tools to assist analysts in performing the most demanding aspects of their tasks. CTA extends traditional task analysis techniques (so-called behavioral task analysis) to produce information regarding the knowledge, cognitive strategies, and goal structures that provide the foundation for task performance (Chipman, Schraagen, & Shalin, 2000). The goal of CTA is to discover the cognitive activities that are required for performing a task in a particular domain to identify opportunities to improve performance by providing improved support of these activities (Potter, Roth, Woods, & Elm, 2000).

Our overall approach for the first phase of our research involved the following steps: (a) review of the intelligence literature, (b) semistructured interviews, followed by (c) structured interviews, and then (d) review of the results by subject-matter experts (SMEs). The second phase of this research, conducted in the summer of 2004, involved developing and comparing several alternative hypotheses based on material presented in a case study. A prototype tool developed to assist the intelligence analyst in comparing alternate hypotheses was introduced and simulated tasks were performed to empirically evaluate the tool's effectiveness. A follow-on study will involve the use of think-aloud protocol analysis while using a more advanced version of this tool. This multiple-phase plan is in line with the approach employed by several successful CTA efforts (Hoffman, et al., 1995; Patterson, Roth, & Woods, 2001). We are using a "balanced suite of methods that allow both the demands of the domain and the knowledge and strategies of domain experts to be captured in a way that enables clear identification of opportunities for improved support" (Potter et al., 2000, p. 321).

Types of activities that typically require the sort of resource-intensive analysis frequently required when conducting a CTA are in those domains that are characterized as involving complex, ill-structured tasks that are difficult to learn, complex, dynamic, uncertain, and real-time environments, and multitasking. A CTA is appropriate when the task requires the use of a large and complex conceptual knowledge base, the use of complex goal–action structures dependent on a variety of triggering conditions, or complex perceptual learning or pattern recognition skills. Intelligence analysis involves all of these characteristics.

When considering which knowledge elicitation technique is most appropriate, the differential access hypothesis proposes that different methods might elicit different "types" of knowledge (Hoffman et al., 1995). Certain techniques are appropriate to bootstrap the researcher and generate an initial knowledge base and more structured techniques are more appropriate to validate, refine, and extend the knowledge base (Hoffman et al., 1995). A direct mapping should exist between characteristics of the targeted knowledge and the techniques that are selected (Cooke, Salas, Cannon-Bowers, & Stout, 2000).

A model of knowledge or reasoning that delineates the essential procedural and declarative knowledge is necessary to develop effective training procedures and systems (Annett, 2000). This entails building a representation that captures the analysts' understanding of the demands of the domain, the knowledge and strategies of domain practitioners, and how existing artifacts influence performance. CTA can be viewed as a problem-solving process where the questions posed to the subject matter experts (SMEs), and the data collected, are tailored to produce answers to the research questions, such as training needs and how these training problems might be solved (DuBois & Shalin, 2000). A partial listing of the types of information to be obtained by conducting a CTA includes factors that contribute to making task performance challenging, what strategies are used and why, what complexities in the domain practitioners respond to, what aspects of performance could use support, concepts for aiding performance, and what technologies can be brought to bear to deal with inherent complexities (see Crandall et al., 2006).

Use of Multiple Techniques

Analysis of a complex cognitive task, such the intelligence analyst's job, requires the use of multiple CTA techniques. When results from several techniques converge confidence is increased regarding the accuracy of the products (Cooke, 1994; Flach, 2000; Hoffman et al., 1995; Potter et al., 2000). Flach and many others recommend sampling a number of experts and using a variety of interviewing tools to increase the representativeness of the analysis. During the initial bootstrapping phase of our research, several CTA approaches were examined with an eye toward determining which approach would be most productive for our domain of interest. The remainder of this section describes two CTA techniques that were used for the initial phase of this research.

Applied Cognitive Task Analysis Method.   Our initial set of interviews drew upon the Applied Cognitive Task Analysis (ACTA) Method (Militcllo & Hutton, 1998; Militello, Hutton, Pliske, Knight, & Klein, 1997) and the Critical Decision Method (CDM; Hoffman, Crandall, & Shadbolt, 1998; Klein, Calderwood, & MacGregor, 1989). The ACTA suite of methods was developed explicitly as a streamlined procedure for instructional design and development (Militello et al., 1997) that required minimal training for the task analysts. (Because of the dual meaning of *analysis* in this chapter, in reference to task analysis and in reference to intelligence analysis, we henceforth refer to "researchers" rather than "task analysts.") ACTA is a collection of semistructured interview techniques that yields a general overview of the SMEs' conception of the critical cognitive processes involved in their work, a description of the expertise needed to perform complex tasks, and SME identification of aspects of these cognitive components that are crucial to expert performance.

The standard ACTA methodology includes the use of three interview protocols and associated tools: (a) the Task Diagram, (b) the Knowledge

Audit, and (c) the Simulation Overview. The ACTA Method uses interview techniques to elicit information about the tasks performed and provides tools for representing the knowledge produced (Militello & Hutton, 1998). Objectives for the ACTA method include discovery of the difficult job elements, understanding expert strategies for effective performance, and identification of errors that a novice might make. The focus for researchers using the ACTA method is on interviews where domain practitioners describe critical incidents they have experienced while engaged in their tasks and aspects of the task that made the task difficult.

Our use of the ACTA method produced valuable data for the bootstrapping phase of this research, where our goal was to learn about the task and the cognitive challenges associated with task performance, as well as to determine what tasks to focus on during ensuing phases of the CTA research. Products typically produced when using the ACTA method include a Knowledge Audit and a Cognitive Demands Table. After conducting this first set of CTA interviews we opted to attempt to use an additional method to capture the essence of the intelligence analyst's job. The intelligence analyst's task places great emphasis on deductive and inductive reasoning, looking for patterns of activity, and comparing hypotheses to make judgments about the level of risk present in a particular situation. We felt it was necessary to broaden the scope of the interview probes used with intelligence analysts, to include probes that are used in the general CDM procedure.

Critical Decision Method. CDM is a structured interview technique developed to obtain information about decisions made by practitioners when performing their tasks. Specific probe questions help experts describe what their task entails. CDM's emphasis on nonroutine or difficult incidents produces a rich source of data about the performance of highly skilled personnel (Hoffman et al., 1998; Klein et al., 1989). By focusing on critical incidents, the CDM has proven to be efficient in uncovering elements of expertise that might not be found in routine incidents, and it can help to ensure a comprehensive coverage of the subject matter.

As in all uses of the CDM, ours involved tailoring CDM probe questions to make them appropriate to the domain under study. We created probes to elicit practitioner knowledge and reasoning concerning how they obtain and use information, the schemata or recurrent patterns they rely on to conceptualize the information, how hypotheses are developed to analyze this information, and the types of products that are developed as a result of their analysis. A strength of the CDM is the generation of rich case studies, including information about cues, hypothetical reasoning, strategies, and decision requirements (Hoffman, Coffey, Carnot, & Novak, 2002; Klein et al., 1989). This information can then be used in modeling the reasoning procedures for a specific domain.

In the remainder of this chapter, we describe our development and use of an adapted version of the CDM and results derived from use of ACTA and CDM.

METHOD

In Study 1, we sought to learn about the intelligence analyst's task and the cognitive challenges associated with task performance, as well as to converge on determined tasks that would be our focus in Study 2, in which we refined our methodology and used a different group of participant intelligence analysts. We conducted our adapted version of the CDM where participants were asked to describe a difficult, strategic analysis problem, in lieu of a critical decision problem.

Study 1

Participants. Participants were six military intelligence analysts who were enrolled in a graduate school program at the Naval Postgraduate School (NPS), Monterey, California. Participants were contacted via e-mail with the endorsement of their curriculum chair and were asked to volunteer for this study. (No compensation was given for participation.) These U.S. Naval officers (Lieutenant through Lieutenant Commander) were students in the Intelligence Information Management curricula at NPS. Though this might imply that the participants were apprentices, in fact they had an average of 10 years' experience working as intelligence analysts. Thus, they were considered experts, or at least journeymen, as the literature generally defines an expert as one who has over 10 years' experience and "would be recognized as having achieved proficiency in their domain" (Klein et al., 1989, p. 462).

Our participant intelligence analysts had a variety of assignments in their careers, however the majority of their experience was predominantly focused on performing analysis at the tactical level. (Tactical-level analysis refers to analysis of information that will impact mission performance within the local operating area, e.g., of the battle group, and generally within the next 24 hours.) During the Study 1 bootstrapping phase of our CTA effort, we learned that there are several career paths for intelligence analysts. These career paths can be categorized as having more of either a technology emphasis (where the focus is on systems, equipment, and managing the personnel who operate and maintain this equipment), or an analytical emphasis where the focus and experience is on performing long-range, or strategic, analysis.

Procedure. After a brief introduction to the study, participants were asked to complete a demographic survey. The CTA process for all participants took place in a small conference room at NPS. In the semistructured interviews, the participants were asked to recall and describe an incident from past job experience. Following the ACTA method, participants were asked to draw a task diagram, describe critical incidents they had experienced on their job, and identify examples of the challenging aspects of their tasks. They were asked to elucidate why these tasks are challenging, and to describe the cues and strategies that are used by practitioners, and the context of the work.

Interviews were scheduled for 1½ hours, and were held at a time that was convenient for each participant. Three interviewers were present for each of the six interviews. The interviews were tape-recorded and transcribed, and the analysis was performed using the transcription and any other materials produced during the interview, for example, task diagrams.

Information gathered during the initial phase served as an advance organizer by providing an overview of the task and helped to identify the cognitively complex elements of the task. The ACTA method produced valuable data for the initial phase of this research. After analyzing the data from the initial set of interviews, we determined that we needed to broaden the set of interview probes and tailor them for the specific domain of intelligence analysis to uncover the bigger picture of how intelligence analysts approach performing their job. Thus, tailored CDM probes were developed specifically for the domain of intelligence analysis, and those were to be used in Study 2.

Study 2

Participants. Concurrent with the decision to use an adapted version of the CDM was the decision to switch to a different group within the intelligence community, specifically, intelligence analysts who had experience at the strategic, or national, level. National-level intelligence is more concerned with issues such as people in positions of political leadership and the capabilities of another country. (In contrast, at the tactical level, the user of intelligence information may be concerned only about a specific ship that is in a particular area, at a certain time; that is, the information will be valid for only a limited time.) Descriptions of experiences at the tactical level did not provide examples of the types of problems or cases that could benefit from the technology envisioned as the ultimate goal for this research.

Four military intelligence analysts from the National Security Affairs (NSA) Department were interviewed in Study 2. In the NSA curriculum, there is a stronger emphasis on analytical procedure and analysis assignments at the strategic level. We were fortunate in that this second group of participants was very articulate in describing assignments where they had performed analysis of critical topics at the strategic level. Several researchers have noted the issue of encountering problems with inaccessible expert knowledge (Cooke, 1994; Hoffman et al., 1995).

Procedure.   CDM interviews were conducted in which the intelligence analysts were asked to recall a strategic analysis problem they had worked on. Following the standard procedure of multiple retrospective "sweeps" used in the CDM, participants were asked to describe what they did step-by-step to gather information and analyze it, and then construct a timeline to illustrate the entire analysis process.

Many CTA techniques have been developed and used for tasks that involve the practitioner making decisions and taking a course of action based on these decisions, for example, firefighters, tank platoon leaders, structural engineers, paramedics, and design engineers. A goal for many CTA techniques is to elicit information on actions taken and the decisions leading up to those actions. However, the intelligence analyst's job does not fit this pattern. One finding that emerged during our Study 1 was that making decisions is not a typical part of the intelligence analyst's task. The major tasks consist of sifting through vast amounts of data to filter, synthesize, and correlate the information to produce a report summarizing what is known about a particular situation or state of affairs. Then, the person for whom the report is produced makes decisions and takes actions based on the information contained in the report.

Thus, our modified CDM had to emphasize processes of comparing alternative hypotheses rather than making decisions and taking action. Thus, interview probe questions provided in the CDM literature were tailored to capture information on intelligence analysts' approaches to gathering and analyzing information. Domain-specific probes were developed to focus the discussion on a *critical-analysis assignment* where the analyst had to produce a report on intelligence of a strategic nature. Examples of such strategic analysis problems might include assessments of the capabilities of nations or terrorist groups to obtain or produce weapons of mass destruction, terrorism, strategic surprise, political policy, or military policy. Interview probes were developed to capture information on the types of information used, how this information was obtained, and the strategies used to analyze this information.

One researcher conducted the initial interviews, each of which lasted approximately 1½ hours. Once the initial interview was transcribed and analyzed, the participant was asked to return for a follow-up interview. All three researchers were present for the follow-up interviews. This approach, requiring two separate interviews, was necessitated by the domain complexity and the desire to become grounded in the case before proceeding with the second interview where our understanding was elucidated and refined. (Hoffman, Coffey, & Ford, 2000, reported that in complex domains, such as their case domain of weather forecasting, CDM procedures sometimes have to be split up into multiple sessions.)

Deepening Probes. For the "deepening" sweep of the CDM procedure, domain-specific probes were developed to capture information on the types of information the intelligence analyst was seeking, the types of questions the analyst was asking, and how this information was obtained. Additional information was collected on mental models used by analysts, hypotheses formulated and the types of products that are produced. Table 12.2 lists the questions posed to the participants. Topics for which participants conducted their analyses included modernization of a particular country's military, whether there would be a coup in the Philippines and the potential impact on the Philippines if there was a coup, and the role for the newly created Department of Homeland Security.

Follow-up Probes. Once the data from the initial interviews had been transcribed and analyzed, participants were asked to return for a follow-up interview. The goal during this second session was to further elaborate our

TABLE 12.2
Adapted Critical Decision Method "Deepening" Probes

Topic	Probe
Information	What information were you seeking, or what questions were you asking?
	Why did you need this information?
	How did you get that information?
	Were there any difficulties in getting the information you needed from that source?
	What was the volume of information that you had to deal with?
	What did you do with this information?
	Would some other information been helpful?
Mental Models/ Schemata	As you went through the process of analysis and understanding did you build a conceptual model?
	Did you try to imagine important events over time?
	Did you try to understand important actors and their relationships?
	Did you make a spatial picture in your head?
	Can you draw me an example of what it looks like?
Hypotheses	Did you formulate any hypotheses?
	Did you consider alternatives to those hypotheses?
	Did the hypotheses revise your plans for collecting and marshaling more information? If so, how?
Intermediate Products	Did you write any intermediate notes or sketches?

understanding of the intelligence analyst's task. The analysts were asked to review the timeline produced during the first interview session and to elaborate on the procedures and cognitive strategies employed. Probes used during the follow-up interview are listed in Table 12.3.

Probes included questions about the participants' goals, whether this analysis was similar to other analysis assignments, use of analogues, and how hypotheses were formed and analyzed. Other probes asked about the types of questions raised during their analysis, methods used, information cues they used to seek and collate information, and the types of tools, for example, computer software, they used to perform their analysis. During this second interview, we went through the same intelligence analysis problem as we had in the first CDM interview, with the goal of obtaining additional details to refine our understanding of the intelligence analysis process. This included the types of information they used, and how they structured their analysis to answer the strategic question they had been assigned.

RESULTS

Study 1

Using the ACTA method, participants focused on providing descriptions of the cognitively challenging aspects of the task. Table 12.4 presents an example of one of the formats used to codify the knowledge elicited using the

TABLE 12.3
Additional Deepening Probes, Used During
the Second Set of CDM Procedures

Topic	Probes
Goals	What were your specific goals at the time?
Standard Scenarios	Does this case fit a standard or typical scenario? Does it fit a scenario you were trained to deal with?
Analogues	Did this case remind you of any previous case or experience? What hypotheses did you have? What questions were raised by that hypothesis? What alternative hypotheses did you consider? What questions were raised by that alternative hypothesis?
Information Cues for Hypotheses and Questions	As you collected and read information, what things triggered questions or hypotheses that you later followed up?
Information Tools	What sort of tools, such as computer applications, did you use? What information source did you use? What difficulties did you have?

TABLE 12.4
An Example Cognitive Demands Table Developed
in the Study 1 Use of the ACTA Method

	Why Is It Difficult?	Cues	Strategies	Potential Errors
Synthesizing Data	Lack of technical familiarity with different types of data.	Difficult to know how to weight different kinds of data and disregard other data.	Emphasize the type of data the intelligence analyst has experience with, and disregard other data types.	Tendency to focus on data types the analyst has experience with and ignore data that are not understood.
	Domain expertise is needed to analyze different classes of data (HUMINT, SIGINT, etc.).			
	No one database can correlate across systems.	Systems produce different "results" (e.g., mensuration produces different latitude/longitude coordinates in different systems).	Different commands rely on different databases in which they have developed trust.	Users develop a comfort level with their system, and its associated database; this can lead to wrong conclusions.
	No one database can correlate all inputs from many different analysts to form one coherent picture.			

	Databases are cumbersome to use; have poor correlation algorithms.	Users do not always understand what the information-processing system presents; too many levels in the system are not transparent.	Reliance on past experience	Overreliance on trend information.
	System produces results that users do not trust; tracks are "out of whack."			
Noticing Data	Time-critical information can be difficult to obtain.	Need to decide whether imagery is current enough to proceed with a strike: How long has the data been there?	Need to rely on other sources to verify data	Failure to refer to other sources to verify data.
	Need to assimilate, verify, and disseminate information in a short time window.			

ACTA method. This Cognitive Demands Table was produced based on analysis of data captured during an interview with one participant. A Cognitive Demands Table provides concrete examples of why the task is difficult, cues and strategies used by practitioners to cope with these demands, and potential errors that may result in response to the challenges inherent in the task.

Table 12.5 presents an example of a Knowledge Audit result, which includes examples of the challenging aspects of the task and the strategies employed by experienced analysts to deal with these challenges. A challenging aspect described by several intelligence analysts was the need for the analyst to understand the capabilities and limitations of the systems employed for data collection. Understanding the systems' capabilities is important because the systems used to collect data and the tools used to process data can generate errorful results due to conflicting databases, complexities of the system that are not transparent to the user and other human–system interaction issues.

Another theme was the constant pressure not to let anything slip by without looking at it. The participants described this aspect of their task as trying to find the "little jewels in the huge data stream," while knowing that 90% of the stream will not be relevant. An issue germane to intelligence analysis, also reported by several of our participants, was the tendency to discount information when they see something they don't expect to see, that is, to look for confirming evidence and to discount disconfirming evidence. An additional pressure experienced by intelligence analysts is the need to ensure the customer will take appropriate action as a result of the report (i.e., you are "almost dictating what the customer is going to do").

The next subsection summarizes what was learned from the Study 1 ACTA interviews.

Cognitive Challenges

The intelligence analyst's task is difficult due to the confluence of several factors, including characteristics of the domain and the cognitive demands levied on analysts. The following paragraphs describe the cognitive challenges involved in performing intelligence analysis.

Time Pressure. The shortening of timelines to produce reports for decision makers is becoming an increasingly stressful requirement for intelligence analysts working at all levels, tactical through strategic. An example at the tactical level is provided by a participant who described how the effect of timeline compression coupled with organizational constraints can sometimes "channel thinking" down a specific path. An example of time pressure at the strategic level is provided by one participant, whom we men-

TABLE 12.5

Knowledge Audit Result for One of the Intelligence Analysts

Example	Cues & Strategies	Why Difficult
Collection Task involves much technical knowledge coupled with expertise.	Start formulating a picture right away and ask: What do I expect to see here? Constantly checking all data coming in.	Collection systems and processors make mistakes: e.g., radar signatures can be similar.
	Constantly think about nature of the collection system.	Not all data are 100% accurate.
	Know what system can do/limitations.	Need to understand systems to assess validity of information.
	Need to question all data for validity: Correlate signals with what is already known.	Deluged with signals in dense signal environment, yet constant pressure not to miss anything.
	Assess validity of information.	Look for incongruent pieces of information.
	Try to extend the area that is monitored to maintain wide-area situation awareness.	Huge amount of raw data, yet analyst has to find the "little jewels" in huge data stream.
	Can't miss the radar contact that is the enemy coming out to conduct reconnaissance, or attack the battle group.	Want to know 10–12 hours ahead of time when the enemy aircraft was coming.
	Look at everything, recognizing that probably 90% is going to be of no use.	Can't afford to let anything slip by without looking at it.
Analysis Task entails multiple ways to obtain certain kinds of information.	Focus on what additional information is needed: Think about what still need to know.	Need some familiarity with different types of sources.
	Potential political ramifications to requesting asset to get something.	Requesting assets to get information may be expensive and conflict with other things.
	Good analyst drives operations; how to present information to customer.	Need to ensure customer will take appropriate action as a result of report.

(Continued)

TABLE 12.5
(Continued)

Example	Cues & Strategies	Why Difficult
	Do not just pass all the information without some level of interpretation included. Analyst brings a lot of knowledge to situation that goes beyond sensor-to-shooter approach. What has occurred in the past week? 2 months: 2 years?	Interpretation can be challenging: Are almost dictating what customer is going to do. Things need to be interpreted in context.
Disseminate Task: providing reports includes pressure to reduce the time to respond.	What does customer need to know? Pick out event-by-event pieces	What is the priority of this target vs. others that are out there? Is it the most important thing to do right now? Need to pass time-critical information right away.
	Simply pushing reports out to people does not always work: See something they don't expect, doesn't fit an established picture.	Time-critical spot reports need to go out to people who need it right away. Need to assess how this fits with what analyst has been observing recently.
	Times when event does not fit in: Try to develop coherent picture based on other things that have been occurring in past 1–2 hours	More likely to discount information if see something you don't expect.
	What do I think will happen in the next hour?	How does the last one event fit in with all the other recent pieces?
	See something outside a pattern of what expected. Always call operator: "We saw X but here is why we don't think it is not necessarily the truth."	Need to watch your back (not look bad). Look for reasons why it might not be correct.
	Pressure on analyst to ensure all high-level decision makers have same picture/information.	High-level decision makers want individual, tailored brief: generates differential exchange of information.

tioned earlier, who had to prepare a report on a matter of strategic importance when he had no prior knowledge of this area and he did not have a degree in political science. The assignment involved the question of whether President Estrada of the Philippines would be deposed, and if so, would there be a coup? This assignment was to include an analysis of what the impact would be on the Philippines. Six weeks was the time he had to gather all the necessary information, including the time needed to develop background knowledge of this area. He began by reading travel books and other ethnographic information. This finding is in accord with those of Patterson et al. (2001), that is, that analysts are increasingly required to perform analysis tasks outside their areas of expertise and to respond under time pressure to critical analysis questions.

Synthesizing Multiple Sources of Information. One aspect of the intelligence analyst's task that is particularly challenging involves merging different types of information—particularly when the analyst does not have technical familiarity with all these types of information. As an example, two analysts looking at the same image may see different things. Seeing different things in the data can occur because many factors need to be considered when interpreting intelligence data. Each type of data has its own set of associated factors that can impact interpretation. In the case of imagery data, these factors would include the time of day the image was taken, how probable it is to observe a certain thing, and trends within the particular country.

Multiple sources of disparate types of data (e.g., open source, classified, general reference materials, embassy cables, interviews with experts, military records, to name a few) must be combined to make predictions about complex, dynamic events—also sometimes in a very short time window. To accomplish the data correlation process, intelligence analysts need to be able to combine seemingly unrelated events and see the relevance. The cognitive challenges involved in synthesizing information from these different sources and distilling the relevance can be especially difficult, particularly when different pieces of data have varying degrees of validity and reliability that must be considered. Furthermore, domain expertise is often needed to analyze each type of data.

Human intelligence, electronic intelligence, imagery, open-source intelligence, measures and signals intelligence can all include spurious signals or inaccurate information due to the system used or to various factors associated with the different types of data. Intelligence analysts described situations where they gave greater weight to the types of information they understood and less weight to less understood types of information. They acknowledged this strategy could lead to incorrect conclusions.

Coping with Uncertainty.   Regarding data interpretation, a strong relationship typically exists between the context in which data occurs and the perspective of the observers. This critical relationship between the observer and the data is referred to as context sensitivity (Woods, Patterson, & Roth, 2002). The relationship between context and the perspective of the observer is an essential aspect of the data interpretation process. People typically use context to help them determine what is interesting and informative, and this, in turn, influences how the data are interpreted. Context is the framework a person uses to determine which data to attend to and this, in turn, can influence how the data are interpreted. This relationship between context and data interpretation is the crux of a problem for intelligence analysts: When high levels of uncertainty are present regarding the situation, the ability to interpret the data based on context sensitivity is likely to be diminished.

High levels of ambiguity associated with the data to be analyzed produce an uncertain context in which the analyst must interpret and try to make sense of the huge data stream. For instance, data that appear as not important might be extremely important in another situation, for example, when viewed from a different perspective to consider a competing hypothesis. In general, people are good at being able to focus in on the highly relevant pieces of data based on two factors: properties of the data and the expectations of the observer (Woods et al., 2002). However, this critical cognitive ability may be significantly attenuated for professionals in the intelligence community, as they may not always have the correct "expectations" while conducting their search through the data due to the inherent uncertainty associated with the data.

High Mental Workload.   One of the most daunting aspects of the intelligence analyst's job is dealing with the high level of mental workload that is entailed when a constant stream of information must be continuously evaluated, particularly when the information often pertains to several different situations. Relevant items must be culled from the continual onslaught of information, then analyzed, synthesized, and aggregated. An additional contributor to the high workload is the labor-intensive process employed when an analyst processes data manually—as is often the case—because many tools currently available do not provide the type of support required by analysts. For example, no one single database exists that can correlate across the various types of data that must be assimilated.

Intelligence analysts often wind up trying to synthesize all the information in their head, a time-consuming process that requires expertise to perform this accurately, and something that is very difficult for a less experienced analyst to do. Moreover, it is stressful to perform the analysis this way because they worry about missing a critical piece of data and doing it correctly: "Am I missing something?" and "Am I getting the right information out?"

Intelligence analysts must assess, compare, and resolve conflicting information, while making difficult judgments and remembering the status of several evolving situations. These cognitive tasks are interleaved with other requisite tasks, such as producing various reports or requesting the re-tasking of a collection asset. A request to gather additional information will often involve use of an asset (e.g., a satellite) that is in high demand. Retasking an asset can be costly and may conflict with other demands for that asset, thus, trade-offs must be made regarding the potential gain in information when retasking the asset to satisfy a new objective. Potential political ramifications of requesting an asset to obtain data to satisfy an objective must also be considered.

Potential for Error. The high level of mental workload imposed on intelligence analysts introduces a potential for errors to influence interpretation. For instance, the potential for "cognitive tunnel vision" to affect the analysis process is introduced by the high cognitive load that analysts often experience (Heuer, 1999). As an example, they may miss a key piece of information when they become overly focused on one particularly challenging aspect of the analysis. Similarly, the analysis process may be skewed when analysts attempt to reduce their cognitive load by focusing on analyzing data they understand and discounting data with which they have less experience. Additionally, discrepancies regarding interpretation may result when decision makers at different locations (e.g., on different platforms, different services) rely on systems that produce different results. Moreover, the sheer volume of information makes it hard to process all the data, yet no technology is available that is effective in helping the analyst synthesize all the different types of information.

Data Overload. Though data overload is actually a relatively new problem for the intelligence community, it is a major contributor to making the task difficult. It was once the case that intelligence reporting was very scarce, yet with technology advances and electronic connectivity it has become a critical issue today. A former Marine Lieutenant General, describing the situation in the 1991 Persian Gulf conflict, commented on the flow of intelligence: "It was like a fire hose coming out, and people were getting information of no interest or value to them, and information that was [of value] didn't get to them" (Trainor, cited in Bodnar, 2003, p. 55). Data overload in this domain is attributed to two factors: (a) the explosion of available electronic data coupled with (b) the Department of Defense emphasis on tracking large numbers of "hot spots." These factors place analysts in a position where they are "required to step outside their areas of expertise to respond quickly to targeted questions" (Patterson et al., 2001, p. 224).

Complex Judgments. Difficult judgments are entailed when decision makers have to consider the plausibility of information, decide what information to trust, and determine how much weight to give to specific pieces of data. Each type of data has to be assessed to determine its validity, reliability, and relevance to the particular event undergoing analysis. Intelligence analysts must also resolve discrepancies across systems, databases, and services when correlation algorithms produce conflicting results or results that users do not trust. Evidence must be marshaled to build their case or to build the case for several competing hypotheses and then to select the hypothesis the analyst believes is most likely. Assessing competing hypotheses involves highly complex processes.

Inadequate Tools. The sheer volume of information makes it hard to process all the data, yet the information-processing and display tools currently available are not always effective in helping the analyst assimilate the huge amount of information that needs to be analyzed and synthesized. Many of the systems and databases available to analysts are cumbersome to use due to system design issues and a lack of "human-centered design" considerations. Systems across and within agencies do not "talk" to one another. Furthermore, analysts are not always able to readily understand the information presented by the system (e.g., when there are discrepancies across system databases) or the system presents results that users do not trust (e.g., tracks that don't make sense). Tools include poor correlation algorithms and have too many levels within the system that are not transparent to the user.

Organizational Context. Several themes related to organizational context emerged from the Study 1 interviews. The first involves communication between the analyst and their "customers" (a term used to refer to the person for whom the report or product is produced, usually higher level decision makers and policymakers). When the customer does not clearly articulate his or her need—and provide the reasons they need a specific item—the analyst has an ill-defined problem. When the analyst does not have an understanding of the situation that merits the intelligence need, this will make it more difficult for the analyst to meet the analysis requirement(s).

A second organizational context issue is that a goal for analysts is to ensure that all high-level decision makers are given the same picture, or information. Yet, high-level decision makers will often demand an individual, tailored brief. This generates a differential exchange of information between the analyst and various decision makers.

Organizational constraints are placed on analysts to maintain the "status quo," such that new information is filtered through a perspective of being considered as not falling outside of normal operations. There is pressure to avoid being "the boy who cried wolf." This is in accord with other findings (Vaughan, 1996) that describe organizations that engage in a "routinization of deviance, as they explain away anomalies and in time come to see them as familiar and not particularly threatening" (chap. 6, this volume). Finally, there is a perception among analysts of feeling unappreciated for their work: Because people often do not understand what is involved there is a perception among intelligence analysts that people ask, "Why do we need you?" This credibility issue results in part because different data in different databases produce discrepancies. Intelligence analysts feel they lose credibility with operational guys because of these system differences.

The initial set of knowledge representations for the intelligence analyst's job provided the basis for the more detailed CTA. We now turn the discussion to present results from analysis of data gathered using the modified CDM.

Study 2

Our modified CDM method was used with a group of analysts who had experience working on analysis problems at the strategic level. The emphasis was on having intelligence analysts describe tasks where the focus was on producing a report to answer a question of strategic interest. Each analyst was asked to describe a difficult case they had been assigned to work on and then we went through the case with them several times to identify what made it challenging, and how they went about completing the assignment. Similar to Study 1, our goals for the use of the modified CDM were to learn about the types of information they used and how they obtained the information, schemata employed, hypotheses formulated, and types of products produced. The length of time our second group of interviewees had devoted to the assignments that they described ranged from 6 weeks to 3½ years. (The latter case involved the fact that the researcher had to work on the project intermittently, while serving on a U.S. Navy ship followed by attending graduate school at NPS.)

Example 1: Likelihood of a Coup in the Philippines

In this example, the intelligence analyst described his task of having to create a brief to answer a political question regarding whether President Estrada would be deposed from the Philippines, whether there would be a coup, and if there was a coup, what the implications would be for the Philippine Islands and for the United States? He was asked to complete this analysis task within a time span of 6 weeks on a topic that was outside his base of expertise (i.e., the geopolitical area).

From the initial search of raw reports, he produced an initial profile of what was known. Many additional searches and follow-up phone calls were conducted to fill in the gaps in his knowledge and to elaborate on what was learned during the initial set of queries. This step resulted in producing a large number of individual documents, one on each political person or key player. These included biographies on approximately 125 people, including insurgency leaders, people in various political groups, people with ties to crime, and so on. The information in these files was then grouped in various ways to consider several hypotheses. Next, he developed a set of questions to use to work backward to review all the material from several different perspectives to answer a series of questions related to the main question of interest: Will there be a coup? Will it be peaceful or not? Will it be backed by the military? Will the vote proceed, or will the military step in, prior to the vote? What is the most likely scenario to pan out?

Schemata

We define a schema as a domain-specific cognitive structure that directs information search, guides attention management, organizes information in memory and directs its retrieval, and becomes more differentiated as a function of experience. Schemata are a way of abstracting the information into a representation. The schema summarizes the external information by abstracting and aggregating information and eliminating irrelevant information. Schemata are structured to efficiently and effectively support the task in which they are embedded.

Figure 12.2 depicts the schema used to represent the dual-problem space of various information sources that the analyst researched to develop a comprehensive understanding of the issue and answer key questions designed to help answer the overarching question regarding a possible coup. The analyst began, in Week 1, by reading general background information to develop knowledge on the history and cultural ethnography of the country and also by examining prior Naval intelligence on the previous history for political turnover in the Philippines. During Week 2 he began contacting intelligence centers and reading U.S. Embassy cables, an important source for this particular topic. Although this step provided valuable information, because this material was from a secondary source it had to be corroborated. Thus the analyst had to decide which of these reports were to be given greater emphasis and in which reports he did not have much confidence.

One way the analyst structured his analysis was to sort people according to whether they were pro-Estrada or anti-Estrada, and who would be likely

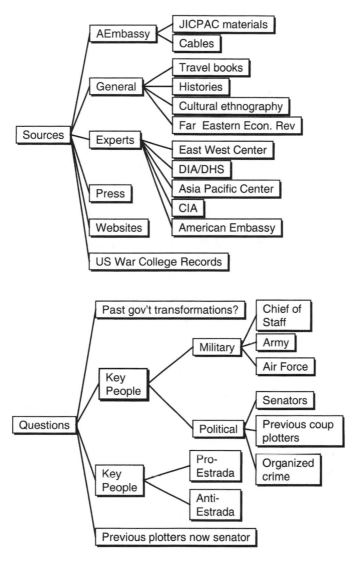

FIG. 12.2. Information foraging in a dual-problem space: schema for analyzing coup question.

to drop allegiance to the constitution, and so on. The analyst structured, and restructured, all the information to see how it might support various scenarios associated with the analysis questions. For example, if the United States invests money, will the country remain stable? How should the United States react? What is the most dangerous potential outcome? Most/least likely?

The analyst had five hypotheses that he used to organize his material. Previous coup attempts that occurred around the time of past-President Aquino were reviewed to examine how the allegiance of these people who were involved in past coup attempts might develop. Voting records provided another way to sort people. For a portion of his analysis, he used nodal analysis software to examine relationships between people. He used a whiteboard to play "20 questions" to come up with new questions to pursue. Relationship diagrams were constructed for each scenario and tables were developed to facilitate comparison of hypotheses. Many other sources were examined, such as political figures' ties to certain newspapers to determine which camp they would fall into. Additional schemata for grouping people to examine where their allegiance might lie involved examining all types of associations that exist between people, including family, military, political, as well as other forms of association such as persons who were classmates while attending school, shared board memberships, business ties, past coplotters who are now senators, and so on.

Multiple ways of grouping people were used by the analyst to consider competing hypotheses on how their allegiance would "fall out" based on their various associations. This analyst grouped key people in both the military and civilian sectors according to their military associations, political, family, geographic region, and various other associations, for example, professional groups and boards they belonged to, to try to ascertain their loyalty. The analyst developed many branches and sequels between people and events in his attempt to examine their affiliations from many different vantage points.

Example 2: Modernization of Country X's Military

This analysis problem evolved as a result of a discrepancy the analyst observed between the stated political military objectives of Country X and the observations made by this analyst during a 6-month deployment on an aircraft carrier. During his time as strike-plot officer he spent a lot of time collecting and sifting through raw message traffic and interpreting its meaning for the Battle Group. He had developed a considerable knowledge base for this part of the world and was aboard the carrier during the EP-3 crisis, in 2001, when it landed on Hainan Island. (An American EP-3 [a long-range Navy search aircraft] was forced down by the communist Chinese. It was forced to land in China and aircraft and crew were held for about a week by the Chinese government.) During the EP-3 crisis, he was able to provide background information on what had been occurring up to that point as well as during the crisis.

When this analyst reported to NPS to focus on Asia he noticed a disconnect between what professors described in terms of this country's political stance and things he had observed while operating in this part of the world. Things discussed in his courses were incongruent with the types of military training exercises he had observed this country engage in, and the types of military equipment acquisitions made by this country. He began with two or three factors that he knew could be used to support a separate hypothesis to explain the incongruity between what the political leaders are saying and what they are doing. His task was to compare the publicly stated policy of Country X regarding their planned military modernization with other possible scenarios for how things might evolve.

This analysis was based on a comparison of this country's officially stated military policy with data collected during detailed observations, and the associated daily reporting, that occurred over a 6-month period while the analyst was onboard the aircraft carrier. Table 12.6 presents a cognitive demands analysis of this analytical problem. The cognitive demands analysis developed using the Study 1 ACTA methodology was modified to represent the process that was used by this analyst. Because intelligence analysis involves an iterative process of data analysis and additional collection, we arranged the table to focus on specific data inputs and outputs. Additional columns include cues that generate processes that operate on data, and the strategies or methods used by the analyst to achieve goals when working with specific inputs and outputs. In addition, the table includes expert assessments of why specific inputs and outputs might be difficult. This provides indications of potential leverage points for system design. Finally, the table records specific examples mentioned by the analyst. These examples might be used as task scenarios to guide design and evaluation of new analyst strategies.

Analysis for this task included building the case for several other possible military scenarios regarding actions that might be taken by this country in the future. A comprehensive analysis of two competing hypotheses was developed to take into account future changes in political leadership, the economy, and sociopolitical factors. Data obtained on factors including economic stability, system acquisitions, and military training exercises conducted were manually coded on a daily basis, placed in a database, and aggregated over larger periods of time to depict trends.

For this intelligence problem, the analyst was looking for evidence to build the case to support several competing hypotheses regarding future political-military scenarios. Several types of information were viewed as indicative of the type of data that could be used to develop and substantiate alternative hypotheses and several methods were used to represent his analysis of the data. For example, a timeline was developed that depicted the following information. (a) location of U.S. forces, (b) geopolitical landscape of the world, and (c) the economy, based on economic decline affecting industry in the country. One scenario depicted a situation where the

TABLE 12.6

Cognitive Demands Table for Case 2: Develop Competing Hypothesis
Regarding Military Modernization Efforts of Country X

Inputs	Outputs	Cues/Goals	Strategy	Why Difficult?	Examples
Observations that support hypothesis that Country X has embarked on a different modernization effort for a number of years.	Data files that depict Country X's trends.	Compare stated modernization policy and economic trends within the country.	Evaluate the political landscape of Country X, by examining economic and cultural shifts in leadership to gain insight into ways they are looking to modernize.	Stated (public) policy says one thing; Observations point to potentially very different goals.	Types of military training exercises, equipment acquisitions.
Observed modernization efforts.	Determine Country X's military capability to conduct precision strike.	Does the political/economic/cultural environment support this operation?	Consider other possibilities beyond their stated military modernization goals.	Build a case for possibilities.	Use observations from exercises, purchases, etc., to see a different perspective, supported with data.
Many prior products: Intelligence sources, e.g., unclassified writings, interviews with political leaders.	Documents describing discrepancies between observed activities and stated policy.	Notice discrepancies between stated policy and observed activity.	Match up things seen in open press with what is occurring militarily.	How do observations relate to each other and to the stated policy?	Stated policy of Country X does not align with activities observed during exercises.

Classified sources; personal observations; anecdotal memories of deployments and experiences from past deployment.	Data files of detailed observations gathered over a 6-month period.	Help operational side of Navy explore a different view that is not based on established norms of thought.	Avoid "group think." Despite the mountain of evidence to the contrary, you don't want to "spool people up."	Difficult to distill the relevance of the information: Take 100 reports and find the gems.	Tendency is to report everything and treat everything as of equal importance.
Read message traffic all day	Two seemingly unrelated events are reported on individually.	Take analysis to next level of what is occurring.	Ask: "Does this make sense?"	Answer question: "Is this relevant?"	Goes against organizational constraints, i.e., events are "not to be considered outside normal routine training activity."
Volume of information is constrained to the geographic area.	Graphs to depict trends of different types of activity.	Factor in Army or ground troop movement in addition to Navy activity.	Classify information as relevant or irrelevant. Maintain databases of activity, e.g., by day/week/months.	Several hours a day sorting through message traffic; If had a crisis would be completely saturated.	Group all different categories of activity, e.g., local activity, aggressive activity, exercise activity.
Read everything can find.	1. Brief for the Commander each day. 2. Daily Intel Analysis Report.	Pick out things that are relevant.	Take raw message traffic (w/o anyone's opinion associated with it).	Databases do not match up (even capabilities listed in them).	Extract what think is relevant and highlight activity thought to be relevant.
Based on observations of activities that did not match up with what others believed.	Form a model of the situation; imagine events over time.	Force people to look at a different possibility.	Build "Perry Mason" clinch argument.	Organizational constraints not to "go against the grain."	Had lots of documented real-world observations.

(Continued)

TABLE 12.6
(Continued)

Inputs	Outputs	Cues/Goals	Strategy	Why Difficult?	Examples
Data on emerging political environment in transition.	Understanding of relationships between important actors.	Paramount to understand who is driving what action.	New leadership person is still "driving" things: Added credibility to thesis that there is a split.	Could not get access to all material (databases) needed for analysis.	Inconsistent capabilities listed in different databases.
Location of U.S. forces; geopolitical landscape; economic decline affecting Country X.	Build timeline to depict more aggressive posture.	West will not have same influence on economy which leads to political unrest: Political rivalry between old/new leadership.	Describe political factors that could set off a change in direction. Set stage for how things could go in a fictional scenario.	Credibility issue: operational guys rarely understand analysis, especially strategic.	When presented brief on threat, operational personnel did not perceive information as representative of a threat.
Difference between what they're saying and what they're doing.	Revised hypothesis.	Initially 2–3 factors that will support a separate hypothesis from the accepted hypothesis on what is transpiring.	Marshal evidence to support alternate hypothesis.	Selecting which pieces of information to focus on.	Fact that found so many pieces to support hypothesis indicates hypothesis has to be considered.

West would not have the same influence on the economy and the fallout will be some political unrest. Political rivalry between the old and new leadership will ensue and the scale will tip to the negative side as a result of political factors that have "gone south." Congressional papers were used, in addition to all the information developed by this analysis, to write a point paper on an assessment of this country's military activity and the kind of threat he saw as a result of his analysis.

With the two previous examples to provide an illustrative context, we now summarize the main results from the Study 2 CDM procedures. These all converge on the notion of sensemaking. Sensemaking is indubitably one of the cognitive processes performed by the intelligence analyst to understand complex, dynamic, evolving situations that are "rich with various meanings (Heuer, 1999, used such terms as "mind-sets" and "schemas"). Klein et al. (chap. 6, this volume) describe sensemaking as the process of fitting data into a frame, an explanatory structure, for example, a story, that accounts for the data, and fitting a frame around the data. The frame that is adopted by the intelligence analyst will affect what data are attended to and how these data items are interpreted. When the intelligence analyst notices data that do not fit the current frame, the sensemaking cycle of continuously moving toward better explanations is activated. Sensemaking incorporates consideration of criteria typically used by intelligence analysts: plausibility, pragmatics, coherence, and reasonableness (chap. 6, this volume).

Sensemaking applies to a wide variety of situations. As Klein et al. (chap. 6, this volume) describe it, sensemaking begins when someone experiences a surprise or perceives an inadequacy in the existing frame. Sensemaking is used to perform a variety of functions, all related to the intelligence analyst's job, including problem detection, problem identification, anticipatory thinking, forming explanations, seeing relationships, and projecting the future.

DISCUSSION

Intelligence analysis is an intellectual problem of enormous difficulty and importance (Wirtz, 1991). We, and others, find abundant evidence that the problems with which intelligence analysts are tasked often involve forcing them to work "out of context," and we have cited examples that illustrate that, such as the analyst who had to work on a problem that involved a country with which he initially knew next to nothing.

Many opportunities for tool development to assist the processes used by intelligence analysts exist. Under such programs as NIMD and programs under the Defense Advanced Research Projects Agency (DARPA), prototype tool development has begun, and will continue in conjunction with the next phase of CTA being conducted by us and other researchers who are working in this area. Because an ultimate goal is to develop computational models of the intelligence analyst's tasks, detailed data must be captured on analysts while they are performing their tasks. Use of process-tracing methods, for example, verbal protocol analysis, in conjunction with the Glass Box software, developed for the NIMD Program (ARDA, 2002), might provide a rich source of data to develop a detailed model of the intelligence analyst's processes. NIMD's Glass Box is an instrumented environment that collects data on analyst taskings, source, material, analytic end products, and analytic actions leading to the end products (Cowley, Greitzer, Littlefield, Sandquist, & Slavich, 2004).

Use of an instrumented data collection environment in conjunction with think-aloud protocol analysis might enable us to gather detailed knowledge about the knowledge and cognition entailed in intelligence analysis. The next phase of this CTA will involve asking SMEs to perform an analysis task while thinking aloud. This technique typically provides detailed data concerning the mental content and processes involved in a specific task.

Identification of an appropriate sample of problems or tasks is essential to ensure sufficient coverage of critical skills and knowledge. The initial set of CTA interviews we conducted was intended to help us develop a foundation of knowledge regarding the intelligence analyst's task domain. During the next phase of this research, additional data will be gathered to further refine the CTA-based model of intelligence analysis. The CTA process is an iterative process that builds on subsequent design activities. New tools and training will impact the cognitive activities to be performed and enable development of new strategies. An additional goal, one that lies in the near future, involves Woods's "envisioned world problem" (Woods & Hollnagel, 1987)—how to predict the impact the technology will have on cognition for the intelligence analyst.

ACKNOWLEDGMENTS

This research was supported by Susan Chipman in the Office of Naval Research.

REFERENCES

Advanced Research and Development Activity (ARDA). (2002). Program on Novel Information from Massive Data (NIMD). Retrieved August 10, 2005, from http://www.ic-arda.org/Novel_Intelligence/mass_data_struct_org_norm.htm

Annett, J. (2000). Theoretical and pragmatic influences on task analysis methods. In J. M. Schraagen, S. F. Chipman, & V. L. Shalin (Eds.), *Cognitive task analysis* (pp. 25–37). Mahwah, NJ: Lawrence Erlbaum Associates.

Bodnar, J. W. (2003). *Warning analysis for the information age: Rethinking the intelligence process* (Report). Washington, DC: Joint Military Intelligence College.

Chipman, S. F., Schraagen, J. M., & Shalin, V. L. (2000). Introduction to cognitive task analysis. In J. M. Schraagen, S. F. Chipman, & V. L. Shalin (Eds.), *Cognitive task analysis* (pp. 3–23). Mahwah, NJ: Lawrence Erlbaum Associates.

Clark, R. M. (2004). *Intelligence analysis: A target-centric approach.* Washington DC: CQ Press.

Connable, A. B. (2001, June). *Open source acquisition and analysis: Leveraging the future of intelligence at the United States Central Command.* Unpublished Masters Thesis, Naval Postgraduate School, Monterey, CA.

Cooke, N. J. (1994). Varieties of knowledge elicitation techniques. *International Journal of Human–Computer Studies, 41,* 801–849.

Cooke, N. J., Salas, E., Cannon-Bowers, J. A., & Stout, R. (2000). Measuring team knowledge. *Human Factors, 42,* 151–173.

Cooper, J. R. (2005, December). *Curing analytic pathologies: Pathways to improved intelligence analysis.* Washington, DC: Center for the Study of Intelligence, Central Intelligence Agency.

Cowley, P. J., Greitzer, F. L., Littlefield, R. J., Sandquist, R. & Slavich, A. (2004). Monitoring user activities in the glass box analysis environment (Panel on Designing Support for Intelligence Analysts). In Proceedings of the 48th Annual Meeting of the Human Factors and Ergonomics Society (pp. 357–361). Santa Monica, CA: Human Factors and Ergonomics Society.

Crandall, B., Klein, G., & Hoffman, R. R. (2006). *Mind work: A practitioner's roadmap to cognitive task analysis.* Cambridge, MA: MIT Press.

DuBois, D., & Shalin, V. L. (2000). Describing job expertise using cognitively oriented task analyses (COTA). In J. M. Schraagen, S. F. Chipman, & V. L. Shalin (Eds.), *Cognitive task analysis* (pp. 317–340). Mahwah, NJ: Lawrence Erlbaum Associates.

Flach, J. M. (2000). Discovering situated meaning: An ecological approach to task analysis. In J. M. Schraagen, S. F. Chipman, & V. L. Shalin (Eds.), *Cognitive task analysis* (pp. 317–340). Mahwah, NJ: Lawrence Erlbaum Associates.

Garst, R. (1989). Fundamentals of intelligence analysis. In R. Garst (Ed.), *A handbook of intelligence analysis* (2nd ed.). Washington, DC: Defense Intelligence College.

Heuer, R. J. (1999). *The psychology of intelligence analysis.* Washington, DC: Center for the Study of Intelligence.

Hoffman, R. R., Coffey, J. W., Carnot, M. J., & Novak, J. D. (2002). An empirical comparison of methods for eliciting and modeling expert knowledge. In *Proceedings of the Human Factors and Ergonomics Society 46th Annual Meeting* (pp. 482–486). Santa Monica, CA: Human Factors and Ergonomics Society.

Hoffman, R. R., Coffey, J. W., & Ford, K. M. (2000). *A case study in the research paradigm of human-centered computing: Local expertise in weather forecasting* (Report on the Contract, "Human-centered system prototype." Arlington, VA: National Technology Alliance.

Hoffman, R. R., Crandall, B., & Shadbolt, N. (1998). A case study in cognitive task analysis methodology: The Critical Decision Method for the elicitation of expert knowledge. *Human Factors, 40,* 254–276.

Hoffman, R. R., Shadbolt, N. R., Burton, A. M., & Klein, G. (1995). Eliciting knowledge from experts: A methodological analysis. *Organizational Behavior and Human Decision Processes, 62,* 129–158.

Johnston, R. (2004). Retrieved August 10, 2005, from http://www.cia.gov/csi/studies/vol147no1/article06.html

Johnston, R. (2005). *Analytic culture in the United States intelligence community: An ethnographic study.* Washington, DC: U.S. Government Printing Office.

Joint Military Intelligence Training Center. (1996). *Open Source Intelligence: Professional Handbook.* Washington, DC: Open Source Solutions.

Josephson, J., & Josephson, S. (1994). *Abductive inference.* Cambridge, England: Cambridge University Press.

Klein, G. A., (2001). *Sources of power: How people make decisions.* Cambridge, MA: MIT Press.

Klein, G. A., Calderwood, R., & MacGregor, D. (1989). Critical decision method for eliciting knowledge. *IEEE Transactions on Systems, Man, and Cybernetics, 19,* 462–472.

Krizan, L. (1999). *Intelligence essentials for everyone* (Occasional Paper No. 6). Washington, DC: Joint Military Intelligence College.

Militello, L. G., & Hutton, R. J. B. (1998). Applied Cognitive Task Analysis (ACTA): A practitioners toolkit for understanding task demands. *Ergonomics, 41,* 164–168

Militello, L. G., Hutton, R. J. B., Pliske, R. M., Knight, B. J., & Klein, G. (1997). *Applied cognitive task analysis (ACTA) methodology* (Tech. Rep. No. NPRDC-TN-98-4). Fairborn, OH: Klein Associates.

Millward, W. (1993). Life in and out of Hut 3. In F. H. Hinsley & A. Stripp, *Codebreakers: The inside story of Bletchley Park* (pp. 19–29). Oxford, England: Oxford University Press.

Moore, D. T. (2003). *Species of competencies for intelligence analysis.* Washington, DC: Advanced Analysis Lab, National Security Agency.

National Commission on Terrorist Attacks. (2004). *Final Report of the National Commission on Terrorist Attacks Upon the United States.* Washington, DC: Author.

Patterson, E. S., Roth, E. M., & Woods, D. D. (2001). Predicting vulnerabilities in computer-supported inferential analysis under data overload. *Cognition, Technology & Work, 3,* 224–237.

Pirolli, P., & Card, S. K. (1998). Information foraging models of browsers for very large document spaces. In *Advanced Visual Interfaces (AVI) Workshop* (AVI '98) (pp. 75–83). New York: ACM Press.

Pirolli, P., & Card, S. K. (1999). Information foraging. *Psychological Review, 106,* 643–675.

Pirolli, P., Fu, W.-T., Reeder, R., & Card, S. K. (2002). A user-tracing architecture for modeling interaction with the World Wide Web. In *Advanced Visual Interfaces* (AVI 2002). New York: ACM Press.

Potter, S. S., Roth, E. M., Woods, D. D., & Elm, W. C. (2000). Bootstrapping multiple converging cognitive task analysis techniques for system design. In J. M. Schraagen, S. F. Chipman, & V. L. Shalin (Eds.), *Cognitive task analysis* (pp. 317–340). Mahwah, NJ: Lawrence Erlbaum Associates.

Vaughan, D. (1996). *The Challenger launch decision: Risky technology, culture, and deviance at NASA.* Chicago: University of Chicago Press.

Wirtz, J. W. (1991). *The Tet Offensive: Intelligence failure in war.* Ithaca, NY: Cornell University Press.

Woods, D. D., & Hollnagel, E. (1987). Mapping cognitive demands in complex problem-solving worlds. *International Journal of Man–Machine Studies, 26,* 257–275.

Woods, D. D., Patterson, E. S., & Roth, E. M. (2002). Can we ever escape from data overload?: A cognitive systems diagnosis. *Cognition, Technology, and Work, 4,* 22–36.

Chapter 13

Human-Centered Computing for Tactical Weather Forecasting: An Example of the Moving-Target Rule

James A. Ballas
Naval Research Laboratory, Washington, DC

Designing intelligent tools is always challenging, but especially so if the larger system is changing in its essential characteristics. This is called the Moving-Target Rule (Hoffman, Coffey, & Ford, 2000) and is stated as follows:

> *The sociotechnical workplace—the environment in which cognitive work is conducted— is constantly changing. Constant change in the constraints that are imposed by the work environment entails constant change in the requirements for cognition, even if the constraints coming from the domain remain constant.*

The rule is clearly relevant to weather forecasting. The weather-forecasting workplace is constantly changing, involving swap-outs of entire workstations as well as software changes (new software suites as well as upgrades to existing ones), and the introduction of new data types and displays. Weather forecasters are almost always having to work out-of-context. As a part of a year-long project on human-centered computing (HCC) for weather forecasting in the U.S. Navy, Hoffman et al. (2000) conducted a number of procedures of cognitive work analysis, only to find themselves standing on shifting sand: The main forecasting workstation, both hardware and software, was changed out, leading to cascades of changes in forecasting procedures, standard operating procedure (SOP) guidance, and so on.

The U.S. Navy weather systems that Hoffman et al. (2000) studied were supporting military operations in the continental United States. In this chapter, I focus on forecasting weather for tactical military operations and how the Moving-Target Rule seems to be in force in this venue as well. Tactical weather forecasting presents some special (out-of-context) challenges including the likelihood that the forecaster will be asked to provide weather for operations in areas for which there are few or no local weather reports and observations. For several reasons, including the high level of U.S. military operations recently, nearly every major aspect of the tactical weather forecasting has been changing. These include changes in the forecasting computer systems, the software in these systems, and changes in the concept of weather-forecasting operations as well as in the military operations that must be supported. The latter type of change is driven by the transformation in the U.S. military, which in turn, drives change in the tactical forecasting process. Because of these changes, HCC computing for tactical forecasting has been an exemplar instance of the Moving-Target Rule.

CHANGE IN SOFTWARE AND COMPUTER SYSTEMS

One of the major changes in forecasting software is the increasing use of high-resolution numerical forecasting models. During the 1999 military campaign in Yugoslavia, the U.S. Air Force deployed a system developed by the Aerospace Corporation that increased the resolution of cloud forecasts from 40 kilometers to 6 kilometers (The Aerospace Corporation, May, 1999). An example of the increased resolution images provided by the system developer is presented in Figure 13.1. A major benefit of these models is that they reflect the effects of local terrain, which can be an important consideration in planning the flight path of a military flight operation that will need to fly close to the ground.

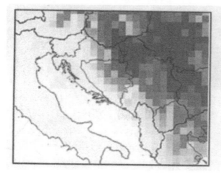 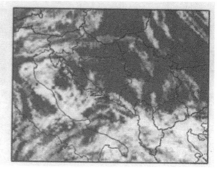

FIG. 13.1.   Images from low-resolution (left) and a high-resolution (right) numerical forecasting model. Actual images used in forecasting rely on use of color. Images courtesy the Department of Defense and The Aerospace Corporation.

However, high-resolution models always have to be nested within a lower resolution model—the computational models begin their calculations with initial "boundary conditions" from the lower-resolution model. Therefore, the accuracy of the nested model depends on the accuracy of the lower resolution model, and one of the forecaster's tasks is to ensure that the accuracy of the nesting model is sufficient to support the high-resolution nested model.

The verification of numerical models represents a significant challenge to the military forecaster. An essential process in tactical forecasting is the daily assessment and comparison of the models. This is often done using another type of software tool—Internet relay chat (IRC) (National Research Council, 2003; Commander, Naval Meteorology and Oceanography Command, 1999). IRC supports real-time, synchronous collaboration among the forecasters who are assigned to support an operation. IRC is used because the support team might not be in the theater of operations, but elsewhere in the world. The chat is not free-form; it is usually structured and led by a moderator. One method of structuring the session is to go through a forecast (which has been posted beforehand on the internet) section by section. In order to stay on the structured discussion, if peripheral topics arise, agreements are made to either stay on afterward or join another session to discuss the peripheral topics.

The IRC discussions might focus on which models appear to be correct and which do not. The "model of the day" can then be used for forecasting. Alternatively, instead of discarding the further use of the incorrect model, an HCC approach would allow the human to make fixes to models that over- or underpredict weather phenomena. This type of human correction is supported in a system called On Screen Field Modification (Hewson, 2004) in the United Kingdom. When a model is judged to be biased, the forecaster can manipulate the numerical fields in the four-dimensional (latitude, longitude, altitude, and time) forecasted data to produce a new set of fields. This approach to numerical modeling lets the forecaster produce output that might more closely match the forecaster's mental model than the unmodified output. For example, a forecaster can graphically move a barometric high to a new location, or make a slight change to the maximum barometric pressure, and this will numerically propagate through the four-dimensional data field. Though such capability simplifies the generation of model-based forecast products that are consistent with the forecaster's mental model, the implications of this direct manipulation of numerical models are controversial (e.g., Scott et al., 2005).

The changes in software in the military weather-forecasting facility come about not only because of advances in the computer technology, but also because the U.S. military has been upgrading its computer systems on ships and ashore. In the Navy, the Information Technology-21 program puts desktop-to-desktop connectivity into ships, using PCs with the same operating system (Windows) and applications (Microsoft Office, Web browsers, etc.) that are used in commercial office environments. Thus we find military weather forecasters using Web-based technology (i.e., browsers, IRC chat, web page services) to distribute and retrieve information.

CHANGES IN MILITARY SYSTEMS SUPPORTED BY FORECASTING

Change in tactical weather forecasting is also being driven by changes in the military systems that must be supported. For example, the introduction of unmanned aerial vehicles (UAVs) has created a need to be able to generate weather forecasts for ultralight air vehicles that operate effectively only in benign weather (Persinos, 2002; Teets, Donohue, Underwood, & Bauer, 1998). Because UAVs do not have an on-board operator, the local weather observations must be done through electronic sensors rather than eyeballs, and the reporting of these observations to the aircraft controller is delayed.

Sensor systems are also rapidly changing, and tactical forecasting must provide information about the capabilities of new air, space, and ocean sensors. I elicited an example of this type of forecasting support by using a critical-incident technique with a senior Navy forecaster. An air-based visual sensor was producing images that seemed blurred, and the forecaster was asked to provide information about the problem. The images were being taken of a particular area, and the forecaster determined that the blurring was not in the image or its processing, but was due to suspended dust produced by winds that were funneled by the terrain at particular times in the day. His recommendation was to reschedule the flights. With this done, the images that came were much clearer. The forecaster in this case had not been trained explicitly about this phenomenon. But he had been trained about the effects of terrain and diurnal weather phenomena, and had the insight to apply this general knowledge to the specific problem.

The way in which military systems are operated ("concept of operations" or CONOPs) is also changing as part of the military transformation process. In particular, planning and decision timelines are being shortened. Tactical weather forecasting must support these shorter timelines, which means that the forecasting process must be shortened. For example, in the Kosovo rescue of a downed F-117A, weather forecasters provided minute-by-minute updates to the aircrews as they flew in to rescue the airman (American Meteorological Society, 1999). This is a departure from the traditional weather support that occurs in the mission planning stage and in a preflight brief.

Addressing the forecasting requirements of new military systems and CONOPs is a formidable challenge to the tactical forecaster. The net effect is that the tactical forecaster may have a large number of potential customers, each having unique forecasting requirements. These customers will want their products quickly. One method of serving these customers that we observed in air strike forecasting (Jones et al., 2002) was the development and distribution of a threshold matrix that summarizes the effects of forecasted weather conditions on missions. The matrix is a type of spreadsheet that illustrates the effects of weather on particular aspects of the mission such as take-off, en-route flight, sensing, in-air refueling, weapon delivery, and so on. The threshold matrix uses the military standard "stoplight" color codes (red for severe effect; yellow for moderate effect; green for no effect) and documents the effect of the weather parameters on the mission by time period and by area. Because the color code provides little information about how the weather effect might be distributed over an area, some versions of this matrix have a link to a visualization that overlays the color coding over the area of interest.

The production of this type of matrix takes tactical weather forecasting from a process involving meteorological cognition into one that includes aspects of tactical decision making. Although there are systems that can produce this matrix through the formal application of threshold rules to weather data, go/no-go decisions have much in common with statistical decisions because the underlying phenomena to which the thresholds are applied have statistical variability. Therefore, in practice, the forecaster has to have some flexibility in applying the rules and be able to use knowledge about the details of the mission. Forecasters are sometimes assigned to the operations rooms (e.g., Clancy, 1999) where they are available to provide rapid weather assessments. In these situations, they will be aware of the details of ongoing operations, and this tactical information will likely be a factor in their weather effects decisions. Under a current project, my colleagues and I are applying HCC notions to support this process by providing:

- A flexible method of editing the rules.
- A summary of the magnitude of the weather field that falls above the threshold.
- A method of easily editing the computed decision and the associated graphic so as to show the threshold effects.

We observed the production of a matrix such as this on the USS *Carl Vinson* in 2001, and found that it could take approximately 2 hours to produce this matrix (Jones et al., 2002). We found that with an improved work flow including software that computes a "first answer," we can significantly decrease this time and enable the forecaster to modify the rules and the outcome (Ballas et al., 2004).

A FRAMEWORK FOR CHARACTERIZING TACTICAL
FORECASTING AND THE MOVING TARGETS

In order to illustrate the larger context that the matrix implies, I have adapted a model of decision making being developed under a NATO project (NATO, 2006). The model coming out of the NATO project is an extension of the OODA (observe, orient, decide, act) loop and is shown in Figure13.2 for tactical weather forecasting. While legitimate questions can be raised about whether people (including military decision makers) actually reason this way, or *should* reason this way (see Klein et al., 2003), the OODA model is sufficient for present purposes.

Weather information is obtained through sensors, satellite, weather reports, and so forth, and is viewed by the forecaster in various visualization forms (Jones et al., 2002; see also chap. 14, this volume). Using this information, the forecaster builds a mental model. The mental-model process in weather forecasting has been described by several researchers (Hoffman, 1991; Pliske et al., 1997; Trafton et al., 2000), all of whom describe the initial stage in similar terms as an examination of the current weather situation. Hoffman describes the subsequent stages as hypotheses testing from the initial conceptual model, with the specific hypotheses driven by factors

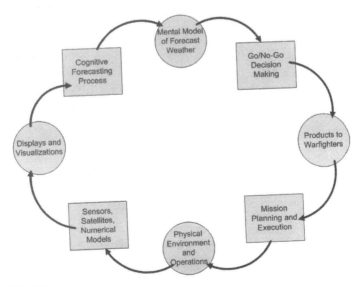

FIG. 13.2.   Decision process model of tactical forecasting within an operational context.

such as the weather situation, the problem of the day, or testing and refining the initial conceptual model (see also chap. 2, this volume). For Pliske et al., the subsequent stages involve construction of an understanding of the dynamics that are causing the current weather and the use of this understanding to predict the future weather. For Trafton et al, the subsequent stages involve construction of a qualitative mental model for an area and verification and adjustment of this mental model.

Using this mental model together with information about mission parameters, operational weather rules, and information from other decision aids, the forecaster engages in one or more decision-making tasks to produce products for the warfighter. These tasks include making go/no-go recommendations about missions, performing quality control of the output of numerical models and if necessary, revising the output, and executing decision-aiding software. The tactical forecasting products are delivered to both mission planners and those commanding an ongoing mission, as illustrated by the minute-to-minute updates in the Kosovo rescue.

This theoretical sketch of forecaster reasoning can serve several purposes. Besides showing how the forecasting process might be represented in its larger context, and how the forecaster's task involves aspects of decision making, it places the types of moving-target changes that I have discussed within a cognitive framework that allows expression of the different forms of the information. As aspects of the overall system change, the processes that must adjust to these changes can be indicated in the framework.

The challenge of relying on any framework such as this is that although it might provide some utility in identifying how change in some aspect of a system will affect other components of the overall process, it also raises a risk that an initial focus on human-centering considerations will derail into technology-driven design. The real challenge posed by the Moving-Target Rule is to stay on-target to the goal of ensuring a system that is human centered in its critical human-implicated components.

CONCLUSION

These examples of how tactical weather forecasting is changing (software and computing systems, and the missions that are supported by forecasting) illustrate only two categories of changes. There are also changes in the CONOPS of the forecasting organizations themselves and the training systems. Together, these moving targets make the challenge of supporting HCC in this domain formidable. One important approach to meet this challenge is to tightly integrate HCC design/analysis/evaluation processes into the larger systems design/analysis/test processes (Bisantz et al., 2003; Hoffman, Lintern, & Eitelmnan, 2004). The integration has to occur by getting cognitive systems engineers on system design teams. The success of this type of integration depends on how well the individuals who are focused on human centering can interact with technology-focused individuals. Some of the principles of HCC, including the Moving-Target Rule (see also Endsley & Hoffman, 2002), should resonate with engineers, and a reliance on these principles should assist in bridging communication and cultural gaps that may make for a gulf between the engineering of intelligent systems and the human centering of such systems.

ACKNOWLEDGMENTS

This research was supported by work order N0001403WR20093 from the Office of Naval Research. The views and conclusions contained in this document are those of the author and should not be interpreted as necessarily representing the official policies, either expressed or implied, of the U.S. Navy.

REFERENCES

The Aerospace Corporation. (1999, May). Aerospace Corporation team helps improve weather forecasting for Yugoslavia operations. The Aerospace Corporation news release. Available at: http://www.aero.org/news/newsitems/forcasting-051699.html

American Meteorological Society. (1999, May). Meteorologists face major challenges in providing forecasts in Kosovo. American Metrological Society Newsletter, 20. Available at: http://www.ametsoc.org/newsltr/nl_5_99.html

Ballas, J. A., Tsui, T., Cook, J., Jones, D., Kerr, K., Kirby, E., et al. (2004) Improved workflow, environmental effects analysis and user control for tactical weather forecasting. In Proceedings of the Human Factors and Ergonomics Society 48th Annual Meeting, Santa Monica, CA: Human Factors and Ergonomics Society, 48, 320–324.

Bisantz, A. M., Roth, E., Brickman, B., Gosbee, L. Hettinger, L., & McKinney, J. (2003). Integrating cognitive analyses in a large-scale system design process. International Journal of Human–Computer Studies, 58, 177–206.

Clancy, T. (1999). Every man a tiger: The Gulf War air campaign. New York: Berkeley Books.

Commander, Naval Meteorology and Oceanography Command. (1999, March 31). Navy weather fills a vital role in JTFEX [News release]. Commander, Naval Meteorology and Oceanography Command, 1020 Balch Boulevard, Stennis Space Center, MS.

Endsley, M., & Hoffman, R. (2002, November/December). The Sacagawea principle. IEEE Intelligent Systems, pp. 80–85.

Hewson, T. (2004). The value of field modifications in forecasting (Forecasting Research Tech. Rep. No. 437). Devon, England: The Met Office. Available at: http://www.met-office.gov.uk

Hoffman, R. R. (1991). Human factors psychology in the support of forecasting: The design of advanced meteorological workstations. Weather and Forecasting, 6, 98–110.

Hoffman, R. R., Coffey, J. W., & Ford, K. M. (2000). *A case study in the research paradigm of human-centered computing: Local expertise in weather forecasting* (Report). Pensacola, FL: Institute of Human and Machine Cognition.

Hoffman, R. R., Lintern, G., & Eitelman, S. (2004, March/April). The Janus principle. *IEEE: Intelligent Systems*, pp. 78–80.

Jones, D., Ballas, J., Miyamoto, R. T., Tsui, T., Trafton, J. G., & Kirschenbaum, S. (2002). *Human systems study on the use of meteorology and oceanography information in support of the naval air strike mission* (University of Washington Tech. Memorandum APL-UW-TM 8-02). Seattle, WA: Applied Physics Laboratory, University of Washington.

Klein, G., Ross, K. G., Moon, B. M., Klein, D. E., Hoffman, R. R., & Hollnagel, E. (2003, May/June). Macrocognition. *IEEE Intelligent Systems*, pp. 81–85.

National Research Council. (2003). *Environmental information for naval warfare*. Washington, DC: National Academies Press.

NATO. (2006). Modelling of organisations and decision architectures (Final report of NATO RTO Task Group IST-019/RTG-006). Paris: NATO Research and Technology Organization.

Persinos, J. (2002, March). New technology: UAVs: pie in the sky? *Aviation Today: Rotor & Wing*, pp. 38–44.

Pliske, R., Klinger, D., Hutton, R., Crandall, B., Knight, B., & Klein, G. (1997). *Understanding skilled weather forecasting: Implications for training and the design of forecasting tools* (Report No. ALFRL-HR-CR-1997-0003). Brooks AFB, TX: Armstrong Laboratory.

Scott, C. A., Proton, V. J., Nelson, J. A., Jr., Wise, R., Miller, S. T., & Eldred, C. (2005, January). *GFE—The Next Generation*. Paper presented at the 21st International Conference on Interactive Information Processing Systems (IIPS) for Meteorology, Oceanography, and Hydrology, San Diego, CA.

Teets, E. H., Donohue, C. J., Underwood, K., & Bauer, J. E. (1998). *Atmospheric considerations for uninhabited aerial vehicle (UAV) flight test planning* (NASA/TM-1998-206541). Edwards, CA: National Aeronautics and Space Administration, Dryden Flight Research Center.

Trafton, J. G., Kirschenbaum, S. S., Tsui, T. L., Miyamoto, R. T., Ballas, J. A., & Raymond, P. D. (2000). Turning pictures into numbers: Extracting and generating information from complex visualizations. *International Journal of Human–Computer Studies, 53*, 827–850.

Chapter 14

Comparative Cognitive Task Analysis

Susan S. Kirschenbaum
Naval Undersea Warfare Center Division Newport, Newport, RI

J. Gregory Trafton
Naval Research Laboratory, Washington, DC

Elizabeth Pratt
University of Connecticut

It is easy to force a weather forecaster to work out of context—simply move him or her to some new locale. It takes at least one full seasonal cycle for forecasters to reacquire expertise. Worse, move a forecaster from the Northern Hemisphere to the Southern Hemisphere. Major things change. Low-pressure systems spiral clockwise, not counterclockwise. Effects of ocean currents, seasonal variations, and effects of land masses change everything. Any knowledge of trends that the forecaster had relied on are now utterly useless.

In the studies we report in this chapter, we did something like this, but the switch involved making *us*, as cognitive systems engineers, work out of context. Work on forecaster reasoning with which we are familiar (e.g., Hoffman, 1991), including on our own research, has involved the study of forecasters in the U.S. Navy and U.S. National Weather Service. We think we have some ideas about how forecasters think (see chap. 15, this volume), but are we sure? How does what we *think* we know transfer to, say, forecasting in Australia? Or does it transfer? What if we were to advise building some new tool, only to learn that it does not help forecasters in regions other than the continental United States?

Weather forecasting is a complex process. The supporting information is multidimensional, distributed, and often uncertain. It includes both "raw" observations (e.g., current temperature, winds, pressure, clouds, precipitation, radar returns, satellite pictures, etc.) and analytic weather models that predict future weather conditions at various scales of space and time. The information that the weather forecaster uses is often downloaded from external Web sites. Local weather organizations use (or build) support tools for displaying downloaded data and images and for building and displaying their own forecasts.

To optimize these tools, consideration must be given to the user-tool-task triad that is central to the principles of human-centered computing (HCC) (Hoffman, Coffey, Ford, & Bradshaw, 2001; Hoffman, Hayes, Ford, & Hancock, 2002). These principles require the designer to build tools that facilitate the task that the user does and accommodate human perceptual, cognitive, and motor functioning. How does the designer incorporate the user-tool-task triad of HCC into this complex and specialized domain? How does the designer gain enough knowledge of the users' tasks and processes to provide useful assistance? And how does the designer disentangle the effects of task, training, teamwork arrangements, and basic human cognition from those of the design of the tools?

The traditional way human factors engineers approach this problem is to perform a task analyses to determine how people operate in a specific domain on a specific task. Cognitive Task Analysis (CTA) is a set of methods that takes into account the perception (e.g., vision), cognition (e.g., decision making), and motor actions (i.e., mouse movements) needed to accomplish a task. In this chapter, we build on CTA methods by suggesting that comparative cognitive task analysis (C2TA) can help solve the aforementioned problems. C2TA is based on replication studies conducted in different environments. Replication is a basic principle of the scientific method, but usually replication aims at duplicating the original conditions. Comparative studies are also a common scientific practice. Within the HCC literature, comparative studies usually employ a traditional experimental design to ask such questions as which device or design is faster (Haas, 1989), more efficient (Haas, 1989), and/or lowers workload (Kellogg & Mueller, 1993). However, CTA is often an exploratory research strategy that focuses on process rather than final performance (Sanderson & Fischer, 1994). C2TA draws on all these traditions, applying elements of replication and comparative methods to the exploratory process approach of CTA. Because it derives data from more than one environment, C2TA provides insight into interface design that single-site studies and individual CTA methods cannot.

There are many versions of task analysis ranging from time-and-motion study (Gilbreth & Gilbreth, 1917) to GOMS (goals, operators, methods, selection rules) analysis (Card, Moran, & Newell, 1983), to ecological interface design (EID) (Vicente & Rasmussen, 1992). Each is best suited to particular aspects of design problems. For example, GOMS analysis is a keystroke-level process for describing human–computer interactions (e.g., mouse and keyboard interactions). EID focuses on how the operator interacts with indicators of physical functioning such as in a power plant or manufacturing control room. CTA is especially useful in situations where the task is heavily dependent on human interpretation and integration of dynamic and highly uncertain data (Schraagen, Chipman, & Shalin, 2000).

Weather forecasters typically deal with large amounts of data over time and space (Hoffman, 1991). Additionally, the information they examine is uncertain on several dimensions (i.e., the predictive weather models that are run may be based on a small number of data points in some areas—like in the middle of the ocean—which necessitates interpolating from the current data, which may cause the final output to be more uncertain). The need for expertise at interpreting the weather prediction models, the dynamic nature of weather, and the uncertainty in the weather models makes weather forecasting an excellent candidate for CTA.

However, most of the data analyzed by CTA methods come from a single source (i.e., most CTA studies have been performed on a single system and/or a small group of people). Although the single approach is adequate in many situations, it may not be as generalizable as it could be. That is, any problems might be traced to the interaction between the person and the system. You may discover, for example, that a specific pointing device is not very effective on a particular system, but you do not know if that is a limitation of the pointing device or the way people (in general) think, or the way people in a particular organization think; you only know that the combination of people and pointing device on the task you are examining is not very effective. By examining different tools (i.e., different types of pointing devices on similar tasks), you can start to dissociate the effects of cognition and those of the tool.

For example, the pen, the typewriter, and the computer keyboard are all tools that can be used for writing a document. The writing process consists of planning, composing, editing, and production (writing/typing). The quantity and sequence of these processes is differentially supported by the three tools. The computer supports longer compositions, however, the writer plans longer before editing with a pen (Haas, 1989). This may be because editing with a pen includes crossing out, rewriting, cutting pages apart and taping them back together, arrows for inserts, and so on, and then repeating the production process (rewriting) on clean paper. Editing on a typewriter uses similar cross-out, cut, glue, and retype processes. With both of these tools, the rewrite (production) process is effortful. However, writers using a computer edit more as they write and new versions do not require redoing the physical production (Kellogg & Mueller, 1993).

The data for the two analyses reported here were collected during two studies in two different locations, a United States Navy (USN) Meteorology and Oceanography (METOC) center in California and a Royal Australian Navy (RAN) METOC facility. These studies employed the methods of cognitive field research and quasi-naturalistic observation in what Hoffman and Deffenbacher (1993) termed a "laboratory field" study. The studies were part of a project to provide improved tools for Navy weather forecasting. Only by understanding current practices and forecasting tools could improvements be suggested that would make the process more efficient while retaining accuracy levels. (Because accuracy data are regarded as sensitive, they are not reported here.) The two studies allowed us to map the information usage of decision makers to information visualization tools, and to compare the USN and RAN forecasters in order to distinguish between effects that are dictated by the tools and training of these specialists and those due to basic human cognition.

The intent of this chapter is to introduce a new approach to answering the questions in the previous paragraph. We need not report a full analysis of the data, but we do present sample results. We first briefly describe the data collection at the two sites. Then we review the results of the C2TA and show how suggestions for the design or redesign of tools flow from the C2TA results. More detailed results from both studies can be found in Kirschenbaum (2002) and Trafton et al. (2000)

THE TWO STUDIES

Study 1: U.S. Navy, 2000

Study 1 took place in San Diego, California, at a Naval meteorological and oceanographic facility. We set up a simulated METOC center with computer access to the tools that the forecasters typically use. Most of the weather information came from meteorological Web sites including military, nonmilitary government, and university sites.

Three pairs of participants consisting of a forecaster and a technician took part in the study. Each pair developed a forecast and prepared a forecast briefing for a (pretend) air strike to take place 12 hours in the future on Whidbey Island, Washington. All actions were videotaped and the participants were requested to "talk aloud" so as to produce a verbal protocol.

Study 2: Royal Australian Navy, 2001

The second study was a naturalistic observation of RAN forecasters working at a Weather and Oceanography Centre at an airbase in eastern Australia. Like their USN counterparts, they were forecasting for 12-, 24-, and 72-hour intervals for air operations. They prepared forecasts and forecast briefings, and used computer-based tools. As with the USN forecasters, most of the forecasting information came from meteorological Web sites. Also as in our study of the USN forecasters, they were videotaped and instructed to "talk aloud" to produce verbal protocols.

By retaining the task (forecasting) and moving to another group of practitioners with different tools (workstations, software), training, and teamwork practices, we might disentangle the effects due to human cognition, versus those due to the organizations, versus those due to the tools used—thereby permitting inferences about how to better support the common forecasting tasks for both groups.

RESULTS

The data were analyzed at two levels of detail. The first is a high-level description of the stages of weather forecasting. The second is a detailed description of the information-processing procedures used during each stage.

Information Use

Comparative CTA can tell two kinds of stories. *Similarities* in classes of information usage that are independent of the tools, training, and teamwork patterns imply basic processes of human cognition. In contrast, we can impute *differences* in information usage patterns as being due to the impact of differences in tools, training, and teamwork. To find either, we must code the verbal protocols to capture the way the forecasters use information. To analyze these data we selected usage encodings that capture what the forecaster did with the information. In other reports, we have examined the format of the information (text, graph, animation, etc.) or the form of the information (qualitative or quantitative) (Kirschenbaum, 2002; Trafton et al., 2000).

The major encoding categories for cognitive activities that we used are described in Table 14.1. Note that, in terms of expertise required and cognitive effort, there is a clear ordering from simplest to most demanding: Record <Extract < Compare < Derive.

The transcripts were encoded using the Table 14.1 categories, and the results for the USN and RAN forecasters were compared. Overall, the results indicated a strong similarity between USN and RAN information usage—the basic processes are the same. There were no methods that were used by one group but not by the other. However, the order, tools used, and relative frequency with which these methods were used did show significant differences in some areas. These areas are indications that the tools differentially support the tasks. They are of interest for C2AT and for the information they provide about opportunities to improve the toolset.

TABLE 14.1
Categories of Cognitive Work Used in the Analysis

Action	Definition	Example
Extract	To read information from any visible source. This occurs when a forecaster examines a visualization and extracts some sort of local or global features that are explicitly represented in the visualization.	"Looks like PVA over the area."
Compare	To use two or more sources and comparing them on any data.	"Radar shows precipitation, but I can't really see anything on the satellite picture."
Derive	To combine information that is available in the visualizations with the forecaster's own knowledge, so as to make inferences and come to a conclusion that differ from what is in the visible source.	"I think that's probably a little fast due to the fact that I don't think the models taking into account the topography of the area."
Record	To write down or copy information for reporting to users. It need not be the final form.	"This is a good picture right here, I'll take this. . . . Just crop this picture a little bit."

Note. The examples come from USN transcripts.

Figure 14.1 indicates differences in the details of how USN and RAN forecasters accomplish their task, using the resources at hand and within their own specific environments (weather, training, and manning). We concentrate on differences during the central tasks of developing and verifying the forecast. (There are no differences in the relative frequency of *record* actions even though specific tools and the pattern of tasks did differ.)

Two observations stand out. The RAN forecasters appear to spend the same proportion of time in *extracting, comparing,* and *deriving* information whereas the USN forecasters spend significantly more of their time extracting information, $\chi^2(3) = 31.31$, $p < .001$. In contrast, RAN forecasters spend virtually as much time *comparing* as *extracting* data. Thus, compared to the USN forecasters, the RAN forecasters spent a significantly larger proportion of their time engaged in comparing information, $\chi^2(1) = 7.28$, $p < .01$.

C2TA reveals the differences between the two groups. However, the analyst must find the reasons for these differences. Candidate causes include task, tool, and training differences. In this case, the goal is the same, predicting weather for naval aviation operations in the 12+-hour time frame. Though training differs between the groups, tool differences appear to be the more likely cause. For example, the RAN forecasters have better support for comparisons because they either use adjacent monitors or adjacent windows on the same monitor. Thus, they can see a satellite or radar pic-

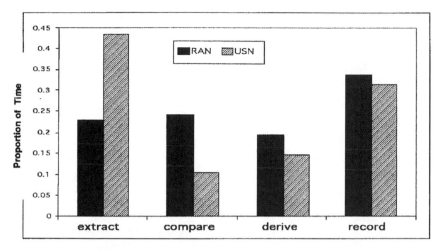

FIG. 14.1.   Proportion of time spent performing cognitive operations.

ture simultaneously or can examine the outputs of two or more computer forecasting models side by side on the same monitor, as shown in Figure 14.2. In contrast, the USN forecasters must extract information from one data source, and then either compare it to information shown at some other workstation, or store it in memory or on paper, and then make comparisons from memory. With the RAN dual view (either on the same or adjacent monitors) the forecaster can make direct *comparisons*. The process of extraction is an integral part of the process while the storage burden is greatly reduced. In Figure 14.2, the forecaster is comparing the outputs of two computer models displayed side by side on the same monitor. Other comparisons observed were comparisons of predictions for the same computer model across time and comparisons of the computer model prediction for current time and current observations (e.g., a satellite image on an adjacent monitor).

Sequences

Another observation from Figure 14.1 is that both groups spend a considerable portion of their time recording information for use in their forecasts. Further insight into this process can be achieved by examining the sequence of processes. Table 14.2 shows the probability of going from one process to another for a USN and a RAN forecaster. For example, given that the RAN forecaster is currently extracting data, the probability of his next action being *comparing, deriving,* or *recording* are $p = .11, .44,$ and $.44,$ respectively. This transition table emphasizes the importance of the two poles, *extract* and *record*. These are the most common transition points for both the

FIG. 14.2.   Forecaster comparing the outputs of two computer models, using screen sectoring.

TABLE 14.2
Representative Transition Probabilities

From \ To	Extract	Compare	Derive	Record
RAN				
Extract	.00	.11	.44	.44
Compare	.40	.00	.20	.40
Derive	.31	.08	.00	.62
Record	.48	.24	.29	.00
USN				
Extract	.08	.25	.42	.25
Compare	.33	.00	.33	.33
Derive	.50	.13	.00	.38
Record	.60	.00	.20	.20

USN and RAN forecaster. Of the three-node transitions, the most common cycles for both were either

   $extractr \rightarrow record \rightarrow extract$

or

   $record \rightarrow extract \rightarrow record.$

For RAN,

extract → *derive* → *record*

was also common. Transitions between *compare* and *derive* were noticeably fewer than those involving the poles.

As with the frequency data, sequence data provide insight into how tools do (or do not) support the cognitive tasks that make up weather forecasting. Design implications from the sequence data suggest the most effective places to automate. For example, as *extract* ™ *record* sequences are common, a semiautomated tool might allow the forecaster who is *extracting* information to *record* the selected data at the press of a button and without having to change screens. This would speed the recording process, eliminate accidental recording errors (typos, memory errors, etc.) and reduce the need to cycle between two tools.

IMPLICATIONS

C2TA is only one of the inputs to inform human-centered tool design. It is, however, necessary because without it, the tools would likely not meet the needs of the user, even if the designer were knowledgeable about the users' tasks. In contrast, with a "traditional" CTA, we could have observed the processes of *extraction, comparison, deriving,* and *recording* during the development of a weather forecast, and studied forecasters' cycle between developing their forecast (*extract, compare, derive*) and recording data. We would not have known whether these processes and cycles are common to other forecasting environments. Furthermore, we would not have learned the important role that the supporting tools play in the *comparison* process.

The cognitive systems engineer and the system developers and designers must work together to exploit these observations to guide the development of better tools. C2TA is only the first step but one that can inform and guide design toward making improvements where they are most needed. Our results suggest a need for better tools to further facilitate the *comparison* process, thus affirming an hypothesis about workstation design from traditional task analyses (Hoffman, 1991) that forecasters have to be able compare multiple data types at a glance. New display approaches and products are coming along to further support forecasting. For instance, it is now possible now to compare the outputs of the differing computer forecasting models. It is possible to superimpose computer model outputs over satellite pictures for current model comparisons.

These are just examples of the kinds of conclusions that can be derived from C2TA and have contributed to the momentum to develop better tools to help forecasters. With a single data set, the designer cannot know if the observed behavior is due to some demand characteristic of the tool set or to some facet of human cognition. With the addition of a second data set, the designer can separate the two and is thus free to develop better ways to support common cognitive processes with new tools.

ACKNOWLEDGMENTS

Susan Kirschenbaum's research was supported in part by the Office of Naval Research (ONR) as Project A424102 and by the Naval Undersea Warfare Center's (NUWC) In-House Laboratory Independent Research (ILIR) Program, as Project E102077. The ILIR program is funded by ONR. Greg Trafton's research was supported by Grants N00014-00-WX-20844 and N00014-00-WX-40002 from the Office of Naval Research.

The views and conclusions contained in this document are those of the authors and should not be interpreted as necessarily representing the official policies, either expressed or implied, of the U.S. Navy.

REFERENCES

Card, S. K., Moran, T. P., & Newell, A. (1983). *The psychology of human–computer interaction.* Hillsdale, NJ: Lawrence Erlbaum Associates.

Gilbreth, F. B., & Gilbreth, L. M. (1917). *Applied motion study.* New York: Macmillan.

Haas, C. (1989). Does the medium make a difference? Two studies of writing with pen and paper and with computers. *Human–Computer Interaction, 4,* 149–169.

Hoffman, R. R. (1991). Human factors psychology in the support of forecasting: The design of advanced meteorological workstations. *Weather and Forecasting, 6,* 98–110.

Hoffman, R. R., Coffey, J. W., Ford, K. M., & Bradshaw, J. (2001, September). Human-centered computing: Sounds nice, but what is it? Paper presented at the Human System Integration Symposium, Crystal City, VA.

Hoffman, R. R., & Deffenbacher, K. A. (1993). An analysis of the relations of basic and applied science. *Ecological Psychology, 5,* 315–352.

Hoffman, R. R., Hayes, P., Ford, K. M., & Hancock, P. A. (2002, May/June). The triples rule. *IEEE: Intelligent Systems,* pp. 62–65.

Kellogg, R. T., & Mueller, S. (1993). Performance amplification and process restructuring in computer-based writing. *International Journal of Man–Machine Studies, 39,* 33–49.

Kirschenbaum, S. S. (2002). Royal Australian Navy meteorology and oceanography operations research report (Tech. Rep. No. 11-346). Newport, RI: Naval Undersea Warfare Center Division.

Sanderson, P. M., & Fisher, C. (1994). Exploratory sequential data analysis: Foundations. *Human–Computer Interaction, 9,* 251–317.

Schraagen, J. M., Chipman, S. F., & Shalin, V. L. (Eds.). (2000). *Cognitive task analysis.* Mahwah, NJ: Lawrence Erlbaum Associates.

Trafton, J. G., Kirschenbaum, S. S., Tsui, T. L., Miyamoto, R. T., Ballas, J. A., & Raymond, P. D. (2000). Turning pictures into numbers: Use of complex visualizations. *International Journal of Human–Computer Studies* [Special issue: *Empirical Evaluations of Information Visualization*], *53,* pp. 827–850.

Vicente, K. J., & Rasmussen, J. (1992). Ecological interface design: Theoretical foundations. *IEEE Transactions on Systems, Man, and Cybernetics, 22,* 586–606.

Chapter 15

Computer-Aided Visualization in Meteorology

J. Gregory Trafton
Naval Research Laboratory, Washington, DC

Robert R. Hoffman
Institute for Human and Machine Cognition, Pensacola, FL

Our topic in this chapter is not so much what happens when experts have to work "out of context," but how cognitive engineering might help weather forecasters, in particular, remain within familiar decision-making spaces by improving on their display technology. Ballas (chap. 13, this volume) makes it clear how weather forecasting, as a workplace, is a constantly moving target by virtue of continual change in displays and data types. Since weather maps were invented around 1816, meteorological data visualization has gone through many dramatic changes (Monmonier, 1999). Weather maps are now displayed and manipulated by computer, even though hand chart work is still generally regarded as a critical activity in the forecaster's trade (see Hoffman & Markman, 2001). Most weather forecasters get data, charts, and satellite images from Internet sources including the World Wide Web. In this chapter, we discuss some of what we know about how weather forecasters use information technology to display and support the interpretation of complex meteorological visualizations. Based on notions of human-centered computing (HCC), we offer some suggestions on how to improve the visualizations and tools.

Norman (1993) offered design guidance that would (hopefully) ensure human-centeredness. Ideally, such guidance should be applied throughout the entire design process, but this does not always happen. In the case of weather forecasting, some of the ways of representing data (such as wind-barbs or iso-pressure lines) were standardized long ago. Ingrained traditions in meteorological symbology and display design force one to keep in mind what we call the Will Robinson principle. This principle is named after a character in the television series *Lost in Space* who was continually cautioned by his robot about impending dangers. The idea of the principle is that it is potentially dangerous for the outsider—a researcher who is analyzing a domain to create new technologies—to inject changes in long-standing traditions, no matter how flawed the traditions may seem at first glance.

One of our first experiences that (eventually) spawned this principle involved studies of expertise in the field of remote sensing in general (Hoffman, 1990; Hoffman & Conway, 1990) and a study of expertise at interpreting infrared satellite imagery (Hoffman, 1997; Klein & Hoffman, 1992). Until color was introduced, infrared weather satellite images were portrayed using a gray scale representing temperatures. On first thought, one might think that cold things would be represented as black and dark gray tones, and relatively warmer things as light tones to white. However, the Earth seen in this tonal scale looks like a surreal marble cake, not even much like a planet. Reversing the scale (warm = black, cold = white) suddenly makes the Earth look like the Earth and the clouds look like clouds. One of the tonal scales (or "enhancement curves") that forecasters found valuable (and still use to map temperatures, though color has been added), uses black to dark gray tones for the relatively warmest temperatures (i.e., the land surface), lighter grays for the lower and relatively warm clouds, but then reverts back to dark gray for the relatively cooler midlevel clouds, proceeding up through the lighter gray shades to white, and then yet again back to dark gray and black for the relatively coolest and highest cloud tops. An example is shown in Figure 15.1.

Taken out of context as a tone-to-temperature mapping scale, it is somewhat mysterious to the outsider. But with practice comes a skill of being able to use the repeating ascending tone scales to perceive cloud height and thereby gain an awareness of atmospheric dynamics. (Higher cloud tops are relatively colder and also more massive, thus representing the presence of greater amounts of moisture that can be associated with storms and precipitation at ground levels. Note the storm cell in the lower left of Figure 15.1.) It was only after a wave of cognitive task analysis with expert forecasters that the value of the enhancement curve became apparent.

With this cautionary tale in mind, the question we ask in this chapter is: "Given the current state of weather visualizations, how can we apply psychological and human-centering principles to improve the forecasting process?" In order to answer this question, we briefly describe two of the "design challenges" of HCC (see Endsley & Hoffman, 2002)—the Lewis and Clark Challenge and the Sacagawea Challenge, and provide examples from studies supported by the Office of Naval Research that suggest how we can apply the notions of human centering to improve the visualizations that forecasters use, and the forecasting process itself, to help forecasters stay "in context."

FIG. 15.1. A representative image generated by applying the "MB enhancement curve" to infra-red radiometric data sensed by GOES (Geostationary
Operational Environmental Satellite) of the National Oceanographic and
Atmospheric Administration. The upper gray scale in the legend at the bottom maps temperatures onto a continuous tonal palette. The bottom gray
scale is the enhancement curve. The upper right corner appears cropped
but is the apparent temperature of space, coded as white (i.e., cold).

TWO PRINCIPLES OF HUMAN-CENTERED COMPUTING

How should information be displayed on a computer screen? The answer
is, of course, "It depends." What we want to focus on is the fact that the easiest way to display information is not necessarily the easiest way for a person
to understand the information. Presenting external information in a way
that people find meaningful is the key to what Donald Norman (1993) calls
"naturalness of representation":

> Perceptual and spatial representations are more natural and therefore to be
> preferred but only if the mapping between the representation and what it
> stands for is natural—analogous to the real perceptual and spatial environment
> . . . the visible, surface representations should conform to the forms that peo-
> ple find comfortable: names, text, drawings, meaningful naturalistic sounds,
> and perceptually based representations. The problem is that it is easiest to pres-

ent people with the same representations used by the machines: numbers. This is not the way it ought to be. Sure, let the machines use numbers internally, but present the human operators with information in the format most appropriate to their needs and to the task they must perform. (pp. 72, 226)

It is a strong claim indeed that displays should use representations that match the user's mental representations, and it by no means clear that Norman's (1993) sweeping generalization is the right way to go. The simplest case, the case lying at the heart of Norman's treatment of this issue, is to present information graphically if people understand it visually or in the form of mental imagery, rather than presenting tables showing the numbers that computers so adeptly process. But beyond this simplest case is a large murky zone in which one cannot always be certain how the human mind represents things, or how weather displays should somehow align with mental representations. A case in point might be the example given earlier, of the tonal enhancement curves for satellite imagery. As Endsley and Hoffman (2002) point out, the core notion for which Norman is reaching seems to be that displays need to present meanings, and do so in a way that is directly perceptible and comprehensible.

But ease, directness, and psychological reality are differing meanings of "natural." What we might call immediately interpretable displays might, but need not necessarily, present information in the same way it is represented in the human mind or manifested in the "real world." On the contrary, displays of data from measurements made by individual sensors may be computationally integrated into goal-relevant higher order dynamic invariants or compound variables.

A good example from meteorology is the "skew-T, log p" diagram. This diagram uses a clever trick—measuring elevation or height in terms of pressure. This makes the y-axis rather like a rubber ruler. When a mass of air has relatively high pressure, the ruler is scrunched; and when a mass of air has relatively low pressure, the ruler is stretched. Temperature is the x-axis, and the interpretation of the diagram involves looking for patterns that appear as changes in temperature (skews) as a function of height in the atmosphere as measured in terms of pressure ("geopotential height"). To those who are unfamiliar with the skew-T diagram, its appearance and interpretation are a mystery. To those who are familiar with it, the diagram provides immediately perceptible clues to atmospheric dynamics.

Stimulated by Norman's guidance, Endsley and Hoffman (2002) offered two related principles of HCC. One is the Sacagawea Challenge, named after the guide for the Lewis and Clark Expedition:

Human-centered computational tools need to support active organization of information, active search for information, active exploration of information, reflection on the meaning of information, and evaluation and choice among action sequence alternatives.

The Lewis and Clark Challenge states that:

> The human user of the guidance needs to be shown the guidance in a way that is organized in terms of their major goals. Information needed for each particular goal should be shown in a meaningful form, and should allow the human to directly comprehend the major decisions associated with each goal.

Before we discuss the application of these design challenges to meteorological visualization, we need to present a primer about forecasting. (For a discussion of the human factors in general remote sensing, see Hoffman, 1990; Hoffman & Conway, 1990; Hoffman & Markman, 2001.)

WEATHER-FORECASTING TOOLS AND TECHNOLOGIES

In a nutshell, the forecaster's job is to use weather data (observations, radar, satellite, etc.) and the outputs of the many weather-forecasting computer models to make accurate products, including forecasts, guidance for aviation, flood warnings, and so on. Observational data available to the forecaster include ground-based observations from specialized sensor suites located at airports and forecasting facilities (winds, precipitation, air pressure, etc.). Data are also provided by balloon-borne sensors that are launched from weather-forecasting facilities.

The computer models for weather forecasting rely on both the statistics of climate and physical models of atmospheric dynamics to go from current observational data (which are used to "initialize" the physical models) to guidance about what the atmospheric dynamics will be at a number of scales of both space and future time. The computer models make forecasts only in a limited sense. No one can be a really good weather forecaster by relying on the computer model outputs unless she or he can forecast the weather without using the computer model outputs. The computer models are not infallible. Indeed, they are often "supervised" (that is, tweaked; see Ballas, chap. 13, this volume). The data used for initialization may not be as timely or reliable as the forecaster might like. The different computer models (based on different subsets of physics) often make different weather predictions and all have certain tendencies or biases (e g , a model may tend to overpredict the depth of low-pressure centers that form over the eastern U.S. coastline after "skipping over" the Appalachian mountains). This has led to the creation of "ensemble forecasts" that integrate the outputs from a number of the individual computer models. On the one hand, this adds to the forecaster's toolkit, but on the other, it adds to the data overload problem.

Satellite images show both past and recent truth and provide the forecaster with the "big picture" of dynamics at a global scale. There are numerous types of satellite data from many platforms including GOES, SEASAT, and others. The image products portray a variety of data types, including visible, infrared, water vapor, and so forth. Radar, especially from the remarkable NEXRAD system, provides the forecaster with a great deal of information, including winds, precipitation, and so on. The NEXRAD is linked into offices of the National Weather Service, where forecasters can request any of a great variety of products from the radar products generator (velocity data, reflectivity, precipitable water, etc.).

New software tools allow the forecaster to combine data types into single displays. For instance, a map showing "significant weather events " (e.g., storms) may have overlaid on it the data from the national lightning-detection network. An image from the GOES satellite may have overlaid on it the data from the NEXRAD radar, and so on.

As one might surmise from this brief presentation, the typical weather-forecasting facility centers around work areas populated by upwards of a dozen workstations displaying various data types. Staff includes forecasters responsible for general forecasting operations but also forecasters responsible for hydrometeorology and for aviation. At any one, time a forecaster might decide to examine any of scores of different displays or data sets (Hoffman, 1991; Trafton, Marshall, Mintz, & Trickett, 2002). Despite the recent introduction of all the new software and hardware systems, the typical weather-forecasting facility still finds need for such things as chart tables, clipboards, and colored markers. It is by no means clear whether, how, and to what degree any of the technologies either supports or handicaps the forecasting process. One might therefore legitimately ask whether the forecasting facilities might benefit from the application of the notions of HCC.

APPLYING THE PRINCIPLES

When a system is designed, how are the types of displays or interfaces chosen? Frequently, decisions are based on what is easiest or most efficient from a programming point of view with little regard to people actually perform the task. The Lewis and Clark Challenge and the Sacagawea Challenge suggest that the goals and mental operations of the human are critical to the success of the computational tools or displays. After determining the human's reasoning, the designer can try to determine how well (or poorly) an interface supports (or prevents) it. The designer can then change or even completely redesign the interface to facilitate both frequent and difficult mental operations, and then engage in an empirical usability evaluation.

A necessary and often difficult part of this task, of course, is revealing and describing the forecaster's reasoning. There are many methods of cognitive task analysis including methods for modeling cognition at the microscale of the reaction times and sequences of individual mental operations (e.g., GOMS). There are several tutorials and descriptions available for conducting cognitive task analysis (e.g., Crandall, Klein, & Hoffman, 2006; Hoffman & Lintern, 2006; Schraagen, Chipman, & Shalin, 2000; Trafton et al., 2002; Vicente, 2000). There are also ways of revealing the perceptual steps a user goes through (e.g., with an eye tracker) in conducting particular tasks. Forecasting also has to be described at a macrocognitive level of high-level strategies and the drive to comprehend of what's going on in the atmosphere. One thing that makes weather forecasting a challenging domain for HCC analysis is that there is no one single model of how forecasters reason. Reasoning methods and strategies depend on climate, season, experience level, and a host of other factors. (A detailed discussion appears in Hoffman, Trafton, & Roebber, 2006.)

The studies we now discuss examined how experienced meteorologists understand the weather and create a weather forecast. The forecasters used a variety of computer tools including internet sources, meteorology-specific tools, and off-the-shelf tools such as Microsoft PowerPoint. They also communicated with other to make sure their forecast and understanding of the current weather conditions was correct.

How might we apply the Sacagawea Challenge to weather forecasting? Norman's (1993) guidance, which we have already qualified previously, suggests that weather-forecasting displays that show graphics (images, charts) would already be "natural," and those that depict quantitative information would be "not-natural." Here we see how Norman's distinction is of little help, because weather charts are anything but "natural" and because forecasters need to know quantitative values (e.g., wind speeds at a height of 700 millibars) and must have some idea of specific likelihoods, certainly by the time they finish a forecast. For example, a forecaster may predict that there will be a 30% chance of rain tomorrow in a given region over a given time period, or the maximum temperature will be 14°C. Because these types of numbers are frequently a part of the final product for the forecaster (after a great deal of work), it seems necessary that the tools used to make forecasts would show or contain a great deal of quantitative information. Indeed, some forecasting tools are built to help the forecaster find the "best" computer model and then extract specific numeric information from that model.

But forecasters do not reason solely, or even primarily, with numbers as they try to understand and predict the weather. Forecasters glean large amounts of information from many weather visualizations and then combine that information inside their heads. An experienced weather forecaster (the type we are concerned with in this chapter) is able to create a mental model of the current atmospheric dynamics and project the likely future weather (Lowe, 1994; Perby, 1989; Trafton et al., 2000). The mental model has a significant qualitative aspect manifested as imagery, but is also "driven" by an understanding of the principles of atmospheric dynamics. Some forecasters, those who grew up on the traditional technology of hand chart work, report that their mental images are like animated charts populated with such graphic elements as isolines and wind barbs. Others report visualization of air masses and air mass interactions. Indeed, the notion of a mental model has for some time been quite familiar to the meteorology community (see Chisholm, Jackson, Niedzielski, Schechter, & Ivaldi, 1983; Doswell & Maddox, 1986) because they have to distinguish forecaster understanding ("conceptual models") from the outputs of the computer models.

The role of mental models has been demonstrated in a series of innovative experiments by Ric Lowe of Curtin University, investigating how novices and experts perceive and conceptualize the sets of meteorological markings that comprise weather charts. In Lowe's research, college student participants and weather experts carried out various tasks that required them to physically manipulate or generate markings from a given weather map. In a task in which people had to group map elements and explain the groupings, meteorologists' groupings involved the division of the map into a northern chunk and a southern chunk, which corresponds with the quite different meteorological influences that operate for these two halves of the Australian continent. Next, weather map markings were organized according to large-scale patterns that corresponded to the location of zones of regional meteorological significance. In contrast, the novices' groupings divided the map into eastern and western chunks on the basis of groups of figurally similar elements that happened to be in close proximity. Such subdivision has no real meteorological foundation.

In another task, participants were shown a map with an extended and unfilled perimeter, and had to attempt to extend the markings in the map. As well as producing significantly fewer markings in the extended region, the novices' markings appeared to have been derived quite directly from the graphic characteristics of the existing original markings by extrapolation or interpolation (e.g., turning simple curves into closed figures, continuation of existing patterns). In contrast, the meteorologists were operating in accordance with superordinate constraints involving a variety of external relations that integrated the original map area with the wider meteorological context. The resulting patterns in markings suggested the progressive clustering of lower level weather map elements into high-level composite structures that correspond to meteorologically significant features and systems of wider spatio-temporal significance.

In the task involving copying weather maps, the meteorologists began by drawing the major meteorological features. The second stage was then to pass through the map again in order to fill in subsidiary elements around this framework. In contrast, the novices tended to make a continuous pass around the map, filling in all elements they could remember in each region as they progressed, influenced primarily by the figural similarity of elements and their spatial proximity.

An especially interesting finding from a task involving the recall of maps was that the meteorologists' recall of the number of barbs on a frontal line was actually worse than that of the novices. This was because the meteorologists were concerned with the meteorologically important aspect of the cold-front symbol (the cold-front line itself) while glossing over the more "optional" aspect of the symbol (the particular number of barbs on the line).

In another task, participants attempted to predict future weather on the basis of what was shown in a map. For the nonmeteorologists, markings on the forecast maps could be largely accounted for as the results of simple graphic manipulations of the original markings; that is, they tended to move markings *en masse* from west to east without regard to meteorological dynamics. In contrast, the meteorologists' predictions showed a much greater differentiation in the way the various markings on the map were treated. Rather than moving markings *en masse,* new markings were added. This shows that meteorologists' mental representation of a weather map extends into the surrounding meteorological context.

In general, novices construct limited mental models that are insufficiently constrained, lack a principled hierarchical structure, and provide an ineffective basis for interpretation or memory. A major weakness of their mental models was the apparent lack of information available regarding the dynamics of weather systems. To quote Lowe (2000), "The expert's mental model would be of a particular meteorological situation in the real world, not merely a snapshot or image of a set of graphic elements arranged on a page" (pp. 187–188).

The consistent pattern of findings suggested a training intervention based on a new display of weather chart data. Animations were developed that portrayed temporal changes that occur across a sequence of weather maps, the idea being that animations would empower novices to develop richer mental models that would include or provide necessary dynamic information. But when novices worked with the animations, Lowe (2000) got a surprise:

Animated material itself introduces perceptual and cognitive processing factors that may actually work against the development of a high quality mental model. . . . When the information extracted by novices was examined, it was found that they were highly selective in their approach, tending to extract material that was perceptually conspicuous, rather than thematically relevant to the domain of meteorology. . . . [For example] for highly mobile features such as high pressure cells, trajectory information was extracted while information about internal changes to the form of the feature tended to be lacking. There is clearly more research required to tease out the complexities involved in addressing ways to help meteorological novices become more adept at weather map interpretation. In particular, we need to know more about the ways in which they interact with both static and dynamic displays. (p. 205)

This finding captures the motivation for the research that we report in this chapter.

A sample display that forecasters use is shown in Figure 15.2. This shows 700-millibar heights, winds, and temperatures, and involves both a graphical (i.e., map) format and individual data points (i.e., color-coded wind

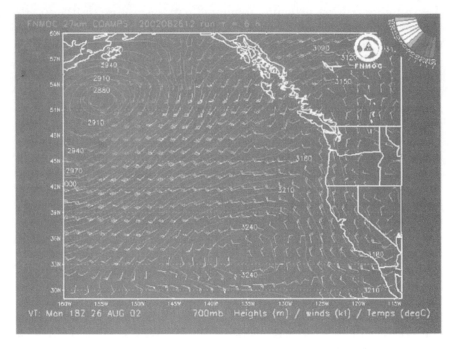

FIG. 15.2. A typical visualization that meteorologists use. Wind speed and wind direction are shown by the wind barbs; temperature is color coded according to the legend on the upper right, and equal pressure is connected by the lines (isobars) connecting the same values. The actual displays rely heavily on color, which we could not reproduce here.

barbs). Meteorological charts of this kind present a selective and somewhat decontextualized view of the atmosphere, and depict some information that is beyond direct or everyday experience. For instance, many weather charts show fronts, which are boundaries between air masses projected onto the Earth's surface. In actuality, air mass boundaries have complex shapes as a function of geopotential height (e.g., they are intersecting and interacting "blobs" of air). "Fronts" are hypotheticals. At higher levels in the atmosphere, meteorologists refer to "troughs," "ridges," and "domes":

> In weather maps these are indicated not by isolated graphical features but rather by patterning. . . . Minor local convolutions that are echoed across a series of adjacent isobars indicate the presence of a meteorologically significant feature. However, this subtle patterning of isobars can be obscured to a large extent by their visually distracting context, and so these features are likely to be overlooked unless given special attention. (Lowe, 2000, p. 189)

Hoffman (1991) and Trafton et al. (2000) have affirmed Lowe's (2000) findings: Expert forecasters do more than simply read off information from the charts or computer model outputs; they go through a process. They begin by forming an idea of the "big picture" of atmospheric dynamics, often by perceiving primarily qualitative information (e.g., "The wind is fast over San Diego" or "This low seems smaller than I thought it would be"); they continuously refine their mental model; and they rely heavily on their mental model to generate a forecast including numeric values (e.g., "The wind speed over San Diego at 500 mb will be 42 knots"). Experienced meteorologists use their mental model as their primary source of *information* (rather than copying *data* directly from the best or a favorite visualization or computer model output). Thus, simply showing a complex visualization, expecting a user to extract the necessary information, and to be finished is an oversimplification of how complex visualizations are used. Figure 15.3 shows a macrocognitive model of the forecaster reasoning process.

VISUALIZATION SUGGESTIONS

In the case of weather forecasting, the application of the Lewis and Clark Challenge and the Sacagawea Challenge is a bit tricky because forecasters must see quantitative information, but they also reason on the basis of qualitative information and their mental models. Thus, the best kind of external representation might be one that emphasizes the qualitative aspects of the data, but where quantitative information can also be immediately perceived. One good example of this is the wind-barb glyph, shown in Figure 15.2. A wind barb shows both wind speed (by the number and length of the

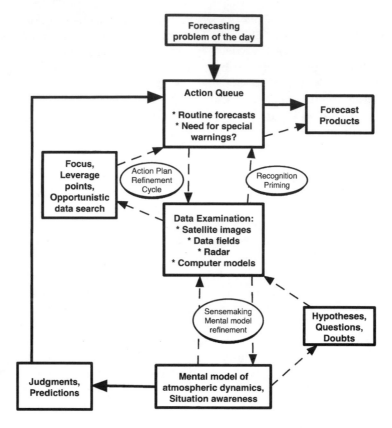

FIG. 15.3. A macrocognitive model of how expert weather forecasters predict the weather (based on Trafton et al., 2000). This diagram includes all of the fundamental reasoning processes that experts engage in: recognition-primed decision making, the action plan refinement cycle, the situation awareness cycle, and mental-model formation and refinement. In the course of cycling through multiple cycles of situation awareness and mental-model refinement, the forecaster will iterate between making products and gathering data.

barbs) and direction (by the direction of the major line). Though collapsing information in a way that might be regarded as efficient, this comes at a cost of legibility and display clutter. On the other hand, because individual wind barbs tend to cluster in ways suggestive of atmospheric dynamics (see the swirl patterns in the upper left and lower middle of Figure 15.2), the display allows the forecaster to see the qualitative aspects of the wind (areas of increasing wind speed, the formation of lows, cyclonic winds, etc.). A wind barb has the added benefit of allowing a forecaster to extract quantitative information from the glyph (e.g., each long barb is 10 knots; each short barb is 5 knots). Creating additional iconic representations that are primarily qualitative but allow quantitative information to be easily extracted from them should enhance visualization comprehension and usage.

What kinds of mental operations do weather forecasters use when predicting the weather and when working with these complex visualizations? Referring to Figure 15.3, we already know that they can make decisions by recognition priming. That is, weather data are perceived, and on the basis of experience, the forecaster knows immediately what the weather situation is and what actions need to be taken. We also know that forecasters rely on their mental models to formulate and test hypotheses and maintain a state of situational awareness. But what other kinds of mental operations are performed? To examine this issue, we developed a framework called *spatial transformations.*

A spatial transformation occurs when a spatial object is transformed from one mental state or location into another mental state or location. Spatial transformations occur in a mental representation that is an analogue of physical space and are frequently part of a problem-solving process. Furthermore, they can be performed purely mentally (e.g., purely within spatial working memory or a mental image) or "on top of" an existing visualization (e.g., a computer-generated image). Spatial transformations may be used in all types of visual-spatial tasks, and thus represent a general problem-solving strategy in this area.

There are many types of spatial transformations: creating a mental image, modifying that mental image by adding or deleting features, mental rotation (Shepard & Metzler, 1971), mentally moving an object, animating a static image (Bogacz & Trafton, 2005; Hegarty, 1992), making comparisons between different views (Kosslyn, Sukel, & Bly, 1999; Trafton, Trickett, & Mintz, 2005), and any other mental operation that transforms a spatial object from one state or location into another.

Because the mental models that meteorologists use to reason about the weather are dynamic and have a strong spatial component, it is not surprising that many spatial transformations are applied while making a forecast. We examined the type and amount of spatial transformations as expert scientists analyzed their data. We found that by far the most common spatial transformation is a comparison: Experts frequently compared and contrasted two different visualizations or compared their qualitative mental model to a visualization (Trafton et al., 2005; see also chap. 14, this volume). In fact, when the scientists used complex scientific visualizations, they performed almost twice as many comparisons as all the other spatial transformations combined. Similarly, each data field or visualization they examined was compared to others, up to four comparisons per visualization. This shows that comparisons are frequent, suggesting that they are extremely important to the forecasting process itself. Table 15.1 shows sam-

TABLE 15.1
Sample Comparisons That Forecasters Made
While Examining Meteorological Displays

Utterance by Forecaster
Location of the low, pretty similar to what, uh, NOGAPS has.
AVN at first guess looks like it has a better handle on the weather than NOGAPS.
I can't believe that radar shows precipitation, but I can't really see anything on the satellite picture.
Fleet numeric COAMPS seem to have a better handle on it than what I've seen based on the 27 kilometer.
18Z (18 Zulu or GMT time) still has a lot of moisture over the area. [At 21Z] there is not a whole lot of precipitation, and then, 00Z has a lot of moisture over the area just off the coast.
And uh, Doppler radar currently showing the precipitation a little bit closer [than I expected].
[This visualization] has a little bit more precipitation than No Gaps [or] AVN.

Note. NOGAPS, COAMPS, and AVN are computer-forecasting models.

ples of what different forecasters said as they were viewing different meteorological visualizations and making comparisons. Notice the very frequent comparisons both between and within displays that are made by the forecaster.

SUPPORT FOR MENTAL-MODEL FORMATION AND REFINEMENT

Having identified the mental model as a linchpin phenomenon in forecaster reasoning, it seems clear that what might be of use to forecasters would be a graphical tool that supports them in constructing a depiction of their four-dimensional mental model (see Hoffman, Detweiler, Lipton, & Conway, 1993). It is certainly within the reach of computer science and artificial intelligence to build a tool that might support the forecaster, for instance, in defining objects or regions within satellite images, radar, or other data types, grabbing them as graphic objects, dragging them onto a window, progressively building up a dynamic or runnable simulacrum of their understanding of atmospheric dynamics—a sketchpad to represent their mental model and spatio-temporal projections from it. Current weather information-processing workstations do not support such an activity.

Having identified the most common spatial transformation, we could also build systems that support the comparison process. Unfortunately, other than an overlay capability there is very little support for forecasters making comparisons between different weather models or to determine how well a weather model matches to a satellite image. Because comparisons are such a vital part of the forecasting process, it makes sense to support them. Specifically, there should be ways of providing forecasters with simpler ways of comparing weather models, and comparing weather models to satellite images, radar, and so forth.

With regard to both of these suggestions, there is a host of theoretical and practical research issues in the cognitive science of weather forecasting that have not yet been addressed. For example, when is it best to provide side-by-side visualizations and when is it best to provide intelligent overlays? If overlaid, how should transparency be handled? One of the difficult technical issues includes time syncing and geo-referencing the visualizations with each other so that comparisons are facilitated; it won't help the forecaster if she or he has to mentally animate one visualization to get it in sync with the other.

SUPPORT FOR THE INTERPRETATION OF QUANTITATIVE DATA

The previous section examined the spatial and reasoning processes in which forecasters engage. This section examines the perceptual processes that forecasters use to comprehend a weather visualization and extract information from it.

Most of the visualizations that forecasters examine have many variables that are used to represent upwards of tens of thousands of data points (see Hoffman, 1991, 1997). One obvious question for the designer is how to represent such an amount of data in a single visualization. There are several standard ways of accomplishing this task: color-coding variables (such as temperature), using various forms of isolines (isotachs, isobars, isotherms, etc.), and using glyphs that can combine information (as do wind barbs). All these can be used to "compress" data. However, some of these graphical tricks force the forecaster to ferret out the needed information from the mass of data. For example, one might have to interpolate between isobars or compare colors at different locations to determine which area is warmer. Such operations have to be deliberative, and can be effortful even for experienced forecasters. If we understood how forecasters perceive meaning from these types of displays, we might be able to build better ones.

Trafton et al. (2002) examined experienced forecasters' eye movements as they inspected meteorological visualizations. They found that interpolating between isobars was about twice as difficult as reading information directly off of a data chart. In the same study, they examined how forecasters extracted information by relying on legends. Extracting information from a legend can be difficult for several reasons, some cognitive, some perceptual, and some dealing primarily with the interface itself. First, the legend itself may be small or otherwise hard to find (see the temperature legend in the upper right of Figure 15.2). Second, the colors can be difficult to differentiate. Third, meteorological visualizations still tend to rely on color palettes consisting solely of highly saturated primary hues. (There are a number of lingering issues of the use of color in meteorological displays. See Hoffman et al., 1993.) Fourth, sometimes the legend is not labeled, making it unclear which of several possible variables it refers to. These factors can make finding the right match of color-to data value quite difficult.

These difficulties can manifest themselves in a variety of ways. For example, forecasters may examine different legends (if there is more than one on a visualization) or miss-guess what a legend refers to if a legend is not labeled. A forecaster may also spend an inordinate amount of time searching for the exact color if the colors are not differentiated well. Using the method of eye tracking, Trafton et al. (2002) found evidence that forecasters do all of these things when a legend is not well defined, labeled, or colored.

Trafton et al. (2002) presented experienced forecasters visualizations such as that in Figure 15.4. These visualizations all came from a forecasting Web site that was familiar to the forecasters. The researchers asked forecasters to extract specific information from these graphs while their eye movements were recorded with an eye tracker. The eye tracker allowed us to record exactly what they were looking at with a temporal resolution of 4 milliseconds. Figure 15.4 shows the relatively large amount of time and number of eye movements (saccades) one forecaster used when trying to determine what the temperature was in Pittsburgh. This back-and-forth eye movement pattern is representative of how forecasters read the legend. It shows that it took a while for the forecaster to find the right color that matched the color at Pittsburgh. Of course, these eye movements are quite fast, but over many visualizations, the time and effort invested in performing color interpretation can become substantial.

Similarly, Figure 15.5 shows a less skilled forecaster trying to figure out what the legend is showing: relative humidity or geopotential height. Notice that because the legend is unlabeled, the forecaster searches all the text in the display in the hope of determining what the legend represents. This forecaster may not have had a great deal of familiarity with this visualization, but even if she or he had, a simple label would have simplified the forecaster's search.

If we apply the Lewis and Clark Challenge and the Sacagawea Challenge to legend design, we can make some suggestions that should improve the readability of these visualizations. First, even experienced graph readers have problems color matching on legends. Most weather visualizations do not show an extremely wide range in temperature, so colors could be sepa-

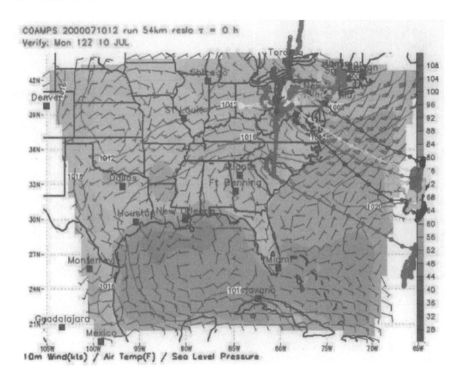

FIG. 15.4.   An example visualization that was used in the study of the eye movements of weather forecasters. The forecaster was asked, "What is the temperature at Pittsburgh, Pennsylvania?" The colored lines that one can see going between the legend and the map show the eye movement tracks. The dots are 4 milliseconds apart and the changes in eye track color represent every 2 seconds, starting at Pittsburgh (obscured by the large eye-tracking trace blob). We show only some of this forecaster's eye movements for clarity.

rated more within the legend. There may still be color match problems, but the color match should be between red and blue rather than between orange-red and red-orange (see Hoffman et al., 1993). Second, we can label each legend. This is perhaps an obvious point, but it is one that even high-traffic Web sites sometimes ignore. Many newspapers, magazines, and even scientific journals do not enforce this rule. By understanding the kinds of processes that people engage in to extract information from a visualization (color matching, search of unknown information), we can attempt to prevent or ameliorate many of these problems and improve the visualizations.

An excellent example of the role that human-centering considerations can play in display design is the recent work of Lloyd Treinish and Bernice Rogowitz at IBM (e.g., Rogowitz & Treinish, 1996; Treinish, 1994; Treinish & Rothfusz, 1997). They have created a rule-based advisory tool for the

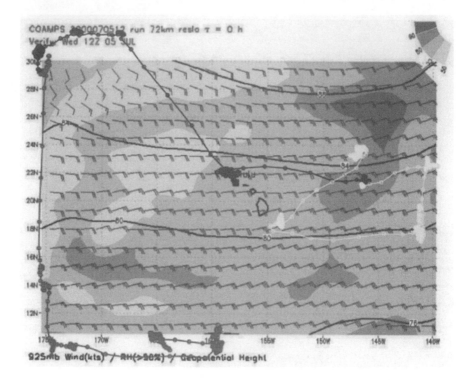

FIG. 15.5.   Eye movement tracks for a forecaster attempting to answer the
question, "What is the relative humidity at Honolulu?" The legend in the upper right-hand corner shows humidity, but because it is unlabeled, the forecaster searches the text and the axes to attempt to determine whether it is
the isolines (solid black lines) or the legend that refers to the relative humidity.

specification of appropriate color-to-data mappings depending on whether
the goal of visualization is exploration or presentation. In addition, they
have relied on a human-centered strategy for visualization based on the
need to preserve the fidelity of the original data, and the need to take into
account known facts about human perception and cognition. Specifically,
they have developed guidelines for how to collapse multiple variables and
data types into individual displays and guidelines to support the user in defining coordinate systems onto which data may registered in space and
time. One of their perspectival displays portrays horizontal winds (using a
color palette of saturation shades of violet), relative humidity (using saturation shades of brown), surface temperature overlaid on the base map (using a two-tone palette of saturation shades of blue and green-blue), and air

pressure (indicated in a semi-transparent vertical plane using saturation shades of blue-violet and green). Also depicted are three-dimensional cloud structures. For all of their graphic products, the use of perspective, depth pseudoplanes, and animation permits the perceptual discrimination of the multiple variables. (Images can be viewed at http://www.research .ibm.com/people/l/lloydt/).

CONCLUSION

In one project in which one of us was involved, the sponsor wanted to create displays to enable nonforecasters to understand the uncertainty of various types of weather data. The researchers who worked at the sponsor's laboratory assumed that these nonforecaster users would benefit by seeing the same sorts of displays that forecasters use to display data uncertainty. We knew intuitively that this might not be the best way to approach the issue. "Meaning" is always relative to the person who is doing the comprehending. Yet, it seems too easy for people, including smart, well-intentioned people, to create systems and displays that force people to work out of context. *The road to user-hostile systems is paved with user-centered intentions* (Woods & Dekker, 2001).

How might we go about improving the forecasting process, and in particular, helping forecasters stay "in context" by creating better displays? Research on the cognition of expert forecasters, compared with novices, has revealed a great deal of useful knowledge concerning the forecasting process and what is required for forecasting expertise. This knowledge can be leveraged in the application human-centering principles. We have discussed two principles of HCC, and have suggested how they might be used to improve existing meteorological products and systems. There is also a need for innovation and revolutionary redesign, especially in the creation of systems that support the forecaster in creating a graphical depiction of their own mental models of atmospheric dynamics.

ACKNOWLEDGMENTS

This research was supported in part by grants N0001400WX20844 from the Office of Naval Research and N0001400WX40002 to Greg Trafton from the Office of Naval Research. The views and conclusions contained in this document are those of the author and should not be interpreted as necessarily representing the official policies, either expressed or implied, of the U.S. Navy. Thanks to Jim Ballas for comments on an earlier draft.

REFERENCES

Bogacz, S., & Trafton, J. G. (2005). Understanding dynamic and static displays: using images to reason dynamically. *Cognitive Systems Research, 6*(4), 312–319.

Chisholm, D. A., Jackson, A. J., Niedzielski, M. E., Schechter, R., & Ivaldi, C. F. (1983). *The use of interactive graphics processing in short-range terminal weather forecasting: An initial assessment* (Report No. AFGL-83-0093). Hanscom AFB, MA: U.S. Air Force Geophysics Laboratory.

Crandall, B., Klein, G., & Hoffman, R. (2006). *Minds at work: A practitioner's guide to cognitive task analysis.* Cambridge, MA: MIT Press.

Doswell, C. A., & Maddox, R. A. (1986). The role of diagnosis in weather forecasting. In *Proceedings of the 11th Conference on Weather Forecasting and Analysis* (pp. 177–182). Boston: American Meteorological Society.

Endsley, M., & Hoffman, R. R. (2002, November/December). The Sacagawea principle. *IEEE Intelligent Systems, 17*(6), 80–85.

Hegarty, M. (1992). Mental animation: Inferring motion from static displays of mechanical systems. *Journal of Experimental Psychology: Learning, Memory and Cognition, 18*(5), 1084–1102.

Hoffman, R. R. (1990). Remote perceiving: A step toward a unified science of remote sensing. *Geocarto International, 5,* 3–13.

Hoffman, R. R. (1991). Human factors psychology in the support of forecasting: The design of advanced meteorological workstations. *Weather and Forecasting, 6,* 98–110.

Hoffman, R. H. (1997, February). *Human factors in meteorology.* Paper presented at the Short Course on Human Factors Applied to Graphical User Interface Design, held at the 77th annual meeting of the American Meteorological Society, Los Angeles.

Hoffman, R. R., & Conway, J. (1990). Psychological factors in remote sensing: A review of recent research. *Geocarto International, 4,* 3–22.

Hoffman, R. R., Detweiler, M. A., Lipton, K., & Conway, J. A. (1993). Considerations in the use of color in meteorological displays. *Weather and Forecasting, 8,* 505–518.

Hoffman, R. R., & Lintern, G. (2006). Eliciting and representing the knowledge of experts. In K. A. Ericsson, N. Charness, P. Feltovich, & R. Hoffman (Eds.), *Cambridge handbook on expertise and expert performance* (pp. 203–222). New York: Cambridge University Press.

Hoffman, R. R. & Markman, A. B. (Eds.). (2001). *Human factors in the interpretation of remote sensing imagery.* Boca Raton, FL: Lewis.

Hoffman, R., Trafton, G., & Roebber, P. (2006). *Minding the weather: How expert forecasters think.* Cambridge, MA: MIT Press.

Klein, G. A., & Hoffman, R. R. (1992). Seeing the invisible: Perceptual-cognitive aspects of expertise. In M. Rabinowitz (Ed.), *Cognitive science foundations of instruction* (pp. 203–226). Mahwah, NJ: Lawrence Erlbaum Associates.

Kosslyn, S. M., Sukel, K. E., & Bly, B. M. (1999). Squinting with the mind's eye: Effects of stimulus resolution on imaginal and perceptual comparisons. *Memory and Cognition, 27*(2), 276–287.

Lowe, R. K. (1994). Selectivity in diagrams: Reading beyond the lines. *Educational Psychology, 14,* 467–491.

Lowe, R. (2000). Components of expertise in the perception and interpretation of meteorological charts. In R. R. Hoffman & A. B. Markman (Eds.), *Interpreting remote sensing imagery: Human factors* (pp. 185–206). Boca Raton, FL: Lewis.

Monmonier, M. (1999). *Air apparent: How meteorologists learned to map, predict, and dramatize weather.* Chicago: University of Chicago Press.

Norman, D. (1993). *Things that make us smart: Defending human attributes in the age of the machine.* Cambridge, MA: Perseus.

Perby, M. L. (1989). Computerization and skill in local weather forecasting. In I. Josefson (Ed.), *Knowledge, skill and artificial intelligence* (pp. 39–52). Berlin: Springer-Verlag.

Rogowitz, B. E., & Treinish, L. A. (1996). How not to lie with visualization. *Computers in Physics, 10*, 268–274.

Schraagen, J.-M., Chipman, S. F., & Shalin, V. L. (Eds.). (2000). *Cognitive task analysis.* Mahwah, NJ: Lawrence Erlbaum Associates.

Shepard, R. N., & Metzler, J. (1971). Mental rotation of three-dimensional objects. *Science, 171*, 701–703.

Trafton, J. G., Kirschenbaum, S. S., Tsui, T. L., Miyamoto, R. T., Ballas, J. A., & Raymond, P. D. (2000). Turning pictures into numbers: Extracting and generating information from complex visualizations. *International Journal of Human–Computer Studies, 53*(5), 827–850.

Trafton, J. G., Marshall, S., Mintz, F., & Trickett, S. B. (2002). Extracting explicit and implicit information from complex visualizations. In M. Hegarty, B. Meyer & H. Narayanan (Eds.), *Diagramatic representation and inference* (pp. 206–220). Berlin: Springer-Verlag.

Trafton, J. G., Trickett, S. B., & Mintz, F. E. (2005). Connecting internal and external representations: Spatial transformations of scientific visualizations. *Foundations of Science, 10*(1), 89–106.

Treinish, L. A. (1994). Visualization of disparate data in the earth sciences. *Computers in Physics, 8*, 664–671.

Treinish, L. A., & Rothfusz, L. (1997). Three-dimensional visualization for support of operational forecasting at the 1996 Centennial Olympic Games. In *Proceedings of the Thirteenth International Conference on Interactive Information and Processing Systems for Meteorology, Oceanography and Hydrology* (pp. 31–34). Boston: American Meteorological Society.

Vicente, K. (2000). *Cognitive work analysis.* Mahwah, NJ: Lawrence Erlbaum Associates.

Woods, D. D., & Dekker, S. W. A. (2001). Anticipating the effects of technology change: A new era of dynamics for human factors. *Theoretical Issues in Ergonomics Science, 1*, 272–282.

Chapter 16

Cognitive Task Analysis of the Warning Forecaster's Task: An Invitation for Human-Centered Technologies

David W. Klinger
Bianka B. Hahn
Erica Rall
Klein Associates, Inc., Fairborn, OH

One of the best ways to challenge a weather forecaster is to confront him or her with a situation in which severe weather might be developing. A study of U.S. Navy and senior civilian forecasters (Hoffman, Coffey, & Ford, 2000) showed that performance at forecasting severe weather distinguishes experts from journeymen forecasters more clearly than skill at forecasting other types of weather events. There are many reasons why forecasting severe weather is difficult, and one of the most powerful factors is local experience. So, if one wants to see what happens when a weather forecaster has to work "out of context," ask him or her to forecast severe weather for a locale with which they are not familiar.

Imagine an inexperienced warning forecaster, freshly trained to use advanced radar algorithms as the primary data source for generating warnings of severe weather. Because this person lacks experience, this forecaster issues warnings every time the computerized system, driven by algorithms, provides a flashing alert. The forecasting organization's measurement of the forecaster's performance is based largely on the numerical count of the storms he or she has missed over the last year, regardless of the storms' severity or the amount of damage sustained in lives or property. On the receiving end of the warnings is a citizen of, say, Oklahoma City, Oklahoma. He has no faith in the warnings issued by the Weather Service because the myriad of inaccurate and unspecific warnings has dulled his desire to notice. He can't make sense of how severe the threat is on a particular day because they all sound alike to him. In short, he mostly ignores warnings. On the day of an intense (F4 supercell) tornado, our citizen does not look for shelter and does not believe that a storm is going to develop a severe tornado that will claim many lives. He goes about his business, unaware that a weather emergency is taking place.

Imagine the same weather situation, but with an experienced warning forecaster in charge who is able to distinguish this storm from all the others he had seen over the years. He realizes that this storm was going to be larger, stronger, and potentially more lethal than anything the area has ever seen before. He knows that he needs to develop a warning that will adequately convey the imminent danger of this storm to the public in order to get a fast response from those in its path. He issues a "Tornado Emergency." This is beyond the normal classification language recommended by the National Weather Service. This dramatic alert results in extensive media reports broadcast over radio and television convincing the population to seek shelter, and increasing the activity on the part of government agencies.

What expertise did this forecaster apply to allow him to make the judgment that this would be a day like no other? How do expert forecasters pluck potential storms out of the wall of data? How do these forecasters gather clues, piece those cues together, and recognize patterns? Only by answering such questions through Cognitive Task Analysis (CTA) can one design truly "intelligent" systems to aid the forecaster. Such technologies should support the expert's decision-making and other cognitive processes, and at the same time decrease mental and physical workload. This article presents a case study in the use of CTA to identify leverage points for the application of intelligent technology to help all forecasters stay *in* context.

RESEARCH METHOD

While members of Klein Associates, we conducted a CTA in order to capture the expertise of warning forecasters at the U.S. National Weather Service (NWS). We interviewed seven meteorologists with experience in warning forecasting, who worked at NWS offices in Alabama, Oklahoma, Missouri, and Texas. Six of the experts had between 12 and 20 years of experience as forecasters. They had advanced through various positions within the NWS and currently held positions as either Science Officers ($n = 4$) or Meteorologists in Charge ($n = 2$). Five of them had worked at offices located in different regions across the country. These experiences exposed them to a variety of weather trends as determined by the geography of the region. Five of the forecasters had extensive experience forecasting tornados associated with severe storm cells, and one forecaster was very experienced with tornados associated with frontal systems. In our estimation, six of the forecasters were experts and one was a journeyman who had just been promoted to forecaster. She had forecasted only two severe weather days, although one of those days had been extremely active in terms of the number of warnings issued.

The Critical Decision Method

The Critical Decision Method (CDM) explores specific incidents that the expert has experienced. (For a detailed description of the procedure and a review of its successful applications, see Hoffman, Crandall, & Shadbolt, 1998.) The detail and specificity of the cognitive events that CDM is designed to uncover require firsthand experience and lead to a deeper description than a generalized account. It is not enough for a person to say, "Usually when this happens you will know it because X." CDM requires a specific case of X and a detailed description of the event from beginning to end—the observations the expert made, the decisions they made, the actions they took.

The initial step guides the participant to recall and recount a relevant incident. We asked for an event that challenged their expertise as severe-weather forecasters. Next, the expert was asked to recount the episode in its entirety starting with the beginning of their shift that day or when they left to go to work. Interestingly, some of the important information that the experts noticed happened outside of their office, often early in the day from their homes or even on their drive to work as they skywatched. This information seemed to set the stage for the additional information gathering and decision making later in the incident. We therefore frequently asked the experts to start telling us about their day even before their shift began.

In the third step of the CDM, "retelling," the expert was asked to attend to the details and sequence in order to correct any errors or gaps in our understanding of the incident. The experts typically offered additional clarifying details on specifics of the technology or data they used. This step allowed the expert and the interviewers to arrive at a shared overview of the incident.

The interviewer then conducted another sweep through the incident in which the expert was asked to structure and organize the account into ordered segments. It was our goal in this sweep to gain a better understanding of the timeframe of the developing severe-weather event, the experts' thoughts as time went on, and the steps the forecaster took to interrogate the storm.

The interviewer then conducted the next step of the CDM, "progressive deepening." The expert reviewed each identified segment of the timeline, while addressing question probes designed to focus attention on particular cognitive aspects of the incident. The purpose of this step is to deepen our understanding of the event, and build a comprehensive, detailed, and context-specific account of the incident from the perspective of the decision maker. In particular, we were looking for the presence or absence of salient cues in the environment, the specific nature of these cues, assessments of the situation and the basis of those assessments, expectations of how the situation might evolve, and options that were evaluated and chosen. We were also interested in the role and value of individual team members to the expert's decision-making process in the incident.

The interviewer then conducted the next sweep through the incident, in which the expert is presented with "what if" queries for each segment of the timeline (e.g., "What might a less experienced person have done in this case?") We used such queries to explore what it meant to "think ahead of the storm." This gave us some insight into the importance of the expert's functional mental models (see chap. 15, this volume).

The results helped us appreciate the need to create technologies that leverage and amplify human expertise at severe weather forecasting, rather than technologies intended to substitute for the human expert.

RESULTS

Data analysis began with team brainstorming to discuss cognitive themes and trends that arose during the interviews. Next, we reviewed and coded the interview notes looking for instances of those themes and trends. We worked through each of the interviews to reframe the incidents in terms of cognitive activities. We describe our results in terms of two main topics: (a) forecaster reasoning, and (b) the cues on which their reasoning is based.

Forecaster Reasoning

Our study confirms other findings concerning the cognition of expert forecasters (e.g., Hoffman, 1991; chaps. 14 and 15, this volume). Warning forecasters often come into the office at the start of their shift already having garnered an overall environmental assessment on their way to work. They sometimes do this by checking data on the Internet from home, yet we also heard interesting stories regarding subtle cues like moisture on metal rails and/or automobile hoods. Once they arrive in the office they quickly come up-to-speed by checking their favorite data sources before they start talking to the forecasters on the previous shift.

Current technology allows the forecasters to configure their individual workstations to suit their personal assessment style. The arrangement of the screen real estate seems to be a personal choice based on the season and the associated potential for severe weather. One strategy that an expert described regarding an uncertain weather day was to keep a variety of data sets available, in order to retain flexibility in making an assessment.

Experts told us that they "interrogate" storms as they develop. They instigate a process of proactive investigation of the characteristics of the developing storm. Like a detective trying to solve a mystery, they access various types of data sets to get different views of the problem (storm in this case), and overlay different data sets in an attempt to identify patterns. During this process, they develop certain questions that they want the data to answer for them. Expert severe-storm forecasters tend to run large sets of data past their eyes and cycle through different systems almost continuously, never sticking too long with any one data set in order to support their assessment of the storm's development. In this process, the individual expert uses a large variety of different products to make his or her assessment.

We also heard many stories from experts regarding their propensity to spend a lot of time projecting what will happen in the near future. This type of projection allows the forecaster to stay ahead of the storm and to be proactive regarding warnings. They talked about how they generate hypotheses regarding the worst-case scenario and play it out in their minds. This hypothesis generation helps them to develop expectations about "where the data might go." When the data line up with their hypothesis, they are ahead of the game. When the data violate their hypothesis, they are quickly able to recognize which data element is "out of line" and quickly develop a new assessment.

Season after season of observing weather patterns has instilled a "sensitivity for severity" for the experts. What they mean by this is that they have developed the ability to identify that a certain weather event is going to be much more intense than any that the forecaster has experienced before. In our interviews, we heard a number of reports about "career-changing" events. Often they were able to realize at some point in the forecasting process that this storm would be larger, more destructive, or faster developing than the usual storms that developed in the given region during that time of year. It was not unusual for our expert forecasters to be the first ones in their office to anticipate this extraordinary level of storm development. Yet, it goes far beyond just anticipating "big storms." We found that our experts were also able to pinpoint the exact microstorm within the larger storm front. Much like finding a needle in a haystack, they were able to identify the exact data elements within the larger storm that needed to be monitored closely.

Without exception, all of the interviewed experts described the importance of determining "ground truth." In each of the events we heard about in our interviews, the informative element of ground truth played an important role in the decision-making process. In some cases, it confirmed predictions about the storm's developments and helped in the follow-on warning decisions. In other cases, it confirmed that a warning that was sent out was the correct choice for the specific region and provided post facto confirmation of a correct warning decision. Experts use all possible means to receive ground truth including HAM radio reports and phone calls that flood the office during severe weather. We also heard forecasters report that they turned on the television and caught an ESPN report about the cancellation of a sports event based on severe rainfall, which in turn was a valuable cue that the weather system was carrying as much water as suspected. There are also cases where the "ground truth" report is far more compelling. We heard that a TV camera on top of a city building happened to capture a large funnel cloud that was moving toward a densely populated downtown area. The forecaster happened to catch the local news report and saw the funnel cloud. This proved to be invaluable as he was able to evaluate the size, location, and potential of this severe weather event. In search of any "eyes on" data, they often call the cell phone of an off-duty forecaster in the hope that the person is somehow out there looking at the storm. Experts seek this type of information to cut down on uncertainty. In essence, they are removing the impersonal data presented by the technology and replacing it with something far more useful, context-based and information-rich data.

Experts tend to be skeptics and maintain that skepticism about any given weather situation for long periods of time. They are not easily swayed by simple patterns in the data that might indicate one weather pattern over another. Experts are often *not* swayed by the opinions of their colleagues until they find that confirming piece of data. We heard in more than one case that the warning forecaster in charge was much more skeptical about the developing weather situation than other people in their own office. In one case, this skepticism was credited with allowing the forecaster to recognize that the developing weather system was not a typical tornado producer but that a flash-flood situation was developing. This forecaster delayed their assessment as he searched for a critical piece of data. In this case, that piece of data was developing and he needed to wait to see "where it would go." It moved in a sudden direction, as he predicted, and he issued a flash-flood warning. The other forecasters thought he was waiting too long to issue a tornado warning.

After a few cycles of looking through radar data, expert warning forecasters develop a mental model of the current state of the atmosphere. Experts use their mental models to "stay ahead of the storm" so that they can make sound and defendable decisions on where and when warnings should be issued. Based on their initial assessment, they develop predictions or hypotheses of what the next data scan would need to look like in order to confirm their mental models. They look through many different data sets, and data overlays, looking for certain cues in the data to confirm their model.

Our expert pool included experts from across the nation and their mental models varied quite drastically based on regional and seasonal differences, as one would predict (see chap. 15, this volume). An expert from one region with sophisticated skills of developing and using his or her mental models will not be equally efficient and successful in applying those if transferred to another region. Experts use their research experience as a basis for the development of their mental models. The knowledge they have gained through either investigation of former severe-weather cases or involvement in research on certain weather patterns allows them to make sense of the data and atmospheric anomalies. We saw experts who had a distinct number of available mental models based on former research of, for example, developing squall lines, and they were able to leverage those during their forecasting process. The preexisting mental models do not lock experts into a rigid framework, but rather provide a general knowledge structure that allows them to take in new and surprising information, such as the rapid development of aspects of a severe storm. Experts are able to learn from the surprises they encounter and use them as additions and refinements of their existing mental models to support future forecasting decisions.

Cues and Cue Perception

We developed a Cue Inventory to represent how the forecasters' expertise supports sensemaking within this data-intensive domain. First, we identified the critical cues and information from each CDM interview. We then asked each expert (via e-mail) to fill in any missing information for his or her incident account. This process not only supplied missing details for our inventory, but also allowed the experts to confirm, or adjust, our interpretation of their incident.

As one might imagine, dozens of individual cues are involved in the warning forecasting task. In fact, the incidents we analyzed resulted in a listing of 43 individual kinds of cues. The list of cues includes:

- *Cues coming from direct perception.* For example, the smell of humid, salty air coming off the Gulf of Mexico implies the presence of moisture to fuel developing storms. Forecasters described subtle cues they noticed during their drive to work, like cloud shapes and movements (i.e., wisps of stratus clouds) suggesting that an air mass is destabilizing.
- *Cues coming from weather data.* For example, satellite imagery shows "towering cumulus" clouds typical of severe weather. Doppler radar might suggest that supercells are developing and indicate whether conditions are favorable for tornado formation. The radar "loop" might show that storms are moving eastwardly very slowly, suggesting increasing rainfall rates;

merging storm cells will also increase rainfall rates. Data from balloon-borne sensors can be critical to show that a layer of relatively warmer moist air is developing, thus suggesting that moisture is being lifted.

- *Cues coming from forecast products.* Examples include severe weather watches issued by the NWS and data from the computer models (e.g., overlap of both strong instability and vertical wind shear parameters is typical in supercell development).

- *Certain kinds of cue* absences. For instance, in one case the forecaster heard no discussion of developing severe weather on the radio during his drive to work, suggesting that a severe weather event would surprise people. In another instance, a forecaster told us that he was looking at a specific geographic point in search of a specific piece of data. If he saw it, it would cause him to discard his current assessment. But, if he didn't, his assessment was right on. So, he was looking to make sure that certain data *didn't* exist.

- *Cues coming from persistence.* At higher resolution scales, weather events can last over several days. One forecaster told us that it is not uncommon to take a look at the sky late in the afternoon and use it as a predictor for the next afternoon. As certain systems grow, stall, and/or settle into an area, much can be gained by observing changes from one day to the next. Hence, persistence can be a powerful tool in anticipating conditions conducive to the formation of severe storms.

- *Cues from other forecasters.* Severe-weather forecasters love what they do. Almost to a person, they described their passion for severe weather. These individuals use their days off to chase and observe severe weather firsthand. These are often the eyes and ears of the forecaster stationed at the office. This expert onsite evaluation of current conditions often proves to be the critical data point for the accurate issuance of warnings.

IMPLICATIONS

Our main focus was on identifying areas that are ripe for further research to benefit the development of intelligent technologies. We present these as invitations or opportunities for the computer science community. These areas include:

- Providing information to the public.
- Team collaboration and decision making.
- Modifications of current technology to provide for a more human-centered system.

- Support for decision making for cases where expertise has the greatest impact.
- Archiving of tough or unusual cases to build up an experience base.
- Expert decision making.

Relationship With the Public and the Issuance of Warnings

One aspect of expertise was the sensitivity of forecasters to their role in protecting the public. They were very aware of storms' proximities to metropolitan areas, and this factor almost always played into their decisions of when and how to warn. They know when public weather awareness is more urgent, and when the public is not concerned "enough" in a particular situation.

Because of this concern for public safety, forecasters take extraordinary measures to alert the public to coming dangers. For example, they might call event organizers when they know there is a major event taking place in the path of a potential storm, or call the Storm Prediction Center (SPC) to request they upgrade their products to reflect the severe-weather potential for that day. A number of forecasters mentioned using unusually ominous language in their warnings to get the attention of the media and the general public. These forecasters have developed sensitivities and strategies to get important weather messages out to the people who need to hear it, to get them to take precautions when necessary, and as a result, to save lives.

All these aspects of forecasting expertise must feed into the creation of aids to help the warning forecaster.

The Team Aspect

It was apparent that expert forecasters develop strategies to take advantage of interactions with other members of the forecasting team in their offices. The more experienced forecasters did not attempt to dominate difficult situations, but rather leveraged opportunities presented within these incidents to mentor and share insights with their colleagues. In more than one incident, we heard that an experienced forecaster took the event of the day as an opportunity to provide on-the-job training to less experienced colleagues. Some of those colleagues had either less experience as forecasters in general, or were recently transferred from a different region of the country and had less experience with the weather patterns specific to that area. Knowledge sharing would therefore be a prime topic for the application of intelligent systems technologies.

In other cases, we heard experts use their colleagues as sounding boards for developing their mental model of the storm. In these situations, the expert communicated to the colleagues what she or he saw on the different screens, how she or he interpreted the weather patterns, and what she or he expected to see in the next data set. The experts sought the input and feedback of their colleagues either to share general opinions about their need for concern about the weather, to confirm suspicions, or to reevaluate their mental model of a developing storm. They valued these interactions as a crucial part of the forecasting process. Thus, collaborative decision making would also be a prime topic for the application of intelligent systems.

The forecasters appreciate the need and importance of developing a shared situational awareness of the developing storm across the personnel in the office. They were strongly aware of the possibility of losing control over a situation if they became too focused on the smaller tasks at hand. In order to counteract this possibility, many offices have a coordinator during busy weather days. This coordinator's responsibility is to keep the big picture in the forecasting area, communicate information to those who need it, find holes in the process, and find ways to fix those holes. This position becomes even more important when the workstations are distributed and lines of communication and awareness of each other's work are easily disrupted. The impact of the distribution of the workstations and personnel on the team's shared awareness, ability to anticipate, generate expectations, coordinate, and synchronize, must be lessened by the coordinator. Yet, no specialized software has been created to aid the coordinator. Decision aiding for the coordinator, and tools to support shared situational awareness would be another topic ripe for the application of intelligent technology.

HUMAN-CENTERING ISSUES: TECHNOLOGY TO HELP FORECASTERS STAY *IN* CONTEXT

Every expert we interviewed believed that math-based prediction algorithms are "not the sacred truth" but are useful as safety nets or as attention management tools. Experts described situations in which they had to focus their attention very closely on real data—the development of one particular storm with the highest probability to turn severe—while giving less time to other less probable or less severe storms. In these situations, programs that run algorithms (e.g., an algorithm that detects vorticity) can be used to alert the forecaster to additional proto-storms or storms that might be increasing in strength. Experts are well aware of the capabilities and also the limitations of the individual algorithms. They discuss the strengths and limitations of the products with which they deal. Rather than referring to programs that are based on algorithms, they refer to the algorithms themselves, regarding them as entities that do things, as opposed to being enumerations of implemented processes. It was not uncommon for experts to cite the rate of false alarms of algorithms. The experts also feel that some algorithms are more trustworthy than others. This is often based on their ability to follow the "reasoning" of the mathematical formulas.

Few radar algorithms allow the forecasters to drill down to the data to confirm that the prediction is trustworthy. Over time, the forecasters have come to distrust a number of tools available to them because of their disability to confirm, trace, or understand the "reasoning" of the technology. Experts are not looking for more algorithms that are designed to replace the human as the decision maker, but rather algorithms that make sense to the decision maker and can therefore be trusted. What are lacking are technologies to help the forecaster understand the meaning of the mathematics behind the algorithms. What are lacking are technologies to help the journeyman understand the bias tendencies, limitations, and strengths of the various computer models. (This points to an issue for meteorology and forecasting that is discussed by Ballas, chap. 13, this volume: Because the computer models are constantly being improved and are therefore a moving target, how can one create human-centered systems to aid the journeyman forecaster in understanding model biases and tendencies?)

We should point out, for fairness sake, that expert forecasters do not mistrust and dislike *all* of their technologies. Most are enthusiastic about trying out new tools. It would be safe to say that most forecasters are hopeful that new technologies will be more user-friendly than many of their current systems and interfaces. (Researchers who are experienced in cognitive task analysis and cognitive engineering would probably not be quite so hopeful. It remains a struggle to get human-centering considerations fully entrenched into the procurement process.)

Using Tough and Unusual Cases to Build Up Experience Base

Experts recognize the importance of building a "lessons learned" knowledge base. For example, one forecaster was able to instantly recognize a subtle signature in the radar data, an unusual cue that a less experienced person might have missed. The forecaster was able to make a nearly instantaneous decision because the pattern he saw on the radar screen matched a pattern he had seen in an investigation years before. In that previous case, he and other forecasters had pored over the data for hours, finding only this one persistent signature to explain the phenomenon.

Building up a repertoire of these unusual cases allows forecasters to more efficiently diagnose situations. Instead of spending valuable time connecting the factors and projecting the causes and effects, they can instantly match the patterns they are seeing to patterns in the current data. The larger the repertoire of these cases, the quicker they can recognize the unusual situations that may face them (Klein, 1993). What is needed is software support, distributed across weather-forecasting offices, allowing the creation of a shared knowledge and experience base particular to weather and weather forecasting.

Expert Decision Making

We identified three types of decisions where expertise has the most impact: (a) identifying "The Big One," (b) predicting which part of a line of storms will accelerate, or how much velocity/reflectivity is enough to warrant the generation of a warning, and (c) recognizing a storm's severity, and communicating the potential danger to the public.

Identifying "The Big One." Forecasters refer to this as the ability to recognize the potential severity of a storm and use that recognition to identify the one storm that distinguishes itself from all others. Severe storms, in this case, are defined as the ones that have the potential to cause extensive damage and pose a high risk to the public. In areas where severe storms are most likely and during specific times of the year, a large weather system usually produces a dozen or more storms. Experts are often able to pick out the one storm that will be most threatening. There are many factors that play into that assessment, including the speed of development of the storm, the direction it is moving (e.g., toward a metropolitan area vs. empty farmland), and the potential severity of the storm. Experts can assess these and other cues fairly early on and make a judgment regarding which storm to attend. This judgment is difficult because there are often fine subtleties in the data that help to predict, for example, if a severe (or soon to be severe) storm is likely to shift in the direction of a highly populated area. Recognizing such nuances is anchored in the expert's experience base. Experts are able to perceive patterns of cues that do not mean much to a novice. They accomplish this by leveraging their mental models and a carefully chosen technological safety net.

Creating intelligent systems to assist in the identification of especially severe storms represents a major challenge for computer science, but it must be based on further CTA efforts.

Predicting Which Part of a Squall Line Will Accelerate, or How Much Velocity/ Reflectivity Is Enough to Warrant the Generation of a Warning. The expert forecaster looks at a line of thunderstorms and is able to make specific predictions as to which part of the line will accelerate and where it will turn and potentially cause damage, allowing the forecaster to make a warning decision for a specific area before the actual event occurs. We heard the same from forecasters whose expertise lies in the prediction of supercell tornadoes. Often they are faced with signatures on their radar that do not clearly indicate a tornado, yet they are able to make an accurate prediction of whether the storm will or will not produce one. In our interview with a less experienced forecaster, we heard how hard it is to know how much velocity and reflectivity is enough to warrant the issuance of a warning on any given day. There are no exact guidelines, because at some times the reflectivity is strong while the velocity is supposedly not developed enough to indicate a tornado, yet a tornado develops. Experts have developed mental models that allow them to play out the individual scenarios and recognize patterns to which journeymen are not yet sensitized, and strategies for querying the data that journeymen have not yet acquired. It is this expertise of knowing when severe is severe enough, recognizing the nuances of the storm's development even in the face of irregularities that allows experts to make early warning decisions and, thus, early warning declarations.

Creating intelligent systems to assist in the identification of tornados associated with squall lines (lines of storms) or supercells represents a major challenge for the computer science field.

Recognizing Storm Severity, and the Need to Make It Explicit to the Public. The recognition that a standard warning message will not be enough is the third high-payoff area. The ability to make this decision encompasses a number of crucial assessments and an overall awareness of the public's "state of mind." The expert has realized, based on the forecasting process he or she has gone through and the data he or she has looked and compared to his or her mental model, that the storm is large and destructive in nature and is moving toward a populated area. This combination of assessments sets off an alarm in the forecaster's mind. The warning needs to go out but the question arises, is the public going to recognize the intense urgency and intrinsic danger of this storm? Experts often are aware of large outside festivals or sporting events in the area that attract a lot of people, and take this into consideration. Experts know to consider what it might take to make the public aware of the severity of the weather event, and they choose effective means to do so. The wording of the warning message can catch the media's attention and reach a larger number of people.

Intelligent systems to aid the journeyman forecaster in making these sorts of assessments would be of great value to the weather-forecasting community.

CONCLUSIONS

Any attempt to overapply technology, including intelligent technology, to fields in which detailed, context-specific expertise is required can be misguided unless it is recognized that the purpose of the technology is to amplify and extend the human's ability to perceive, reason, and collaborate. Although this point may be obvious to those who advocate the paradigms of Naturalistic Decision Making and human-centered computing (HCC), we have heard many in the weather-forecasting community yearn for intelligent systems that enhance and support their abilities.

The technology that is introduced must amplify the decision-making processes of the forecasters. Simply providing more tools (algorithms, image enhancements, etc.) and faster processors or faster radar updates is not the answer. Enhancements cannot be based on capabilities that technologists find possible and that system designers believe the forecasters need. System designers must acquire a deep understanding of how the forecasters look for data, what data are important, and which data are noise. Designers must understand when in the forecasting process the forecasters need certain pieces of information and when they don't, how they use the algorithms and when they simply turn them off, and so on. It is the drilling-down into the cognitive processes of the end-user that provides the necessary foundation for technology development. Without this step, money will be wasted.

Little attention has been paid in the weather-forecasting domain to create technologies that support *team* collaboration and *team* decision making. Here, too, the paradigm of HCC comes into play. Cognitive systems engineers have been involved in design efforts for command posts, nuclear power plant control rooms, and information centers. A general finding is that it is difficult to balance the design of a decision center for both steady-state operations and crisis operations (Klinger & Klein, 1999). Designing a decision center requires an understanding of the decisions, but also the context surrounding those decisions, the information flow (this includes receiving the right information in the right format at the right time), and the coordination with other team members (both internal and external). The paradigm of HCC, especially The Triples Rule (Hoffman, Hayes, Ford, & Hancock, 2002) regards the technology as one of the team members. As decisions regarding technology are made, one must consider the impact the technology has on team performance—this is the Envisioned World Problem (Dekker, Nyce, & Hoffman, 2003). The ultimate goal would be to minimize scramble time, to make available to the experts the information they need when they need it, to support decision making so that products are delivered to the public as efficiently as possible, and to facilitate the ramp-up of journeymen to the level of the expert.

ACKNOWLEDGMENT

The research reported here was supported by a contract from the Office of Climate, Water, and Weather Services of the National Weather Service, Norman, Oklahoma.

David W. Klinger is now with SRA, Inc.; Bianka Hahn is happily at home raising her son and daughter, and Erica Rall is at Vanderbilt University

REFERENCES

Dekker, S. W. A., Nyce, J. M., & Hoffman, R. R. (2003, March/April). From contextual inquiry to designable futures: What do we need to get there? *IEEE Intelligent Systems,* pp. 74–77.

Hoffman, R. R. (1991). Human factors psychology in the support of forecasting: The design of advanced meteorological workstations. *Weather and Forecasting, 6,* 98–110.

Hoffman, R. R., Coffey, J. W., & Ford, K. M. (2000). A case study in the research paradigm of human-centered computing: Local expertise in weather forecasting (Report on the contract, "human-centered system prototype." Arlington, VA: National Technology Alliance.

Hoffman, R. R., Crandall, B. W., & Shadbolt, N. R. (1998). Use of the critical decision method to elicit expert knowledge: A case study in cognitive task analysis methodology. *Human Factors, 40,* 254–276.

Hoffman, R. R., Hayes, P., Ford, K. M., & Hancock, P. A. (2002, May/June). The Triples Rule. *IEEE: Intelligent Systems,* 62–65.

Klein, G. A. (1993). A recognition-primed decision (RPD) model of rapid decision making. In G. A. Klein, J. Orasanu, R. Calderwood, & C. E. Zsambok (Eds.), *Decision making in action: Models and methods* (pp. 138–147). Norwood, NJ: Ablex.

Klinger, D. W., & Klein, G. (1999). Emergency response organizations: An accident waiting to happen. *Ergonomics In Design, 7,* 20–25.

Part V

TEAMS OUT OF CONTEXT

Applying NDM to World Security Needs: Perspectives of a U.S. Military Practitioner

Lt. General Frederic J. Brown
McLean, VA

THE WORLD SITUATION

Terrorists have been quite clear in expressing their goals and intent (e.g., statements by Osama bin Laden; see http://idaho.indymedia.org/news/2002/11/207.php), including a threat to use chemical and radiological weapons of mass destruction (WMDs). Arguably, some sort of WMD attack, even on U.S. soil, is no longer a *whether*, but a *when*. Many government agencies and departments are currently in the midst of substantial change in decision making, within the U.S. military and many of its national and international partners. The challenges of countering world terrorism have focused leaders at every level and echelon, and across all functions, on the goal of forming effective command and staff action teams in complex joint, interagency, and multinational organizations, including *ad hoc* teams assembled, on very short notice, to conduct any of a very wide variety of operations.

On the one hand are the effects of global terrorists' actions. Among them is an increase in the difficulty and complexity of decisions. On the other hand are the information and knowledge sharing opportunities of the Internet and the World Wide Web, which offer exceptional supporting capabilities for leaders. In the past few years, decision-making capabilities have been made less difficult to manage in some ways, but have become more complex in other ways. Though challenges of effective decision making have grown with the diversity of complex threats, so have the capabilities of the supporting tools provided by the Internet and various technology research communities, including the Naturalistic Decision Making (NDM) community.

The various threats that the world faces, and the national and international responses, establish the framework for capabilities required by the national security community and its many decision makers. Requirements for these capabilities were broadly perceived before the attack of September 11 (e.g., Shinseki, 2000), and emphasized joint forces (i.e., units of employment composed of army, naval, and air elements), international formations, and a variety of operations including urban warfare civil engineering, peacekeeping, and humanitarian missions. And there is the ever-present potential for conflicts in which WMDs are used.

The U.S. Army must be prepared to conduct combat operations with *ad hoc* hybrid organizations composed to ensure dominant force whatever the military requirement across the full spectrum of operations. These capabilities requirements can be expected to expand even more as doctrinal preparation proceeds for the future Army, which will be transforming even as it fights across the spectrum of conflict from high- or midintensity operations to counterinsurgency to counterterrorist to stability or support operations "composed of modular, scalable, flexible organizations for prompt and sustained land operations" (Riggs, 2003, p. I). With regard to the composition of teams and methods of decision making, we need not only to bring about change, but also prepare for changes that are not currently predictable. At least three explicit new land power missions have appeared since 9/11. They are preemption of WMD attack, support to homeland security (national, state, and local), and mending failed states that might become WMD operations bases directly threatening the United States. These challenges cannot be met if considered solely in the context of the known challenges of decades of postwar rebuilding of Germany and Japan—as has been evident in operations in Southwest Asia.

These are the macro challenges to military decision makers, and they are paralleled by increasing complexity at the micro level of the tactical units. The implications of all these challenges for decision-making requirements and evolving practices seem quite substantial. Afforded the opportunity to observe all this recently at tens of Army Combat Training Center rotations, I offer some of my insights in this chapter. Mine is the perspective of decades of practice in decision making within the military, as well as an abiding intellectual curiosity with respect to "futures" developments. My goal is to suggest some points for the NDM community to consider as it reaches for new methods and support for decision-making professionals.

SOME IMPLICATIONS FOR DECISION-MAKING THEORY AND PRACTICE

All Soldiers (Corporal and Above) Will Have to Be Adaptive Leaders. The combination of highly competent, motivated volunteer soldiers and increasingly distributed tactical data and information (Tactical Internet) are driving task performance responsibilities lower and lower in the chain of command. The future Army will routinely bring digital data and information capabilities to the level of the Infantry Squad. Hence, corporals will be expected to master tasks formerly expected of much more senior noncommissioned officers who in turn have assumed many responsibilities formerly expected of officers. The challenge then becomes conversion of data and information to knowledge and understanding by effective leaders and leader teams practicing much more the art of command than the science of control.

Soldiers Will Perform Under Remarkable and Rapidly Changing Forms and Patterns of Stress.   Brutal, hostile, decisive fighting, whether using stand-off or close assault, will always characterize close combat. Added to this, however, are the threat of WMDs, and the uncertainty of who is, and who is not, a hostile combatant, in genuinely complex tactical environments.

Combat operations will be joint, interagency and multi-national (JIM). Most operations already are international, and although the decision-making focus here may be on practices of the US military, there are far broader applications to other nations and organizations through JIM operations.

Leaders Will Also Operate as Members of Teams.   Teams will operate vertically, forming chains of command down the echelons: Division, Brigade, Battalion, Company, and Platoon. Other vertical teams of leaders are chains of coordination, which are needed when JIM operations are added to chains of command, and chains of functional support (noncommissioned officers or commanders and staff officers supporting a single function such as intelligence or fire support; Brown, 2002, pp. III 1–30.) Teams will also operate horizontally, in the form of groups of peers who are responsible for such areas as cross-staff coordination. At every intersection of this matrix, there is at least one leader, more often several leaders, who are in near-continuous communication vertically with both seniors and subordinates, and horizontally with peers, and often both. These individual leader and leader team interactions, enabled by near-continuous communication, will become pervasive.

Teams Will Need to Be "Expert Teams" That Are Flexible and Able to Cope With Change, Both Predicted and Unpredicted.   Preparation of individual leaders is necessary but insufficient. Teams must also be prepared to practice good decision-making *as teams*. As NDM research has demonstrated, effective teaming involves more than team leadership (i.e., shared vision [purpose], shared trust, shared competence, and shared confidence). The skills, knowledge, and attributes needed by individual leaders within teams can be distinguished from those needed in order for a team as a whole to be high performing (see chaps. 19 and 22, this volume). For example, the individual's confidence does not directly translate to team confidence—the shared belief that together, we are all confident in our team's performance. Add to this is the fact that team membership may change as casualties occur or as new *ad hoc* organizations evolve. Furthermore, teamwork and team decision making must be sustained despite fear, uncertainty, fatigue, and disruption. Thus, "continuity of command" requirements mandate practice at teamwork by vertical leader teams of decision makers.

To date, there has been some institutionalization of teamwork practice, especially in some chains of coordination. Specifically, methods and requirements of the Joint Interagency Task Forces, the Joint Counter Terrorism Task Forces, the NATO Combined Joint Task Force, and Partnership for Peace associations rely on an understanding of teamwork requirements. They provide signposts for what will be required for broader multinational teamwork. However, clearly more research and development is required in this area of high-performing leader team decision making in dynamic chains of coordination.

Chains of Authority and Responsibility Will Shift.

Another chain appears if JIM operations are involved, such as Stability and Support Operations. If, for example, a MultiNational Division has been interjected, it is no longer an Army chain of command with its well-understood responsibilities and authorities—it becomes a chain of coordination. Yet another chain appears if the decision making is interagency, as when the Army National Guard supports federal, state, and local authorities in homeland defense, or when the Army is committed in operations with the CIA or FBI. Each of these new chains will involve new sets of horizontal and vertical leader teams. In those, the explicit and implicit authorities and responsibilities of the traditional chain of command will no longer be in effect. Each chain of coordination might present unique authorities and responsibilities dependent on the joint, interagency, or multinational representation.

Forms and Processes of Decision Making Will Change.

Although expressed here largely in terms of Army practice and structure of the immediate past, it seems reasonable to assert that team decision-making theory and practice need to develop new and distinctive alternative vertical and horizontal frameworks to meet the demands of future decision making. Successful processes of decision making under combat conditions exist and are practiced routinely at the Combat Training Centers. Now, responding to new diverse threats in the global war on terrorism, traditional detailed command processes such as the Military Decision Making Process (MDMP) are supplanted with increased reliance on more intuitive decision processes (Department of the Army, 2003; Klein, 2003). The art of command is now being addressed as rigorously as the science of control has in the past (Brown, 2000, pp III 1–3). Furthermore, the Tactical Internet is designed to train as well as to support the military decision-making process, which is also formalized in current NATO agreements. However, inadequacies remain in our understanding of decision making at the level of interagency and multinational decision making when international coalitions morph to respond to terrorist threats. More research seems clearly required.

Expansion of the Internet. The potential of the Internets (public and DoD) and potential of the Semantic Web are increasingly evident. For instance, we now have Internet-based methods of knowledge management and knowledge sharing. Less appreciated perhaps is the recent formation of military-oriented Communities of Practice (CoP) that address various important professional issues (Companycommand.army.mil). In time, there seem certain to be expanding families of CoPs, where concerned professionals can share data, information, knowledge, and understanding as they serve in units or other organizations. These are emerging today in the Army in the rapidly expanding Battle Command Knowledge System supported by Army Knowledge Online. Inevitably, much effort will focus on decision making. There would be seem to be many opportunities to stimulate individual leader and leader team acquisition of the skills of decision making for chains of command, chains of coordination, and chains of functional support.

To that matrix we would now add horizontal peer CoPs. This interwoven interaction of vertical and horizontal exchanges of data, information, and knowledge is termed "double knit" (Wenger, McDermott, & Snyder, 2002). In double-knit, each leader has "dual citizenships" in various combinations of horizontal CoPs and vertical chains of command or support, forming nested, distributed teams. In addition to the sharing of data, information, knowledge, and understanding, there would also appear to be opportunities to draw on double-knit to support more rapid preparation of high-performing leader teams—to accelerate the development of shared trust and shared vision within the chain. Clearly this all should support better individual and team decision making, but an answer to "how?" awaits the findings from the research community.

Fortunately there are current and emerging tools available to support rapid leader team creation.

There Will Be New Tools to Prepare Decision Makers. Several tools have been developed to prepare leaders as decision makers both as individuals and, increasingly, as members of teams. The tools form a part of the infrastructure of the U.S. Army's Combat Training Centers. Perhaps the best example of focused leader learning has been the various Mission Rehearsal

Exercises conducted for units that were about to deploy to the Balkans, Afghanistan, and currently Iraq. Unit transitions into the theater of operations were seamless for both active and reserve units. That is, conditions postdeployment were successfully replicated in leader-intense learning environments predeployment such that new units were not surprised by conditions "in country."

Another example is an innovative electronic decision game that has been created by the Army Research Institute/Army Research Laboratory for precommand preparation. The "Think Like a Commander" tool is now used in Pre-Command Instruction at the School of Command Preparation at Ft. Leavenworth, Kansas. It draws the commander into increasingly challenging situations, while simultaneously suggesting appropriate decision-making considerations. It will soon be available in distributed mode, and hopefully will be a useful tool available to multiple CoPs. Similar tools will be provided for leader teams to improve decision-making practices through the Battle Command Knowledge System. Effective decision games for school use have even been exported to various leader chains that are having to adapt to new requirements.

But there remain significant advanced experiential learning and team-building opportunities—all nascent or forthcoming—designed such that they can be employed in both school and unit for either peer or hierarchical leader teams use in developing necessary decision-making practices. Increasingly, the learning vehicle appears to be the Battle Command Knowledge System, but design of appropriate effective individual leader and leader team learning programs remains incomplete.

CONCLUSION

Hopefully these thoughts may stimulate continuing interest in the NDM community to support improved decision making by military leaders and leader teams. I see a need for proposals for practices for developing high-performing leader teams—involving team leadership training, teamwork training, and training for team decision making—for chains of command, chains of coordination, and functional support. Priority is suggested for the study of hierarchical chains of coordination. These are the least understood of the team types, and yet are the most likely organization of leaders in security in a world at war against global terrorism.

Related to the aforementioned is a need also for remedial training and new forms of training, in teamwork and decision-making practices. Finally, there is an outstanding need to extend and leverage the notion of double-knit—to apply to it in the creation of counterterrorist leader teams composed from very diverse JIM organizations that are themselves modular, scalable, and flexible.

REFERENCES

Brown, F. J. (2000). *Preparation of leaders* (Institute for Defense Analyses Report No. D-2382). Washington, DC: Institute for Defense Analysis.

Brown, F. J. (2002). *Vertical command teams* (Institute for Defense Analyses Report No. D-2728). Washington, DC: Institute for Defense Analysis.

Department of the Army. (2003, August). *Mission command: Command and control of Army Forces* (FM 6-0). Washington, DC: Department of the Army.

Klein, G. (2003). *Intuition at work*. New York: Doubleday.

Riggs, J. (2003). *The objective force in 2015* (Department of the Army Report). Washington, DC: Department of the Army.

Shinseki, E. (2000). The Army vision. (*Department of the Army Weekly Summary*, April 7, 2000). Washington DC: Department of the Army.

Wenger, E., McDermott, R., & Snyder, W. M. (2002). *Cultivating communities of practice*. Boston: Harvard Business School Press.

Chapter 18

A Dynamic Network Analysis of an Organization With Expertise Out of Context

John Graham
Cleotilde Gonzalez
Mike Schneider
Carnegie Mellon University

The Department of Defense is rapidly transitioning from historically large and hierarchical organizations into smaller, more mobile, and more distributed network-centric organizations. In the traditional hierarchy that characterizes the legacy force, tasks are handled within each formal, functional staff subgroup and functional solutions are then devised at the top of the hierarchy. If there are conflicts susceptible to functional solutions, they are resolved after each task has undergone independent analysis. A disadvantage of this traditional organizational model is that conflicts needing solution are sometimes identified late in the problem-solving process (Cebrowski & Garstka, 1998). In additional, hierarchical organizations are more static, difficult to change and adapt to new changing conditions of a task. Our understanding the communication and decision-making processes for the network-centric organization has become critical (Alberts, Garstka, & Stein, 2001; Cebrowski & Garstka, 1998).

In 2003, a large group of officers with 10 to 30 years each of legacy-force experience were told to role-play as members of a notional network-centric organization. For 2 weeks, these experts were to ignore all of their experience and training with respect to hierarchical organizations and hierarchical decision-making processes and instead adopt a free-flowing network organization design. The exercise was conducted at the Fort Leavenworth, Battle Command, Battle Lab (BCBL). Army officers were required to learn (a) the concepts behind the hypothetical organization, (b) a new method to make decisions in the hypothetical organization, (c) their role in the structure of the hypothetical organization, and (d) how to use the simulation software. The exercise involved the execution of a simulated battle, using simulation software and battling against role-playing enemy officers. The Army officers gathered information and input actions on the battlefield via the computer simulation.

Researchers from the Dynamic Decision Making Laboratory at Carnegie Mellon University were invited to the BCBL exercise to provide support for testing a network-centric organization. Support included five observers and analysts as well as custom-developed data collection software. Our goal was to use network analysis techniques to describe the behavior of a network-centric battle staff. We also intended to use our analysis to gain insight as to the empirical strengths and weaknesses of network organizations. In addition, because there would be follow-up exercises, the Army wanted us to provide feedback on the design of the exercise, and on the use of legacy experts as futuristic role players. Our group applied social network analysis to the data collected during the exercise.

This chapter reports the results from the dynamic network analysis performed in this organization. This chapter describes a selected set of social network analysis measures used in the Army exercise and explains how these measures were implemented in the command-and-control context. This chapter also provides details of the network organization design, and describes the findings from this exercise. The chapter concludes with a discussion of the strengths and weaknesses of the network organization and the managerial challenges it poses, along with the implications of using legacy-force experts as exercise role players.

DYNAMIC NETWORK THEORY AND MEASURES

In any organization, people exchange ideas and information (Borgatti, Everett, & Freeman, 2002). Social network analysis is a technique that seeks to quantify the relationships among people in an organization. People and organizations are represented as nodes in a network, and the relationships (e.g., information flows) between people are represented as lines drawn between these nodes. Thus, a social network is a graph consisting of individuals and connections among them, where each connection is associated with some form of communication or relationship between the nodes (Borgatti, 1994; Scott, 1992). An example appears in Figure 18.1. Social network theory allows for the quantification of dyadic ties that exist between team members. Aggregating these dyadic ties yields characteristics of the overall organization.

As with social network analysis, dynamic network analysis allows researchers to concurrently perform individual-, group-, and organizational-

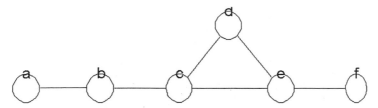

FIG. 18.1.   A simple network graph. From Borgatti (1994). Copyright 1994
by INSNA. Reprinted by permission.

level analysis. Dynamic network theory is a temporal form of traditional social network analysis (Carley, 2003): It examines the trends in a network
over a time span. Dynamic network analysis also seeks to account for an organization's current state based on characteristics of the organization's previous state.

Communication data among an organization's members can be gathered from shared e-mail headers, chat room traffic, instant messaging, and
phone calls, or by surveying the individuals (Wasserman & Faust, 1994).
Though each of these communication media has different qualities, measure relevance is determined by the organizational context and collaborative tool characteristics. For instance, in one study, we found that early in an
organization's life, people are more comfortable with communication conducted face-to-face, but as they become comfortable, including comfort
with their collaborative tools, they migrate their important communications to those tools. As researchers, we would need to track both communication media until a full transition to one or the other occurs. In the studies
reported in this chapter, we describe both chat room and self-reported
data.

Selected Set of Social Network Analysis Measures

Table 18.1 presents selected social network analysis measures defined and
translated for a military command-and-control context for the purposes of
our data analyses. Subsequently, the measures are translated for an Army
brigade command-and-control context.

To use any of the social network analysis measures, communications data
must be collected to be able to construct a network graph. Network software can help represent communications data into network graphs and obtain values for the measures described in Table 18.1. There exists general
network analysis software, such as UCINET (Borgatti et al., 2002) or ORA
(Reminga & Carley, 2003), that were used in this research.

Figure 18.2 is a UCINET-produced network graph representing 90 minutes of communication relationships in a 10-person command-and-control

TABLE 18.1
Network Measures and Concepts

Measure	Organizational concept
Network density	Network density is the number of actual links observed among the members of an organization divided by the number of all possible links (Freeman, 1979). A fully dense network or organization would have every person (node) linked to every other person. Organizations with high density share information quickly, but the higher the number of connections, or the greater the density, the greater the workload.
Network distance	Network distance, often referred to as the geodesic, is the shortest relationship path between any two members of the organization (Borgatti, 1994)—or, operationally, the shortest path distance separating two people within a communications network.
Physical distance	This is the actual metric distance between two members of the network. Shorter physical distance improves coordination. Coordination over greater physical distance may be partially facilitated by collaborative tools.
Self-forming team	This is a group of three or more network members who work together and reciprocate information sharing. Members of ad hoc, self-forming teams seek one another out in order to integrate expertise required to solve impending tasks.

cell of a network organization. The 10 are members of a Command Integration Cell (CIC) designed to coordinate the activities of other functionally oriented cells, and constitute a subset of the 56-member prototype network organization that was the focus of this research. We used this subset to further explain the measures in Table 18.1 into this exercise context.

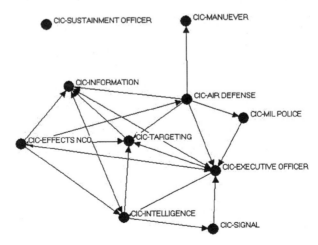

FIG. 18.2.   Command Integration Cell (CIC) of a prototype network organization. Links external to the CIC are not represented in this figure.

Network Density

Network density provides a gross level measure of the connectedness of organization members. In the CIC example, there are 10 members and 90 possible links if each member has two possible links (a "to" and a "from") with each other member. As it is evident from counting the links in Figure 18.2, only 19 relationships were observed during the session depicted. The resulting density of the organization is 21.1%.

Each organization could have a 100% network density, but density is not a static measure. Fluctuations from 100% are a result of numerous factors tied to organizational and environmental dynamics. For example, changes in density could be attributed to how comfortable organization members are in their roles, to organizational procedures, and to changes in task workload.

Network Distance

The network distance is the shortest path distance separating two people within a network. A person's network distance from other members is a function of whom they choose to communicate with in the organization. In the CIC example, the Executive Officer has a path of length value of one. This is, the shortest connection in the graph between the Executive Officer and the other members of the network is one. He has a path length of two to the Maneuver Officer, meaning that the Executive Officer does not communicate directly to the Maneuver Officer, but rather he communicates to the Air Defense Officer who in turn communicates to the Maneuver Officer. He has a zero path length with the Sustainment Officer (from whom he is disconnected).

Physical Distance

Research has found that team members in close proximity function better than team members that are not colocated (Herbsleb, Mockus, Finholt, & Grinter, 2000). Physical distance is also an important indicator of the situational awareness any two members have in common and of how well they function as a team (Graham, Schneider, Bauer, Bessiere, & Gonzalez, 2004). In the CIC example, all of the members were positioned at a round table facing one another, and they were all categorized to have a physical distance of zero. Members of groups outside the CIC were categorized to be at a distance of one.

Self-Forming Team

Ad hoc, self-forming teams are unique to the network-centric organization. They consist of three or more members of the organization who form a group to solve some problem, engage in reciprocal communications, and may exist only until the problem is resolved. Traditional social network analysis characterizes teams of this type as *cliques* (Scott, 1992). We assume that this strong form of a team normally arises only in the presence of more difficult problems that require extensive collaboration and negotiation.

In Figure 18.2, the only subteam that met this definition consisted of the Executive Officer, the Targeting Officer, and the Intelligence Officer. From our notes, we know that this subteam focused on selecting and confirming enemy targets for the commander. This task apparently required reciprocal collaboration and communications in order to coordinate activities. All other potential subteams failed to pass the reciprocal-communications test, as no other member reported reciprocating the communications of another during this collection period.

In what follows, we explain the exercise conducted at the BCBL, Ft. Leavenworth. The data collection and analyses were done using the network measures explained previously. Furthermore, the data analyses conducted looked into dynamic network analyses in which we analyzed the change of the network measures over time. A dynamic network analysis considers the organizational form to be a living entity, capable of shifting form and structure, and it analyzes the change of this organizational communication (Carley, 2003).

DYNAMIC NETWORK–BASED MILITARY EXERCISE

The Army gathered 56 officers to serve as role players for a futuristic network organization command-and-control staff. The exercise design involved a distributed five-cell Army organization responsible for a brigade-size unit. Each functional cell involved three to eight role players. The role players gathered information, coordinated with appropriate staff members as they needed in order to execute a simulated battle successfully, and entered battlefield actions into the battle simulation software.

Individuals were also provided with communication software tools, including audio conferencing, instant messaging, a computer-based chat room, shared whiteboard, file transfer capabilities, and application-sharing software. Partitions or walls separated the five cells, so that a participant could talk directly to members of his own cell, but could communicate directly with members of other cells but only through the use of communication tools.

Communication data were collected over 5 days following a 1-week training period. We did not collect data during the training period. Communication data were collected on three sessions per day over 5 days. The training provided individuals with an explanation of network organizations and attempted to reduce the reliance on the legacy-force procedures during the exercise. A process in which the organization planned and then executed the scenario in the computer simulation was used in this exercise. Experienced military observers were placed within each cell to capture locally observable phenomena, such as face-to-face communications within each cell that wasn't captured through the communication technology. Our group was allowed to collect data from all role players every 60 to 90 minutes for 16 sessions, using an online questionnaire networked to each of the participant's workstations.

Though we would have preferred to log all communications regardless of medium, the computer system for this particular exercise would not support an automated logger; as a result, a self-report questionnaire was employed as a simplified means of collecting data on all communications between members regardless of the communication method used. A questionnaire asked participants to report the people with whom they had communicated during the time elapsed since the previous questionnaire. They could choose up to 10 responses by selecting participants from pull-down menus; the responses were ordered by the frequency of communication during the previous session.

The communications data set was used to construct network graphs. We created the network graphs within minutes of data collection in a way that the military observer and controllers could have immediate feedback on the critical trends. As a consequence of observing the patterns of communication, the military controllers were able to alert the observers of "hidden" communications on the part of the role players under observation.

Using the networks constructed from the communications data, we calculated the network measures described in the previous section. The calculation of these network measures over the course of the military exercise allowed us to draw conclusions about the dynamic behavior of the organization. In addition to the traditional network measures, we created a new measure that we call the workload congruence. This new measure as explained next helped understand how shared understanding changed over time. The results from these measures are presented next.

Network Density

In the "Measures" section we explained that network density is the sum of active links in an organization divided by all potential links between members (Freeman, 1979). Figure 18.3 shows the density for this exercise over

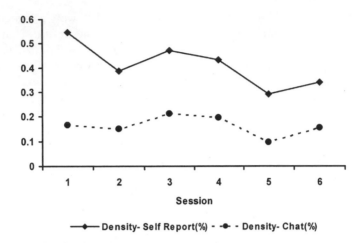

FIG. 18.3.   Density from self-reported communications and logged chat room communications.

the last six consecutive sessions. We calculated the network density using the self-reported communications survey (density-self) and the logged chat room communications (density-chat). Based on these data, and given that the chat room was only one of the communications media available (face-to-face, chat room, instant messaging, or e-mail), the self-report is more representative of combined communication media quantity (higher density was found with the self-report). Although self-reports suffer from a number of biases and memory failures when they are collected sporadically, we attempted to reduce the memory effect by increasing the frequency with which we collected the self-report data to three times a day.

By collecting network density data throughout the exercise, we could construct a temporal graph of network density. Intuitively, one would expect the fluctuations to indicate activity in the organization due to the evolving number and types of tasks created by the scenario, and, potentially, the normative communication characteristics of the organization.

Figure 18.4 shows the change in network density over the course of the 16 sessions. After testing the linear trend of the network density, we verified that it significantly increased over the sessions ($F(11) = 10.81$, $p = .007$): density starts at approximately 10% and finishes at approximately 20%. This is a significant increase in magnitude, considering the lack of familiarity of the role players on network organizations and the short period of time they have worked together in this new organizational design.

Multiple possible reasons exist for the increase, including individual role development, requirements of the scenario or task, and improved under-

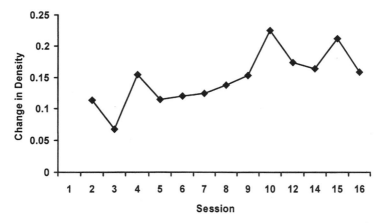

FIG. 18.4.   Network density change over the course of 16 sessions.

standing of communication technologies. Using this density data, it is not possible to discern the real cause for the increase. However, during the interviews, the role players spoke of establishing procedures and developing a set of coordination rules. Chat rooms were identified for specific functions and tasks. Essentially the organization was learning, as the legacy-force expertise that did not apply to this network organization was discarded and replaced through collaborative trial and error.

Observer reports indicate that the variations starting at time period 10 can be attributed to changes in the scenario task load. During that period and supported by the feedback provided through the networks, the commander asked the team to pay special attention to the team members they should be communicating with, and asked them to concentrate on the execution of the simulated battle. Variations prior to time period 10 can more likely be attributed to organizational learning as individuals learned their own and others' roles and responsibilities. In a newly formed organization, norms have not yet been established (Moreland, 1999). For instance, people do not know who has expertise on different topics and whom they can or should contact when attempting to solve a specific problem and we would expect this to be reflected in the communication patterns. The participants in these studies were operating as a team for the first time. We expect a new organization to start out with a relatively sparse network, to gradually increase linkages as the members explore and establish necessary communication channels, and then to peak at some level of density. Once organizational norms are established, we expect that variations in behavior are attributable to factors such as organizational task load and changes in the external environment.

Self-Forming Teams

In a traditional, legacy-force hierarchy, problem-solving teams are functional in nature and do not cross functional boundaries; it is not until functional solutions ascend up the hierarchy that multifunctional teams are formally designated to coordinate their solutions. Any cross-functional teams occurring at lower levels are approved by each of the functional leaders affected. This formal process exists to prevent subordinates from incurring workload or shifting task priorities without the knowledge of their staff supervisors.

This deliberate, formal process is discarded in the network organizational design we tested. Instead, members are able to assign themselves to problems and to accept workload. Furthermore, they are able to freely coordinate across functional boundaries without the express approval of their respective supervisors. In network organizations, these interactions result in ad hoc teams—which are not part of the formal organizational structure, but come into being nonetheless. We wanted to know whether the role players would adjust to the network organizational concept and form ad hoc teams. If the role players were not able to shed their legacy-force behaviors, we expected to see stabilization of a few teams reflecting the typical organizational ties in a traditional hierarchy.

We analyzed the communication data to look for the groups of three or more network members who worked together and reciprocated information sharing. These groups were counted as the ad hoc, self-forming teams. Figure 18.5 shows the increase in ad hoc team formation over time. The number of ad hoc teams significantly increased ($t[12] = 6.68$, $p < .01$) as the organization members spent more time working within the network organization model. However, if the ad hoc teams found at each time period were persistent for a long period, it would be possible that the group members established a hierarchy so that the teams would no longer really be ad hoc. In the data, not one team stabilized and persisted over the duration of the exercise.

Figure 18.6 shows that only one team lasted for more than half of the sessions, whereas the majority of the ad hoc teams (204) were unique to a single session and never recurred in the communications network. Therefore, the organization members appear to have steadily adapted their collaborative behavior to that expected in a network organization.

Whereas the role playing members were adopting network behaviors, the role players assigned to staff leadership roles were losing oversight of their staff. The most prolific ad hoc team consisted of the CIC-Executive Officer, the MSC-Operations Officer, and the BSC-Sustainment Officer. At the end of the exercise, we queried all role players, and none were aware of this team's existence except for the members of the team itself. Krackhardt

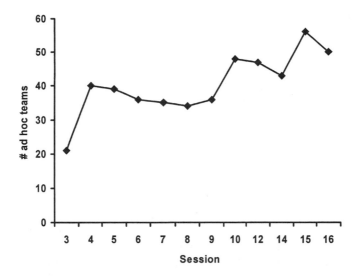

FIG. 18.5.   Number of ad hoc teams formed per session.

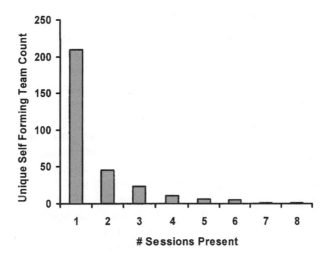

FIG. 18.6.   Unique self-forming team count by number of sessions present.

(1987) found that managers exhibited about 80% accuracy in identifying the leaders and work teams in their organization. In a geographically collocated command-and-control organization, a leader has but to listen and watch to be aware of the number of ad hoc teams. In the geographically distributed network organization, however, the most prolific ad hoc team was invisible to the leader and other members of the organization.

Workload Congruence

All the traditional network measures provide information based on the links formed by the communication patterns. But none of the traditional network measures provide information about the shared understanding or knowledge of individuals that communicate. To understand how each of the participants' understandings of other players changed as they adapted to the network organization, we created a measure of each participant's workload shared awareness.

We measured the workload of each individual and their predictions of others' workload. This measure was derived from Entin (1999). We gathered estimates of workload using NASA TLX (Task Load Index; Hart & Staveland, 1988), comprising six workload parameters on a Likert scale. During the exercise, the role players were asked to rate themselves, as well as five other team members randomly selected from among the other organization role players. This allowed us to sample the workload measure at multiple time periods throughout the scenario. When rating other people, the role players had the option of selecting "Don't know" in answer to each of the seven questions. In a typical laboratory exercise consisting of college sophomores, "Don't know" would not have been an option and the participants would be expected to make their best guess. Working with experts requires different methods, however. Experts tend to know when they do not have sufficient knowledge to accurately answer a question. Adhering to a traditional experimental design and forcing the experts to blindly guess would have created frustration and decreased response validity.

The workload congruence was determined by comparing the workload estimations with respect to a particular role player with that role player's self-reported workload. This measure was computed by summing the absolute differences between the ratee's self-reported ratings and the raters' estimations. Congruence scores could range from 0 (no differences between the ratee's ratings and the rater's estimations, indicating perfect congruence) to 36 (corresponded to the possible maximum possible difference in each of the six TLX workload items measured, indicating perfect incongruence). If a role player chose "Don't know," the workload congruence was assigned a score of zero.

Figure 18.7 is a graph of workload congruence over all the sessions in the exercise. The upper line indicates the mean congruence for organization members who reported that they communicated directly during that particular session (direct communicators). The lower line indicates the mean congruence between all organization members, whether they communicated directly or not (entire organization). Note that both congruence means initially decrease. Overall, however, the workload congruence for direct communicators held approximately constant at a mean of 59.5%,

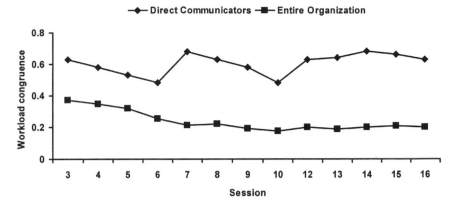

FIG. 18.7. Mean workload congruence by session and the comparison of between organization members in direct communication and all members of the organization.

whereas the overall workload congruence suffered a significant decrease over the life of the organization ($F[34, 2128] = 24.94$, $p < .01$).

Throughout all sessions, the participants were relatively consistent in their workload estimations for people with whom they reported they communicated with directly during that session. The interaction may have provided information about one another's workload. In fact, it is surprising that the direct communicators did not have better workload awareness over time. Nevertheless, the significant and continuous relative decrease in overall workload estimation is indicative of a loss of congruency of the workload awareness.

It appears that when operating in the new organization, these individuals had a high workload congruence at the start of the exercise. At that point, the organization was still operating with some legacy-force behaviors, and the role players' organizational mental models could produce somewhat accurate workload congruence. As the members explored and adapted to the network organization, however, their work procedures changed. Therefore, as the organization became more networklike, the role players' existing schemata did not apply, and their workload congruence deteriorated.

Network Distance and Physical Distance

Network organizations are highly dependent on members that are distributed across the battlefield, working and coordinating effectively with one another. At the outset of this exercise, a number of legacy-force experts expressed the opinion that "there is no substitute for face-to-face" interaction in battle staff coordination. In fact, much of the laboratory and field research has supported the claim that there is no substitute for physical proximity in team performance (Clark & Brennan, 1991; Olson & Olson, 2000). We were concerned that the distributed nature of the military organizations being tested would hinder their effectiveness.

Furthermore, given that research indicated that direct communications were more likely to produce an accurate workload congruency, we wanted to know what defined the boundaries of a typical role player's workload congruency. That is, how far, in terms of both physical distance and network distance did a role player's workload congruency extend in the organization—and, as either measure of distance increased, to what extent did that workload congruency deteriorate?

Physical distance is a geographical measure whereas network distance reflects the shortest network path between two role players. We were seeking to determine whether the accuracy of the role players' workload congruency extended only over the organizational subteam that they were physically collocated with, or whether it was more dependent on their communications with organizational members who were distributed across the organization.

First, we compared network distance to physical distance over the course of the exercise. We found that the two distance measures were weakly correlated at $r = .247$. This supports the hypothesis that the members at a short network distance from each other and those at a short physical distance from each other are not necessarily the same. In fact, we found that 53% of direct communications were with physically collocated members, whereas 47% were with members external to the role player's immediate location. That is to say, direct communications were nearly evenly distributed between members within the same cell and those in different cells.

Next, we wanted to understand whether direct communications with physically collocated members produced a better workload congruency than direct communications to those with physically distributed members. We found that, overall, the workload congruence of physically collocated members was 11% better than that of geographically distributed members ($F[32, 267] = 16.75$, $p < .01$). However, direct communications are not the only determinant of workload congruency accuracy. Observation, direct or virtual, can provide enough information to increase workload congruency. We first tested the hypothesis that physical distance is a predictor of workload congruency. Overall, we found that it was ($F[1, 24] = 211$, $p < .0001$), controlling for the effect of network distance, rater, and session.

We then tested the hypothesis that network distance is a predictor of workload congruency. Results are shown in Figure 18.8. We found that, overall, the effect of network distance on workload congruency is significant ($F[1, 34] = 302$, $p < .0001$), controlling for the effect of physical distance, rater, and survey period.

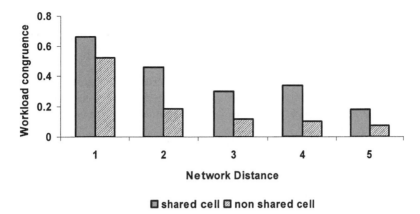

FIG. 18.8.   Workload congruence by network distance.

Both social network distance and physical distance, then, were predictors of workload congruency. Figure 18.8 indicates that role players' initial workload congruency extended equally whether the person they were rating was collocated or not. Initially, organizational behavior reflected the role players' training in traditional hierarchies. In this environment, the role players could apply their legacy-force expertise to achieve equivalent workload congruency in both local and geographically distributed contexts. However, as the role players adopted network organizational behaviors, they were no longer capable of applying their expertise. As a result, they lost workload congruency with respect to organizational elements not collocated.

We found that geographical distribution had the least effect at a network distance of one and a progressively greater effect at greater distances. The role players were able to use their expertise to achieve workload congruency in this new organization. They were dependent on direct communication to achieve their best workload congruency; and as the exercise continued, they required physical collocation to achieve their best workload congruency.

CONCLUSION

This chapter has sought to understand expertise out of context through the study of role players operating outside of their traditional organizational design into a new network-centric organization. We collected communication data from a large military organization and used social network analysis to draw conclusions about the dynamic nature of the organization. Traditional social network analysis measures used included network density, network distance, physical distance, and self-forming teams. In addition to the traditional social network analysis measures, we also proposed a new measure, the workload congruence, which helps describe shared knowledge in the organization.

Clearly, the role players adapted to the new network organization behaviors in a relatively short period of time. The network density was high for self-reported communication data. The analysis of the change of density indicates an increase on fluctuations of network density over time, a sign of adaptation to the network organization. There was an increase in self-forming teams over time and most of those self-forming teams were unique to single sessions over the course of the simulation exercise.

However, in future research it is important to identify and categorize the effects of all possible causes for the increase in density and the need for self-forming teams. Also, it is important to further investigate shared understanding and situational awareness in network organizations. Our attempt to create a shared measure, workload congruency, yielded only limited success.

Although workload congruency initially seemed to increase, overall the congruency seemed to decrease for the overall organization and remain mostly stable for those individuals that communicated directly. Workload congruency was predicted by the network and physical distance. Higher congruency was found for those individuals physically collocated than for those belonging to different teams in the organization. Also higher congruency was found for those individuals that communicated more often than for those that communicated less frequently.

Research on network-centric organizations is ascendant now. As network organizations become more prevalent (Miles & Snow, 1992; Nohria & Eccles, 1992), workload congruence will become more critical as a design tool for organizational structure and collaborative tool selection. Validation of the workload congruency measure created in this research and the creation and validation of new measures of shared understanding are necessary. Dynamic network analyses is at its beginnings and measures of organizational change, and in particular measures of dynamics of shared understanding are expected to play a key role in future research. Finding ways to shorten the "effective network distance" through the use of shared visual space, and to increase workload congruence without high retraining costs, are worthwhile areas for future research.

It should be noted that large-scale network organization exercises are extremely expensive. Consequently, available replications would be rare; and it is difficult to accord general validity to observations that result from a small number of replications. These early findings, however, will shape both the organizational design and the methodologies of future exercises.

ACKNOWLEDGMENTS

The work reported here was conducted with the support of the Advanced Decision Architectures Collaborative Technology Alliance; sponsored by the U.S. Army Research Laboratory (DAAD19-01-2-0009). We thank personnel involved in the exercise conducted at the BCBL, Ft. Leavenworth. In particular, we thank Diane Unvarsky, who played a key role in the success of data collection presented in this research.

REFERENCES

Alberts, D. S., Garstka, J. J., & Stein, F. P. (2001). *Network centric warfare: Developing and leveraging information superiority* (2nd, Rev. ed.). Washington, DC: National Defense University Press.

Borgatti, S. P. (1994). A quorum of graph theoretic concepts. *Connections, 17*(1): 47–49.

Borgatti, S. P., Everett, M. G., & Freeman, L. C. (2002). *UCINET 6 for windows: Software for social network analysis.* Natick, MA: Analytic Technologies.

Carley, K. M. (2003). Dynamic network analysis. In R. Breiger, K. Carley, & P. Pattison (Eds.), *Dynamic social network modeling and analysis: Workshop summary and papers* (pp. 133–145). Washington, DC: National Research Council, Committee on Human Factors.

Cebrowski, C., & Garstka, J. (1998). Network-centric warfare: Its origins and future. U.S. Naval Institute *Proceedings, 124*(1), 28–35.

Clark, H. H., & Brennan, S. A. (1991). Grounding in communication. In L.B. Resnick, J. M. Levine, & S. D. Teasley (Eds.), *Perspectives on socially shared cognition* (pp. 127–149). Washington, DC: American Psychological Association

Entin, E. E. (1999). Optimized command-and-control architectures for improved process and performance. In *Proceedings of the 1999 Command-and-control Research and Technology Symposium* (pp. 116–122). Washington, DC: U.S. DoD Command and Control Research Program.

Freeman, L. C. (1979). Centrality in social networks: Conceptual clarification. *Social Networks, 1,* 215–239.

Graham, J. M., Schneider, M., Bauer, A., Bessiere, K., & Gonzalez, C. (2004). Shared mental models in military command-and-control organizations: Effect of social network distance. In *Proceedings of the 47th Annual Meeting of the Human Factors and Ergonomics Society* (pp. 439–443). Santa Monica, CA: Human Factors and Ergonomics Society.

Hart, S. G., & Staveland, L. E. (1988). Development of NASA-TLX (Task Load Index): Results of empirical and theoretical research. In P. A. Hancock & N. Meshkati (Eds.), *Human mental workload* (pp. 139–183). New York: North-Holland.

Herbsleb, J. D., Mockus, A., Finholt, T. A., & Grinter, R. E. (2000). Distance, dependencies, and delay in a global collaboration. In *Proceedings, ACM Conference on Computer-Supported Cooperative Work* (pp. 319–328). New York: ACM Press.

Krackhardt, D. (1987). Cognitive social structures. *Social Networks, 9,* 109–134.

Miles, R. E., & Snow, C. C. (1992). Causes of failure in network organizations. *California Management Review, 34,* 53–71.

Moreland, R. (1999). Transactive memory: Learning who knows what in work groups and organizations. In L. Thompson, D. Levine, & J. Messick (Eds.), *Shared cognition in organizations: The management of knowledge* (pp. 3–31). Mahwah, NJ: Lawrence Erlbaum Associates.

Nohria, N., & Eccles, R. G. (1992). *Networks and organizations: Structure, form and action.* Boston: Harvard Business School Press.

Olson, G. M., & Olson, J. S. (2000). Distance matters. *Human–Computer Interaction* [Special issue: New Agendas for Human-Computer Interaction], *15*(2–3), 139–178.

Reminga, J., & Carley, K. M. (2003). *Measures for ORA (Organizational Risk Analyzer).* Paper presented at the NAACSOS Conference, Pittsburgh, PA.

Scott, J., (1992). *Social network analysis.* Newbury Park, CA: Sage.

Wasserman, S. & Faust, K. (1994). *Social network analysis: Methods and applications.* Cambridge, England: Cambridge University Press.

Chapter 19

Preparing for Operations in Complex Environments: The Leadership of Multicultural Teams

C. Shawn Burke
University of Central Florida, Orlando, FL

Kathleen Hess
Aptima, Inc., Woburn, MA

Eduardo Salas
Heather Priest
University of Central Florida, Orlando, FL

Michael Paley
Aptima, Inc., Washington, DC

Sharon Riedel
Army Research Institute, Ft. Leavenworth, KS

An Army patrol in Bosnia arrives in a village not particularly happy with the American presence—suspicious glares tend to be the norm. Upon their arrival, a crowd runs toward the vehicle yelling and screaming. The patrol leader and his interpreter exit their vehicle and are immediately crowded to the point that they cannot get back in or draw a weapon for self-defense. The Lieutenant then sees his interpreter disappear into the crowd. The Lieutenant becomes nervous as the two main options—driving away or using his weapon—are no longer available. Thinking he has led his patrol into an ambush with the two most likely courses of action deterred, the Lieutenant improvises and yells into the crowd asking if anyone speaks English. In response, a man emerges from the back of the screaming crowd. Quickly the Lieutenant asks the man to tell the crowd to back off. Surprisingly, the "hostile" crowd complies immediately. Upon asking why everyone was so excited and crowding around the jeep, the Lieutenant learns that a young boy fell into a well and people, hoping that the patrol could help get

403

him out, were yelling to get their attention. After helping to rescue the trapped boy, the Lieutenant realized that, if he had followed his initial instincts to drive away or shoot into the crowd, someone—either the boy in the well or some innocent bystander seeking assistance—might have died.

The preceding incident, recounted to one of the coauthors on a trip to Bosnia, begins to illustrate the complexity inherent within multicultural teams, where experts' expectations of a course of action are challenged by novel, unexpected, or unpredictable situations. In the preceding scenario, the Lieutenant made the right decision despite his initial instincts to do otherwise. Why this decision was made may never be fully understood; however, steps need to be taken to ensure that leaders are better prepared for the difficult situations that often present themselves within multicultural environments as these situations often stretch their leadership expertise by requiring adaptations to traditional methods of operation.

It is imperative that we create training tools that will promote the development of effective leadership within multicultural teams. The complexity and ambiguity illustrated in the previous example is not uncommon for multicultural teams, and originates not only from the operational context, but also from the process requirements needed for effective team performance and adaptability. Though it is widely acknowledged that teams are capable of great synergy, within multicultural teams this synergy is often not realized. Instead multicultural teams are often viewed as frustrating in that they experience tremendous amounts of process loss due to: (a) inadequate cohesion, (b) low levels of trust, (c) misinterpretations and loss of communication, and (d) a reliance on inappropriate stereotypes to assign attributions (Adler, 1997; Distefano & Maznevski, 2000; Horenczyk & Berkerman, 1997; Thomas, 1999; Triandis, 2000).

Challenges also arise because there are often latent and culturally bounded assumptions about how social and cognitive interactions should be conducted (e.g., treatment of women, appropriateness of challenging superiors and manner in which this is done, and importance of social interaction; Sharon Riedel, 2005, personal communication). In order for multicultural teams to be effective, these assumptions and beliefs must be identified so that the leader can appropriately structure team interaction. The challenges inherent in overcoming process difficulties, combined with the fact that multicultural team leaders often have to make decisions using varied input sources with little feedback from the environment, makes it challenging to achieve effective team performance. For example, while damage assessment within war-fighting environments provides an indication of effectiveness, successful peacekeeping operations (where multicultural teams are common) may be marked by the *absence* of action within an area of responsibility.

Given these complexities, effective operation within multicultural teams does not happen automatically. Training is essential. Though leaders in multicultural teams often are leadership "experts," their skills must be applied differently and their strategies adjusted when they operate within these complex interactions. Despite the need for training for multicultural teams, this topic is so new that we have yet to develop and test training methods. For example, though many organizations, including the military, have invested heavily in team or leadership training, the training typically focuses on behaviors and characteristics needed regardless of the cultural/ national profile of the team or the environment within which the team is to perform. Additional evidence of the need for better preparation for leaders operating within multicultural environments is evidenced by interviews and observations of U.S. forces transitioning from war-fighting to peacekeeping operations where a lack of skill in multinational teamwork was specifically identified as a weakness (Klein & Pierce, 2001; Pierce, 2002; Pierce & Pomranky, 2001). This state of affairs is not acceptable for men and women who are increasingly being required to engage in mission-critical tasks. Therefore, the purpose of this chapter is to delineate the components that would be included in a training program for multicultural team leaders. In doing so, the theoretical drivers of such a training program are highlighted, based on what research to date suggests. Second, these drivers are used to assert a series of grounding tenets for such a training program. Finally, an approach is illustrated (i.e., an exemplar) concerning how these various elements might be integrated within a training program for multicultural team leaders.

THEORETICAL BASIS

Given the goal of delineating the basic components that should comprise a training approach for multicultural team leaders, the literature on multicultural team leadership should serve as a theoretical driver. Although there is a large literature on the topics of teams, leadership, and cultural diversity, there have been few efforts to integrate these literature bases to form theoretical models, frameworks, or even individual propositions that might guide research or training development for multicultural team leadership (see Salas, Burke, Wilson-Donnelly, & Fowlkes, 2004). For example, the last two decades have witnessed large developments in understanding the factors contributing to team effectiveness. However, most of the work has been conducted within the United States and therefore does not address the impact that diverse national cultures may have on teamwork processes. Similarly, in looking to the leadership literature a parallel state of affairs is found in that most of the work that explicitly examines culture and leadership has been conducted from the perspective of individual leadership. Though both of these areas may provide some ideas, they are incomplete, or at least non-optimal for understanding multicultural team leadership.

It is only by identifying and examining the convergence of these literatures (e.g., team, leadership, culture) that leverage points for training multicultural team leaders might be identified. Once these leverage points are determined they, in turn, help in forming a basis for identifying training requirements for leaders. As these leverage points have yet to be identified, a review of the literature on culture, teams, and team leadership was conducted. Psychology databases were searched (e.g., PsycInfo, Academic Search Premier, Business Source Premier) as well as military Web sites and databases (e.g., Army Knowledge Online, Center for Army Lessons Learned, Defense Technical Information Center). Next, we briefly describe some of the major themes that arose from this review.

Team Leadership

The first literature that serves as a theoretical driver is that on team leadership. Team leadership has been argued by many (Cannon-Bowers, Tannenbaum, Salas, & Volpe, 1995; Day, Gronn, & Salas, 2004; Kozlowski, Gully, Nason, & Smith, 1999; Zaccaro, Rittman, & Marks, 2001) to be especially important for promoting the dynamic processes that comprise teamwork and facilitate adaptive performance. Therefore, key information was extracted from the literature pertaining to what team leadership within the current context might entail (e.g., What are the competencies that underlie team leadership?).

Theme 1: Team Leadership Is Multifaceted. Drawing from the functional approach, team leadership within complex operational domains (i.e., where leaders face challenging and novel problem contexts) can be described as a series of steps that are accomplished through the leader's response to social problems: problem identification and diagnoses, generation of solutions, and implementation of a chosen solution. These responses are generic and can be captured in four broad categories: (a) information search and structuring, (b) information use in problem solving, (c) managing personnel resources, and (d) managing material resources (Fleishman et al., 1991). The first of these four dimensions, information search and structuring, highlights the leader's role as a boundary spanner in his or her effort to seek out, acquire, evaluate, and organize goal-supporting information. The second dimension, information use in problem solving, describes a leader's application of a "best-fitting" solution to the problem at hand. As the team leader applies a chosen solution to a social problem, both personnel and material resources are called upon and utilized in the process. Finally, team leaders dynamically manage both material (the third dimension) and personnel resources (the fourth dimension). These four dimensions serve to illustrate the multidimensionality of team leadership.

Further evidence of the multidimensionality of team leadership are the competencies delineated by the U.S. Army within their Field Grade Leader Competency Map. Specifically, the Army has identified a set of seven leader competencies (interpersonal, conceptual, technical, tactical, influencing, operating, and improving). These competencies are further broken into skills (e.g., decision making, team building, critical reasoning, establishing intent), behavioral requirements (e.g., encourage initiative, anticipate requirements, develop teams), and supporting requirements (e.g., active listening, negotiating, gaining consensus). For a full breakdown and corresponding mapping, the interested reader is referred to Army Field Manual 22-100: Army Leadership (Department of the Army, 1999).

Theme 2: Team Leadership Involves the Cyclical Application of Two Distinct Roles. While completing requisite leadership functions, team leaders iteratively switch between task-oriented and developmental roles. Within the "task" role, the leader serves as a boundary spanner to gather information and create an enabling performance environment by ensuring that the team has the necessary material and personnel resources to accomplish the task at hand. It is within this role that team leaders structure and regulate team processes in order to meet shifting internal and external contingencies. Functional behaviors such as information search and structuring, information use in problem solving, and managing material resources are subsumed within this role. Conversely, when adopting the "developmental" role, the leader ensures that the team develops and maintains the necessary shared knowledge, behaviors, and attitudes that enable interdependent coordinative action (Kozlowski, Gully, Salas, & Cannon-Bowers, 1996). In doing so the leader functions as a mentor, instructor, coach, or facilitator dependent on the stage of team development (Kozlowski et al., 1996, 1999).

Culture

A second theoretical driver in developing a training system for multicultural team leaders is the work in cross-cultural social science. The goal in examining this literature base is twofold: (a) Leverage against the work detailing the impact of culture on individuals and (b) identify those instructional strategies that have proven effective within this arena. In this effort, several key themes have been identified.

Theme 3: Culture Impacts Team Member Attitudes and Behavior. Culture has been defined as consisting of the shared norms, values, and beliefs of a nation that serve to provide structure for individual action (Dodd, 1991;

Helmreich, 2000). The values and beliefs that comprise culture have been shown to have a substantial impact on team member attitudes, with between 25% and 50% of variation in attitudes being explained by national culture (see Gannon, 1994). In our review of the literature, approximately 43 distinct conceptualizations of cultural variables were identified (see Salas et al., 2004), with a predominant number of them focusing on the attitudes that guide behavior. Specifically, cultures have been shown to differ in terms of their attitudes and corresponding behavior about human relations (e.g., Erez & Earley, 1993; Hofstede, 1980; Schwartz, 1999), power relations (e.g., Triandis, 2000; Trompeanaars & Hampden-Turner, 1998), rules orientation (e.g., Hofstede, 1980; Parsons & Shils, 1951), time orientation (e.g., Hall & Hall, 1990; Trompenaars & Hampden-Turner, 1998), rules governing affect (e.g., Parsons & Shils, 1951; Triandis, 2000), orientation to nature (e.g., Schwartz, 1999; Triandis, 2000), rules governing the transmission of context (e.g., Hall & Hall, 1990; Triandis, 2000), and a series of other less easily categorized areas (e.g., masculinity–femininity, views of time, activity and space orientation). These differences have been shown to impact teams through the impact on behaviors such as aggression, conflict resolution, social distance, helping, dominance, conformity, obedience, decision making, and leadership (Hambrick, Davison, & Snow, 1998; Hofstede, 1980; Schneider & DeMeyer, 1991; Shane, 1994; Triandis, 1994).

Theme 4: Culture Impacts Cognition and Corresponding Behavior. Culture also appears to guide the cognitive approaches used, as well as individual choices, commitments, and standards of behavior (Erez & Earley, 1993; Klein, 2003, 2004). The manner in which culture impacts cognition is especially important for two reasons. First, given that team leadership is a form of social problem solving and is highly dependent on the cognitive skills and structure of the leader's mental models pertaining to the task, team, and environment, the manner in which culture may impact cognition is especially important. Second, research has shown that team cognition is a key driver of team performance within complex domains (Burke, 1999; Marks, Zaccaro, & Mathieu, 2000; Mathieu, Heffner, Goodwin, Salas, & Cannon-Bowers, 2000).

With the preceding discussion as a context for importance, it has been shown that cultures differ in the extent to which they tend toward the use of an "analytic" or a "holistic" cognitive style (Choi & Nisbett, 2000). Individuals from cultures that enact a holistic style tend to focus on the entire problem field and assign causality to interactions between the objects or events in focus and their broader context. Conversely, individuals from cultures that primarily follow an analytic style tend to focus more narrowly on the object or events in focus. Assignment of causality and meaning within this cognitive style can be described as being guided by logic or rules.

Furthermore, cognitive styles have been shown to impact differences in how communication is structured. For example, cultures vary in whether communication is structured such that the mass of information is explicitly stated within the message or whether communication is structured in a manner where most of the information is already known, with very little being laid out explicitly in the transmitted message (i.e., high–low context; Hall & Hall, 1990). These differences in cognitive style not only impact the cues to which team members attend, including the leader, but will also impact the meaning that is assigned. This makes the need for clear, concise communication even more essential within multicultural teams.

Theme 5: Skill-Based Training Programs for Multicultural Environments Are Few and Far Between. A review of the literature revealed that when multicultural training has been provided, it has a variety of goals and forms (e.g., cultural awareness, positive attitudes toward cultural differences, understanding one's own culture and biases, and identifying links between cultures; Elmuti, 2001; Grahn & Swenson, 2000). With few exceptions (Wentling & Palma-Rivas, 1999), skill-based training has not been attempted. Specifically, most of the training strategies primarily focus on increasing multicultural awareness and do not discuss or attempt to promote strategies by which to cope with the challenges and potential misinterpretations that can occur within a multicultural environment (see Littrell & Salas, 2005, for more detail). Programs tend to vary in the degree to which they are experientially based, but few were found that focused on the more specific topic of preparation for work in multicultural teams, and almost none on preparing leaders to deal with the potential process difficulties within these types of teams.

Though familiarization and related approaches may be beneficial and necessary (although few training effectiveness studies have been performed), it is doubtful that meaningful behavioral changes in team functioning will occur solely as a result of such interventions. This conclusion is based on what is known about the nature of effective teamwork as well as what is known about human learning. The need to move beyond familiarization to more skill-based programs is essential. Not only do the process challenges within multicultural teams often stretch leaders' current expertise due to their complexity, but the interaction between the various forms of culture adds uniqueness to team leadership within these problem contexts.

Theme 6. One Size Does Not Fit All. By their very nature, multicultural teams are complex. As such, there are several variables that may impact the type of multicultural training that is most appropriate (e.g., stage of multicultural awareness, nature of team participation). For example, Bennett

(1986) argued that individuals may be at different stages of awareness in terms of interacting with cultures different from their own. Therefore, it is likely that different training strategies will be more/less appropriate depending on the cultural composition of the team in terms of the leader's and team members' level of multicultural awareness. In addition, the majority of programs train individuals who are going to live and work in another country (i.e., expatriates). However, information collected from military sources indicates that current multicultural team operations can include teams composed of individuals with diverse cultural or national backgrounds, or teams performing in different cultural or national environments. It is possible that these types of teams need different types of training strategies and delivery methods because there are differences in the requisite competencies.

Based on the themes extracted from the literature review, we believe there are certain tenets that must be applied to the design of this type of training.

TENETS

In order to train leaders for the complexities of military operations in multinational or multicultural environments, we believe that training must be practical, relevant, experiential, scenario based, skill based, and multifaceted. The following section expands on these propositions.

Tenet 1: Practicality and Relevance

Perhaps the most important tenet of any training program, regardless of its focus, is that it is practical in its application and relevant in its content. This requirement becomes even more vital when the training is being devised to assist teams working in operational environments where they are faced with mission-critical tasks, because their time is often extremely precious. Therefore, training must be practical in terms of both time and ease of use. Developers should seek to devise training systems that offer a vehicle for optimally focusing trainee experiences to facilitate learning. Often, these teams do not have months to acquire the complex knowledge and skills needed to effectively operate within and lead multicultural teams. Training should target areas of high payoff for multicultural team leaders, that is, leverage points—those team leadership behaviors that are: (a) critical for performance and (b) most likely to be adversely affected by cultural diversity. By targeting these leverage points, training should facilitate the team leader's decision-making processes that form a necessary precursor to appropriate leader intervention.

Although these leverage points represent a practical training target, developers also need to ensure that training is practical in terms of its method and implementation time. Interviews and focus groups conducted with subject-matter experts (SMEs) housed within the particular operational environment can assist in ensuring that this dimension of practicality is maintained.

Though practicality is a necessary feature of training for multicultural team leaders to be effective, it is not a guarantee of effectiveness. An additional requirement is that the content and learning objectives must be relevant to the anticipated needs within the operational context. To assist developers in identifying the needed competencies and corresponding training content, a search of literature and archival information on multicultural team leadership is recommended. This literature search should be grounded by talking with SMEs within the domain of interest. This grounding is especially important within domains, such as multicultural team leadership, where the direct theoretical basis within the literature is weak at best. Additionally, when utilizing SMEs to assist in the delineation of needed competencies (knowledge, skills, attitudes), it is important to involve SMEs with experience as leaders. Research has shown that although there exists a generic set of leadership skills, each set varies in its functionality according to the level at which leadership is required (Jacobs & Jacques, 1986).

Tenet 2: Experiential Training

Arising out of the functional approach to leadership is our third tenet—training for multicultural team leaders should be experiential or practice based. Smith (2001) argues that experiential learning has been conceptualized in two primary ways. First, it describes the sort of learning where participants are given a chance to acquire and apply knowledge, skills, and feelings in an immediate and relevant setting. Second, it has been used to refer to learning that is achieved through reflection on everyday experience. Experiential learning is learning by doing. It involves the direct application of behavior rather than merely thinking about the phenomena (Borzak, 1981).

The rationale for asserting that training for multicultural team leaders should have an experiential base arises out of the functional approach to leadership. Whereas the functional approach to leadership acknowledges a set of generic leadership functions, the manner in which these functions are applied is context dependent. Therefore, training methods should be grounded in the operational context of interest and should allow the practicing of skill-based behavior in a relatively safe context where constructive feedback is possible. A second rationale for this tenet is based on the idea that within complex environments, teams and their leaders must treat every opportunity as a learning experience. Incorporating experiential learning into training tools should facilitate team reflection in actual operational contexts.

One instructional method that relies on such an experiential approach (concrete experience, observation/reflection, formation of abstract concepts, and testing for transfer to new situations) is scenario-based training (SBT) (see Kolb & Fry, 1975). Figure 19.1 depicts the SBT development cycle (adapted from Cannon-Bowers, Burns, Salas, & Pruitt, 1998). The cycle begins by determining what competencies (i.e., knowledge, skills, and attitudes—KSAs) will be the focus of the training (Circle 1) using techniques such as training needs analysis (see Salas & Cannon-Bowers, 2000). Next, to guide a particular training event, a subset of the competencies is selected to form the basis of training objectives (Circle 2). The training objectives, in turn, guide the development of events to embed in the scenario. The events provide opportunities to observe performance reflecting the competencies of interest. In this way, training (or measurement) opportunities are not left to chance. Moreover, the training opportunities are directly linked to the competencies to be trained. Using SBT, measurements are taken to assess trainees' responses to the events. The feedback that follows (Circle 5) has direct relevance to the training objectives, thus maintaining and reinforcing the links from competencies to feedback. Finally, the information collected during a specific training exercise can be incorporated into a historical database (Circle 6). Such a database can then be used to determine the next direction that training should take (i.e., what are the next competencies to be trained) based on performance during the previous exercise (i.e., trainee's/team's strengths and weaknesses).

This SBT approach to training should be applicable to developing training methods for multicultural team leaders, because the technique is very adaptable, is context dependent, has been used in highly complex environments, represents a systematic approach to the development of training,

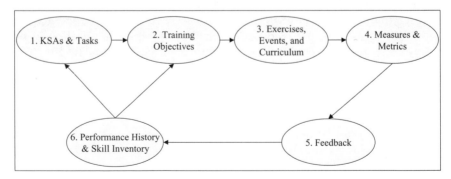

FIG. 19.1.   Scenario-based training life cycle. From Cannon-Bowers, Burns, Salas, and Pruitt (1998). Copyright 1998 by the American Psychological Association. Adapted by permission.

and results in programs that are theoretically sound and able to be evaluated (Cannon-Bowers et al., (1998). Moreover, instructors find the measurement tools that derive from this technique simple to use as a method for providing highly specific feedback. For more information on this method see Burke, Salas, Estep, and Pierce (chap. 23, this volume).

Tenet 3: Skill-Based Training

Training should be designed to meet learning objectives by focusing on the specific knowledge, skills, and attitudes (KSAs) that are required. Though knowledge is a necessary precursor to the effective application of behavioral skills within multicultural environments, training tools must move beyond cultural awareness training. Knowledge does not equal application; trainees need to be provided with the skills that allow this knowledge to be applied in a practical manner.

Because there is a lack of research that explicitly ties team leadership to key cultural variables, we sought to determine those leadership behaviors and corresponding skills that are critical for team performance within this domain, and are most likely to be affected by diversity (i.e., a leverage point). Based on the methodology prescribed earlier, three skill areas were identified that represent an intersection of military and organizational leadership dimensions: interpersonal skills, decision-making skills, and team-building skills. It was believed that these skill dimensions would be most impacted by cultural diversity within dynamic operational environments characterized by complexity.

For example, because the literature reveals that the factors that are detrimental to team performance include failures in communication and the inappropriate reliance on stereotypes, interpersonal skills were felt to be especially important. Decision-making skills are specified in the list based on the idea that leadership within complex domains involves discretion and decision making under conditions of uncertainty. In addition, it is known that cognitive style is a factor that differs across cultures and can impact the method by which decision making occurs. Finally, team-building skills were chosen because this is part of the process by which trust and shared cognitive frameworks are built, which has, in turn, been shown to be an important consideration for multicultural teams.

Training 4: Multifaceted

The literature review indicated that individual team members may be at different stages of development and awareness with regard to operating within a multicultural team or environment. Therefore a sixth tenet is argued for—training for multicultural team leaders should be multifaceted. A review of the literature found a lack of training programs that incorporate multiple methods. Most training programs typically incorporate one primary method (e.g., lecture, role play). Because the environments in which multicultural teams operate are often ambiguous, training tools should be multifaceted in the sense of relying on methods that instill flexibility and operational practicality, to ensure leaders can perform appropriately within these novel environments.

Therefore, in developing a training tool for multicultural team leaders, it is important to expand and integrate methods and instructional strategies from diverse domains of team performance or application that could be applied to the training of multicultural team leaders. In an attempt to do this, a matrix was created that depicts different instructional strategies, learning objectives, assessment methodologies, and typical competency type trained using these strategies. This is presented in Table 19.1.

This matrix is rooted in the literature and examines a number of key variables that are necessary for the development of effective training, while taking into account what has been done in cross-cultural training, what needs to be done, and what can realistically be done given certain restraints (e.g., time and money). This matrix provides a foundation for identifying the appropriate instructional strategies to be integrated for the training of multicultural team leaders.

The training interventions listed in the matrix vary according to the level of expertise of the participant, from novice to expert. Some training interventions have been used extensively in cultural training mostly for individuals who will work in foreign nations (i.e., expatriates). These have been discussed extensively in the literature (Bhawuh & Brislin, 2000; Grove & Torbiön, 1986; Gudykunst, Guzley, & Hammer, 1996; Kealey & Protheroe, 1996). These interventions include self-learning, awareness training, and role playing. Other interventions have been used more in training that has not been focused on cultural or multinational issues: guided facilitation, Situation Awareness Global Assessment Technique (SAGAT), and SBT. It is believed that all these methods merit consideration for multicultural training. Specifically, SAGAT-like training and SBT may be important because they offer practice-based training and feedback, which are not as prevalent in traditional cross-cultural training (e.g., awareness training).

The continuum of delivery and instructional strategies documented in Table 19.1 vary on a number of factors: (a) position on a novice–expert continuum based on experience with and knowledge of multicultural issues, (b) KSAs to be taught, (c) learning objectives, and (d) expense or involvement on the part of the trainer. From this list one can select a variety of instructional strategies that might be embedded within multinational training for military leaders. The determination of the "best" set of instructional strategies is dependent on the particular combination of the vari-

TABLE 19.1

Continuum of Training Strategies for Multicultural Teams and Their Leadership

| NOVICE ◄───► EXPERTISE | | | | | | |
Self-Learning	Awareness Training	Role Play	Guided Facilitation	SAGAT-Like Facilitation	SAGAT-Like Guided Practice	Scenario-Based Training
Training Description • This technique would provide trainees with a "take home" training package. • No instructor involvement. • Training is self-paced. • Geared toward individuals.	• This technique is conducted within a classroom setting. • Instructor-led training. • Pace of training is dictated by instructor. • Geared toward delivery to multiple trainees at once.	• This technique involves trainees acting out simulated roles. • Provides an opportunity to explore and experience solutions to on-the-job problems. • Active, but scripted involvement of trainee. • Instructor involvement consists of setting up the scenario and providing feedback at the end of the role play.	• Using this technique instructors construct a priori vignettes using critical incidents. • Trainees view these critical incidents and the corresponding responses. • Vignettes presented here might be longer and more complex than those presented within role play. • The learning process is guided within this technique	• This technique constructs a priori vignettes representing critical incidents. • Trainees view the incidents and are asked what they would do in each situation. • Trainee responses are not scripted. • After the trainee describes their response the instructor uses probe questions to determine things such as:	• This technique uses a scenario-based approach to training. • Events are a priori embedded into training scenarios. • Similar to SBAT, participants are actively involved in the scenario exhibiting attitudinal, behavioral, and cognitive competencies of interest. • At certain points the simulation is cued to stop.	• This technique uses a scenario-based approach to training. • Trainees are embedded within the simulation. • Participants are actively involved in the scenario exhibiting attitudinal, behavioral, and cognitive competencies of interest. • Trainees' behavior is not scripted. • The scenario is the training curriculum.

(Continued)

TABLE 19.1
(Continued)

NOVICE ◄───► EXPERTISE

	Self-Learning	Awareness Training	Role Play	Guided Facilitation	SAGAT-Like Facilitation	SAGAT-Like Guided Practice	Scenario-Based Training
		•		in that instructors stop the vignette at key points and explain why things happen, what the key cues are. • Feedback and guidance provided throughout. • Intelligent tutor could be used to assist instructor.	(a) what was perceived, (b) what should happen next, (c) why.	• When the scenario stops the instructor asks questions pertaining to situation awareness, cue recognition, and likely behavioral responses. • Trainee responses are not scripted and instructor feedback is intermittent. • Intelligent tutor could possibly be used to augment.	• Instructor provides feedback in the form of AAR at end of each scenario.
Learning Objectives	• Increase diversity literacy • Build cultural awareness • Awareness of own cultural	• Increase diversity literacy • Recognize own and others cultural values • Accept diversity	• Get sensitized to culture. • Create awareness • Learn skills • Development	• Guide to biases • Understand how, why, when biases occur • Understand that biases are	• Create situation awareness • Promote diagnosis and analysis of situation • Create middle	• Demonstrate cognitive skills • Create situation awareness • Promote diagnosis and analy-	• Demonstrate cognitive, attitudinal, and behavioral skills • Build tacit knowledge

Learning Objectives	Key Training Components	Appropriate KSAs
• biases • Awareness of foreign culture biases • Could also provide an awareness of impact of culture on teamwork • Focus on mastery goals	• Information	Knowledge based; attitude based; cognitive based (development of cognitive framework); increases team orientation
• in team members • Create an understanding and respect for cultural differences • Develop common ground • Focus on mastery goals	• Information • Demonstration	Knowledge based; attitude based; cognitive based; increases communication; increases team orientation; improves SMM
• of dissonance • Change Attitudes • Primarily used for analyses of interpersonal problems, attitude change, development of human relation skills	• Information • Demonstration • Feedback	Attitude & skill based; improve communication skills; Build cognitive skills to take on viewpoint of host national and respond to situation as if member of culture; communication skills; professional skills; and personal skills such as listening ability, political astuteness, and relationship building
• natural • Gain knowledge of own biases	• Information • Demonstration • Feedback	Skill based; cognitive learning; improves SMM; improves closed-loop communication
ground b/t analytic and holistic thinking • Express opinion	• Information • Demonstration • Feedback	Knowledge and skill based; attitude skills; improves situational awareness (SA); improved backup behavior; improved mutual performance monitoring (MPM)
sis of situation • Express opinion • Create middle ground between analytic and holistic thinking	• Information • Demonstration • Practice • Feedback	Application of cognitive and behavioral skills; improves closed-loop communication; improves MPM; improves SA; improves backup behavior
• Learn to self correct	• Information • Demonstration • Practice • Feedback	Application of cognitive and behavioral skills; traditionally greater focus on behavioral skills; improves leadership; improves closed-loop communication; improves performance monitoring; improves backup behavior; improves adaptability

(Continued)

417

TABLE 19.1

Continuum of Training Strategies for Multicultural Teams and Their Leadership

NOVICE ◀──▶ EXPERTISE

	Self-Learning	Awareness Training	Role Play	Guided Facilitation	SAGAT-Like Facilitation	SAGAT-Like Guided Practice	Scenario-Based Training
Instructional Strategies	Information-based printed or Web-based materials; video; information booklets on host country; reading case studies; culture assimilator (self-learning tool); reading case studies; Web-based medium (example CultureSavvy.com); self-assessments (designed to determine attitudes and perceptions)	Instructor-led lecture; may also involve video vignettes to illustrate differences; casual conversations with returned expatriates; structured information sessions with experts, host-country nationals, and former expatriates; seminars; audiovisuals; attribution training; value orientation training; value-ranking charts; culture contrast technique	Behavioral scripts; instructor feedback; situation exercises; behavior modeling; simulation games; video playback; contrast culture technique; incident method; drill and practice; modeling; T-group	Computer or video presented vignettes; culture assimilators	Computer, video, or pen and paper vignettes; assignment to microcultures; instructional games; coaching; intercultural communication workshops	a priori events embedded within a simulation; examples of simulations include BAFA BAFA, Barnga, and Albatross; bicultural communication workshops	A priori events embedded within a simulation

418

Diagnostic Tools	Paper and pencil and/or card sort to assess declarative knowledge	Paper-pencil or card sorts to assess declarative knowledge	Paper-pencil or card sorts to assess declarative knowledge; some sort of attitude assessment	Assessment of declarative and underlying, conceptual knowledge	Assessment of declarative and underlying conceptual knowledge / Knowledge elicitation tools	Assessment of the application of conceptual, procedural, and strategic knowledge; assessment of application of behaviors (process and outcome) / Knowledge elicitation tools	Assessment of he application of conceptual, procedural, and strategic knowledge; assessment of the application of behaviors (process and outcome) / TARGETs-type tools / Mental-model assessment
Sources	Bhawuh & Brislin (2000); Kealey & Protheroe (1996)	Bhawuh & Brislin (2000); Gudykunst, Guzley, & Hammer (1996)	Brewster (1995); Grove & Torbión (1987)	Bhawuh (2001); Harrison (1992)	Bennet et al. (2000); Brislin & Bhawuh (1999)	Gudykunst, Guzley, & Hammer (1996); Levy (1995)	Cannon-Bowers, 2001; Hitt, Jentsch, Bowers, Salas, & Edens (2000)
Cultural Issues for Design, Delivery and Practice	Some cultures may prefer teacher centered learning; delivery method	Explanation of source of expertise of instructor; some cultures prefer a more expert instructor with a more structured instructional style	Structure and format of feedback; feedback delivered in group or individual setting; research suggests differences in communication between cultures due to difference in cognitive styles and cultural values	Structure and format of feedback; feedback delivered in group or individual setting; culture assimilators either need to be general or culture specific	Structure and format of feedback; feedback delivered in group or individual setting; cognitive style	Structure and format of feedback; feedback delivered in group or individual setting; cognitive style	Structure and format of feedback; feedback delivered in group or individual setting; willingness to self-correct

ables identified along the axes of Table 19.1, and is itself a topic meriting research and evaluation.

A TRAINING TOOL FOR MULTICULTURAL
TEAM LEADERS

Up to this point, the theoretical basis and corresponding tenets for forming a training tool to facilitate the development of leaders of multicultural teams has been described. Next, an illustration is offered of one manner in which the tenets described earlier can be applied in the development of a training program.

An Example of Training Program Development

Though tool practicality and relevancy are two aspects that must be considered, a precursor to both of these is the determination of the target audience and the purpose of the training. Such information is often determined by the customer and has implications for the type of databases that should be searched and type of SMEs who should be consulted. For the present case, it was decided by the sponsoring agency that the training target was midlevel officers (e.g., Captains, Majors) and that the purpose was to provide them with skills for leading military teams from diverse cultures.

Once the target audience and tool purpose had been defined, the content and methods that would drive the training were developed. To ensure relevancy, data were collected from the literature, SMEs, researcher experience, and operational constraints. Specifically, the literature review served to guide the initial development of content. Interviews and focus groups then served to further populate the training materials and provide comments on content.

Researchers traveled to three Army posts in the United States to conduct focus groups and individual interviews with Lieutenants, Captains, Majors, and Lieutenant Colonels who had experience either leading culturally heterogeneous teams or leading American teams in multinational environments. The interviewer explained the conceptual aspects of the training and the targeted competencies. The SMEs were then asked to answer a set of questions and relate stories of their experiences in multinational teams or leading teams in multinational environments. These stories served as a basis for experiential learning materials and scenario development, and the feedback from the questions was incorporated into training development. The focus groups followed a similar format, and provide further insight on form, content, and practical realities that the training would have to promote. In particular, the focus groups clarified time limits and the level at which the instruction needed to be presented.

Based on information gained from the literature review and focus groups, the training procedure was designed to facilitate the acquisition of team leadership skills (i.e., interpersonal, decision making, and team building) within multicultural teams. Each of the leadership skill sets was delineated into behaviors that are specified in Table 19.2. It is the intersection of these team leadership behaviors with culture that was the focus of the instruction. Correspondingly, seven cultural dimensions were chosen as the areas of focus and served as our leadership leverage points. These are specified in Table 19.3.

Next, the matrix of multicultural training strategies (see Figure 19.1) was examined to identify a combination of instructional features that would allow the training procedure to be flexible and multifaceted. In deciding on a particular set of instructional features, we took into account our target audience, the purpose of the training, and the identified content areas. It was decided that the training would combine elements of self-learning (e.g., community links), skill training, role playing, guided facilitation, and SBT. In addition, to further ensure practicality, the training was designed to be SCORM (Sharable Content Object Reference Model) compliant. This standardizes content display, the collection of information from the user (e.g., via multiple choice), and adaptation of content based on user response. Following this structure, multimedia, diagnostic feedback, and practice guide participants through a series of video-based vignettes designed to tar-

TABLE 19.2
Team Leadership Skills of Interest

Skill Dimension	Corresponding Behaviors
Interpersonal Skill	• Active Listening • Negotiating • Conflict Resolution • Communication • Empathy
Decision-Making Skill	• Envisioning • Obtaining/Factoring Media Support • Shaping the Environment • RPM (recognition-priming model)—a decision-making protocol
Team-Building Skill	• Ongoing Feedback/Correction of Team Behavior • Developing People and Leaders (includes mentoring) • Motivating • Building Team Roles • Transformational vs. Transactional Leadership (as culture appropriate)

TABLE 19.3
Cultural Dimensions of Interest

Cultural Dimension	Definition
Power Distance	"The extent to which a society accepts the fact that power in institutions is distributed unequally" (Hofstede, 1980, p. 45). High-power distance cultures accept this and social exchanges are based on this fact. Low-power distance cultures do not see a strict hierarchy among social exchanges.
Individualism/ Collectivism	"A loosely knit social framework in which people are supposed to take care of themselves and of their immediate families only" (Hofstede, 1980, p. 45). A tight social framework "in which people distinguish between in-groups and out-groups, they expect their in-group to look out after them, and in exchange for that they feel they owe absolute loyalty to the in-group" (Hofstede, 1980, p. 45).
Masculinity/ Femininity	"The extent to which the dominant values of society are 'masculine'— that is, assertiveness, the acquisition of money and things, and not caring for others, the quality of life, or people" (Hofstede, 1980, p. 45).
Time Orientation	"LTO stands for the fostering of virtues orientated towards future rewards, in particular perseverance and thrift"; "STO the fostering of virtues related to the past and present, in particular respect for tradition, preservation of 'face' and fulfilling social orientations" (Hofstede, 2001, p. 359).
Cognitive Style	Analytic cultures pay attention primarily to the object and the categories to which it belongs, use rules and formal logic to understand its behavior. Holistic cultures attend to the entire field and assign causality to interactions between the object and the field, making relatively little use of categories and formal logic; rely on dialectical reasoning (Choi & Nisbett, 2000; Nisbett, Peng, Choi, & Norenzayan, 2001).
Uncertainty Avoidance	"The extent to which a society feels threatened by uncertain and ambiguous situations and tries to avoid these situations by providing greater career stability, establishing more formal rules, not tolerating deviant ideas and behaviors, and believing in absolute truths and the attainment of expertise" (Hofstede, 1980, p. 45). Ranges from high to low.
Context	In high-context cultures, communication involves messages "in which most of the information is already in the person, while very little is in the coded, explicit, transmitted part of the message" While low context the mass of the information is vested in explicit code (Hall & Hall, 1990, p. 6).

get the specific skills and behaviors discussed previously and in Tables 19.2 and 19.3.

The result is a training program consisting of six modules (two modules per skill), each of which consists of three scripted vignettes that follow a single overarching storyline. Targeted within each vignette was the relationship between a subset of the identified cultural dimensions and one or two of the team leadership behaviors (i.e., conflict resolution, active listening) that fall under each identified team leadership skill set (i.e., interpersonal, team building, decision making). Each module is structured such that the first vignette that the participants' view illustrates a situation in which their current training on the targeted skill is effective. This "good" example shows a leader within a nationally homogeneous team (i.e., an American military team) dealing with a situation and having to direct team members. The example targets one or two behaviors that make up each of the skills. The purposes of this first vignette are to: (a) grab trainee attention and (b) illustrate an example of the effectiveness of their training in certain situations. At the conclusion of the first vignette, participants hear a narrator describe in general terms what happened in the previous vignette and what behaviors the leader used to accomplish his or her goal.

Participants then see a second scripted video vignette that illustrates when their current training on the targeted leadership behaviors may not work. This "bad" example illustrates a leader within a nationally heterogeneous team (e.g., a team with both American and Spanish team members) faced with the same situation shown in the "good" example, applying the same strategies, targeting the same behaviors, but in this case the trainee sees how this same strategy may not work out. Following this example, participants receive "lessons learned" from the first two examples.

Following the lessons learned, the participants are shown a final vignette. This final vignette also targets the intersection of the targeted skill behaviors and culture, but illustrates a different situation than the first two vignettes. Participants are then asked to "choose your own" ending from a list of multiple-choice answers. Answers to the "choose your own ending" vignette are based on interviews with SMEs and our knowledge of the impact of culture on the targeted skill behavior. Answers are designed such that no one answer is entirely correct or incorrect, and all the answers differ by sophistication of understanding of multicultural effects. Constructive feedback is given for each chosen response (e.g., this was good because . . . , this was bad because . . .), followed by more "lessons learned," and recommendations for further training needs.

Although the structure of the vignettes within each of the six modules is designed to apply the tenets of content relevancy and experiential learning, they have also been designed with practicality in mind. Specifically, each module is scheduled to take approximately 15 minutes to ensure training is timely and does not tax leaders, with the entire training lasting about 1½ hours. To further illustrate the tenets of practicality and flexibility, the training also contains "community" links, including message boards, chat rooms, and external links for military and country-specific information.

Efforts such as this should result in computer-based training that will support multiple training methods. For example, a leader can receive traditional training on specific topics related to teams or multicultural issues (e.g., readings, written scenario-based situations with questions), or a leader can participate in a computer-simulated team situation with diverse multicultural team members "played" by the computer. This is seen not only as a training interface, but also as a reference that can be used by leaders to more thoroughly prepare for multicultural team situations and thereby mitigate some of the novelty of these types of problem contexts.

CONCLUSION

There is no escaping the fact that the reliance on multicultural teams will expand. Moreover, as the military increasingly becomes involved in support and stability operations around the world, and industry barriers to global partnerships continue to be mitigated through technological advances, the need to understand effectiveness within multicultural teams becomes of paramount importance. Though multicultural teams pose process challenges beyond those of culturally homogeneous teams, when multicultural teams are able to overcome the challenges, in part through effective team leadership, teams can be more productive than their culturally homogeneous counterparts (Adler, 1997). Moreover, for many organizations the use of such teams is no longer a luxury, but is becoming a necessity.

Despite the argument for the important role that leaders play within teams, a thorough understanding of how to train leaders for operation within multicultural teams has yet to be generated. This is evidenced by the fact that both industry and the military have voiced concerns about the lack of such training. In addition, one only has to look at the failure rate of expatriates to see that leading multicultural teams often taxes the leadership expertise gained in the home country when transplanted to novel, culturally diverse settings. Based on this information, we sought to provide theoretical drivers and guidance for the development of training that is flexible in providing experiential learning and will serve to facilitate the training of multicultural team leaders. It is our hope that the work discussed in this chapter will serve to provide "food for thought" on this subject, and will challenge researchers and practitioners alike to push the science forward.

ACKNOWLEDGMENTS

This work was supported by funding from the Army Research Institute, Contract # DASW01-04-C-0005. The view, opinions, and/or findings contained in this report are those of the author(s) and should not be construed as an official Department of the Army position, policy, or decision.

REFERENCES

Adler, N. J. (1997). *International dimensions of organizational behavior* (3rd ed.). Cincinnati, OH: International Thomson.

Bennett, J. M. (1986). A developmental approach to training for intercultural sensitivity. *International Journal of Intercultural Relations, 10,* 179–196.

Bennett, R., Aston, A., & Colquhoun, T. (2000). Cross-cultural training: A critical step in ensuring the success of international assignments. *Human Resource management, 39,* 239–250.

Bhawah, D. P. S. (2001). Evolution of culture assimilators: Toward theory-based assimilators. *International Journal of Intercultural Relations, 25*(2), 141–163.

Bhawuh, D. P. S., & Brislin, R. W. (2000). Cross-cultural training: A review. *Applied Psychology: An International Review, 49*(1), 162–191.

Borzak, L. (Ed.). (1981). *Field study: A sourcebook for experimental learning.* Beverly Hills, CA: Sage.

Brewster, C. (1995). Effective expatriate training. In J. Selmer (Ed.), *Expatriate management: New Ideas for international business* (pp. 57–71). Westport, CT: Quorum.

Brislin, R. W., & Bhawah, D. P. S. (1999). Cross-cultural training research and innovations. In J. Adamopoulos & J. Kashima (Eds.), *Social psychology and cultural context: Contributions of Harry Triandis to cross-cultural psychology* (pp. 205–216). Thousand Oaks, CA: Sage.

Burke, C. S. (1999). *Examination of the cognitive mechanisms through which team leaders promote effective team processes and adaptive team performance.* Unpublished doctoral dissertation, George Mason University, Fairfax, VA.

Cannon-Bowers, J., Burns, J., Salas, E., & Pruitt, J. (1998). Advance technology in scenario-based training. In J. A. Cannon-Bowers & E. Salas (Eds.), *Making decisions under stress: Implications for individual and team training* (pp. 365–374). Washington, DC: American Psychological Association.

Cannon-Bowers, J. A., Tannenbaum, S. I., Salas, E., & Volpe, C. E. (1995). Defining team competencies and establishing team training requirements. In R. Guzzo & E. Salas (Eds.), *Team effectiveness and decision making in organizations* (pp. 333–380). San Francisco: Jossey-Bass.

Choi, I., & Nisbett, R. E. (2000). Cultural psychology of surprise: Holistic theories and recognition of contradiction. *Journal of Personality and Social Psychology, 79,* 890–905.

Day, D., Gronn, P., & Salas, E. (2004). Leadership capacity in teams. *Leadership Quarterly, 15*(6), 857–880.

Department of the Army. (1999). *Field manual 22-110, Army leadership: Be, know, do.* Washington, DC: U.S. Government Printing Office.

Distefano, J. J., & Maznevski, M. L. (2000). Creating value with diverse teams in global management. *Organizational Dynamics, 29*(1), 45–63.

Dodd, C. (1991). *Dynamics of intercultural communication* (3rd ed.). Dubuque, IA: Brown.

Elmuti, D. (2001). Preliminary analysis of the relationship between cultural diversity and technology in corporate America. *Equal Opportunities International, 20*(8), 1–16.

Erez, M., & Earley, P. C. (1993). *Culture, self-identity, and work.* New York: Oxford University Press.

Fleishman, E. A., Mumford, M. D., Zaccaro, S. J., Kevin, K. Y., Korotkin, A. L., & Hein, M. B. (1991). Taxonomic efforts in the description of leader behavior: A synthesis and functional interpretation. *Leadership Quarterly, 2*(4), 245–287.

Gannon, M. J. (1994). Understanding global cultures: A metaphorical journey through 17 countries. *Journal of International Business Studies, 25*(3), 662–666.

Grahn, J. L., & Swenson, D. X. (2000). Cross-cultural perspectives for quality training. *Cross Cultural Management: An International Journal, 7,* 19–24.

Grove, C. L., & Torbiön, I. (1986). A new conceptualization of intercultural adjustment and the goals of learning. In M. Paige (Ed.), *Cross-cultural orientation: New conceptualizations and applications* (pp. 70–110). Lanham, MD: University Press of America.

Gudykunst, W. B., Guzley, R. M., & Hammer, M. R. (1996). Designing intercultural training. In D. Landis & R. S. Bahgat (Eds.), *Handbook of intercultural training* (pp. 61–80). Thousand Oaks, CA: Sage.

Hall, E. T., & Hall M. R. (1990). *Understanding cultural differences: The Germans, French and Americans*. Yarmouth, ME, Intercultural Press.

Hambrick, D. C., Davison, S. S., & Snow, C. C. (1998). When groups consist of multinationalities. *Organizational Studies, 19*(2), 181–206.

Harrison, J. K. (1992). Individual and combined effects of behavior modeling and the cultural assimilator in cross-cultural management training. *Journal of Applied Psychology, 77,* 952–962.

Helmreich, R. L. (2000). Culture and error in space: Implications from analog environments. *Aviation, Space, and Environmental Medicine, 71*(9-II), 133–139.

Hitt, J. M., II, Jentsch, F., Bowers, C. A., Salas, E., & Edens, E. S. (2000, November). *Scenario-based training for autoflight skills.* Paper presented at the Australian Aviation Psychology Association conference, Sydney, Australia.

Hofstede, G. (1980). *Culture's consequences: International differences in work-related values.* Beverly Hills, CA: Sage.

Hofstede, G. (2001). *Culture's consequences: Comparing values, behaviors, institutions, and organizations across nations* (2nd ed.). Thousands Oaks, CA: Sage.

Horenczyk, G., & Berkerman, Z. (1997). The effects of intercultural acquaintance and structured intergroup interaction on ingroup, outgroup, and reflected ingroup stereotypes. *International Journal of Intercultural Relations, 21*(1), 71–83.

Jacobs, T. O., & Jacques, E. (1986). Leadership in complex systems. In J. Zeidner (Ed.), *Human productivity enhancement: Vol. 2. Organizations, personnel, and decision making* (pp. 7–65). New York: Praeger

Kealey, D. J., & Protheroe, D. R. (1996). The effectiveness of cross-cultural training for expatriates: An assessment of the literature on the issue. *International Journal of Intercultural Relations, 20*(2), 141–165.

Klein, G., & Pierce, L. G. (2001). Adaptive teams. In *Proceedings of the 6th ICCRTS Collaboration in the Information Age Track 4: C2 Decision-Making and Cognitive Analysis.* Retrieved December 2, 2004, from http://www.dodccrp.org/6thICCRTS/

Klein, H. A. (2003, April). The cultural lens model: Understanding cognitive differences and aviation safety. Paper presented at the 12th International Symposium on Aviation Psychology, Dayton, OH.

Klein, H. A. (2004). Cognition in natural settings: The cultural lens model. In M. Kaplan (Ed.), *Cultural ergonomics* (pp. 249–280). Oxford, England: Elsevier Science.

Kolb, D. A., & Fry, R. (1975). Toward an applied theory of experiential learning. In C. Cooper (Ed.), *Theories of group process* (pp. 33–54). London: Wiley

Kozlowski, S. W. J., Gully, S. N., Nason, E. R., & Smith, E. M. (1999). Developing adaptive teams: A theory of compilation and performance across levels and time. In D. R. Illgen & E. D. Pulakos (Eds.), *The changing nature of work and performance: Implications for staffing, personnel actions, and development* (pp. 240–292). San Francisco: Jossey-Bass.

Kozlowski, S. W. J., Gully, S. M., Salas, E., & Cannon-Bowers, J. A. (1996). Team leadership and development: Theory, principles, and guidelines for training leaders and teams. In M. Beyerlein, S. Beyerlein, & D. Johnson (Eds.), *Advances in interdisciplinary studies of work teams: Team leadership* (Vol. 3, pp. 253–292). Greenwich, CT: JAI.

Levy, J. (1995). Intercultural training design. In S. M. Fowler & M. G. Mumford (Eds.), *Intercultural source book: Cross-cultural training methods* (pp. 1–15). Yarmouth, ME: Intercultural Press.

Littrell, L. N., & Salas, E. (2005). *Twenty-five years of cross-cultural training research: A critical analysis.* Unpublished manuscript.

Marks, M. A., Zaccaro, S. J., & Mathieu, J. E. (2000). Performance implications of leader briefings and team-interaction training for team adaptation to novel environments. *Journal of Applied Psychology, 85*(6), 971–986.

Mathieu, J. E., Heffner, T. S., Goodwin, G. F., Salas, E.,& Cannon-Bowers, J. A. (2000). The influence of shared mental models on team process and performance. *Journal of Applied Psychology, 85*(2), 273–283.

Nisbett, R. E., Peng, K., Choi, I., & Norenzayan, A. (2001). Culture and systems of thought: Holistic versus analytic cognition. *Psychological Review, 108*(2), 291–310.

Parsons, T., & Shils, E. (Eds.). (1951). *Toward a general theory of action.* New York: Harper & Row.

Pierce, L. G. (2002). *Preparing and supporting adaptable leaders and teams for support and stability operations.* Paper presented at Defense Analysis Seminar XI, Seoul, Korea. (*11th ROK-US Defense Analysis Seminar Proceedings (Manpower Policy, Session 4),* pp. 97–129.

Pierce, L. G., & Pomranky, R. A. (2001, September). *The Chameleon Project for Adaptable Commanders and Teams.* Poster presented at the 2001 Human Factors and Ergonomics Society annual meeting, Minneapolis, MN. (*Proceedings of the 45th Human Factors and Ergonomics Society Anuual Meeting,* pp. 513–517).

Salas, E., Burke, C. S., Wilson-Donnelly, K. A., & Fowlkes, J. E. (2004). Promoting effective leadership within multicultural teams: An event-based approach. In D. Day, S. J. Zaccaro, & S. M. Halpin (Eds.), *Leader development for transforming organizations: Growing leaders for tomorrow* (pp. 293–323). Mahwah, NJ: Lawrence Erlbaum Associates.

Salas, E., & Cannon-Bowers, J. A. (2000). The anatomy of team training. In S. Tobias & J. D. Fletcher (Eds.), *Training & retraining: A handbook for business, industry, government, and the military* (pp. 312–335). New York: Macmillan Reference.

Salas, E., & Cannon-Bowers, J. A. (2001). The science of training. A decade of progress. *Annual Review of Psychology, 52,* 471–499.

Schneider, S. C. & DeMeyer, A. (1991). Interpreting and responding to strategic issues: The impact of national culture. *Strategic Management Journal, 12,* 307–320.

Schwartz, S. H. (1999). A theory of cultural values and some implications for work. *Applied Psychology: An International Review, 48,* 23–47.

Shane, S. (1994). Cultural values and the championing process. *Entrepreneurship Theory and Practice, 18*(4), 25–41.

Smith, M. K. (2001). David A. Kolb on experiential learning. *The encyclopedia of informal education.* Retrieved February 11, 2005, from http://www.infed.org/b-explrn.htm

Thomas, D. C. (1999). Cultural diversity and work group effectiveness. *Journal of Cross-Cultural Psychology, 30*(2), 242–263.

Triandis, H. C. (1994). *Culture and social behavior.* New York: McGraw-Hill.

Triandis, H. C. (2000). Culture and conflict. *International Journal of Psychology, 35*(2), 145–152.

Trompenaars, F., & Hampden-Turner, C. (1998). *Riding the waves of culture: Understanding cultural diversity in business.* New York: McGraw-Hill.

Wentling, R. M., & Palma-Rivas, N. (1999). Components of effective diversity training programmes. *International Journal of Training and Development, 3,* 215–226.

Zaccaro, S. J., Rittman, A. L., & Marks, M. A. (2001). Team leadership. *Leadership Quarterly, 12*(4), 451–483.

Chapter 20

Computational and Theoretical Perspectives on Recognition-Primed Decision Making

Walter Warwick
Alion Science and Technology, MA&D Operation, Boulder, CO

Robert J. B. Hutton
Klein Associates Inc., Applied Research Associates, Dayton, OH

A theme of chapters in this volume is the question of how experts cope with challenging, rare, or unusual problems, whether their expertise can transfer to novel situations, and how expertise can accommodate dynamic and uncertain situations. The goal is to understand how expertise is manifested when experts have to work "out of context." In this chapter, we take the theme at a 90-degree angle, as we try to understand expertise not just out of context, but out of body. Rather than discuss expertise stripped from familiar circumstances, we discuss expertise stripped from the expert. The discussion focuses on our attempts to develop a computational model of the recognition-primed decision (RPD) (Klein, 1993, 1998).

Our sense of disembodiment follows from the fact that we're trying to identify and computationally implement some of the invariant aspects of a Naturalistic Decision-Making (NDM) process. The ultimate goal of the computational work is not to develop a representation of this person making that decision, but rather, to develop a general architecture that can be applied to the representation of a variety of decision makers in a variety of real-world contexts. In turn, these representations should allow us to simulate decision making more realistically and in a manner with greater psychological plausibility than the probabilistic and rule-based representations in common use today in computational modeling.

From a theoretical perspective, one goal is to address the question of whether it is even possible to develop a computational model of an NDM process. Indeed, some might argue that naturalistic models of decision making are interesting precisely because their noncomputational nature seems not to fit with microscale representations (Klein, Ross, et al., 2003), and that it simply makes no sense to try to coerce macrocognitive phenomena into a computational framework. Second, if we assume it is possible to model such processes computationally, we must fix the level of abstraction at which we claim correspondence between the theoretical and computational model and we must clearly delineate the sense in which the resulting computational models are "naturalistic." As a third goal, we must establish a reciprocal relationship between the two models, not just one where the theoretical model informs the computational, but rather, where the computational model helps us explore aspects of the theoretical description that might remain underspecified.

In this chapter, we report on our progress toward these three goals. We begin with a brief overview of the theory and the computational model. Next, we turn to the question of how well the computational model represents the theory. We describe several successive attempts we have made to model one of the several "by-products" of recognition posited by the RPD theory. Each iteration of this process has been instructive, in terms of both raising theoretical questions and identifying limitations in the computational representation. Finally, we describe the implications we see in this work for understanding the relation between the theory and computational model. This relationship has turned out to be more subtle than we expected, and it has caused us to rethink some of our most basic assumptions about the limitations of both the theory and the computational model.

OVERVIEWS OF THE THEORY
AND THE COMPUTATIONAL MODEL

The RPD Theory

The RPD is a description of a certain phenomenon of expert decision making (Klein, 1989, 1993; Klein, Calderwood, & Clinton-Cirocco, 1988). The RPD theory grew out of attempts to understand how decision makers generate and compare multiple courses of action when under extreme stress, uncertain and dynamic environments, and high stakes. Given the burden that such analytic strategies would seem to impose on the decision maker, it was unclear how experts would be able to make consistently good decisions in real-world environments. The answer hinged on the finding that expert decision makers do not employ analytic decision-making strategies. Instead, Klein and his colleagues found that experts typically recognized a single course of action based on their experience. Typically, the first recognized course of action would be deemed workable and immediately implemented; but if, for some reason, a shortcoming was detected, another course of action would be generated and considered. Thus, a serial view of decision making emerged, with experience of the decision maker rather than his or her analytical skill driving the quality of the decision making.

These observations took the form of a description of decision making, often presented as a block diagram showing causal relations among events and processes, as in Figure 20.1. The initial "simple match" description evolved to include variations that emphasize diagnosis of the situation in those cases where the initial assessment is unclear (sensemaking), as well as the mental simulation of a candidate course of action in order to evaluate whether it can satisfy the current requirements of the situation. These elaborations were first detailed in Klein (1989), and were described more recently in Klein (1993). Figure 20.1 shows this elaboration. The vertical path at the right going all the way from recognition to action is recognition-primed decision making. The other paths bring hypothesis testing into the process.

The RPD notion stands in contrast to normative-choice theories of decision making. First, it is descriptive and as such its aim is to inform our un-

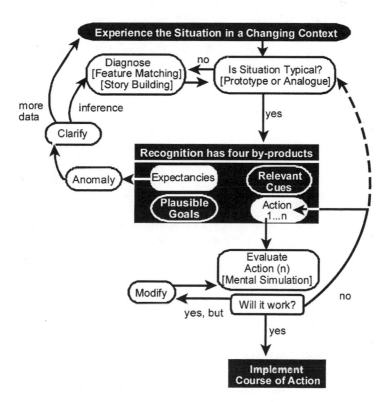

FIG. 20.1. The "integrated" RPD (after Klein, 1993).

derstanding of what decision makers actually do rather than what they should do. Second, the theory suggests a continuous, more holistic view of the decision-making process: Rather than seeing a discrete decision point as the culmination of an abstract utility analysis, the RPD posits a decision maker who is continuously engaged in monitoring the environment, reassessing the situation, and trying to understand what is going on until a decision is required (knowing when to act is often as critical as what to do). The RPD is not a one-time shot at the decision point, but part of a continuous, dynamic decision cycle or decision event (Hutton, Thordsen, & Mogford, 1997).

Third, and most important from the perspective of this volume, the RPD grew out of an attempt to understand decision making outside the laboratory and in the messy, real-world situations where experts actually make decisions. The RPD is very much a description of decision making in context. Indeed, Klein (1998) has even suggested the RPD as a theory of situational awareness, further reinforcing the connection between perception and action—the inexorable link between the decision maker and the environment (Neisser, 1976). In this light, it is unclear how the RPD of an expert can be considered apart from the environment in which particular decisions are made. But this is precisely what we have attempted to do by developing a computational model of the RPD, a model that reduces the environment (the context) to a handful of cues and actions.

The Computational Model

Figure 20.2 depicts the decision-making process of our computational model at a functional level. In the most basic terms, the computational model takes variables from a simulated environment as inputs and produces action as outputs to be implemented in the simulation. But before anything can happen, the computational model must be "populated" with the cues, by-products (such as "expectancies," as described by the RPD description) and courses of action that characterize a particular decision in a particular situation. Although the computational architecture is generic in the sense that it can represent a variety of decisions, it is specific in the sense that individual instantiations of the model must be created for each decision being represented.

After the computational model has been populated to represent a particular agent making a decision, the flow of control during simulation is as follows: The variables from the simulation environment that represent cues are continually passed to the computational model and buffered until the decision is to be made by the simulated agent. At that time, the buffer con-

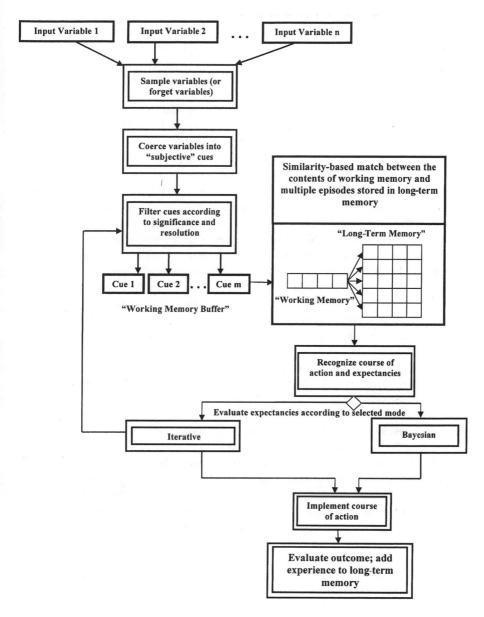

FIG. 20.2.   Flow of control through the computational model of the RPD.

tents are pared down according to the "attention management strategy" and memory decay rates specified by the modeler. This initial filtering is intended to represent the limited capacity of working memory. The values of the remaining variables are cast into ranges that better reflect the subjective terms meaningful to the decision maker. For example, where a variable that represents the distance to a target might be given in terms of meters in the simulation environment, it will be represented in the computational model in "subjective" terms such as "near" and "distant" (or in-range and out-of-range). This representation supports fuzzy judgments of cues such as distance and allows for situations to appear ambiguous to the decision maker. After the cues have been cast into subjective terms, they are then further filtered according to the significance the decision maker attaches to each cue and his or her ability to discriminate each cue more or less finely. We thus transform a subset of the original simulation variables into the set of cues that will prompt recognition.

Recognition occurs when the contents of this "working memory" are compared to the contents of "long-term memory." Our model of long-term memory comes from Hintzman's (1984, 1986) multiple-trace model. The essential idea behind a multiple-trace model is that each experience leaves behind a trace in long-term memory and thus the sum of experience is quite literally a collection of episodes, rather than an abstract schema. Recognition in Hintzman's model is a process of comparing a given trace against every trace in long-term memory. The comparison results in similarity values for each trace, which in turn are used to generate an emergent "echo" from long-term memory. In our model, each trace represents an individual decision-making experience in terms of the cues that prompted the recognition, the by-products of that recognition and the outcome of the decision. Following Hintzman, we use a similarity-based routine to compare the cues of the current situation to cues from past experience and to generate the by-products of recognition as the echo.

The by-products of recognition depend on the type of decision being modeled. The user can represent the "simple match" variation of RPD (Klein, 1993), in which case by-products of recognition consist only of a course of action. In the case of a more complex recognition, the user can model expectancies, in one of two ways (depicted as parallel paths through the flow of control in Figure 20.2). In either case, the course of action is returned to the simulation environment, implemented, and then evaluated with respect to its outcome as either a success or a failure according to criteria specified by the modeler. Finally, the entire decision-making episode, including the cues that prompted recognition, the course of action that was implemented, and the outcome of that decision, is added into long-term memory so that the experience can influence subsequent decisions.

COMPARING THE THEORY
AND THE COMPUTATIONAL MODEL

Whether we are describing the cognitive processes that characterize the RPD or the software mechanisms that constitute our computational model, it is important to remember that in both cases we are dealing with a set of abstractions that we impose to facilitate explanation, description, and prediction. Although this observation might seem obvious, we mention it to guard against a natural tendency to expect the computational model to somehow "resemble" the theory at every level of description. For instance, note that RPD by itself does not assert that recognition involves comparing a list of cues to a memory record of all prior experiences. Conversely, the computation model is mute concerning ways in which experts might perceptually integrate individual cues into patterns or into what, from the perspective of the memory, might be "chunks."

We have often found ourselves struggling against this tendency in the past—defending this or that aspect of the computational model as naturalistic despite the requirement that computational implementation must nail down many things that cognitive theories often leave unspecified. But now we no longer see a privileged perspective from which to judge the correspondence between the theoretical and computational model, as if some ultimate measure of verisimilitude could be established merely by inspection.

The real issue here is whether the abstractions we entertain within both the theory and computational model are commensurate. That is, given some level of description of the processes postulated by the theory, does the computational model serve as a reasonable analogue at a comparable level of description? We recognize that "reasonable" and "comparable" might be in the eye of the beholder, but that should not and cannot prevent us from adopting *some* basis for comparison and making assessments. For instance, given that we are willing to ignore discussion of the actual mechanisms that underlie the cognitive processes posited by the RPD theory, we should likewise be willing to consider the intended relationships captured by a computational model without laboring over the fact that underlying implementation is (necessarily) computational. This position is reminiscent of the cliché that nothing actually gets wet inside the computer simulation of a hurricane; that is, we can talk about what is and isn't represented in a computational model even if the phenomenon being modeled isn't itself computational by any stretch. Conversely, however, we cannot simply ignore the plainly computational aspects of a model just because we declare it "naturalistic" by fiat. Indeed, there is something unsatisfying about calling the application of a rule in an expert system "recognition" just because the antecedent conditions of the rule were satisfied before a specific action was taken.

So, returning to the question at hand, how well does the computational model compare with the theoretical model? One of our primary goals was to capture the role of experience in decision making. It is interesting to note here that the RPD model grew out of a failed attempt to uncover analogical reasoning in fireground command. Klein recalls, "We expected to see heavy uses of analogous cases. We found very little. Never was there an entire fire that reminded a commander of a previous one. The people we studied had over twenty years of experience, and all of it had blended together in their minds" (Klein, 1998, p. 12). In our computational model, the recognition process is driven by a similarity-based routine that reflects the sum of a decision maker's experience rather than his or her ability to recall a single episode. In this respect, our computational approach captures an important aspect of the RPD insofar as individual decision-making episodes are "blended together" during recall.

As to the nature of experience itself, the RPD theory suggests that expert decision making is not a matter of knowing and following the right rules, but rather one of gaining enough experience in particular domain so that the recognition of a suitable course of action will emerge rapidly. Side-stepping questions about the nature of rule-governed versus rule-following behavior, such a view of experience suggest that a straightforward expert–systems approach to the RPD, where rules specifying conditions and actions are elicited from a domain expert and hard-coded into a production system, is misguided. By dovetailing our multiple-trace model with a simple reinforcement routine, we avoid the need for a priori specification of relationships between situations and courses of action. Instead, the modeler merely defines the available cues and courses of action and then at run time the computational model stores the outcome of each new decision-making episode along with the situation (i.e., the specific values of the available cues) that prompted the decision. In this way, the model learns how to associate situations and actions according to the feedback it receives from its environment. Moreover, changes to that environment can lead to changes in the decisions the model will make, even if the available cues and courses of action that characterize the decision remain constant. Although hardly novel from the perspective of machine learning, this approach captures some of our intuitions about the RPD; the decision-making performance of the computational model depends directly on experience it gains making decisions.

Finally, the mechanics of the multiple-trace memory model and recall have allowed us to implement a grab bag of naturalistic intuitions. For instance, we represent the decision maker's attention management strategy and working memory capacity as it relates to a workload. Although workload issues and working memory capacity are not at the center of the theoretical focus, the RPD theory does suggest that part of expert decision making lies in recognizing which cues to attend to and which can be ignored in a high-pressure environments. We have found that workload-induced shifts along the analogous dimensions in the computational model can produce qualitative differences in decision-making performance. Similarly, the RPD theory suggests that expert decision makers can adjust the significance that they attach to a cue in response to changing circumstances. By filtering the set of cues that are used to calculate similarity in the computational model, we can represent such adjustments in terms of both the importance attached to each cue and the decision maker's ability to make more or less fine-grained discriminations when "perceiving" a cue. In this way, perception of the environment is dynamic and sensitive to the details of the situation being perceived.

We have also implemented functionality to capture the intuition of a "sentinel event" in long-term memory, whereby a single decision-making episode can have a disproportionate impact on subsequent decision making. This can be used to bias decision making behavior—for example, making a model "risk averse" by disproportionately reinforcing failures of a particular kind—and, similarly, to influence what a model actually learns from experience. Finally, although the theory doesn't emphasize how an individual decision maker might vary in his or her ability to make good decisions from time to time, there are parameters in the computational model we can use to represent how confident the decision maker must be in his or her recognition of the situation and, similarly, how precisely he or she will recall past situations. These too influence how the model learns from experience.

The structure of long-term memory together with the similarity-based recognition routine support much of the "naturalistic functionality" of our computational model and, in turn, provide some fairly direct analogues of the decision-making processes postulated by the RPD model. Like the theory, the computational model emphasizes the central role of experience and the "blended" recall that leads to recognition in the decision-making process. Furthermore, unlike an expert–systems approach, populating the computational model is not a matter of eliciting and codifying the rules that relate situation and actions in a given domain. Rather, the modeler defines the cues that characterize a situation and the manner in which they will be "interpreted" and lets the model figure out how to relate situations and actions. Given the functionality of the computational model, we have been able to simulate human decision-making behavior that is satisfying at both qualitative and quantitative levels (see Warwick & Hutchins, 2004; Warwick, McIlwaine, Hutton, & McDermott, 2001).

This is not to suggest that there is nothing more to be said about the comparison between the theoretical and computational models. Quite the contrary, having fixed the level of comparison in terms of the experience and recognition that drive decision making, questions recur at deeper levels of analysis having to do with the nature of experience and recognition themselves. For instance, the computational model suggests that experience can be quantified in a rather direct manner—simply count up the number of decision-making episodes in long-term memory adjusting for those episodes considered as sentinel. Is the same true of humans? Could we simply estimate the number of decision-making episodes of a certain type from, say, the years of experience on the job? Would this tell us anything interesting about the nature of expertise? And if it did, should the performance of the computational model by judged in part by comparing how long it takes the computational model to gain humanlike proficiency?

As to the nature of recognition, knowing what to look for in a situation is certainly one aspect of recognition and is represented in the computational model insofar as the modeler specifies the cues that will be made available to the model and even how those cues will be "interpreted." Still, the question arises whether the variables that constitute the simulation environment have psychological validity. Indeed, it is one thing to recast a variable such as distance into "subjectively meaningful" terms, but it is quite another if a real-world decision doesn't depend so much on a crude measure of distance as it does, say, on a more complex affordance of "reachableness," for which no obvious analogue exists in the simulation environment. Conversely, it is possible to trivialize complex decisions with a sufficiently (and artificially) rich cue set, but how do we judge when that threshold has been crossed?

These are difficult questions to answer. But they concern the concepts and processes that underlie the theory itself rather than the adequacy of the mapping between the theory and the computational model. Until a more complete theoretical account is given of the psychological constructs that underlie the RPD, there is no reason *not* to entertain the resemblance of the theoretical and computational model at the "macro" level of description (cf. Klein, Ross et al., 2003).

The same cannot be said for the role of expectancies in the computational model. Here there is real work to be done sorting out our theoretical intuitions and the computational representations meant to capture those intuitions.

DEALING WITH EXPECTANCIES IN THE COMPUTATIONAL MODEL

Up to this point, we have been discussing the adequacy of the computational model as it relates to the "simple match" variation of the RPD. As we mentioned earlier, the initial RPD hypothesis was elaborated to include the expert decision maker's ability to detect and diagnose anomalies in a recognized situation. Whereas it was relatively straightforward to implement the computational models of experience and recognition that underlie the simple-match variation of the RPD, it was unexpectedly difficult to represent diagnosis. The difficulties followed from the fact that when the RPD theory was developed, Klein and his colleagues consciously eschewed the information-processing descriptions—as one might see in an algorithmic state machine diagram—in favor of a more "holistic" representation (Klein, 2003, personal communication), which we now understand better in terms of the distinction between macrocognition and microcognition (Klein, Ross, et al., 2003).

The computational representation of experience and recognition was straightforward precisely because their role in the theory was so general. In fact, any number of machine-learning and "soft" recognition routines could be composed to implement the kinds of naturalistic processes we discussed previously. But in the case of expectancies, the "holistic" nature of the theory obscured a host of complexities that were revealed when they were exposed to the rigor required of a computational implementation.

For starters, where our first pass at the simple-match variation required only that we represent some of the impacts of experience on recognition without much concern for the order of operations in the decision-making process, implementing any representation of anomaly detection required that we decompose the otherwise holistic decision cycle into discrete, ordered phases. Given the cyclical nature of the description given in RPD (see Figure 20.1), ambiguities arise as to what goes in and what comes out of the recognition process. Figure 20.1 includes a box labeled "Is the situation typical?" followed by a box labeled "By-products of recognition" (implying the result of recognition) including: relevant cues, plausible goals, expectancies, and actions. However, no hint is provided in terms of what exactly is recognized. In an earlier version of the model (Klein et al., 1988), several factors were singled out as being compared to the prototype: goals, perceptual cues, and causal factors. The result of this comparison is a match to a prototype resulting in expectancies and an action queue (presumably with several possible courses of action in order of most feasible first, ready to be evaluated serially). In this case, a prototype consists of classes of situations characterized by their goals, perceptual cues, and causal factors. (One could understand this in terms of the idea that the prototype plus the actions and expectancies is a "mental model" of typicality, which is subsequently used in course of action evaluation by monitoring expectancies and mentally simulating courses of action.) In a dynamic decision-making context, such processing loops mean that we have to confront a problem of prioritizing chickens and eggs. What does the decision maker perceive about the situation that is compared to experience, and what is the result of that comparison? Are relevant cues the input to or the output from recognition? Do expectancies prompt the recognition or follow from the recognition?

The original motivation for thinking of recognition as a holistic process was rooted in the intuition that input–output relationships belied the nature of the recognition as reported by the subject-matter experts. This holistic viewpoint with its emphasis on recognition of solutions rather than elements of the situation itself suggests a direct perception (Gibson, 1966, 1979) or immediate perception process where the affordances of the situation are recognized based on experience and attuning to the appropriate information. This explanation also reinforces the perceptual-cognitive emphasis on expertise in the RPD model (Klein & Hoffman, 1993). Unfortunately, these theoretical explanations did little to help us decide how to anchor the order in which expectancies might be evaluated in the computational model.

In more concrete terms, our questions about the place of expectancies in the decision cycle took the following forms:

- Are expectancies just future states of the current cue set used by the model?
- Are they a different set of cues that have to be sampled to see if they concur with an expected value, trend, or relationship?
- When is an expectancy "evaluated"?
- How often is it checked?
- How do we define a violation?
- What types of checks can we conduct?
- What is the effect of a violated expectancy?

Answering these questions entailed several iterations in the development of the computational model. Each iteration of the computational model led to additional ruminations on the nature of expectancies. We describe each of these iterations in the sections that follow and discuss how well each satisfied our theoretical intuitions.

Expectancies as Fixed By-Products of Recognition

Before we could make any progress implementing expectancies in the computational model, we had to resolve the ambiguity of expectancies as by-products of recognition as opposed to cues that prompt recognition. Taking the RPD theory at face value, we opted to treat expectancies as by-products of recognition. Thus the computational model would take in a set of cues to prompt the recognition of a provisional course of action and a fixed set of expectancies. We treated the expectancies themselves as Boolean-valued variables that described aspects of the decision-making situation some fixed amount of time after initial recognition. So, recognition of expectancies in the computational model would consist of recalling whether each condition from some fixed set should be asserted or denied, and checking to see whether in fact each condition was satisfied (i.e., recalling that a condition should be denied/asserted and discovering that it is indeed denied/asserted). If even a single condition was not satisfied, the computational model would resample the cues and repeat the recognition–evaluation process; otherwise, the model would implement the recognized course of action and record the actual state of the expectancies at the time the course of action was implemented.

Though crude, this first pass at expectancies satisfied some basic theory-based intuitions. First, the recognized course of action was provisional insofar as no course of action could be implemented until all its attendant expectancies were satisfied. Second, one of the feedback loops implicit in the RPD theory was reflected at least in part in the resampling of the cues and rerecognition of the situation in the face of an unsatisfied expectancy. Third, although the conditions that defined the expectancies had to be defined a priori by the modeler, recalling whether those conditions should be satisfied in any situation was learned by the model at run time. That is, in the same way the model learned to associate the available cues and courses of action, it also learned to associate expectancies by situation. Finally, because expectancy evaluation occurred some fixed time after recognition (as specified by the modeler), the time it took to make a decision could vary from situation to situation depending on how many times an expectancy failed, and thus we saw a simple analogue for a decision maker struggling with a tough decision.

On the downside, our first implementation of expectancies was wanting in several respects. Chief among these was the static nature of the evaluation and feedback it provided; the same set of expectancies would be evaluated at the same time, every time a decision of that type was made. There was no facility for representing an expectancy that might be evaluated early and often as opposed to another that might take some time to be evaluated only once. Likewise, the impact of any one expectancy on the confirmation or diagnosis of a situation was no different than any other. Finally, nothing new was learned by the failure of an expectancy—failure led directly to reassessment of the situation using the same cue set as was used to prompt the initial assessment (i.e., recognition) of the situation.

These shortcomings led us to consider a more flexible implementation of expectancies.

Expectancies as a More Dynamic By-Product of Recognition

The first modification was simple: We allowed the modeler to specify "cycle times" for each expectancy that would control how long after the initial assessment of the situation each expectancy would be evaluated (allowing that one expectancy might be evaluated multiple times during the single evaluation cycle of another expectancy). This allowed expectancy evaluation to be an interleaved process.

The second modification was more substantial. Based on our data from studies using the Critical Decision Method, we knew that a violated expectancy was often something of a surprise to the decision maker that could lead to an elaboration of the assessment (at least) or a complete change or shift in the assessment (Klein et al., 1988). Subsequent reassessment is informed by knowledge of the violated expectancy, which can lead the decision maker to shift the relative importance the decision maker attaches to the cues, and may even prompt the decision maker to look at other related sets of cues (i.e., seek more and different information).

The computational mechanisms to reflect such dramatic shifts in the reassessment strategy were twofold. First, we modified the feedback from the evaluation cycle so that the value of the failed expectancy would be appended to the cue set used to reassess the situation. In this way, reassessment in light of a failed expectancy would include information that wasn't originally available to the model. Moreover, we provided functionality to allow the modeler to specify how much significance would be attached to a failed expectancy when used as a cue and further how both the significance and ability to resolve the each of the original cues might shift in light of the failure. Thus, the model might reassess the situation using different cues considered at different levels of detail in light of the failed expectancy.

These modifications allowed us to capture some of the intuitions that were not captured in the initial implementation, but something was still missing. Although we had introduced a more dynamic evaluation cycle, the expectancies themselves represented little more than delayed cues—just something the decision maker had to look at later rather than sooner. The theoretical commitment to expectancy evaluation as a rich activity of prediction and diagnosis was clearly lacking from our computational model.

Expectancies as the Product of a Causal Model

Our attempts to represent expectancies had focused almost entirely on their role as triggers for reassessment; we gave little consideration to the content of the expectancies themselves. On the front end of the evaluation process, the predictive content of the expectancies was determined entirely by the recognition process; depending on the model's experience, some subset of the predefined expectancies would be recognized and evaluated. Conversely, on the back end of the evaluation process, the impact of a failed expectancy was registered mostly in terms of the process-level effects during reassessment of the situation. The failed expectancy was treated simply as an additional cue for subsequent recognition cycles whereas the real emphasis rested on the induced shifts in cue resolution and significance.

Although some of our theoretical ruminations emphasized the role of expectancies as "trip wires" in the assessment process (e.g., supporting shifts in attention management strategy, pointing out that there are shortcomings in the original assessment of the situation, etc.), much of the theorizing focuses on the role of expectancies as substantive elements in the decision maker's "mental model" of the situation, to support process such as "story building" and "sensemaking." At the very least, we came to the view that the predictive and diagnostic impact of a failed expectancy should be more than just "try looking at the situation again."

Philosophers of science have long recognized the role of abduction, or inference to the best explanation. More recently, the subject has come under computational study as researchers have attempted to formalize patters of causal reasoning and abductive inference (see, e.g., Josephson & Josephson, 1996; Pearl, 1988, 2000). These computational efforts provided us a mechanism for representing a more substantive account of the prediction and diagnosis implicit in many of our discussions of expectancies.

In particular, we looked to the notion of a causal Bayesian network (Pearl, 2000) for a representation of expectancies and their evaluation. Informally (and quite briefly), such networks provide a graphical depiction of the causal relationships between variables (where nodes represent variables and the links between them represent causal relationships). Beliefs about the strength of the causal relationships between variables are locally encoded at each node in terms of the probabilities that a particular value of a child node will obtain given the value of the parent node. Global beliefs about a situation (i.e., beliefs about the state of all the variables) can be determined by several-well-know algorithms, given either changes in prior beliefs about the values of some variables that might obtain or actual information about the state of the world.

The study of these networks is a subject (and an industry) unto itself. But even this admittedly brief overview should be enough to ground what we see as a natural representation for the complementary relationship between prediction and diagnosis. We grafted a causal Bayesian network to our computational model in the following way: As before, cues about the situation still prompt a soft recognition of the available courses of action.

However, expectancies are no longer recognized at this point. Instead, exploiting the fuzzy nature of the similarity-based recognition routine, we treat each course of action as a root-level node in a causal belief network (where the topology of the network is defined in advance, at the same time as the rest of the computational model is populated), and we push the relative strength of recognition of each course of action as prior beliefs about the situation. In turn, these prior beliefs about the courses of action "trickle down" through other nodes in the network, each of which can represent an intervening causal relationship, until nodes at the leaf level are reached.

These nodes represent the expectancies about the situation. The expectancies are multivalued and the belief that any one value should obtain is a product of the initial recognition of the situation and the causal relationships specified between courses of action and expected (i.e., predicted) states of the world. At this point, the expectancies can be evaluated against the actual state of the (simulated) world. Note that every expectancy need not be evaluated (given thresholds, supplied by the modeler for how strongly a belief must be asserted in order to warrant evaluation). Diagnostic information about the world now "trickles up" through the network to the root level where beliefs about the courses of action might change. As before, not every change at the root level need be deemed significant, again depending on how the modeler specifies belief thresholds. This cycling between top-down prediction and bottom-up diagnosis continues until either a belief threshold is met for a course of action or no new information can be supplied to the network (i.e., all the expectancies have been checked).

This approach satisfies some of the intuitions that were not captured in our earlier implementations. First, expectancies need no longer be Boolean valued (i.e., either satisfied or denied); they can be multivalued and a particular value of an expectancy (e.g., "the smoke should be gray, not black or white") may be believed in a given situation. Second, there is a more robust sense of surprise built into this representation insofar as the impact that a failed expectancy has depends on both how far its predicted value diverges from its actual value and whether its fellow expectancies are checking out. The entire situation must be considered before the impact of a failed expectancy can be determined. Finally, and most important, this representation elevates expectancies from mere triggers for process-level effects, to more substantive reflections of the decision maker's causal understanding of the situation. More generally, the computational implementation here suggests an unambiguous view of what it means to hold an expectancy. Namely, to hold an expectancy is to hold a belief that a certain causal relationship holds between situations of a type—where situations are individuated by course of action—and some set of additional variables.

RECONSIDERING THE RELATIONSHIP BETWEEN
THE THEORETICAL AND COMPUTATIONAL MODEL

As we have explained, each iteration in our implementation of expectancies has supported a more robust and detailed representation of our theory-based intuitions. At the same time, these iterations have placed a greater burden on the both the modeler and the descriptive model of the RPD to supply more detail. Can we use task analytic techniques to elicit causal information from the human decision maker in sufficient detail to define network topology within the computational model? For that matter, can we reliably elicit any of the information we need to populate the computation model—for example, the cues and courses of action that define the structure of long-term memory, the reinforcement schedule we use to provide useful experience to the model, or the details of the shifting attention management strategy? In practice, this has turned out to be harder than we expected. As we noted earlier, we have had some success, both qualitative and quantitative, in representing simple-match decision making with the computational model. We are not yet sure whether the increasingly detailed representation of expectancies will lead to similar results (and hence we allow the modeler to decide whether to use the static implementation, the dynamic implementation, the causal Bayesian implementation, or none at all).

At the same time we've struggled with these questions about the computational model, our theoretical view of expectancies has broadened considerably. During the course of this effort, we began to document the various types and functions of expectancies. We summarize some of these next:

Diagnosis. Expectancies serve the purpose of helping the decision maker to make sense of events, especially when the situation is not obvious. What's going on right now? How did the situation get here (past) and how is it evolving (near-term future)? (For further treatment of sensemaking, see Klein et al., chap. 6, this volume; Klein, Phillips, Battaglia, Wiggins, & Ross, 2002; Klein, Phillips, et al., 2003.)

Story Building. This can be thought of as the reverse engineering of the current situation and making sense of events, and is related to diagnosis. How did this situation evolve? By understanding the way the situation emerged, the decision maker can now assess the current situation, whether it is typical or atypical, if it is "recognized" or not, and how the situation may continue to evolve.

Confirming/Disconfirming Assessments of the Current Situation. Expectancies help the decision maker to assess whether the situation is anoma-
lous or whether it is similar to previous experiences. Thus, situations can be
compared to previous situations based on the expectancies generated, and
a judgment as to whether those expectancies are violated or supported.

Detecting Problems. The generation of expectancies is critical to problem detection, a critical cognitive skill in complex domains. Is this situation
typical or does an anomaly exist? Will this course of action work? They also
help to set trip wires that can be used to monitor a situation over time and
detect whether the situation is still as originally assessed, and that nothing
significant has emerged or changed. (For further discussion of the largely
neglected issue of how people detect problems, see Klein & Crandall,
1997.)

Mental Simulation. Expectancies serve a role as both inputs and products of mental simulation. Mental simulation is an aspect of recognition-
primed decision making used to verify or evaluate a candidate course of action. Expectancies are used as inputs to the mental model that serves as the
basis for evaluating the course of action. They can provide the values that
certain parameters may take given the current situation, and the way that
the situation is thought to have evolved (see Trafton and Hoffman, chap.
15, this volume). As a result of this process, new expectancies may be generated based on insights provided by the mental simulation (Klein &
Crandall, 1995).

Coordinating Action. One function of generating expectancies is to anticipate and prepare for upcoming events, for example, visualizing a course
of action and the actions necessary to coordinate with other resources.
These expectancies are usually a result of the mental simulation of a plausible course of action.

Managing Uncertainty. Expectancies are a way for decision makers to
overcome some uncertainties in a situation by providing data to the mental
simulation of future events and plausible courses of action. They also help
the decision maker to "fill in" some of the areas of missing, ambiguous, or
unreliable data by taking the "typical" situation as represented in the mental models, and extrapolating or filling in some of those spurious data.
These are assumptions of typicality that help the decision maker make informed and educated decisions based on past experience. They also provide the decision maker with a feel for the degree of uncertainty and
whether it is tolerable, based on the nature, amount, and degree of "filling
in" required (see Schmitt & Klein, 1996).

Managing Resources. Related to coordinating action, expectancies from mental simulation and mental models of typicality help the decision maker project trends and rates of usage, and help the operator monitor and control resources to achieve the required goals. Expectancies can serve as trip wires for potential problems or envisioned bottlenecks.

Enhancing the Perceptions of Affordances of Situations. Expectancies provide the decision maker with a sense of what actions are plausible given the situation. They add information to the situation and provide the decision maker with a better sense of what the situation affords (Greeno, 1994). What can or could I do here and what would I expect to happen if I did that? Where are the leverage points? How do I take advantage of them?

Managing Attention. One of the more important roles of expectancies may be in the area of attention management. The anticipation of events or behaviors of various aspects of the environment allows the decision maker to improve his or her focus of attention and "cue sampling" strategy. Key elements of this include: What to scan or pay attention to? Where to scan? How often to scan? When to scan? How closely to scan, or can I skim?

Controlling Stance. Stance is the disposition of the decision maker with regard to what is good, appropriate, valuable, or necessary. Should I be defensive? Offensive? More alert? Relax a little? Expectancies serve a purpose in terms of helping the decision maker feel his or her stance appropriately for the situation now, and the expected situation at some later time. This also serves a stress or workload management function.

Judging the Timing and Sequence of Events. Related to coordinating action, a critical aspect of decision making and acting is knowing when to do something. The course of action may be obvious, but the critical judgment may be when to implement the action. Expectancies, based on experience and a sense of typicality, provide critical information about the timing of events and how they will unfold, providing critical information for the timing and sequencing of actions.

There is a long way to go before our computational model reflects all these functional aspects of expectancies. Furthermore, in terms of reconciling the abstractions of the two models, it might also seem that theory offers a comparatively rich account of expectancies against a rather impoverished representation of the computational model. If this were the end of the story, one might conclude that the computational model compares poorly with the theory, and leave it at that. But there is more to say.

First of all, given our progress to date, there is no reason not to think that given enough time and effort we would could implement an ever more complex (and presumably, more psychologically plausible) computational model with an ever more sophisticated representation of expectancies. As we indicated earlier, each iteration of expectancies in the computational model led to new theoretical ruminations, and the questions that came up were not so much about the feasibility of a computational approach per se but whether thinking of expectancies as trip wires, or expectancies as cues fed back into recognition, or expectancies as the product of causal reasoning would do justice to theoretical intuitions.

Although it might seem that a method of trial and error should bring us closer and closer to representing some fact of the matter, each iteration has revealed an otherwise unexplored aspect of the theory that a model should somehow embrace. Rather than peeling away layers of the onion, so to speak, we have found the theoretical view of expectancies becoming increasingly ramified. But not rarified—each implementation cycle has resulted in at least a nominal correspondence between model and theory only to reveal additional aspects of the theory that could or should be considered.

Despite these difficulties, we see real benefit in trying to cast our theoretical intuitions in terms sufficiently concrete to support a computational model; nothing cuts through a tangle of intuitions like the computational implementation of a theoretical model. Still, although there is no doubt that the RPD is deceptive in its apparent simplicity, the slippage we have experienced in attempting to represent expectancies suggests that there is something systematically askew in both our computational and theoretical perspective (i.e., the problem is deeper than just a question of representing additional detail in the computational model).

In particular, when we review the various aspects of expectancies, we are struck by intuitions of continuous control, monitoring, reacting, adjusting, attuning, and so on. From an NDM perspective, it is easy to view such intuitions as reflections of a macrocognitive, or even a noncomputational process and, by extension, to assume that such processes cannot be represented computationally. But this may miss the point. Again, the fact that a process is somehow, or inherently non-computational (e.g., a hurricane) does not entail that it cannot be modeled computationally. Rather, our theoretical intuitions suggest a process that is not just non-computational, but a process that defies description even in the more abstract terms of a knowledge-level system (cf. Newell, 1982, 1990). To the extent that we conceive of expectancies as constituting continuous, dynamic control within a complex decision-making process (i.e., a process in which, when modeled as a box diagram, everything feeds back into everything else; see Klein, Ross, et al., 2003), we move away from the view of an agent whose knowledge is rationally applied in pursuit of a goal. And yet the theoretical descriptions of expectancies—expectancies as trip wires, as newly recognized cues, as the product of causal reasoning—are couched in essentially those terms.

As a reaction to the received view in cognitive science, our naturalistic view of expectancies does not go far enough. Indeed, the significance of a naturalistic view of decision making isn't just that it undercuts the traditional view of decision making as a rational, analytic process but, rather, that it calls into question whether it makes sense to talk about decision making as a disembodied, symbolic, knowledge-level process in the first place. There are, in fact, several issues lurking here. The first is a matter of degree. Where Newell marginalizes the role of "control" in cognition (Newell, 1990, p. 45), we see expectancies as a potential example of high-level cognitive behavior where control is paramount. Second, even granting that a single phenomenon might be explained at many different levels of abstraction, our experience representing expectancies suggests that a knowledge-level explanation can be counterproductive. Finally, where a knowledge level of explanation of expectancies emphasizes the knowledge "inside the head" of the decision maker, thinking of expectancies in terms of control highlights the delicate relationship between the decision maker and the environment and reminds us that the two cannot be easily decoupled.

So if we're not going to think about expectancies at the knowledge level, how should we think about them? Resorting (or resigning, as the case may be) to metaphor, if we think of expectancy evaluation in terms of signal processing, our intuitions suggest that it might be better understood as an analogue process than as a knowledge-level process—where the output of the process is not determined by some manipulation of abstract representations of the situation but rather by the unconscious blending and filtering that occurs across a continuous spectrum, where both the process and the environment in which it occurs must be understood. Perhaps this is where the theory, and hence the computational model, runs into difficulty. In describing expectancies in terms of cues, trip wires, and finite resources and outcomes to be managed, we implicitly commit ourselves to a knowledge-level description—as if we could reduce the whole process to the identification of a critical set of cues and the knowledge needed to diagnose the situation. At this level of description, the process is (somewhat) easily implemented, and, consequently, largely unsatisfying. If, instead, we focused on what leads to the development of the right kind of "circuitry" to recognize expectancies—the environmental constraints, the cognitive constraints, and the interplay between them—we might end up with a more satisfying account. That is, the meat of the problem might lie in understanding how it is that an expert is able to transform the flux of experience into a cue or pattern we can call an expectancy rather than understanding what he or she does with that cue once it has been given.

This brings us back to the question of expertise out of context. Obviously, there is much more that should be said about the development of an analogue model of expert decision making, but even at this stage it is possible to sketch the relationship between our modeling efforts and context. Namely, to the extent we talk about our models at the knowledge level, we always run the risk of disembodying the decision-making process and trivializing the context. It no longer matters whether we're talking about this decision maker or that one, so long as we can specify the input to and output from the process. Likewise, when we identify a set of discrete cues as germane to decision making, and thereby individuate situations along corresponding dimensions, we reduce the complexity of the situation to a handful of variables, and run the risk of missing the patterns or family resemblances, especially the patterns that form across data types.

This in itself is not *necessarily* a bad thing. In fact, it is exactly this ability to abstract that can make modeling exercises worthwhile. At the same time, however, if we lose sight of these abstractions when we return to the theory, we're likely to find ourselves unsatisfied with the products of our theorizing. We believe this has happened with our attempts to model expectancies. By rethinking the relationship between the theory and the computational model, we find ourselves in position to speculate about NDM at a deeper level. Perhaps this approach will even bring us to a point where decision making will be modeled as a set of continuous equations reflecting the interactiveness and dynamics of the decision-making processes-in-context, rather than as a disembodied, decontextualized information processor operating over a set of cues abstracted from the environment.

REFERENCES

Gibson, J. J. (1966). *The senses considered as perceptual systems.* Boston: Houghton Mifflin.

Gibson, J. J. (1979). *The ecological approach to visual perception.* Boston: Houghton-Mifflin.

Greeno, J. G. (1994). Gibson's affordances. *Psychological Review, 101,* 336–342.

Hintzman, D. L. (1984). MINERVA 2: A simulation model of human memory. *Behavior Research Methods, Instruments & Computers, 16,* 96–101.

Hintzman, D. L. (1986). *Judgments of frequency and recognition memory in a multiple-trace memory model* (Report). Eugene, OR: Institute of Cognitive and Decision Sciences.

Hutton, R., Thordsen, M., & Mogford, R. (1997). En route air traffic controller decision making and errors: An application of the recognition-primed decision model to error analysis. In *Proceedings of Ninth International Symposium on Aviation Psychology* (pp. 721–726). Columbus: Ohio State University Department of Aviation.

Josephson, J. R., & Josephson, S. G. (Eds.). (1996). *Abductive inference: Computation, philosophy, technology.* Cambridge, England: Cambridge University Press.

Klein, G. (1989). Recognition-primed decisions. In W. B. Rouse (Ed.), *Advances in man–machine systems research* (Vol. 5, pp. 47–92). Greenwich, CT, JAI.

Klein, G. (1993). The recognition-primed decision model: Looking back, looking forward. In C. Zsambok & G. Klein (Eds.), *Naturalistic decision making* (pp. 285–292). Mahwah, NJ: Lawrence Erlbaum Associates.

Klein, G. (1998). *Sources of power: How people make decisions.* Cambridge, MA: MIT Press.

Klein, G., Calderwood, R., & Clinton-Cirocco, A. (1988). *Rapid decision making on the fireground* (Tech. Rep. No. DTIC AD-A199 492). Alexandria, VA: U.S. Army Research Institute.

Klein, G., & Crandall, B. W. (1995). The role of mental simulation in naturalistic decision making. In P. Hancock, J. Flach, J. Caird, & K. Vicente (Eds.), *Local applications of the ecological approach to human–machine systems* (Vol. 2, pp. 324–358). Mahwah, NJ: Lawrence Erlbaum Associates.

Klein, G., & Crandall, B. (1997). *Characteristics of problem detection* (Report to the Japan Atomic Energy Research Institute). Fairborn, OH: Klein Associates.

Klein, G., & Hoffman, R. (1993). Seeing the invisible: Perceptual/cognitive aspects of expertise. In M. Rabinowitz (Ed.), *Cognitive science foundations of instruction* (pp. 203–226). Mahwah, NJ: Lawrence Erlbaum Associates.

Klein, G., Phillips, J. K., Battaglia, D. A., Wiggins, S. L., & Ross, K. G. (2002). *Focus: A model of sensemaking* (Interim Report–Year 1, prepared under Contract No. 1435-01-01-CT-31161 [Dept. of the Interior] for the U.S. Army Research Institute for the Behavioral and Social Sciences, Alexandria, VA). Fairborn, OH: Klein Associates.

Klein, G., Phillips, J. K., Rall, E. L., Thunholm, P., Battaglia, D. A., & Ross, K. G. (2003). *Focus year 2 interim report* (Interim Report prepared under Contract No. 1435-01-01-CT-3116 [Dept. of the Interior] for the U.S. Army Research Institute for the Behavioral and Social Sciences, Alexandria, VA). Fairborn, OH: Klein Associates.

Klein, G., Ross, K. G., Moon, B. M., Klein, D. E., Hoffman, R. R., & Hollnagel, E. (2003). Macrocognition. *IEEE Intelligent Systems, 18,* 81–85.

Neisser, U. (1976). *Cognition and reality: Principles and implications of cognitive psychology.* San Francisco: Freeman.

Newell, A. (1982). The knowledge level. *Artificial Intelligence, 18,* 87–127.

Newell, A. (1990). *Unified theories of cognition.* Cambridge, MA: Harvard University Press.

Pearl, J. (1988). *Probabilistic reasoning in intelligent systems: Networks of plausible inference.* San Francisco: Morgan Kaufmann.

Pew, R. W., & Mavor, A. S. (Eds.). (1998). *Modeling human and organizational behavior: Application to military simulations.* Washington, DC: National Academy Press.

Schmitt, J. F., & Klein, G. (1996). Fighting in the fog: Dealing with battlefield uncertainty. *Marine Corps Gazette, 80,* 62–69.

Warwick, W., & Hutchins, S. (2004). Initial comparisons between a "naturalistic" model of decision making and human performance data. In *Proceedings of the 13th Conference on Behavior Representation in Modeling and Simulation* (pp. 73–78). Arlington, VA: SISO.

Warwick, W., McIlwaine, S., Hutton, R., & McDermott, P. (2002). Developing computational models of recognition-primed decision making. In *Proceedings of the Tenth Conference on Computer Generated Forces* (pp. 323–331). Norfolk, VA: SISO.

Chapter 21

Analysis of Team Communication and Performance During Sustained Command-and-Control Operations

Donald L. Harville
Air Force Research Laboratory, Brooks City-Base, TX

Linda R. Elliott
Army Research Laboratory, Fort Benning, GA

Christopher Barnes
Michigan State University

United States Air Force (USAF) command-and-control (C2) warfighters face increasingly complex and novel environments that represent the essence of Naturalistic Decision Making (NDM): multiple demands for enhanced vigilance, rapid situation assessment, and coordinated adaptive response (Cohen, 1993; Klein, 1993; Mitchell & Beach, 1990; Orasanu & Connolly; 1993; Orasanu & Salas, 1993; Rasmussen, 1993). In tactical C2 situations, the focus is on dynamic battle management and time-critical targeting. Coordination demand is high—reconnaissance and resource allocation depend on close coordination between ground and air forces in a distributed network system of systems. Situations requiring close coordination and adaptive replanning are increasingly prevalent and challenging.

It is clear that challenges within these battle scenarios are critically important to air and ground superiority. Much effort has been focused on the development of advanced technology to provide and represent time-critical information during mission execution. These capabilities are needed to facilitate, even enable, situation awareness and coordinated response in conditions of information complexity and time pressure. However, technology can only support, not replace, the role of the war fighters. In fact, it can easily be argued that technology sometimes (perhaps often) increases the role and demands of the human decision maker.

C2 operators must face these ever-increasing demands in conditions that are not ideal. It is, and will be, quite likely that operators will be chronically tired and sleep-deprived. Such is the nature of battlefield operations. During the early stages of actual operations, members of the command center are often up for several days. Over time, chronic fatigue will affect everyone, and the likelihood of error will increase (Bonnet, 2000; Hursh, 1998). This is particularly relevant to C2 situations, which require constant monitoring, even when events are still.

Though interventions for fatigue exist (Eddy & Hursh, 2001) and have proven to ameliorate effects, additional interventions (e.g., in information display, monitoring, decision support, and alerting mechanisms) can also be developed. To facilitate their development, we need controlled studies to determine what kinds of errors are made, by whom, and when.

Whereas extensive data are available on effects of sleep loss on physiological, attitudinal, and cognitive function (Kryger, Roth, & Dement, 2000), very few studies have reported data regarding sleep loss effects on particular aspects of information processing in complex team performance or decision-making tasks. A few preliminary studies provide some introductory results (Coovert et al., 2001; Hollenbeck, Ilgen, Tuttle, & Sego, 1995; Mahan, 1992, 1994; Mahan, Elliott, Dunwoody, & Marino, 1998).

There is a need for in-depth study of the effects of fatigue on team performance, based on operational scenarios and participants. As a result, the Fatigue Countermeasures Branch at Brooks City-Base (San Antonio), Texas, initiated a program of research, to identify effects of sleep loss and circadian rhythm on information processing, communication, coordination, and decision making in team-based sustained C2 simulation-based task environments. In this chapter, we describe a baseline study, with a focus on issues related to elicitation and assessment of team communications. We predicted that communication-based measures of information transfer, coordinating behavior, and encouragement would decline with fatigue. Participants were expected to tunnel in, not notice the predicaments of others, not care, and not realize when new bandits arise, or when additional friendly resources are available. These predictions are consistent with findings concerning performance under stress in general (Cannon-Bowers & Salas, 1998; Driskell & Johnston, 1998; Klein, 1996). As a result, they were also expected to be less effective in sequencing activities.

METHOD

Our overall approach was to: (a) develop operational C2 scenarios that elicit desired aspects of team performance, (b) develop an array of measures, (c) obtain military research participants comparable to USAF C2 operators, (d) train participants to a high level of performance, and (e) have them perform the scenarios overnight, after having been awake all day.

Development of C2 Scenarios

Scenario development is essential to valid investigations of complex NDM situations. Whereas many experts are situated in environments not easily replicated in the lab, C2 offers an ideal situation, in that C2 operators "in the wild" work in a realm dominated by tables, computer screens, and computer-mediated communications. Therefore, we have the opportunity to study expert behavior in a "naturalistic" setting. However, that is not to say that scenario development is easy. It is not. First, scenarios must be carefully constructed to ensure operational relevance, they must elicit the performance of interest, and they must enable good measures. These things must be achieved for any study, but investigations of sustained operations pose greater challenges, because multiple scenarios are necessary.

The goals for our scenario development were to: (a) capture operational relevance to C2 (content fidelity), (b) identify and assess aspects of individual and team performance (construct fidelity), and (c) develop scenarios that were equivalent in difficulty (cognitive fidelity), but (d) were not exact replicas of each other. If participants performed the same scenario over and over, then effects of fatigue would potentially be confounded with effects of practice and performance based on recognition of superficial or literal features. Scenarios had to be demanding, operationally relevant, equivalent in difficulty, and yet distinct from each other (Elliott, Coovert, Barnes, & Miller, 2003). Although the scenarios needed to include the same "deep" or conceptual features, they could not repeat the same surface features, or we would have to tease out an effect of practice that was not the immediate concern (i.e., performance improvement due to repetition of literal features).

Scenarios were developed to reflect USAF C2 tactical operations and require coordinated action, decision making, and adaptive response to time-critical situations. Time-critical retargeting was chosen as the operational theme of all scenarios. This is currently a critical issue in actual USAF operations. In previous operations, battle damage assessments revealed that enemy forces accomplished several methods to construct decoys that were targeted. In order to assess this threat, Uninhabited Aerial Vehicle (UAV) assets are now deployed to verify suspected targets, just prior to actual targeting. When a decoy is identified, the asset that had been deployed toward that target needs to be retargeted immediately, for effective use of resources. This requires a great deal of communication, coordination, and time-critical decision making.

Scenarios were constructed to ensure a reasonable degree of equivalence in task demand and difficulty (Elliott et al., 2003). This challenge entails much consideration and planning; therefore, it is described in detail here. A prototype scenario was constructed, with designated targets and decoys that appear at intervals throughout the scenario, to help assure equivalence in cognitive demand. Friendly assets (e.g., UAV assets; bomber, jammer, and fighter aircraft) were assigned to friendly roles that were played by the participants. Additional assets appeared at intervals in the scenario. Enemy assets included surface-to-air missile (SAM) sites and fighter aircraft (both varied in threat level). Additional enemy assets appeared at intervals, in the form of "popup" SAM sites and additional fighter aircraft. Participants in the scenario had to identify and verify SAM targets, and coordinate their resources to form "strike" packages. Enemy SAM sites needed to be jammed, so that friendly bombers and fighters would not be shot down. At the same time, friendly bombers had to be protected by friendly fighter aircraft, against enemy fighter aircraft. When assets are distributed across the participants, there is a high need for coordinated action.

The primary mission was always the same: Find and destroy hostile targets. The same number of targets (20) were presented in each scenario. Ten were presented within the first 5 minutes of the scenario, whereas the remaining 10 were "popup" targets that appeared systematically throughout the scenario. In all scenarios, half of the targets were always decoys. All targets were placed about the same distance from the friendly weapon assets. All targets appeared in similar task tempos. All friendly assets were equivalent in each scenario. In each scenario, all friendly roles were started with the same type and number of assets, and were presented with additional assets that were equivalent in type and timing across the scenarios. Thus, for every 5-minute increment in each scenario, participants owned the same number of assets and faced the same number of hostile targets— except for any losses in targets and assets that were a function of their own performance. Throughout the scenario, the participants had to monitor the fuel and armament states of their controlled assets and perform aerial refueling when necessary. Friendly assets varied with regard to fuel states, such that some needed immediate refueling, whereas others would have to be refueled sometime during the scenario.

In addition to equivalence in mission and task demands, the scenarios required similarity of actions to meet these demands. For example, in the premission planning phase, certain topics needed to be discussed and orders to assets given, prior to mission execution. During the first few minutes, the UAV assets must be effectively employed (covering enemy territory) to discriminate true targets from decoys. Some friendly assets were located far away, such that orders must be given right away in order to have the assets when needed. These events were documented and used as a behavioral guide for observer-based assessment of performance.

Once the prototype scenario was developed, alternates were constructed based on the same underlying structure. In each scenario, similar assets were assigned, at similar times, creating similar events. To minimize recognition of this underlying structure, "surface" characteristics of the scenario were changed. Scenarios differed in geographical and political context that was plausible for hostile conflict. Participants were given briefing materials that explained each situation. Scenarios also differed in the nature of the enemy targets (e.g., aircraft, ships).

Scenarios were created for a PC-based synthetic team task environment that was developed for investigations of C2 team performance. It is essentially a scaled microworld that strives to meet criteria of representativeness and ecological validity for controlled naturalistic research, as discussed by Brehmer and Dorner (1993) and Bowers, Salas, Prince, and Brannick (1992). The Airborne Warning and Control System (AWACS) Agent-Enabled Decision Group Environment (AEDGE™) is a PC-based federation of intelligent agent–based functions that enable C2 scenario construction, with multiple roles and numerous entities (Hicks, Stoyen, & Zhu, 2001; Petrov & Stoyen, 2000). The intelligent-agent technology enables decision support to each role and utilization of synthetic computer-driven role players. These PC-based scenarios look and feel much like a networked videogame. However, data collection is much more extensive. In addition, the roles in the scenario that are not played by participants are "played" by "intelligent" decision aides, following decision rules that are both generalizable and reasonable. In this way, scenarios can be "played" without any human participants. This also provides enemy roles that are more intelligent than scripted entities, and more consistent than humans playing enemy roles.

Complexity in scenarios used in this study was enhanced by the addition of an AEDGE™ software agent to play the role of a fourth friendly team member, whose responsibility was to maintain high-value assets such as tankers and airborne reconnaissance platforms. All decisions and actions taken by each participant were documented in output. Thus, AEDGE™ scenarios enable systematic investigation of expert team decision making in a naturalistic environment.

Measures

An array of measures of AEDGE™ performance were taken. These included observer-based assessments, many indices taken from AEDGE™ output, and capture/coding of communications. AEDGE™ output records and time-stamps every action and decision made by each participant; therefore it can yield indicators of individual and team decision making. It also records and summarizes every outcome. In addition, a number of other measures were taken, including computer-based performance on basic cognitive tests, body temperature, and subjective reports.

Assessment of Team Performance.   For this analysis, we wanted to relate indices of communication to overall mission outcome, in order to identify
patterns of communication that typified good-performing versus poorly
performing teams, as well as changes that might have occurred over time in
the sessions. Thus, we used measures of overall mission performance, based
on mission outcomes.

Raw measures of mission outcome and team process were captured and
time-stamped by the simulation. Mission outcome scores were represented
by the type, number, and relative value of assets that were lost, by "friendly"
and "hostile" roles. Friendly assets included air bases, cities, SAM launchers,
uninhabited aerial vehicles, tanker aircraft, high-value reconnaissance aircraft, fighter aircraft, and bomber aircraft.

Each asset was given a relative score value, generated by our weapons director expert, and validated by other experienced weapons directors. The
loss of any friendly asset detracted from the score of the friendly team, and
added to the score of the enemy. The loss of hostile assets added to the
score of the friendly team, and detracted from the score of the hostile. The
overall mission outcome score was based on the point value obtained after
subtracting all friendly "losses" from the total hostile "losses."

Assessment of Communications.   All communications, during the permission planning session, mission execution, and debriefing, were digitally recorded. In addition, all e-mail communications were captured, along with
other communications such as requests for asset transfers. As one can imagine, coding was a very laborious and time-consuming process. Several approaches can be taken, such as event-based analyses (i.e., communication
related to particular events), analyses for certain dynamics (e.g., identification of indications of "double loop" learning), or a variety of coding
schemes that classifies individual utterances.

We chose to transcribe all communications, no easy task. We considered
several existing schemes that related communication dynamics to C2 performance (Artman & Granlund, 1998; Colquitt, Hollenbeck, Ilgen, LePine,
& Sheppard, 2002; Hutchins, Hocevar, & Kemple, 1999; Kanki & Foushee,
1989). We also reviewed other coding schemes related to team or group
processes as related to performance (Bales, 1950; Mulder & Swaak, 2000).
There is great diversity in approaches to communication analyses, each in
accordance to their particular research goal. However, there are also similarities with regard to coding categories, given similar research goals. Several include aspects of information exchange.

We crafted our coding scheme, illustrated in Figure 21.1, upon core
aspects of communication that were common to several approaches.
Each category was chosen through consideration of: (a) relevance to per-

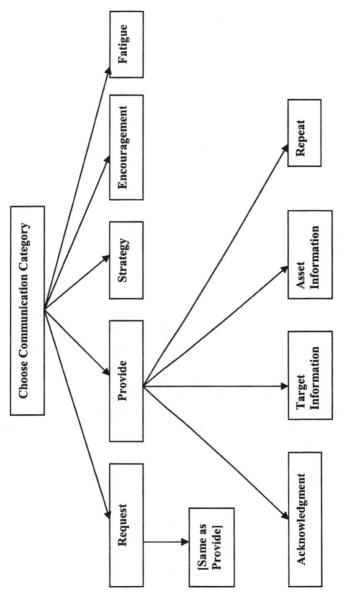

FIG. 21.1.   Communication coding scheme.

formance of interest (i.e., team decision making, team coordination, and fatigue) and (b) distinctiveness of category (in some schemes, coding categories appeared overlapping and difficult to distinguish). Specifically, we wanted to capture aspects of information exchange as they relate to identification of threats and coordination of action. At the same time, we also wanted to capture efficiency of communication (acknowledgments, requests for repeats), encouragement of others, and expressions of fatigue.

We chose this framework for its usefulness in analyses relating to decision making. The breakdown of "provide" versus "requests for" information, along with some other categories (acknowledgment and encouragement) have been used as indicators of leadership and communication effectiveness (Bales, 1950). One can see that this scheme allows one to assess total counts of different types of communication, in terms of information transfer, coordinating activities, acknowledgments, encouragement, and fatigue. It also allows one to compare the proportions of provide versus requests. When a team is providing task-related information without requests, it is an indication of implicit coordination. Implicit coordination has been related to highly performing teams (Serfaty & Entin, 1995; Serfaty, Entin, & Deckert, 1994).

Each utterance was first coded as to who provided it and whether the utterance: (a) requested information, acknowledgment, or repeat; (b) provided information, acknowledgment, or repeat; (c) concerned strategy (i.e., coordinating activities among the participants, such as you get that target, while I get this one); (d) was encouragement (e.g., "attaboys"); or (e) expressed fatigue (e.g., "Man, I'm tired," loud yawns, etc.). Subclassifications used for request and provide were the same. Thus, all utterances were coded as to whether the utterance was task related or not. If request or provide was the communication's category, the utterance was further categorized as to whether it was referring to a particular target (e.g., threat level, location), a particular asset (e.g., location, type), acknowledging a prior comment (e.g., roger that; did you get that?), or asking for a repeat of a communication.

Explicit definitions and examples were provided for each category. Two researchers independently used this initial scheme to code the same set of utterances. They then compared results, discussed discrepancies, and refined the definitions. This was repeated until high agreement was consistently achieved (95% or greater). Each utterance was coded as to whether researchers agreed or disagreed. Once high agreement was established, we trained additional coders using the refined category distinctions and examples. Examination of their ratings also yielded high agreement (95% or above).

Participants

Research participants were drawn from a pool of USAF officers awaiting Air Battle Management Training at Tyndall AFB, Florida. A total of 10 three-person teams participated in this study (23.3% female, mean age = 26, *SD* = 3.1 years). Four of the teams were all male, and six were mixed gender. All of the participants had already attended the Aerospace Basics Course, and so they were very familiar with USAF C2 concepts and principles. Although they were not experienced warfighters, their performance on training scenarios was equivalent to that of experienced AWAC Weapons Directors who participated in a previous study.

Training

Each participant participated in a 40-hour training session occurring during a 1-week period. The training was conducted by an experienced, retired warfighter, who previously served as an AWACS Weapons Director Instructor. They were primarily trained on C2 assets, capabilities, and tactics, along with AEDGE™ interface functions (30 hours). All of the participants were trained in three distinct C2 functional roles: Intelligence, Surveillance, and Reconnaissance (ISR); SWEEP; and STRIKE. The ISR role owned assets related to ISR functions, such as UAV. The STRIKE role owned assets such as air-to-ground bombers and airborne jammers, whereas the SWEEP role owned assets such as air-to-air fighter aircraft. Training consisted of initial description and discussion of roles and tactics, followed by hands-on training. These practice sessions began with interface functionality, then simple scenarios where each participant had all assets (thus lowering the need for coordinated action). These were followed by simple scenarios with more specialized, interdependent roles (ISR, SWEEP, and STRIKE) that progressed to more complex scenarios that were comparable to the experimental sessions. Their performance was closely monitored and guided by the subject-matter expert instructor. Debriefs after each scenario contributed greatly to learning.

Participants also learned how to adjust and effectively utilize the ergonomic chairs and workstations. The training also included administrative processing (1 hour) and training on the Automated Neuropsychological Assessment Metric (ANAM; Reeves, Winter, Kane, Elsmore, & Bleiberg, 2001) cognitive test battery to reach specified performance levels (9 hours). During the week of training, each participant wore a wrist actigraph, which records physical activity and is typically used to track sleep patterns. This enabled identification of participants who had erratic sleep patterns.

Procedure

The experimental session began at 6 p.m. on the last day of training (always a Friday) and ended at 11 a.m. the following morning. Each participant chose a specific role, which was constant throughout the session. During the session, they participated in eight 40-minute team-based C2 decision-making scenarios, with 20 additional minutes for scenario mission planning and debriefing. Every other hour, between each scenario session, they performed on the ANAM cognitive test battery that assesses reaction time, working memory, simple mathematical processing, and multitasking. At intervals through the experimental sessions, they provided oral temperature and self-reports on mood state and sleepiness.

ANALYSES AND RESULTS

Analysis of mission outcome scores indicated that teams performed least well during Session 6, improving somewhat for Sessions 7 and 8. This points to effects due to circadian rhythm as opposed to fatigue. Because this trend was evident after the second team, we decided to focus on coding and analysis of the first and sixth sessions. Table 21.1 provides descriptives and paired t-test results for each variable.

We expected that, in general, all task-related communication and encouragement would decrease, and references to fatigue would increase. All measurements differed in the expected direction. Four of the differences were not statistically significant, but this may be due to low statistical power (i.e., low N).

One unexpected finding was that both providing and requesting information on targets did not statistically decline between Sessions 1 and 6. We speculate that, given information regarding targets is essential for performance, team members maintained communication of most essential information when tired, and simply reduced all other types of communications (except expressions of fatigue and total information requests).

Despite the small sample size, several comparisons illustrated in Figure 21.2 were statistically significant, or approached significance. Expressions of fatigue increased, as expected ($p = .002$). Also, participants reported they felt sleepier in self-reports (means of 1.90 and 4.80, $p = .000$), so that we can assume the experimental manipulation did succeed at having them feel tired. Expressions of encouragement also declined (from a mean of 4.00 to 2.56, $p = 0.208$).

Multiple aspects of task-related communication also differed significantly. Requests for asset-related information decreased ($p = .049$), as did information on assets and strategy ($p = .002$). As illustrated in Figure 21.2,

TABLE 21.1
Descriptives and Paired-Sample t Tests
for Communication Variables by Session

Variable	Session	M	N	SD	t	df	p
Total Communications	1	164.778	9	72.057	3.410	8	
	6	108.333	9	56.628			.009**
Total Task	1	160.333	9	68.642	4.102	8	
Communications	6	92.222	9	50.734			.003**
Provide Information	1	69.556	9	24.136	2.313	8	
	6	48.111	9	23.846			.049**
Provide Target	1	25.333	9	9.962	1.024	8	
Information	6	22.444	9	12.300			.336
Provide Asset	1	44.222	9	17.050	2.686	8	
Information	6	25.667	9	14.739			.028**
Request Information	1	20.556	9	11.359	1.799	8	
	6	13.778	9	10.183			.110
Request Target	1	3.444	9	2.068	−.924	8	
Information	6	4.778	9	5.740			.383
Request Asset	1	17.111	9	9.880	2.324	8	
Information	6	9.000	9	5.937			.049**
Strategy	1	70.222	9	37.439	5.043	8	
	6	30.333	9	18.934			.001**
Information on Assets	1	131.556	9	58.728		8	
and Strategy	6	65.000	9	37.383			.002**
Encouragement	1	4.000	9	3.937	1.368	8	
	6	2.556	9	3.321			.208
Fatigue	1	0.444	9	0.727	−4.523	8	
	6	13.556	9	8.748			.002**

$*p < .10.$ $**p < .05.$

the first variable represents the total number of communications regarding assets and strategy. Total communications decreased ($p = .009$), as did total requests for information ($p = .110$), total provides for information ($p = .049$), and total task-related communications ($p = .003$).

The trends shown in the data are further illustrated in Figure 21.3, which represents the total of all communications, the total of all task-related communications, total information regarding assets and strategy, total provides, total requests, and total target information. It is interesting that whereas task-related communication decreased significantly, the total requests and total target information did not change as much. This further indicates that communications were focused on target information when participants were tired.

Communication variables were examined for relation to broad outcome variables. The mission outcome variables in Table 21.2 reflect outcomes based on the number of assets remaining at the end of each mission: (a)

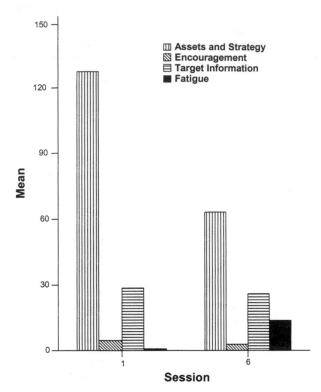

FIG. 21.2. Number of communications related to (a) assets and strategy, (b) encouragement, (c) target information, and (d) expressions of fatigue, at Sessions 1 and 6.

the number of hostile assets killed by friendly assets and (b) the number of friendly assets killed by hostile assets.

Paired *t* tests indicated statistically significant differences between means for both outcome variables. Hostiles killed by friendlies decreased (p = .005), whereas friendlies killed by hostiles increased (p = .079). These are the types of differences expected with increased fatigue (i.e., lower performance in the later session).

For Session 1, the two outcome variables had a correlation of –.689 (p = .040). As shown in Table 21.3, none of the correlations between the two outcome variables and the communication variables were significant in Session 1. With the range restriction evidenced by having a mode of 0 and a maximum of 2 for self-reports of fatigue, there was little fatigue in Session 1. A .581 correlation (p = .101) between level of fatigue and hostiles killed by friendlies was the correlation closest to significance.

For Session 6, the two outcome variables had a correlation of –.728 (p = .026). Table 21.4 shows that provide target information was the only com-

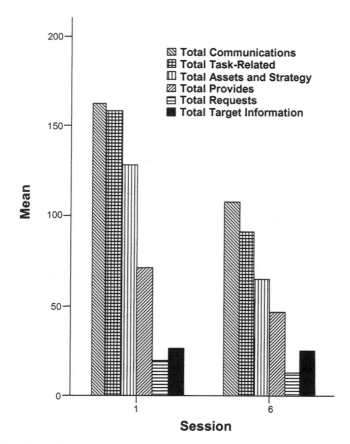

FIG. 21.3.   Total communications and task-related communications at Sessions 1 and 6.

munication variable that was significantly correlated with hostiles killed by friendlies for this session ($p = .081$). In contrast, five of the correlations between the communication variables and friendlies killed by hostiles were significantly negative. Provide and request variables that concerned target communications were significant. In addition, total communications, provide information, and encouragement were significant.

DISCUSSION

Implications outside the military arise from this chapter. Long working hours causing signs of worker burnout are increasingly prevalent. It is becoming more common for employees to take less time off from work. Due in part to cell phones, e-mail, and remote-access voice mail, workers are in-

TABLE 21.2
Descriptives and Paired-Sample *t* Tests for Overall Mean
Outcomes of Hostile Assets Killed by Friendly Assets
and Friendly Assets Killed by Hostile Assets by Session

Variable	Session	M	N	SD	t	df	p
HKBF	1	43.556	9	5.854	3.879	8	
	6	32.444	9	6.044			.005**
	Total	38.000	18	8.124			
FKBH	1	4.778	9	3.930	−2.016	8	
	6	8.222	9	4.684			.079*
	Total	6.500	18	4.554			

Note. HKBF = hostile assets killed by friendly assets; FKBH = friendly assets killed by hostile assets.
$*p < .10.$ $**p < .05.$

TABLE 21.3
Correlations Between Communication Variables
and Outcome Variables for Session 1

	HKBF	FHBH
Total Communications	.415	−.245
Total Task Communications	.405	−.247
Provide Information	.441	−.133
Provide Target Information	.236	−.078
Provide Asset Information	.486	−.143
Request Information	.322	−.207
Request Target Information	−.054	−.032
Request Asset Information	.381	−.231
Strategy	.361	−.305
Information on Assets and Strategy	.435	−.275
Encouragement	.418	−.129
Fatigue	.581.	−.224

Note. HKBF = hostile assets killed by friendly assets; FKBH = friendly assets killed by hostile assets.
$*p < .10.$ $**p < .05.$

creasingly tethered to their work. American productivity is up, but so is the financial cost of coping with worker stress (Robinson, 2003).

The primary contribution of this chapter to issues of concern within this book, and to the field of NDM and cognitive engineering in general, is not the results regarding fatigue effects per se, but rather the methodology used to enable the study of team communication and performance in a complex naturalistic setting (Chaiken et al., 2004). This methodology transfers to the investigation of numerous issues of concern to NDM—be it cognitive analysis techniques, modeling of decision making, investigation and training of critical thinking, or applied issues related to equipment design.

TABLE 21.4
Correlations Between Communication Variables
and Outcome Variables for Session 6

	HKBF	FKBH
Total Communications	.375	−.631*
Total Task Communications	.317	−.566
Provide Information	.465	−.634*
Provide Target Information	.611*	−.826**
Provide Asset Information	.243	−.336
Request Information	.111	−.429
Request Target Information	.230	−.644*
Request Asset Information	−.031	−.112
Strategy	.204	−.489
Information on Assets and Strategy	.194	−.398
Encouragement	.428	−.587*
Fatigue	.385	−.507

Note. HKBF = hostile assets killed by friendly assets; FKBH = friendly assets killed by hostile assets.

$*p < .10.$ $**p < .05.$

The primary challenge to achieving this study was the creation of realistic and demanding scenarios reflecting expert performance in an operational setting. This is the foundation from which all inferences can be made. One cannot study what is not properly elicited, and for the study of expert performance, realism is key. Our task was made easier in that the operational task is basically computer based. Even so, scenarios must reflect the same task and cognitive demands. Certainly one can create scenarios where even novices would excel, or where experts would fail. They may reflect operational realism, but would not allow study of the cognitive processes of interest. Here, mission scenarios were developed systematically to (a) reflect operational credibility (i.e., they were constructed, reviewed, and approved by operational Weapons Directors), (b) elicit high but not unreasonable workload (e.g., a proficient operator, with effort, could do well), and (c) elicit the same cognitive and decision-making processes. We described the process of scenario development in detail. Our challenge was made even greater in that several different scenarios had to be developed that were equivalent. In this situation, it was to investigate performance over time. However, the same issues apply to investigation of performance in general—one cannot repeat the same scenario for different manipulations, yet one must achieve equivalence in task demands.

A second methodological contribution regards assessment of communication, decision making, and performance in this context. Particular effort was taken to create scenarios with specific decision-making events that are linked to performance outcomes. The simulation utilized for study tracked all operator actions, and algorithms were constructed to elicit measures from this data log. The functional coding of communications was chosen to be consistent with existing taxonomies of communication, and also to generate hypotheses specific to the operator context (e.g., providing and requesting information can be directly linked to awareness and decision making) and to expected effects of fatigue (e.g., less explicit coordination). The level of effort and attention to experimental control enables present authors to argue that success was met in eliciting performance that was complex, naturalistic, and operationally relevant, while maintaining equivalence in workload and minimizing practice effects.

The transcription and coding of communications were achieved through much systematic effort, with very high interrater reliability (over 95% after training). This speaks to the clarity of the communication coding scheme. In addition, the communication scheme allowed researchers to track information exchange that was both explicit (responses to requests for information) and implicit (provision without request). At the same time, the classification of content, with regard to decision cues versus communication, related more directly to teamwork (e.g., sequencing of activities). This communication appeared to be more affected by fatigue, which indicated a "tunneling in" effect.

Though great care was taken with regard to development of scenario, measures, simulation platform, and expertise/training of operators, this study had limitations typical of a more operational field study. Our report is based on nine teams ($N = 9$), a low sum for degrees of freedom. Though scenarios were very carefully constructed, insights were gained during the study in terms of how to improve timing and sequencing of task demands and capture of operator actions. For example, in follow-up studies we would want even greater demand for coordination with regard to time pressure, so that consequences of failure to coordinate will be more immediate and certain. We would also want more uncertainty with regard to unexpected events. Our care in creating equivalent scenarios may have contributed toward operator expectations regarding timing of events and workload management.

Insights with regard to fatigue effects are limited but promising. Results indicate trends that will be further investigated. It is very likely that the physiological state and consequences of fatigue were not fully elicited. Participants were young, healthy, and very motivated to do well. The loss of one night of sleep did not appear to affect them as strongly as it affected the more senior researchers who observed them, even though the researchers worked in shifts. Fatigue research usually involves a stronger manipulation, often involving sleep deprivation over several nights. Follow-on studies will increase the sleep deprivation by (a) having participants experience some sleep deprivation over several nights (e.g., 4 hours per night) and/or (b) extending the period of continuous performance.

Though the sleep loss manipulation may need to be stronger, preliminary results do indicate the need for further investigation. It was evident that researchers did elicit participants' subjective experience of fatigue, based on communications and outcome measures (see Table 21.1 and Figure 21.2). In addition, researchers found evidence of fatigue in participants' communication patterns and performance. Participants definitely communicated less when tired. However, they maintained communications about certain aspects of information, such as core critical information regarding hostile targets. Reductions in communication had to do with discussion of their assets and coordinating strategies, information that was not as critical to mission success. It is speculated that team members maintained communication regarding targets even when tired because it was essential for performance, and simply reduced all other communications (except expressions of fatigue).

The reduction in communication regarding coordinating strategies and the assets of other team members, indicated possible detriments to team processes, particularly as mission outcomes were reduced. On the other hand, reduced communications, as indicated by higher levels of implicit coordination, have been associated with better performance in C2 situations (Elliott, Dalrymple, & Neville, 1997; Serfaty et al., 1994). When a team is providing task-related information without requests, it is an indication of implicit coordination, and this has been related to high-performing teams (Serfaty & Entin, 1995; Serfaty et al., 1994). As teams in a study by Elliott et al. became more practiced, they learned to provide critical information and execute coordinating strategies with less effort and explicit communication. If this were so in this study, the streamlined process may have reduced the effect of fatigue on performance. This would be consistent with findings and the argument that in some cases, stress can enhance performance (Klein, 1996).

Certainly, change in communications is an area for further study. It has been argued and demonstrated that stress can be an effective context for training (Driskell & Johnston, 1998; Johnston & Cannon-Bowers, 1996). Perhaps in the same way, performance in fatiguing conditions can enhance or accelerate learning of team coordination. In this study, participants had in-depth training of coordinating strategies. This kind of training has been demonstrated effective for performance under stress (Serfaty, Entin, & Johnston, 1998).

Our findings describe diminished communications during sustained operations, which may inhibit the sharing of expertise between team members. As team members become fatigued, they seek and share less information. Those who need to access the expertise of others are less likely to do so, and those who have expertise are less likely to disseminate it. Thus, teams whose collective members may have the requisite knowledge and expertise to succeed in their task may fail in their endeavors, due to the fact that the knowledge and expertise of its members are not matched to the demands of the task. Heterogeneity in the expertise of team members, which may normally not inhibit team functioning, may become a problem when this expertise is not ported to the team level. These findings suggest the potential negative results of diminished team communications when expertise is distributed.

Our current goals are to model and predict, to the extent possible, what kinds of error will occur, by whom, when, and why. An underlying goal is the identification and development of countermeasures to ameliorate the effects of fatigue. These will likely focus on training, information visualization, decision aiding, and sleep/wake management. Potential research topics are numerous. We hope that this chapter communicates the operational relevance, methodological challenges, and implications of team communication performance research under conditions of sustained operations.

REFERENCES

Artman, H, & Granlund, R. (1998). Team situation awareness using graphical or textual databases in dynamic decisionmaking. In T. R. G. Green, L. Bannon, C. P. Warren, & J. Buckley (Eds.), *Proceedings of European Association of Cognitive Ergonomics (ECCE), European Conference on Cognitive Ergonomics* (Vol. 9, pp. 151–156). Limerick, Ireland: University of Limerick.

Bales, R. F. (1950). *Interaction process analysis: A method for the study of small groups.* Reading, MA: Addison-Wesley.

Bonnet, M. H. (2000). Sleep deprivation. In M. Kryger, T. Roth, & W. Dement (Eds.), *Principles and practices of sleep medicine* (pp. 53–68). Philadelphia: W. B. Saunders.

Bowers, C., Salas, E., Prince, C., & Brannick, M. (1992). Games people play: A method for investigating team coordination and performance. *Behavior Research Methods, Instruments, and Computers, 24*(4), 503–506.

Brehmer, B., & Dorner, D. (1993). Experiments with computer-simulated microworlds: Escaping both the narrow straits of the laboratory and the deep blue sea of the field study. *Computers in Human Behaviour, 9,* 171–184.

Cannon-Bowers, J., & Salas, E. (1998). Individual and team decision making under stress: Theoretical underpinnings. In J. Cannon-Bowers & E. Salas (Eds.), *Making decisions under stress: Implications for individual and team training* (pp. 17–38). Washington, DC: American Psychological Association.

Chaiken, S., Elliott, L., Barnes, C., Harville, D., Miller, J., Dalrymple, M., et al. (2004). Do teams adapt to fatigue in a synthetic C2 task? In *Proceedings of the 9th Command and Control Research and Technology Symposium.* Washington, DC: CCRP Press. Retrieved August 11, 2005, from http://dodccrp.org/events/2004/CCRTS_San_Diego/CD/papers/121.pdf

Cohen, M. S. (1993). The naturalistic basis of decision biases. In G. Klein, J. Orasanu, R. Calderwood, & C. Zsambok (Eds.), *Decision making in action: Models and methods* (pp 51–99). Norwood, NJ: Ablex.

Colquitt, J. A., Hollenbeck, J. R., Ilgen, D. R., LePine, J. A., & Sheppard, L. (2002). Computer-assisted communication and team decision-making accuracy: The moderating effect of openness to experience. *Journal of Applied Psychology, 87,* 402–410.

Coovert, M., Riddle, D., Gordon, T., Miles, D., Hoffman, K., King, T., et al. (2001). The impact of an intelligent agent on weapon directors behavior: Issues of experience and performance. In *Proceedings of the 6th International Command and Control Research and Technology Symposium.* Washington, DC: CCRP Press. Retrieved August 11, 2005, from http://dodccrp.org/events/2001/6th_ICCRTS/Cd/ Tracks/Papers/Track4/033_tr4.pdf

Driskell, J., & Johnston, J. (1998). Stress exposure training. In J. Cannon-Bowers & E. Salas (Eds.), *Making decisions under stress: Implications for individual and team training* (pp. 191–217). Washington, DC: American Psychological Association.

Eddy, D. R., & Hursh, S. R. (2001). *Fatigue avoidance scheduling tool* (AFRL-HE-BR-TR-2001-0140). Brooks City-Base, TX: Air Force Research Laboratory.

Elliott, L., Coovert, M., Barnes, C., & Miller, J. (2003). Modeling performance in C2 sustained operations: A multi-level approach. In *Proceedings of the 8th International Command and Control Research and Technology Symposium.* Washington, DC: CCRP Press. Retrieved August 11, 2005, from http://www.dodccrp.org/events/2003/8th_ICCRTS/pdf/023.pdf

Elliott, L. R., Dalrymple, M. A., & Neville, K. (1997, September). *Assessing performance of AWACS command and control teams.* Paper presented at the Human Factors and Ergonomics Conference, Albuquerque, NM.

Hicks, J., Stoyen, A., & Zhu, Q. (2001). Intelligent agent-based software architecture for combat performance under overwhelming information inflow and uncertainty. In *Proceedings of the International Conference on Engineering of Complex Computer Systems* (Vol. 7, pp. 200–210). Piscataway, NJ: IEEE Computer Society.

Hollenbeck, J. R., Ilgen, D. R., Tuttle, D. B., & Sego, D. J. (1995). Team performance on monitoring tasks: an examination of decision errors in contexts requiring sustained attention. *Journal of Applied Psychology, 80*(6), 685–696.

Hursh, S. R. (1998). *Modeling sleep and performance within the integrated unit simulation system (IUSS)* (Natick/TR-98/026L). Natick, MA: United States Army Soldier Systems Command; Natick Research, Development and Engineering Center.

Hutchins, S., Hocevar, S., & Kemple, W. (1999). Analysis of team communications in "human-in-the-loop" experiments in joint command and control. In *Proceedings of the 4th Command and Control Research and Technology Symposium.* Washington, DC: CCRP Press. Retrieved August 11, 2005, from http://www.dodccrp.org/events/1999/1999CCRTS/pdf_files/track_1/101hutch.pdf

Johnston, J., & Cannon-Bowers, J. (1996). Training for stress exposure. In J. Driskell & E. Salas (Eds.), *Stress and human performance* (pp. 223–256). Mahwah, NJ: Lawrence Erlbaum Associates.

Kanki, B., & Foushee, C. (1989). Communication as a group process mediator of aircrew performance. *Aviation, Space, and Environmental Medicine, 60,* 402–410.

Klein, G. (1993). A recognition-primed decision (RPD) model of rapid decision making. In G. Klein, J. Orasanu, R. Calderwood, & C. Zsambok (Eds.), *Decision making in action: Models and methods* (pp. 139–147). Norwood, NJ: Ablex.

Klein, G. (1996). The effects of acute stressors on decisionmaking. In J. Driskell & E. Salas (Eds.), *Stress and human performance* (pp. 49–88). Mahwah, NJ: Lawrence Erlbaum Associates.

Kryger, M., Roth, T., & Dement, W. (2000). *Principles and practices of sleep medicine* (3rd ed.). Philadelphia: Saunders.

Mahan, R. P. (1992). Effects of task uncertainty and continuous performance on knowledge execution in complex decision making. *International Journal of Computer Integrated Manufacturing, 5*(2), 58–67.

Mahan, R. P. (1994). Stress induces strategy shifts toward intuitive cognition: A cognitive continuum framework approach. *Human Performance, 7*(2), 85–118.

Mahan, R. P., Elliott, L. R., Dunwoody, P. T., & Marino, C. J. (1998). The effects of sleep loss, continuous performance, and feedback on hierarchical team decision making. In *Proceedings of RTA-NATO Conference on Collaborative Crew Performance in Complex Operational Systems RTO-MP-4* (pp. 29/1–29/8). Neuilly-sur-Seine, France: Research and Technology Organization.

Mitchell, T. R., & Beach, L. R. (1990). ". . . Do I love thee? Let me count . . ." Toward an understanding of automatic decision making. *Organizational Behavior and Human Decision Processes, 417,* 1–20.

Mulder, I., & Swaak, J. (2000). *How do globally dispersed teams communicate? Coding technology-mediated interaction processes* (Report No. TI/RS/2000/040). Enschede, Netherlands: Telematica Instituut. Retrieved June 30, 2000, from https://extranet.telin.nl/docuserver/dscgi/ds.py/Get/File-9069).

Orasanu, J., & Connolly, J. (1993). The reinvention of decision making. In G. Klein, J. Orasanu, R. Calderwood, & C. Zsambok (Eds.), *Decision making in action: Models and methods* (pp. 3–20). Norwood, NJ: Ablex.

Orasanu, J., & Salas, E. (1993). Team decision making in complex environments. In G. Klein, J. Orasanu, R. Calderwood, & C. Zsambok (Eds.), *Decision making in action: Models and methods* (pp. 327–345). Norwood, NJ: Ablex.

Petrov, P., & Stoyen, A. (2000). An intelligent-agent based decision support system for a complex command and control application. In *Proceeding of the IEEE International Conference on Engineering of Complex Computer Systems* (Vol. 6, pp. 94–104). Piscataway, NJ: IEEE Computer Society.

Rasmussen, J. (1993). Deciding and doing: Decision making in natural contexts. In G. Klein, J. Orasanu, R. Calderwood, & C. Zsambok (Eds.), *Decision making in action: Models and methods* (pp. 158–171). Norwood, NJ: Ablex.

Reeves, D., Winter, K., Kane, R., Elsmore, T., & Bleiberg, J. (2001). *ANAM 2001 user's manual* (Special Report No. NCRF-SR-2001-1). San Diego, CA: National Cognitive Recovery Foundation.

Robinson, J. (2003). *Work to live: The guide to getting a life.* New York: Berkley.

Serfaty, D., & Entin, E. E. (1995). Shared mental models and adaptive team coordination. In *Proceedings of the International Symposium on Command and Control Research and Technology* (Vol. 1, pp. 289–294). Washington, DC: CCRP Press.

Serfaty, D., Entin, E. E., & Deckert, J. C. (1994). Implicit coordination in command teams. In A. H. Levis & I. S. Levis (Eds.), *Science of command and control: Part III coping with change* (pp. 87–94). Fairfax, VA: AFCEA International.

Serfaty, D., Entin, E, & Johnston, J. (1998). Team coordination training. In J. Cannon-Bowers & E. Salas (Eds.), *Making decisions under stress: Implications for individual and team training* (pp. 211–246). Washington, DC: American Psychological Association.

Chapter 22

Training for Multinational Teamwork

Helen Altman Klein
Wright State University

Debra Steele-Johnson
Wright State University

U.S. military personnel are among the best trained and equipped in the world. Recent global changes have, however, entailed new missions (see Brown, chap. 17, this volume). For example, the U.S. military was asked to take on the challenges presented by multinational peacekeeping operations forces at the Headquarters of NATO's Stabilization Forces (HQ SFOR) in Bosnia-Herzegovina and at many other world trouble spots. Beyond the difficulties introduced by nation building in the face of ethnic conflict, military personnel must work with individuals and groups who differ in behavior, social role expectations, and cognition. The research reported here describes the development of a prototype for training that acknowledges the situational constraints imposed in these difficult settings.

In the past, the U.S. military occasionally assumed leadership in multinational operations or worked in coalitions with well-marked responsibilities requiring little direct collaboration. Reflecting its main agenda, the U.S. military prepared personnel for unilateral action against long-term adversaries during traditional warfare. The last decade has brought a shift toward multinational operations for both adversarial and humanitarian missions. Multinational collaboration has economic, technical, and political advantages, but it also introduces complexity and difficulties. Partners can have different operational norms, leadership expectations, values, interpersonal styles, and cognition. These differences permeate many aspects of organizational behavior, including teamwork (e.g., Adler, 1997; Granrose & Oskamp, 1997; Helmreich & Merritt, 1998; Lane & DiStefano, 1992) and can impede joint efforts that require coordination, judgment, and authority relationships (e.g., Thomas, 1999). National differences can hinder collaboration and the expression of expertise during complex multinational operations such as peacekeeping.

The research described in this chapter was directed at the challenges experienced during multinational peacekeeping. Military personnel need to understand and anticipate the differences in judgment and decision making encountered in these multinational operations. This is important when multinational teams must make high-stakes decisions under time pressure and stress. It is easy to assume that others think as we do, but this assumption can lead to misunderstanding, frustration, and mission failure. In the second section of this chapter, we look at the contributions made by cognitive engineering toward understanding performance in complex domains. We then use the Cultural Lens Model (CLM) to describe how national differences influence cognition and how these differences can disrupt collaboration. Finally, we describe the rationale for this research. Recognizing the pressures facing multinational peacekeeping forces, we first developed and tested prototype training material with business students in a laboratory setting. The success of this initial work led to support from the U.S. military and access to field personnel. We then developed a scenario-based training prototype to overcome the challenges of national differences presented during multinational collaboration at HQ SFOR.

COGNITIVE ENGINEERING IN COMPLEX DOMAINS

Cognitive engineers have helped develop technologies that enable practitioners to perform more effectively. Cognitive Task Analysis (CTA) is an interview technique aimed at discovering how people understand and perform complex tasks. The knowledge gleaned from CTA can provide a window into the thinking of domain practitioners. It has transformed design, procedural, and training applications (e.g., Klein, Orasanu, Calderwood, & Zsambok, 1993).

As cognitive engineering applications have proven their worth, they have been applied to multinational issues and settings. This has exposed a troublesome problem: People around the world differ in cognition and in the social context of cognition (e.g., Klein, 2004; Wisner, 2004). Applications based on careful and comprehensive research in one nation are not always effective in other nations (Kaplan, 2004).

Initial attempts to address the problem of national differences in performance focused on differences in customs and social behaviors (e.g., Kluckhohn & Strodtbeck, 1961; Triandis, 1994). Many guidebooks provide advice on customs and behavior (e.g., Morrison, Conaway, & Borden, 1994).

Early acknowledgment of the importance of national and cultural variables for ergonomics, for example, mentioned anthropometry, language, physiology, and customs but mentioned psychology only with regard to simple learning technology and perception. These differences are important but not sufficient. National differences in thinking and reasoning styles also create formidable barriers to understanding and collaboration. These cognitive differences are less visible and have received less attention than differences in behavior, language, and customs. This neglect, no matter how benign, has created gaps in the ability of practitioners to function effectively in multinational settings.

Peacekeeping operations present complex cognitive challenges that must be met. Even highly competent and dedicated military personnel can have difficulty working with their equally competent and dedicated counterparts from other nations. The difficulties extend beyond differences in equipment and regulations to the underlying cognitive processes that differentiate national groups. To address the gaps and help insure effective performance in multinational environments, we look to the CLM (Klein, Pongonis, & Klein, 2000). This model describes the nature and origins of differences in cognition and in the social context of cognition, providing a framework for the impact of culture and an approach for increasing performance effectiveness.

THE CULTURAL LENS MODEL

The CLM starts with the assumption that all people are born with similar endowments and potentials. These include the capacity for behavioral plasticity in response to the developmental context (Agrawal, 2002). Accumulated early experiences in families and in the broader social and physical world shape the way people think, relate, and act (e.g., Berry, 1986; Segall, Dasen, Berry, & Poortinga, 1990; Triandis, 1994). These experiences provide a "lens" that focuses the beliefs and values a person holds, as well as the nature of a person's self-construct. A person's cultural lens shapes his or her emotional reactions, relationships with other people, and cognition. Because people from the same culture are likely to share early experiences and socialization, they are likely to share a worldview. This is not to say that culture can predict individual patterns. Rather, culture describes useful tendencies and reflects ways in which experience and socialization shape reasoning, judgment, and beliefs. These common tendencies help members of a group interpret and react to each another.

During multinational interactions, practitioners may encounter others who have a worldview that differs from theirs. The present research identified differences that influence expectations and behaviors during collaboration and crafted training for increasing task-related cultural understanding. The model of a cultural lens captures the need to "decenter"— to see as if through the worldview of a person from another culture. When people can decenter, they are better able to understand and interact with different people.

Before practitioners can accommodate differences, they must see and understand them. To apply the CLM, we must first identify the dimensions that distinguish groups and that can compromise mission effectiveness. Cultures may differ, both subtly and drastically, on many important dimensions. Triandis (1994) suggested 19 dimensions associated with perceptual differentiations, information use, and evaluation. Wise, Hannaman, Kozumplik, Franke, and Leaver (1997) identified 919 behaviors that together provide a structured, topical taxonomy of cultural information (e.g., understand/recognize/react appropriately to religious activities, gestures, members of the opposite sex, authority, sense of private space, and greetings). Salas, Burke, and Wilson-Donnelly (2004) catalogued over 40 cultural dimensions of values, attitudes, and behaviors covering a range of human differences. These extensive lists, gleaned from the research literature, show limited overlap. We used a naturalistic approach to choose mission critical dimensions from those proposed.

The dimensions included in the CLM (Klein, 2004) are ones that emerged during observations of practitioners in complex domains and with multinational personnel and/or challenges. Domains have included civil aviation (Klein, Klein, & Mumaw, 2001a, 2001b), multinational collaborations (Klein & McHugh, 2005), military command and control (Klein et al., 2000), and conceptual modeling of foreign command decision processes (Lannon, Klein, & Timian, 2001). A concern of these studies has been with discovering the source of functional difficulties and identifying conceptual links and overlaps. Whereas many dimensions have been observed, a subset that appears most frequently in the research is provided in Table 22.1.

The list not is exhaustive and has two caveats:

1. Because different domains have different task demands and different task demands may be sensitive to different dimensions, observations in other domains may reveal additional dimensions, and
2. Because groups of people may differ in as yet unexplored ways, additional dimensions may be needed to describe interactions with yet unstudied groups.

Table 22.1 lists dimensions that emerged from observations in a considerable but finite set of domains with a broad but finite representation of national groups. Three of the dimensions are especially important for the research reported here and are described next in more detail.

TABLE 22.1
Representative Cultural Lens Model Dimensions

Dimension	Definition	Reference
Achievement vs. Relationship	Prime focus of work	Kluckhohn & Strodtbeck, 1961
Time Horizon	Planning for short- vs. long-term goals	Kluckhohn & Strodtbeck, 1961
Mastery vs. Fatalism	Assumption of personal efficacy vs. external control	Kluckhohn & Strodtbeck, 1961
Independent vs. Interdependent	Concept of self and assumed unit of analysis	Markus & Kitayama, 1991
Power Distance	Acceptance/expectation of power distribution	Hofstede, 1980
Tolerance for Uncertainty	Comfort vs. stress with uncertainty	Hofstede, 1980
Hypothetical vs. Concrete Reasoning	Reasoning; dependent on past cases or on future projections	Markus & Kitayama, 1991
Holistic vs. Analytic Thinking	Incorporates a broad range of contextual considerations or focuses on key component	Nisbett, 2003
Attribution	Cause attributed to dispositional vs. situational factors	Choi, Nisbett, & Norenzayan, 1999
Synthesizing vs. Contrasting	Solutions reached by integrating many ideas or by selecting the best approach	Peng & Nisbett, 1999

The *Tolerance for Uncertainty* dimension reflects a person's reaction to uncertainty and ambiguity. Whereas some people experience uncertainty as stressful and take action to avoid it, others are comfortable with and accepting of it (Hofstede, 1980). People with low Tolerance for Uncertainty may follow ineffective rules to alleviate the associated emotional discomfort of uncertainty. They show fear of failure and are troubled by a lack of information. The lower the tolerance, the greater tends to be the desire for specified plans. The lower the tolerance, the longer the time it takes to make and implement plans. In contrast, those with high Tolerance for Uncertainty are comfortable with ambiguity and incomplete information. This is reflected in their planning. Tolerance for Uncertainty affects how ready a person is to respond and adapt to unexpected changes (Hall & Hall, 1990; Helmreich & Merritt, 1998; Lane & DiStefano, 1002).

The *Holistic–Analytic Thinking* dimension is reflected in attribution and in synthesizing versus contrasting differences (Nisbett, Peng, Choi, & Norenzayan, 2001). They are consistent with Confucian versus Aristotelian logic. On the one hand is a tendency to see things as integrated, related, and inseparable; on the other is the tendency to draw distinctions and clear-cut boundaries of cause and effect. Attribution describes the way causality is assigned—situationally or dispositionally (Choi, Nisbett, & Norenzayan, 1999; Ji, Peng, & Nisbett, 2000). Holistic thinkers look at the broader contextual forces and contributions. In explaining behavior or events, holistic thinkers see responsibility in the interactions between objects or people and the situation. In contrast, analytic thinkers tend to see the dispositions of objects or people as the primary causal agents (Choi et al., 1999). Holistic and analytic thinkers may identify different problems in the same situation.

Holistic and analytic thinkers also tend to differ in synthesizing versus contrasting. This describes the customary approach to making sense of complex and contradictory information (Chu, Spires, & Sueyoshi, 1999). Synthesizers evaluate ideas by seeking connectedness. They believe that all perspectives contain some truth. Seeming contradictions are resolved by reconciling perspectives (Ji et al., 2000; Peng & Nisbett, 1999). Contrasting thinkers manage seeming contradictions by separating components and evaluating their individual qualities. They tend to polarize perspectives and select the one they believe to be best (Peng & Nisbett, 1999). Synthesizing and contrasting are different paths to resolving conflicts and making decisions. Differences in reasoning can hinder a team as it seeks resolution.

The *Achievement–Relationship* dimension describes the way in which members of a group approach life, work, and relationships. It has implications for cognition in the areas of planning, decision making, and implementation. Those with an Achievement orientation view their work as the primary and desired focus of activity and their accomplishments as their defining features. They strive to achieve and emphasize accomplishments that are concrete and measurable. They place more emphasis on performance than on the overall team experience. Cultures with a Relationship orientation view social relationships as a person's defining features. Generally, they place more emphasis on learning from the experience, nurturing interpersonal relationships, and growing as a person. This means they are comfortable allowing change to happen without rushing or pushing (Adler, 1997; Kluckhohn & Strodtbeck, 1961).

National culture is important because it provides a functional blueprint for individual behavior, social functioning, and cognition. It provides rules for communication and guidelines for emotional expression. Culture provides cognitive tools for making sense out of the world. Because it is a part of the physical and social ecology, these common behavioral, social, and cognitive patterns confer a survival advantage. This concept of culture is consistent with Berry's ecocultural framework (Berry, 1986; Segall et al.,

1990) in that both describe culture as a dynamic system that adapts to changing contextual demands. Culture is an integrated rather than a haphazard collection of behaviors, social roles, and cognition, providing an integrated vision. Consequently, changes in the physical and social ecology can have repercussions throughout the cultural system. The interdependent and codefining nature of cultural components also means that some cultural elements regularly occur together. Industrial nations, for example, are likely to show social and cognitive similarities.

The dynamic nature of culture suggests that interactive methods might be best for training decentering—sensitizing people to alternate cognitive and social patterns. Children rely on experience, modeling, and perspective taking during development. They monitor people in order to learn how to predict behavior. They watch for clues of upcoming reactions and events so that they can respond. Children also take the perspective of others as they play "make believe." The dynamic mechanisms of childhood may also help adults extend their understanding of cognitive differences and expand their capacity to decenter in multinational interchanges.

IMPLICATIONS FOR TRAINING

Military personnel working in multinational settings may be most effective when they appreciate cultural variations in cognition and the social context of cognition. They also need to understand the impact of these differences on collaboration. Training based on the CLM might assist them in seeing the world through the eyes of someone from a different nation or culture. This decentering might help people appreciate how others perceive and reason, develop a working knowledge of differences, and generate effective approaches to interactions. The CLM was used as the basis of our scenario-based training, which was designed to increase understanding and improve the management of cultural differences that could impact the peacekeeping mission.

Two features may contribute to the usefulness of the CLM for practitioners in operational domains. First, each domain is expected to have only a small set of critical dimensions. Training can target these high-payoff dimensions. Second, specific national groups may differ on a limited set of cognition dimensions. The United States, for example, may match its European allies in most ways and differ in only two or three. Training can concentrate effectively on the dimensions on which groups diverge.

The development of training reported here was intended to provide the knowledge needed to identify dimensions that are barriers at HQ SFOR, to predict future actions in order to influence behavior, and to foster the acceptance and management of differences. Assuming perspective taking to be a mechanism for decentering, we hypothesized that perspective taking would serve also as a mechanism for acquiring the "corrective lens" needed to interact effectively in multinational teams. The long-term goal was to develop a training prototype for increasing effective international coordination and adaptation during multinational peacekeeping operations.

Peacekeeping operations present yet another difficulty: Professionals in natural settings have little time or patience for untested concepts because the stakes are too high. In the short term, it was necessary to accommodate the practical constraints imposed by the multinational peacekeeping environment. To overcome this barrier, a student sample was initially used in a laboratory setting. The work was extended then to a multinational peacekeeping setting. The research strategy described next moved from the laboratory to the field (see Hunt & Josslyn, chap. 2, this volume).

OVERVIEW OF THE RESEARCH

The research included four studies. The first two were designed to test the efficacy of scenario-based training for increasing awareness, acceptance, and accuracy in predicting key national/cultural differences. We started with business students in a classroom setting (Klein & Steele-Johnson, 2002). In Study 1, we used the Critical-Incident Technique (CIT) to identify difficulties faced by business, military, and nongovernmental organization (NGO) professionals during multinational interchanges. In Study 2, we used the CLM to develop, implement, and evaluate scenario-based decentering training for the difficulties and cognitive dimensions gleaned during Study 1. We addressed two influential cognitive differences—Tolerance for Uncertainty and Holistic–Analytic Thinking—with training designed to increase awareness, acceptance, and skill in predicting the actions of others. Together these studies assessed the scenario-based training and provided the foundation for fieldwork.

This laboratory-based training research provided the foundation for the next two studies with international military peacekeepers at HQ SFOR in Bosnia-Herzegovina (Hahn, Harris, & Klein, 2002). Study 3 used the CIT to identify cognitive differences troublesome to military personnel during international peacekeeping. The difficulties identified were Tolerance for Uncertainty and Achievement–Relationship. Study 4 used these dimensions to develop scenario-based training for the peacekeepers.

STUDY 1: LABORATORY DIMENSIONS ON SCENARIOS

To develop a CLM training protocol, we first needed to identify dimensions that present difficulties during intercultural interactions. The dimensions were identified from interviews with informants who had extensive experience outside their nation of birth. The interviews also yielded incidents that were used to create training scenarios. Because the training was to be conducted with business students, we needed incidents that would be interesting to the intended trainees.

Method

We interviewed 11 informants, each of whom had spent at least 3 years working, traveling, and/or residing outside the United States. As shown in Table 22.2, these informants included U.S. Army and Navy personnel with international field experience, former Peace Corps members with international business support experience, NGO personnel with field experience in public health and community development, and researchers who had been at institutions overseas. Three of the informants also had spouses from other nations.

Interviews for Studies 1 and 3 used CIT, often focusing on a specific incident to elicit information. We probed for difficulties experienced in incidents during extended intercultural work supervision, professional collaborations, or personal interchanges. We borrowed from CTA techniques to probe for the dynamics of the incident (Crandall & Getchell-Reiter, 1993; Hoffman, Crandall, & Shadbolt, 1998; Klein, Calderwood, & MacGregor, 1989). These interviews were designed to reveal what guided decision making and sensemaking during the incident. They also elicit important cues, choice points, options, and action plans, as well as details regarding the role of experience in judgment and decision making. Because of the training goal, the interviews focused on cultural differences. Informants described incidents during which they realized they could not just treat foreign nationals like U.S. citizens. We looked for examples of confusion, surprise, or misunderstandings that would not happen with U.S. citizens. We paid close attention to the sources of frustration and tension in these intercultural interactions. Of particular interest were times when the interviewee found new and successful approaches when interacting with foreign nationals. All interviews were recorded and transcribed.

Results of Study 1

Two judges independently reviewed all transcripts for incidents that expressed conflicts, discrepancies, or surprises during encounters between the U.S. interviewees and people from other nations. The judges scored the statements in terms of the cognitive dimensions that we described earlier in this chapter. A third judge reviewed incidents where the two judges disagreed.

TABLE 22.2
Description of Study 1 Interviewees

	Position/Title/ Organization	Experiences	National Groups
ROTC	Naval Officer ROTC Instructor	Peacekeeping training operations with multinational forces	Bosnia England
	Army Officer ROTC Instructor	Peacekeeping operations with multinational forces	Italy
Active Duty	Army Officer	Joint Operations	Japan Korea Panama England
	Army Officer	Joint Operations Liaison	Germany Saudi Arabia
	Army Officer	Joint operations with multinational forces	Korea
	Army Officer	NATO operations with multinational forces	Saudi Arabia Bosnia Hungary Turkey
International Organizations	Peace Corps International Development Organization	Consultant to small business Economic development worker MBA	Chile Togo
	Nongovernmental Organizations Private Voluntary Organizations or Governments	Health care delivery (RN)	Jordan Australia Bangladesh Angola Kosovo
Academic	Director, University Center for International Programs	Anthropological research Advising and supervision of foreign national students	North Africa France Scandinavia
	Director, Foreign Language Institute	Program development, supervision, and counseling for language students. External experience teaching or recruiting internationally	Europe Asia Africa South America
	University Center for International Education	Coordinate international programs Support international students in the United States Spouse and child of foreign national	Italy Germany Belgium Argentina

482

We found instances of all of the cognitive dimensions. We decided to focus on the dimensions of Tolerance for Uncertainty and Holistic–Analytic Thinking because these two presented particularly frequent problems during multinational interactions. They also represented different kinds of challenges during collaboration.

Examples from the transcripts illustrate how difficult it is for high–Tolerance for Uncertainty professionals who value flexibility, spontaneity, and last-minute decisions to work with low–Tolerance for Uncertainty people, who prefer firm, committed plans of action. It is difficult for those who want complete information to work with those who are comfortable with incomplete data. Contrast high–Tolerance for Uncertainty responses with low–Tolerance ones, in Table 22.3.

The statements in tows 1 and 2 of Table 22.3 show that the individuals from these two nations/cultures have very different concepts of teamwork.

TABLE 22.3
Example Study 1 Transcript Statements
That Show a National/Cultural Contrast

	High Tolerance for Uncertainty	*Low Tolerance for Uncertainty*
1	"They can drive you crazy with endless planning! When we get an assignment, we pick up and plan as we go."	"They are loose cannons. They don't take the time to figure out what they're doing before they set off to do it."
2	"The task was new so I wanted to talk to key people and plan as we learned. Details can wait. They wanted a plan set in concrete!"	"When our team meets, I wanted to develop detailed plans, define roles, and agree on a timeline from the start. They wanted to wing it!"
3	"They want to schedule a dinner out days ahead! We just get together and go. What's the big deal? These things always work out one way or another."	"Everything is last minute. So many times we've just stood around because we have no reservations and everyone wanted to do something different!"
	Holistic Reasoning Style	*Analytic Reasoning Style*
4	". . . I look at all the ideas and try to integrate the best of each. They pick one and ignore the others . . ."	"When you have a job to do you have to prioritize! Ignore unimportant information and focus on key items . . ."
5	"The first thing they do is find someone to blame. They never look at the big picture. They simplify everything and never really solve the problem."	"The first time, I was amazed! They considered every possible thing that might have contributed. It would not have surprised me if they had mentioned El Niño!"
6	"They simplify the situation and jumped to conclusions. Then they ignore everything else. What good does it do to solve a part of the problem? Why throw out a good option?"	"They can't focus or make a decision. They just put everything in the pot and talk on and on! We identify the options and decide."

In the incidents we recorded the conflicts generated considerable consternation. Even in social interactions, differences contributed to discord and frustration, shown in the two statements in tow 3 of Table 22.3.

Parallel conflicts exemplified differences in the components of Holistic–Analytic reasoning. The statements in row 4 of Table 22.3 show that individuals from these two nations/cultures have very different concepts regarding how to work effectively. Row 5 in Table 22.3 illustrates conflicts involving Attribution and Synthesizing versus Contrasting reasoning, which are aspects of Holistic–Analytic reasoning. Comments such as those in row 5 were common. The preferred approach to a solution is very different for the Dispositional attribution person versus the Situational attribution person. As the row 5 entries suggest, differences in Contrasting and Synthesizing also led to conflict. Interviewees reported that it was frustrating to work with people who differ. Examples are in row 6 of Table 22.3.

The interviews provided a rich set of incidents for each of these two cognitive dimensions. They also provided detailed descriptions of how the informant resolved the difficulty. The incidents and their points of conflict provided the basis for the scenarios developed for training.

Scenario Development

The construction and evaluation of scenarios was iterative. The interview transcripts included approximately 30 stories. From these stories, we identified those that captured conflicts and frustrations associated with each of the two dimensions. Incidents were rewritten as scenarios in which a business or management theme was emphasized. Aspects of the incidents were merged into the scenarios, for example, to capture the context of one incident and the conflict of another. Each scenario was written to present positions from each perspective of a dimension. Textboxes 22.1 and 22.2 present portions of scenarios used for Tolerance for Uncertainty and Holistic–Analytic Reasoning, respectively.

Two researchers, engaged in the study of cross-cultural cognition, reviewed the drafts of the scenarios. When they had questions about the expression of a dimension, the scenario was rewritten or discarded. A third researcher, also engaged in the study of cross-cultural cognition, affirmed the correctness of the final selections. The Tolerance for Uncertainty incidents were easy to identify showing high rater agreement. The Holistic–Analytic incidents were more difficult. For this reason, we confirmed the construct validity of the Holistic–Analytic incidents by consulting five individuals from East Asian nations considered to be holistic. Each informant read the scenarios and evaluated them with questions such as "What I would expect of people from my country?"

**Textbox 22.1.   Tolerance for Uncertainty Scenario:
How Much Is Enough?**

Sara was the Head of Purchasing of a manufacturing firm and ensured that the plant always had parts. She received compliments on her skills and financial recognition from the corporation. Sara's supervisor, the company's VP, then took another position. A new VP from outside the company was appointed. While Sara and the new VP both wanted a smooth transition, the issue of stockpiling supplies caused problems. The steel blanks that the workers milled came in various sizes. They came from different suppliers. The plant typically milled 1,000 blanks every 3 days. Sara had always ordered parts when the blank stock ran below 5,000. This ensured adequate raw material even with delays in shipping. She had successfully used this technique. Sara felt nervous without extra parts in case of unforeseen delays.

The new VP had come from an operation that used "just-in-time" supply acquisition, and approach to reduce inventory and increase room for new opportunities. He thought that keeping 5,000 blanks was excessive. He said that many parts manufacturers in the area were using a "just-in-time" system successfully. They stocked enough parts for a single day's production. The VP liked the flexibility of daily orders. He thought that even if delays occurred, production would not be significantly altered.

Sara felt that the new VP didn't seem to look ahead and didn't plan systematically. His policies could put production at risk if anything disrupted supplies. As she described it, "I don't mind change, but it makes me anxious if it isn't carefully considered based on data. The corporation shouldn't rush into change!" The disagreement really made her very uneasy.

The new VP felt things were getting off to a good start. He felt that he could work with Sara although she seemed rigid. He felt that the new methods were worth a try. Sara seemed to need rules for everything. He was prepared to proceed slowly and even switch back if the new methods didn't work. It didn't seem like a big deal to try a new approach.

Input from these informants allowed revisions of the scenarios for inclusion in the training material. The scenarios were included only if the researchers and the East Asian informants agreed on their typicality and their interpretations (i.e., the content validity).

Final evaluations for interest, understandability, and readability were undertaken with undergraduate psychology students from the United States. First, each of 14 students read and rated three or four of eight scenarios for interest. They also reported their understanding of the scenarios so we could gauge comprehension. Based on their ratings and reports, the sce-

---

Textbox 22.2. Analytic–Holistic Scenario: Earthquake.

The U.S. Army was sent on an international humanitarian mission to help after a devastating earthquake. Fallen buildings still harbored victims, food and shelter needs were urgent, and medical and sanitation services were essential to prevent disease. The U.S. commander asked two aides to each develop a plan for rescue, rebuilding and support. One plan was community centered and one task centered. One aide had seen relief efforts fail because community needs were not addressed, causing alienation. His plan used autonomous teams in each affected community to provide each household with all needed services. The other aide had seen poor technical understanding of complex tasks end in failure. Her plan organized the work by task with separate teams engaging in each needed task nationwide. The medical team, for example, included public health experts, nurses, and physicians who coordinated service across the nation. Each household would have the benefit of highly skilled people. Personnel could move to handle shifting needs.

The commander's staff reviewed the plans to be sure that each covered all needed services and support. They determined that resources were adequate to undertake each plan. The commander was pleased that he could offer local authorities a real choice. To the commander's dismay, the local leader didn't decide between the plans. He thought both plans were great and saw no need to pick between the advantages of community autonomy and specialized teams. He felt the plans were not really different or incompatible and could be combined into an integrated plan.

The American commander saw this pieced-together plan as unnecessarily complex and disjointed. He was disturbed that the local leader didn't just weigh advantages and disadvantages and decided what was best. Both plans could do the job and selecting one would have allowed the work to begin immediately.

The local leader was surprised with the reaction. He explained, "Why pick one approach when the advantages from both plans can be used?" Local autonomy is very important, but so is technical expertise." Both plans were good, but as he looked at the whole picture, he felt sure that together they would better serve the community.

---

narios were rewritten. A final evaluation of the most highly rated scenarios was conducted with nine different students. Each read several of the scenarios and completed a rating sheet. Reading times for the scenario were also surreptitiously collected.

Scenarios rated as interesting were retained; less interesting ones were discarded or modified. We discarded or rewrote those scenarios that were confusing or that had longer reading times. The final scenarios to be used in training and testing were balanced for word length, average sentence length, and approximate reading level.

Discussion of Study 1

The final scenarios, presenting conflicts in Tolerance for Uncertainty or Holistic–Analytic reasoning, were ready to use in training. Taken together, the two dimensions had advantages as initial training targets. Tolerance for Uncertainty appeared to be conceptually simple, based on the ease with which we could explain it to undergraduate subjects. We expected that it would be relatively easy to train. In contrast, Holistic–Analytic Thinking is conceptually difficult because it includes both Synthesizing-Contrasting and Attribution. It was difficult for the U.S. undergraduates to see these as aspects of a single distinction. It might therefore be more difficult to understand, recognize in others, and accommodate. The training intervention was undertaken in Study 2.

**STUDY 2: DEVELOPMENT AND EVALUATION
OF THE INTERVENTION**

The goal of this study was to test the efficacy of training derived by using the CLM. A training program for the dimensions of Tolerance for Uncertainty and Holistic–Analytic Reasoning, consistent with the model, was designed to:

- Increase recognition of cultural dimensions (awareness).
- Increase the appreciation of the differences (acceptance).
- Increase the ability to judge the likelihood of future actions (prediction).

We used basic training development procedures to create workshop material and procedures (e.g., Gagne, Briggs, & Wager, 1992; Goldstein & Ford, 2002). The training materials used scenarios to acquaint participants with the two dimensions. The instructor defined the dimensional constructs. Participants completed surveys with rating scales designed to provide practice applying the concepts. They participated in instructor-facilitated discussions. Focused questions and role-taking exercises were incorporated to help participants understand their own and other students' perspectives.

Method of Study 2

A total of 30 college students, ranging in age from 20 to 46 years, served as participants. They were recruited from international business or international management courses at Wright State University. Five additional students were excluded because English was not their first language or because they did not complete training. Participants had little experience with other national groups. All of participants were in at least their second year, with most being seniors completing their final year. Training was provided in groups with 18 or fewer members to support effective group interaction and discussion.

The materials included: (a) a scenario-based pre- and posttest to assess the training's effect on awareness, acceptance, and prediction, (b) a training module for each dimension, (c) a participant information sheet, and (d) a training program evaluation sheet.

The scenarios used in the pre- and posttests and in the training came from Study 1. The pre- and posttests consisted of a scenario for each dimension. Each test included the question: "Describe the reason behind the difference in how [character's name] and [character's name] behave in the story." Pre- and posttest assessment tools tracked changes in awareness, acceptance, and prediction ability over training. Iterative pilot trials ensured that the pre- and posttest material was understandable and unambiguous.

The three assessment tools enabled us to evaluate training effectiveness by asking participants to make judgments about characters in the scenarios. In the Awareness Scale, trainees rated the two Tolerance for Uncertainty scenario characters on four pairs of opposing characteristics, and rated the two Holistic–Analytic scenario characters on four different pairs of opposing characteristics, for a total of 16 ratings for awareness. The opposing characteristics for Uncertainty, for example, ranged from "Seeks rules" to "Likes flexibility," and from "Willing to take risks" to "Hesitates to take risks." The opposing characteristics for the Holistic–Analytic characteristics ranged from "Trusts analysis" to "Trusts intuition," and from "Accepts contradictions" to "Seeks the best answer." A 5-point choice scale was used for all of these.

The Acceptance Scale measured whether trainees would be accepting of those who differed from themselves. Two scenario characters were each evaluated on three 5-point scales. Items included: "To what extent would you want to interact with [character's name]?" and "How reasonable do you think [character's name]'s approach was in handling the situation?"

The Prediction Scale assessed whether a trainee could predict actions of the scenario's characters in new situations. Each trainee responded to four Tolerance for Uncertainty items and four Holistic–Analytic items. Each item related to a character whose characteristics had been described in the scenario. The Tolerance for Uncertainty assessment included: "[Character name] would be comfortable if a planned business trip changed into a teleconference at the last minute." The Holistic–Analytic assessments included: "Was [character's name] likely to carefully review information about a new job applicant rather than picking one on intuition or impulse?"

The training started with an introduction to individual differences, drawing on familiar personality differences such as active versus relaxed and patient versus eager. The training modules, one for each dimension, were built on scenarios and included descriptions, markers, and definitions as well as typical psychological and behavioral responses. The description for low Tolerance for Uncertainty for example was:

Some people find uncertainty stressful and work to avoid it. These people:

- Prefer rules and specific plans to provide structure.
- Avoid taking any but minimal risks and view risk taking as a dangerous option.
- Attentively adhere to plan and are disconcerted by change.
- Feel threatened by disagreement or dissent in others.

After reading a training scenario, the participants were asked to think of people they knew who were like the scenario character. Then they decided which character they were most like. There were individual rating exercises and opportunities to pick out key points. The modules guided participants through small-group exercises of role playing in which they assumed the role and defended the position of a character. The training was designed to increase perspective taking and provide opportunities to recognize the strengths and limitations in the different ways of thinking.

An information sheet elicited demographic information including age, year in school, foreign travel, work experience, and relevant course work. A program evaluation sheet was used at the end of training to evaluate perceived effectiveness of the program. Trainees were also given cards listing the characteristics of each of the two dimensions. The cards served as memory aids as they worked through the exercises.

Training spanned two 3-hour sessions spaced 2 days apart. The first session started with an introduction and the pretest exercises. The Tolerance for Uncertainty module followed. The second session provided a parallel module for Holistic–Analytic reasoning. After the second session, the posttest was completed. Finally, trainees evaluated the impact and usefulness of the training and provided demographic information.

Results of Study 2

Did this brief training have an impact? Participants in the group that received the training showed more accurate awareness from pre- to posttest for 14 of the 16 Awareness Scale items. Most of the differences were statistically significant, but some were small. Trainees showed increased acceptance on all of the 12 items. Finally, of the four items for each of the two dimensions assessing predictions, three of them changed, reflecting greater prediction skill. Predictions were significantly more accurate for all of the Tolerance for Uncertainty items. Predictions for the Holistic–Analytic items were in the direction of increased conceptual accuracy but were not significant. The results of the analysis are provided in Table 22.4.

After training, trainees rated the usefulness of the training in preparing them for the multinational work they intended to pursue following graduation. The mean ratings were 3.86 (out of 5), confirming an overall favorable evaluation. High ratings were given to: "After this workshop, I better understand how culture can influence thinking, behavior, and decision making" (4.37). Also rated highly was, "After this workshop, I better understand how people who are high/low on a particular cultural dimension could help me/hinder me in a business situation" (4.06), and, "After this workshop, I can list more of the potential situations I am likely to face in other countries" (4.06).

TABLE 22.4
Study 2 Outcomes

	Pretest (1)		Posttest (3)		Difference			
Awareness	Mean	SD	Mean	SD	Mean	SD	t	p
Low T for U	4.38	.46	4.55	.40	0.175	0.52	1.84	0.04*
High T for U	1.74	.50	1.67	.47	−0.067	0.59	−0.62	0.27
Holistic	3.44	.74	3.73	.91	0.297	0.66	2.47	0.01**
Analytic	2.41	.69	2.03	.84	−0.375	0.95	−2.16	−0.02*
Acceptance								
Low T for U	2.94	.64	3.12	.73	0.178	0.61	1.59	0.06
High T for U	3.29	.98	3.63	.95	0.344	0.65	2.89	0.004*
Holistic	3.31	1.03	3.61	.80	0.299	0.81	1.98	0.03*
Analytic	3.03	.88	3.21	.82	0.172	0.85	1.09	0.14
Predictions								
Low T for U	2.03	.52	1.80	.53	−0.233	0.51	−2.51	0.01*
Holistic	3.38	.82	3.54	.82	0.167	0.74	1.23	0.1

*Significant at .05 level. **Significant at .01 level.

One question remained: Might the pretest itself alter participant's responses to the posttest? To check this possibility, we looked at the impact of the pretest alone. A separate group of 30 participants completed the pretest but did not receive training. Two days later, they took the posttest. For all but one measure, differences were smaller than those found in the experimental group. The four awareness measures and three of four assessment measures showed no statistically significant differences comparing pretest to posttest. Participants were more accepting of high Tolerance for Uncertainty and more accurate on the Prediction Scale from pre- to posttest. Many participants in this sample left the pretest talking about the scenarios, perhaps fostering spontaneous role taking. Even so, the training was more effective.

This training intervention was arguably brief, the sample was small but consistent with those commonly used in initial educational-interventional studies, and the trainees were students, not professionals who need the job-related skills offered in the training. Despite these limitations, the training appeared effective. After training, participants were more aware of the cultural differences and made more positive assessments of characters described in the scenarios. There were also some increases in the accuracy with which they could predict likely future outcomes and actions. This CLM-based training improved performance from the pretest to the posttest, the trainees were engaged with the material, and rated they it positively. This training, completed with students in a classroom setting, had an impact on important components of interpersonal interactions. Based on these increases in awareness, acceptance, and predictions, we then asked if these same gains could be achieved with practitioners working in a natural intercultural setting.

For our next step, we hypothesized that training would be effective when provided to personnel who were preparing for multinational interchanges. They would be expected to be quite highly motivated to learn. We also expected that training would be more effective using scenarios that would be viewed as important to the trainees. U.S. military personnel, for example, know that peacekeepers are likely to work on multinational teams in order to deescalate citizen unrest, collect weapons, and plan for transition government. Our next studies allowed us to explore these expectations. In Studies 3 and 4, we performed a first evaluation of this CLM-based training, this time with the participation of international personnel deployed at NATO's HQ SFOR.

STUDY 3: FIELD DIMENSIONS AND SCENARIOS

The SFOR mission was to coordinate rebuilding, foster and support local governance, and provide safety for the citizens of Bosnia-Herzegovina. This effort was difficult because there are over 30 participating nations ranging from the English-speaking nations of the United States, Canada, and the UK to Southern Europe, Western Europe, Scandinavia, Warsaw Pact nations, and Turkey. Personnel screen complex information, make high-stakes decisions, and execute time-sensitive operations. The U.S. Army is cognizant of cultural barriers in these activities. With their support and cooperation, we developed and evaluated CLM-based training at SFOR. Parallel to the work of Study 1 discussed earlier, in Study 3 we collected incidents to identify key dimensions and to provide training scenarios. In Study 4, an initial evaluation of the training was undertaken. This field research extended the laboratory-based research to develop specific training for newly deployed personnel that would be effective and "user-friendly."

Method for Study 3

In 2002, we interviewed military personnel in order to identify the dimensions that are most challenging during multinational teamwork. Two researchers were present at each interview to ask questions and record responses.

In a first visit, we interviewed 23 individuals, including 5 Generals, 17 Colonels, and 1 civilian, from Canada, Czechoslovakia, Denmark, France, Germany, Hungary, Italy, Poland, Spain, Turkey, the United Kingdom, and the United States. All had extensive military experience. Some had extensive training and experience in multinational operations prior to their current deployment whereas others had none. Interviews probed challenges encountered during multinational exchanges. A second set of interviews confirmed and extended the outcomes with another 20 interviewees, ranging in rank from Major to Colonel, from Canada, Czechoslovakia, Denmark, France, Germany, Hungary, Italy, the Netherlands, Poland, Spain, Turkey, the UK, and the United States.

Using a CIT, we asked probe questions about points of stress, frustration, and conflict experienced at HQ SFOR. Situations that caused surprise or misunderstanding were interesting because they indicated cognitive and social context differences. We asked probing questions about incidents of planning, coordination, decision making, and implementation. We looked at the interpretation of cues, resolution of points of difference, weighing of options, and planning. We queried the role of authority and experience in judgment and decision making. We asked for examples where the interviewees realized that they could not deal with allies in their usual way. The examples that were identified provided dimensions of conflict and stories for use as training scenarios.

Results of Study 3

Interview notes recorded incidents of conflict, discrepancy, frustration, and surprise. Three raters independently reviewed the interview notes and highlighted statements that reflected the CLM dimensions. Their classifications were assessed for reliability and consistency. We were interested in identifying dimensions that caused conflicts and stress at HQ SFOR, reflecting training needs. National/cultural differences were evident for most of the CLM dimensions. To support effective multinational operation, we needed to separate differences that were interesting but less important from a pragmatic view from ones that could be distracting, stressful, or even dangerous. The two dimensions were Tolerance for Uncertainty and Achievement–Relationship.

Examples from the interviews illustrate the difficulties encountered. See Table 22.5. Tolerance for Uncertainty was described as stressful during information assessment, planning, and decision making. Faced with variations in Tolerance for Uncertainty, several interviewees expressed their frustration at the haphazard planning of some whereas others were frustrated by what they saw as the paralysis of analysis (row 1, Table 22.5).

The Achievement–Relationship dimension influenced performance during complex or prolonged interchanges when differences hampered coordination and increased frustration (row 2, Table 22.5). Achievement-oriented individuals, focusing on their tasks, can be perceived from another cultural lens as cold, unsociable, unapproachable, and unapprecia-

TABLE 22.5
Example Contrasts for Study 3

	High Tolerance for Uncertainty	Low Tolerance for Uncertainty
1	"I can't tell you how much time they waste 'getting ready.' We could have the whole job done before they finish their planning!"	"Before they've even looked at all the data, they are putting together a plan. They change directions as they go along. It's a waste of time and resources!"
	Achievement-Oriented Individuals	Relationship-Oriented Individuals
2	"They think nothing of taking 2 hours for lunch. You'd think it was a summer camp!"	"Sometimes, it seems Americans don't want to really know other people . . . [They] come in: 'How are you doing, when are you leaving' . . . all in 15 seconds and then they start working the problem. We're not like that. We get to know people, to trust them."

tive of the commitment and personality of others. Relationship-oriented people, who emphasize building ties to others while discussing tasks rather than "working," may be perceived as lazy, uncommitted, and incompetent. Such conflicts and "tacit tensions" can seriously compromise morale, willingness to collaborate, and team performance.

Scenario Construction

We reviewed the transcripts to find incidents that captured problems illustrating the two selected training dimensions—High versus Low Tolerance for Uncertainty and Achievement–Relationship. Those incidents were then used to develop, pre- and posttest material, examples, and four training scenarios for each of the training dimensions. In order to have useful scenarios, each incident was rewritten to illustrate the extremes of one of the two dimensions. The incidents were adapted to capture problems and solutions in the HQ SFOR context. Each of the resulting scenarios presented a conflict related to one of the dimensions during a multinational interchange. For example, a Tolerance for Uncertainty scenario started with the text shown in Textbox 22.3.

**Textbox 22.3. The Opening Statement
in One of the Study 3 Scenarios**

"You are assigned to a task force to recommend training for new personnel. You have one month to complete your work. At the first meeting, several of the officers want to draw out a detailed plan for the task force: 'We need to set a timeline, assign tasks, and make sure everyone knows exactly what will be expected of them.' Though many of the officers are in agreement with this course, several others prefer a different approach: 'We need to develop an understanding of the problem and that understanding will tell us how to proceed. You have to let the pieces fall into place.'"

Subsequent frames referenced this situation to explore differences in Tolerance for Uncertainty during mission planning. Of interest were time management, the different strengths of the two approaches, and techniques for working with people who differ on this dimension.

The Achievement–Relationship dimension influences the social context of cognition and collaboration. Whereas U.S. personnel tend to be pragmatic and objective, focusing on task completion and performance enhancement, others valued and nurtured relationships with coworkers, seeing interpersonal connectedness as vital for effective work. How an individual interpreted and evaluated the actions and behaviors of cowork-

---

**Textbox 22.4. Scenario Expressions of Two Views
on the Achievement–Relationship Dimension**

"We have an amazing tool in technology. I don't have to make calls to other officers or even worse, run downstairs if I want to coordinate our actions. Everyone at SFOR is as close as my computer. We could move fast IF everyone would just use the technology we have available. I can't tell you the number of times someone will walk up here to settle something that could have been done in one tenth the time by e-mail."

Versus

"Technology is good, but sometimes it just doesn't work. You need to have face-to-face time to see what they are actually thinking. When I have walked over to the person I need to work with, I can see what they are really willing to do. When I can watch their faces, I know what's troubling them. We can work out problems rather than waiting for the plan to fail."

---

ers influenced interactions and posed barriers to collaboration. One scenario used communication technology to explore differences in the Achievement–Relationship dimension. Based on interview notes, we presented two very different views. These are shown in Textbox 22.4.

The construction and evaluation of scenarios was iterative, as in Study 1. Scenarios were written to accommodate the varied level of skill in English of the intended users by making the scenarios easy to read and understand. Before inclusion in the training, U.S. military officers who were familiar with the SFOR environment and with multinational challenges reviewed the scenarios to confirm that they represented the situations found in multinational environments. They were able also to help in simplifying the language and incorporating military jargon. Based on this input, we revised the scenarios before beginning construction of the training tool. The construction and implementation of this tool are discussed in Study 4.

Discussion of Study 3

Most interviewees reported that they found the interviews to be interesting and the research problem important. The questions and probes elicited many issues and insights. Military officers, like most experts, often enjoy telling their experiences. During our interviews and our stay at SFOR, personnel from the United States and other nations spontaneously reported their own lack of preparation for multinational operations. Most interviewees voiced concerns about interactions with at least one nationality. We heard many complaints about U.S. personnel, who were perceived as being culturally insensitive. This shortcoming was usually attributed to more limited international experience. A typical statement was, "I can drive 200 miles, and I'm in a different country with a different language. You Americans can grow up never seeing a foreigner." Interviewees were generally appreciative of efforts to improve SFOR effectiveness.

The focus of Study 3 was on using the CLM to elicit information about collaboration challenges experienced by multinational units within the peacekeeping environment. The revised and refined scenarios from this study represented the difficulties encountered during multinational operations. The next section on Study 4 describes how these scenarios were used to design a training program for military personnel.

STUDY 4: DEVELOPMENT AND EVALUATION OF A FIELD INTERVENTION

Parallel to Study 2, Study 4 used the dimensions and incidents to develop CLM-based training. Study 4 was based on the dimensions, scenarios, and guidelines of Study 3. Scenario-based training was developed to deliver a self-paced, computer-based training tool. We needed to test the feasibility of using such training to address the cultural obstacles to multinational collaboration and mitigate the challenges identified at HQ SFOR. The training was designed to increase awareness, recognition, and accommodation skills on the two CLM dimensions identified in Study 3: Tolerance for Uncertainty and Achievement–Relationship.

Headquarters personnel are committed to the peacekeeping mission and its success. They differ, however, in how they manage uncertainty, structure, plans, and interact with others. Such differences can create misunderstanding and increase conflict. The U.S. Army wanted to counter the potential weakness of these differences and build on the strengths in order to facilitate teamwork and improve performance during multinational collaborations.

Methods of Study 4

The training format had to fit with the dynamic and time-pressured demands facing personnel. The computer-based training tool provided easy access for individual and team training prior to deployment or while stationed at SFOR. The material could be completed at one time or paced as other demands permitted. The interactive capabilities allowed trainees to receive individualized feedback and support. Trainees who might be at any place on the dimensions of Tolerance for Uncertainty and Achievement–Relationship could receive responsive information and could monitor their own progress. The training was designed for native English speakers, but reading level and word selection were crafted so that others might also benefit from use.

The PowerPoint format selected for this phase of training development began with on-screen instructions for navigating through modules. The first slides presented an overview of multinational teamwork at HQ SFOR as well as of the purpose and expected benefits of the training. The subsequent modules were designed to increase awareness of and to develop skills for managing differences in Tolerance for Uncertainty and Achievement–Relationship. An additional module, developed for native English speakers, addressed effective communication with team members for whom English was not a first language. Each module had a pretest and posttest to assess progress. The format could be expanded to include additional modules and different languages of presentation.

The modules used scenarios to provide practice in identifying, understanding, and managing differences. One scenario in the Tolerance for Uncertainty module, "The Case of Operation Harvest," describes the views of two leaders who are directing a program to collect weapons from civilians in an effort to decrease violence in Bosnia-Herzegovina. The goal is to collect the weapons without disrespecting the citizens or endangering the troops. Students were presented with the strategies of two very successful leaders, described in Textbox 22.5.

Each unit started with a pretest. Trainees then reviewed definitions and descriptions of the Tolerance for Uncertainty dimension and identified the

---

Textbox 22.5. Two Contrasting Leadership Styles

Leader 1 plans each day down to the smallest detail. Each person in the unit knows his role and follows it carefully. Leader 1 finds it uncomfortable when people disagree because it increases confusion and potential errors in the field. He even has defined procedures for heavy rain and other deviations from a standard day. As Leader 1 always says, "We need to know exactly what will happen so that there will be no surprises. A boring day is a safe day."

Versus

Leader 2 is very different. He does not have a well-defined plan with alternatives for emergencies. He wants his men to be able to handle whatever comes up. Leader 2 believes that you can never have all the information anyway. He would find it difficult to have everything planned. Rigid planning does not leave room for surprises along the way. He believes he has been successful because he has been willing to be flexible and listen to ideas from other people.

leader with whom they would most like to work. They identified strengths and limitations of both leaders. A set of examples highlighted the strengths of each as well as the advantages of teams composed of people who varied on the dimension. The trainees were guided as they explored approaches to improving interactions when team members differed. Interactive exercises allowed practice with and feedback on the concepts. Questions were followed by response choices for which the trainee responded on the screen. Feedback provided correct responses, additional information, or examples as appropriate. Trainees could review the scenario as needed. Finally, the posttest assessed changes with training.

The additional scenarios in the Tolerance for Uncertainty module addressed challenges created by the differences in planning, training, and other mission-critical areas. One scenario, for example, presented the conflicts that have emerged in planning computer skills instruction. Whereas some national groups want flexible and informal training, others want structured curricula and well-defined procedures. In working through this conflict in Tolerance for Uncertainty, officers can see the broader issues faced in day-to-day interactions.

The Achievement–Relationship module was structured similarly to the Tolerance for Uncertainty module. A pretest scenario was followed by a set of scenarios, each with examples, definitions, and descriptions. The module described the impact of this dimension on planning, team building, giving orders, and the use of communication technologies. One scenario, for example, looked at expectations for the first day at one's assigned com-

---

**Textbox 22.6. Two Contrasting Styles
on the Achievement–Relationship Dimension**

Col J: "For coordination, it's the informal relationships that count. I try to meet people before the formal meeting. Often, we solve the problem before we get there. You have to know people and that doesn't happen at formal meetings. You eat and drink with them, you learn about their families. Good relationships build consensus. Col K wants people to run like machines but they don't!"

Versus

Col K: "We're here to plan this visit not to socialize with our buddies. I work as long as it takes to do the job. That may mean a quick lunch at my desk. It may mean a late night. I do it. We're professionals, we work with the staff assigned to the task. I don't have to like them and they don't have to like me. We all just have to do our jobs. Col J's socializing is not part of getting the job done."

mand. For Relationship people, this might include time to meet informally with all of the staff, a long social lunch with sharing of histories and personal lives, and invitations for dinner and a tour of Sarajevo in the evening. An Achievement person would typically be eager to get to the work at hand and learn the job. The trainee could explore the impact of mismatches between commander and staff. A second scenario described two staff members, both competent and respected, who are working to coordinate the visit of an international delegation. The effort is behind schedule and both blame the other, as described in Textbox 22.6.

The module led the trainee to see the strengths and weaknesses of these differing positions and provided approaches by which individuals might find a middle ground, build on different strengths, and expand capacities. Interactive exercises provided opportunities to practice identifying and working with differences. Finally, a posttest assessed mastery of the concepts. The training material thus developed was then available for evaluation onsite.

Evaluation Methods for Study 4

The initial training trials were undertaken at HQ SFOR with 11 trainees ranging in rank from Captain to Colonel, and coming from Argentina, Canada, France, Germany, Italy, the Netherlands, the UK, and the United States. Based on response patterns and posttraining debriefings, we further refined and revised the text and interactive exercises. The training modules then underwent final training trials with eight individuals having ranks ranging from Major to Colonel. They were from Canada, the Netherlands, Portugal, Turkey, the UK, and the United States.

The modules were tested with the first set of trainees to observe performance and improve training. Each trainee received a description of the study and its purpose. They worked through the computer-based training tool on a PC in a computer laboratory. The trainee started with the introductory module. Following this introduction, trainees worked through one or more of the modules. Trainees responded to a pretest assessment. They then completed the module and exercises described previously. Finally, they completed the posttest assessment. Two researchers observed trainees during each of the individual sessions. This allowed us to observe the technical aspects of the tool.

As the trainees completed the modules, the researchers observed their performance and asked questions to detect problems. After completing the training, the trainees provided feedback on the content, usability, and usefulness of the modules. They reported potential problems and judged the value of the material. The procedure allowed the researchers to continually review performance and revise the modules. We made changes that were needed to clarify language and increase functionality. Feedback directly influenced the inclusion of additional examples, as well as the use of military vernacular to present the information. These elements significantly enhanced the suitability of the training.

The revised background information and modules were then tested in the final training trials with the second set of trainees. At the conclusion of training, feedback on the training program was elicited.

Results for Study 4

Feedback from both samples of trainees was quite positive. Most trainees stated that they could easily relate to the examples in the modules. The scenarios appeared to be realistic and relevant to multinational peacekeeping. At many points, trainees would laugh and indicate that they had experienced the problem described in a scenario. We heard of incidents of misunderstandings and frustrations similar to those included in the training scenarios. This suggests that the materials had good face validity. The trainees told us they wish they had had the opportunity for such training prior to deployment. Several officers asked if they could use the training immediately with their current teams. Others asked that the material be sent to them, if possible, so that they could share it with peers or with personnel in their command. They reported the value of the training in their current rotation. Trainees indicated that cultural training is urgently needed either predeployment or soon after arrival. The most common comments reflected the trainees' assessment of the value of their own predeployment training. Typical was, "I sure wish I had seen this before I was deployed. I've learned a lot here, but I've left some bruised feelings behind me."

A common comment from non-U.S. personnel was, "It's great to see that you guys are really looking at these things. In Europe, we have had to face culture problems for a long time."

Discussion of Study 4

Our observations suggest that CLM training can be prepared and delivered as a computer-based training tool. The testing and revision sessions showed that officers judged the training to be appropriate and useful. The accessibility and interactive benefits expected of a Web-based format were observed for the training. Training increased understanding of the cultural differences. It also increased confidence in and hopefully also, ability to manage these differences during collaboration.

Subsequent to our research, the Army Research Laboratory undertook an evaluation of the training concept (for a detailed description, see Sutton, 2003). In the evaluation, 60 participants each completed one module and a usability survey. A small sample ($n = 10$) also completed a post-training scenario-based assessment. Participants judged the modules to be realistic, relevant, and easy to complete. They indicated that they were able to understand the implications of lessons learned for working in a team environment and that the training would be most effective if available prior to deployment or soon after arrival. Also positive were ratings of confidence in ability to work well with multinational personnel and perceptions of module usefulness. The results from the learning assessment were more tentative because of the small sample. Knowledge transfer was described as somewhat encouraging. The findings suggest the viability of this approach and support the value of further development of this approach for improving multinational mission effectiveness.

GENERAL DISCUSSION

A next step should be to evaluate the training without a facilitator. Our training materials were designed so that they might be generalized and customized to the growing number of multinational forces deployed worldwide. The training must allow the individual to complete the training prior to deployment without a facilitator.

In future research, it will be imperative to look at behavioral changes in field settings. Does the training transfer to actual performance, in the sense of skills and strategies for multicultural teamwork? Military leaders are looking to what is being called "effects-based operations" to ensure that personnel are better at accomplishing strategic and operational objectives. It is not enough to increase performance on rating scales. Specific missions with well-defined goals, such as Operation Harvest, are good candidates for quantification. Measures of teamwork and performance will also be needed.

Consistent with the continuing goal of the U.S. Army, these four studies support efforts to enhance collaboration and skill-based actions in multinational forces. Cognitive engineering and the strategies of Naturalistic Decision Making served as a starting point. Because the goal was to support multinational peacekeepers during complex collaborative operations, we could not depend on Western psychology alone.

Future work should build on the strength of cognitive engineering and should reflect sensitivity to national differences in cognition and the social context of cognition. It is valuable because of its focus on settings with ill-defined problems, high stakes, and multiple players. Many domains of professional practice are increasingly multinational and collaborative and this introduces additional complications. Current understanding of collaboration and cognitive engineering are both grounded almost entirely on research undertaken in Western nations—the English-speaking world and

Western Europe—but cognition and the social context of cognition varies over national boundaries.

At each stage of this research, from the interviews with experienced informants to the observations at SFOR, we saw the power of national differences during multinational collaborations. Personnel reported difficulties when working with others who differed in Tolerance for Uncertainty and Achievement–Relationship. They also reported difficulties with regard to differences in Holistic versus Analytic Thinking, Concrete versus Hypothetical Reasoning, Synthesizing versus Contrasting, and Dispositional versus Situational Attribution. Coordination and cooperation were troublesome when team members differed on the Long versus Short Time Horizon, Mastery versus Fatalism, High versus Low Power Distance, and Independence versus Interdependence. A useful science of cognitive engineering must incorporate these differences.

U.S. researchers typically use the notion of a team to describe the efforts of groups working toward a shared goal. Existing team models speak, for example, about mutual performance monitoring and compensating for the workload and stress of others (McIntyre & Salas, 1995; Sims, Salas, & Burke, 2004; Zsambok, Klein, Kyne, & Klinger, 1993). This research confirms that such processes are far from universally understood or practiced. Indeed, best practices in one nation are sometimes judged as being inappropriate in others. One interviewee told us that you shouldn't monitor others because it would show that you did not trust them. Another interviewee explained that to do someone else's job is like "taking their bread"—their feeling of self-efficacy and responsibility. For a U.S. citizen, this might sound like how children react to overly controlling parents rather than as grounded in a different and fairly widespread concept of self.

Western-led team research frequently addresses communication patterns among team members as one of the factors ensuring high performance results (Klinger & Klein, 1999; MacMillan, Entin, & Serfaty, 2002). The free flow of information, shared understanding, and, monitoring are viewed as critical for high-functioning teams. Among some national groups, teams function very differently (Klein & McHugh, 2005). Information may flow hierarchically, sharing of knowledge may be limited, and monitoring others may be unacceptable. The communication patterns among team members of such nations will look very different from those of less hierarchical national groups. In multinational team settings, differences in expectations can be great obstacles to collaboration creating conflicts. This can be particularly troublesome when team members are not aware of the existence or nature of differences or when they lack tools for managing communication differences effectively.

Whenever people assume that others involved in a collaborative relationship think as they do, there can be problems. As we move into international collaborations, we will need to expand the conceptual and empirical basis for describing collaboration. The CLM provides one conceptual framework for this expansion. Effective interactions between members of a multinational collaboration depend on participants being able to take the perspective of others. Whereas practitioners may not be able to describe the dimensions of every potential ally, the CLM can capture differences critical for particular natural settings. More important, it provides people with knowledge about the specific nature of the differences and how they might impact performance. The training described here provides a mechanism for translating knowledge of cognitive differences into training interventions.

Because of the importance of collaboration and teamwork in real-world settings, it will be important that cognitive engineering researchers and practitioners incorporate a broader and more international perspective on cognition. We cannot depend on data collected with U.S. and Western nationals to describe cognition around the world. We can no longer use Western teamwork models without accommodating the vulnerabilities introduced by national differences in cognition. Because collaboration and teamwork are not universals, effective collaboration demands that we see the world through the cultural lens of others.

Work at HQ SFOR and other multinational settings is demanding and stressful. Officers must carefully manage their time, attention, and effort. Even for training experts, the primary demands can be overwhelming. With 6-month deployments common, there is little time to get personnel up to speed. With staggered arrivals, class training is not always possible. With workload sometimes intense, there is not time for extended training. Mistakes may cost human lives and they most certainly cost international understanding and cooperation. It is dangerous and counterproductive to undertake exploratory research in high-stakes, time-pressure environments. The present research describes the value of preliminary laboratory explorations to generate material and methods. It allows researchers to glean initial research concept support before using critical time onsite. The demonstrated training effectiveness of the laboratory intervention provides needed support for the time commitment and risk at HQ SFOR. More important, the initial trials provided experience in scenario, assessment, and perspective-taking exercise development. Whereas there is no substitute for fieldwork, preliminary laboratory studies can make an important contribution.

ACKNOWLEDGMENTS

The work reported here was conducted under ARL Federal Laboratory Prime Contract DAAD19-01-C-0065 as part of the Technology Transfer for the Collaborative Technology Alliance program. It was also supported by the U.S. Army Research Institute under Purchase Order 84334 issued by the Department of the Interior. The authors wish to thank Bianka B. Hahn, and Danyele Harris-Thompson, Mei-Hau Lin, and Sterling Wiggins for their contributions.

REFERENCES

Adler, N. (1997). *International dimensions of organizational behavior* (3rd ed.). Cincinnati, OH: Southwestern College.

Agrawal, A. A. (2002). Phenotypic plasticity in the interaction and evolution of species. *Science, 249,* 321–326.

Berry, J. W. (1986). The comparative study of cognitive abilities: A summary. In S. E. Newstead, S. H. Irvine, & P. L. Dann (Eds.). *Human assessment: Cognition and motivation* (pp. 57–74). Dordrecht, Netherlands: Martinus Nijhohh.

Choi, I., Nisbett, R., & Norenzayan, A. (1999). Causal attribution across cultures: Variation and universality. *Psychological Bulletin, 125,* 47–63.

Chu, P., Spires, E., & Sueyoshi, T. (1999). Cross-cultural differences in choice behavior and use of decision aids: A comparison of Japan and the United States. *Organizational Behavior and Human Decision, 77,* 174–170.

Crandall, B., & Getchell-Reiter, K. (1993). Critical decision method: A technique for eliciting concrete assessment indicators from the "intuition" of NICU nurses. *Advances in Nursing Sciences, 16,* 42–51.

Gagne, R. M., Briggs, L. J., & Wager, W. W. (1992). *Principles of instructional design* (4th ed.). Fort Worth, TX: Harcourt Brace.

Goldstein, I. L., & Ford, J. K. (2002). *Training in organizations: Needs assessment, development, and evaluation* (4th ed.). Belmont, CA: Wadsworth.

Granrose, C., & Oskamp, S. (1997). *Cross-cultural work groups.* Thousand Oaks, CA: Sage.

Hahn, B. B., Harris, D. S., & Klein, H. A. (2002). Exploring the impact of cultural differences on multinational operations (Final Report under Prime Contract #DAAD19-01-C-0065, Subcontract No. 8005 004.02 for U.S. Army Research Laboratory). Fairborn, OH: Klein Associates.

Hall, E., & Hall, M. (1990). *Understanding cultural differences.* Yarmouth, ME: Intercultural Press.

Helmreich, R. L., & Merritt, A. C. (1998). *Culture at work in aviation and medicine: National, organizational, and professional influences.* Aldershot, England: Ashgate.

Hoffman, R. R., Crandall, B. W., & Shadbolt, N. R. (1998). Use of the critical decision method to elicit expert knowledge: A case study in cognitive task analysis methodology. *Human Factors, 40,* 254–276.

Hofstede, G. (1980). *Culture's consequences: International differences in work-related values.* Newbury Park, CA: Sage.

Ji, L.-J., Peng, K., & Nisbett, R. E. (2000). Culture, control, and perception of relationships in the environment. *Journal of Personality and Social Psychology, 78,* 943–955.

Kaplan, M. (2004). *Cultural ergonomics.* Oxford, England: Elsevier.

Klein, G., Calderwood, R., & MacGregor, D. (1989). Critical decision method for eliciting knowledge. *IEEE Transactions on Systems, Man, and Cybernetics, 19,* 462–472.

Klein, G., Orasanu, J., Calderwood, R., & Zsambok, C. E. (Eds.). (1993). *Decision making in action: Models and methods.* Norwood, NJ: Ablex.

Klein, H. A. (2004). Cognition in natural settings: The cultural lens model. In M. Kaplan (Ed.), *Cultural ergonomics* (pp. 249–280). Oxford, England: Elsevier.

Klein, H. A., Klein, G., & Mumaw, R. (2001a). *A review of cultural dimensions relevant to aviation safety.* Technical report completed for the Boeing Company under General Consultant Services Agreement 6-1111-10A-0112.

Klein, H. A., Klein, G., & Mumaw, R. (2001b). *Culture-sensitive aviation demands: Links to cultural dimensions.* Technical report completed for the Boeing Company under General Consultant Services Agreement 6-1111-10A-0112.

Klein, H. A., & McHugh, A. P. (2005). National differences in teamwork. In W. R. Rouse & K. R. Boff (Eds.), *Organizational simulation* (pp. 229–251). New York: Wiley.

Klein, H. A., Pongonis, A., & Klein, G. (2000, June). *Cultural barriers to multinational C2 decision making.* Paper presented at the Proceedings of the 2000 Command and Control Research and Technology Symposium (CD-ROM), Monterey, CA. Available at: http://www.dodccrp.org/events/2000/CCRTS_Monterey/cd/html/pdf_papers/Track_4/101.pdf

Klein, H. A., & Steele-Johnson, D. (2002). *Training cultural decentering* (Final Report for Order No. 84334, Department of the Interior/Minerals Management Service). Fairborn, OH: Klein Associates.

Klinger, D. W., & Klein, G. (1999). Emergency response organizations: An accident waiting to happen. *Ergonomics In Design, 7,* 20–25.

Kluckhohn, F., & Strodtbeck, F. L. (1961). *Variations in value orientations.* Evanston, IL: Row, Peterson.

Lane, H., & DiStefano, J. (1992). *International management behavior: From policy to practice.* Boston: PWS-Kent.

Lannon, G., Klein, H. A., & Timian, D. (2001, May). *Integrating cultural factors into threat conceptual models.* Paper presented at the 10th Computer Generated Forces and Behavioral Representation, Norfolk, VA.

MacMillan, J., Entin, E. E., & Serfaty, D. (2002). From team structure to team performance: A framework. In *Proceedings of the Human Factors and Ergonomics Society 45th Annual Meeting* (pp. 408–412). Santa Monica, CA: Human Factors and Ergonomics Society.

Markus, H., & Kitayama, S. (1991). Culture and the self: Implications for cognition, emotion, and motivation. *Psychological Review, 98,* 224–253.

McIntyre, R., & Salas, E. (1995). Measuring and managing for team performance: Emerging principles from complex environments. In R. Guzzo & E. Salas (Eds.), *Team effectiveness and decision making in organizations* (pp. 149–203). San Francisco: Jossey-Bass.

Morrison, D., Conaway, W. A., & Borden, G. A. (1994). *Kiss, bow, or shake hands.* Avon, MA: Adams Media.

Nisbett, R. E. (2003). *The geography of thought: How Asians and Westerners think differently . . . and why.* New York: The Free Press.

Nisbett, R. E., Peng, K., Choi, I., & Norenzayan, A. (2001). Culture and systems of thought: Holistic versus analytic cognition. *Psychological Review, 108,* 291–310.

Peng, K., & Nisbett, R. (1999). Culture, dialectics, and reasoning about contradiction. *American Psychologist, 54,* 741–754.

Segall, M. H., Dasen, P. R., Berry, J. W., & Poortinga, Y. H. (1990). *Human behavior on global perspective.* New York: Pergamon.

Sims, D. E., Salas, E., & Burke, C. S. (2004). *Is there a "Big Five" in Teamwork?* Poster presented at the 19th Annual Meeting of the Society for Industrial and Organizational Psychology (CD-ROM), Orlando, FL. Available at: http://www.siop.org/Conferences/CDs.htm

Sutton, J. L. (2003). Validation of cultural awareness training concept. In *Proceedings of the Human Factors and Ergonomic Society 47th Annual Meeting* (pp. 2015–2018). Santa Monica, CA: Human Factors and Ergonomics Society.

Thomas, D. C. (1999). Cultural diversity and work group effectiveness: An experimental study. *Journal of Cross-Cultural Psychology, 30,* 242–263.

Triandis, H. C. (1994). *Culture and social behavior.* New York: McGraw-Hill.

Wise, J. C., Hannaman, D. L., Kozumplik, P., Franke, E., & Leaver, B. (1997). *Methods to improve cultural communication skills in special operations forces.* Misenheimer: North Carolina Center for World Languages & Cultures.

Wisner, A. (2004). Towards an anthropotechnology. In M. Kaplan (Ed.), *Cultural ergonomics* (pp. 215–221). Amsterdam: Elsevier Science.

Zsambok, C. E., Klein, G., Kyne, M., & Klinger, D. W. (1993). *Advanced team decision making: A model for high performance teams* (Report on Contract MDA903-90-C-0117, U.S. Army Research Institute for the Behavioral and Social Sciences). Fairborn, OH: Klein Associates.

Chapter 23

Facilitating Team Adaptation "In the Wild": A Theoretical Framework, Instructional Strategies, and Research Agenda

C. Shawn Burke
Eduardo Salas
Sarah Estep
University of Central Florida

Linda Pierce
Army Research Laboratory, Aberdeen Proving Ground, MD

It can be argued that if ever there were a need for teams that can adapt to new and challenging circumstances, it is at the present time. This is particularly the case for teams operating in new and complex environments under ever-shifting goals and responsibilities. In these conditions, the lack of team adaptability may result in a loss of productivity, or worse, the loss of human life. This notion has most recently been manifested in the challenges facing the U.S. military in the countries of Bosnia, Afghanistan, and Iraq. Soldiers who have been trained for combat are finding that their postwar roles of peacekeeping and providing humanitarian assistance (e.g., support and stability operations [SASO]) are unfamiliar and more confusing than they could have imagined (Williams & Chandrasekaran, 2003). This frustration and uncertainty carries the potential for lowered team morale and an overall reduced effectiveness.

Generally, military teams are composed of individuals who have been trained for combat, so when faced with SASO duties, they must adapt to new roles. When this occurs, difficulties arise because those new roles involve negotiating and cross-cultural competencies on the part of military personnel, as opposed to war-fighting skills (Burke, Fowlkes, Wilson, & Salas, 2003). In addition to ambiguous and/or changing roles, SASO teams must also contend with unclear and shifting goals. In this setting, it is usually impossible for the team to generate a sufficient number of contingency plans for each and every situation that may come up; therefore, teams must make and modify decisions as information becomes available. Because a principal measure of SASO success is the absence of any problems or conflicts in a given area, teams may find it difficult to determine which actions taken actually contributed to achieving their goal (Klein & Pierce, 2001). This is particularly the case when there is a delayed response or effect. Each of these aspects of SASO creates an environment that requires teams to have the ability to alter their behaviors, cognitions, and emotions in order to successfully reach their goals.

Though adaptation is important for teams within many military contexts, the need extends beyond this context to nonmilitary organizations. Factors within the internal and external environments of organizations require teams to adapt to changing market conditions and changing customer demands. For example, the series of hurricanes that hit Florida during the summer of 2004 drastically impacted how many organizations and the teams within these organizations conducted business. For a time, many types of teams (e.g., utilities, local governments, emergency management, etc.) were forced to forego the technology that they had become accustomed to and were reliant on, in completing projects. They had to reengineer alternative ways to accomplish the same tasks. A case in point is the crews that work for the various power companies within the state. These crews had to adapt to many contingencies such as: working with new crew members (i.e., crews from other states), a multitude of hazardous conditions interacting at once (heat, rain, standing water, long hours, numerous downed power lines), differing state regulations (crews coming in to help from other states had to adapt to Florida regulations and local utility safety regulations), and often reengineering work due to a shortage of materials/ supplies and in an effort to get as many customers back online as quickly as possible in a safe manner. The adaptations that teams were required to make were sometimes slight (requiring modification of existing strategies) whereas at other times they were large (requiring innovation of new strategies).

As a result of the need for adaptive capacity within teams, it is imperative that practitioners and researchers alike closely investigate the processes of adaptive team performance and the resultant team adaptation so that its role in operational settings can be better understood. This information can then be utilized to delineate methods by which to create and maintain adaptive teams.

The primary objective of this chapter is to provide an overview of the nature and importance of team adaptation for teams operating "in the wild," and use this conceptualization to provide practical guidance for the creation of adaptive teams. To this end, team adaptation and adaptive team performance are defined. We then briefly describe a cyclical model of team adaptation and discuss the role of expertise within that model. Next, this model is used as a basis for delineating a training approach and instructional features that can be embedded within this approach to facilitate the creation and maintenance of adaptive capacity. Finally, a research agenda is delineated in the hopes of stimulating research on team adaptation.

TEAM ADAPTATION: WHAT DOES IT MEAN?

Although argued by many to be important, team adaptation has received little attention by researchers. Though the study of team adaptation is still in its infancy there are some established principles that can serve to form a foundation from which to operate. Using this foundation, Burke, Stagl, Salas, Pierce, and Kendall (2006) have recently put forth a conceptual model of team adaptation (see Figure 23.1). The model integrates work from multiple disciplines to delineate the cyclical processes and emergent states that constitute adaptive team performance and the resulting outcome of this process, team adaptation. Laying the foundation for the components within the model is the definition of team adaptation: "a change in team performance, in response to a salient cue or cue stream, which leads to a functional outcome for the entire team. Team adaptation is manifested in the innovation of new, or modification of existing structures, capacities, and/or behavioral or cognitive goal directed actions" (Burke et al., 2006). The antecedents to team adaptation advanced within Figure 23.1 comprise adaptive team performance, individual characteristics, and job design characteristics. Adaptive team performance forms the core of the conceptual model and is defined as follows:

> [Adaptive team performance is] an emergent phenomenon which compiles over time from the unfolding of a recursive cycle whereby one or more team members utilize their resources to functionally change current cognitive or behavioral goal directed action or structures to meet expected or unexpected demands. It is a multilevel phenomenon which emanates as team members and teams recursively display behavioral processes and draw upon and update emergent cognitive states to engage in change. (Burke et al., 2006)

Adaptive team performance is depicted as a four-phase cycle consisting of individual and team processes and the resulting emergent states. Emergent states characterize properties of the team; they describe the cognitive, motivational, and affective state of the team (Marks, Mathieu, & Zaccaro, 2001). The four phases of this cycle include: (a) situation assessment, (b) plan formulation, (c) plan execution (via individual and team-level interaction processes), and (d) team learning. As the adaptive cycle forms the core of the model, it provides the avenue through which training and design

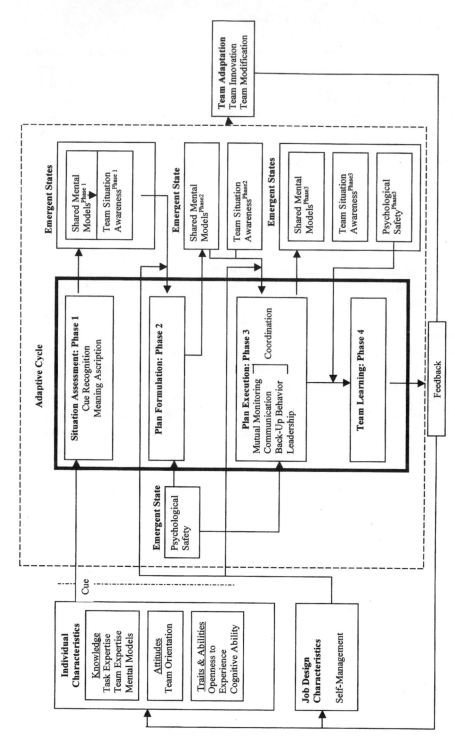

FIG. 23.1. Model of a team adaptation (Burke et al., 2006).

510

guidance can be extracted. Therefore, this portion of the model is briefly reviewed.

The first phase of the adaptive cycle is composed of the situation assessment process and is dependent on members' abilities to notice environmental cues, classify the cue(s), and interpret the possible impact on team success. The culmination of this individual process is two emergent states, mental models and situation awareness. Within teams, the process is a bit more complex; once team members have interpreted a cue's possible impact this information is relayed to the rest of the team, in turn providing the input for emergence of two team-level cognitive states (shared mental models, team situation awareness [TSA]). These cognitive states provide an initial base for adaptation by promoting shared or compatible understandings of the equipment, task, and how team member roles, responsibilities, and characteristics should intersect in context.

To illustrate this first phase of the adaptive cycle, one can look at an airplane cockpit crew. The pilot of an aircraft may slowly begin to notice certain environmental cues such as dryness of mouth or continuous ear popping. After noticing these cues, he or she begins to classify what may be responsible and comes to the conclusion that the plane could be slowly losing pressurization. Experience tells the individual that the meaning of this cue can have consequences and could ultimately impede the success of the crew (i.e., arrive at destination safely). As a team, the next step would be to communicate the information to the rest of the crew to ensure that the individual has interpreted the cue correctly; if the team agrees that the cues have been appropriately interpreted, they can begin the next phase of the adaptive cycle, plan formulation. Thus, the aircraft team has effectively noticed the environmental cue (i.e., dryness of mouth), classified it, assigned meaning to the cue, and finally communicated with the rest of the team. As a result, the team will have enhanced both of the emergent states that lead to adaptive performance (i.e., shared mental model and TSA).

The cognitive states that result from the process of situation assessment serve as the input to plan formulation. For example, TSA serves to provide the context around which the plan is developed. A lack of TSA may lead to misinterpretation or nonrecognition of mission-essential cues that serve as input for how the plan should be formulated. TSA has also been argued to provide the ability to forecast future events (Endsley, 1995) thereby facilitating the formulation of plans that are both proactive and adaptive. In addition to cognitive states, affective states such as psychological safety are also depicted as impacting plan formulation. Edmondson (1999) argues that psychological safety enables interpersonal risk taking through a shared belief that the team is "safe" (i.e., free from reprimand, retribution, or ridicule). A psychologically safe environment in which team members can interact, provide and receive feedback, and state their opinions will help the team to reach an appropriate plan by utilizing each member's expertise and/or knowledge for the situation at hand.

Using the cockpit crew example, after identifying the environmental cues in phase one of the adaptive cycle (i.e. situation assessment), the next step for the team would be to formulate a plan to resolve the situation. Because the team would have already participated in information sharing, their collective shared mental models and TSA are on point for developing a plan. In order for the team to maximize their options, they must maintain TSA and establish a psychologically safe environment. The senior pilot can help to ensure these variables are sufficiently met by communicating with the rest of the team and encouraging members to actively participate in possible solutions. Facilitating such an environment ensures that all team members have access to the same information; it will also aid in utilizing the individual knowledge and expertise from all of the team members by encouraging their interaction in plan formulation.

In the event of a plane losing pressurization, a team might come up with a plan to: (a) release oxygen masks for the passengers and crew, (b) contact senior officials via radio and report the problem, (c) take the plane off of auto-pilot, (d) begin to descend to a safer altitude, (e) locate the nearest airport where the plane can land, and finally (f) radio the tower to approve an emergency landing. Effective plans such as that just listed are a product of the team's shared mental models and are directly impacted by both TSA and psychological safety. By formulating a sound plan, the team is then able to transition to the next phase of the adaptive cycle, plan execution.

Once the plan has been developed and shared, it creates a framework that establishes meaning to commands and information requests during plan execution (Orasanu, 1990). Plan execution is depicted as the third phase within the cycle and consists of team members interdependently engaging in verbal and/or behavioral activities in order to attain established team goals (Marks et al., 2001). Within the model depicted in Figure 23.1, it is argued that individual (i.e., mutual performance monitoring, backup behavior, communication, and leadership) and team-level behaviors (i.e., coordination) will facilitate adaptive behavior during plan execution. Specifically, leaders and corresponding leadership processes set the stage for team adaptation by often coordinating operations, serving as a link to external influences, forming the team's goal or direction (Zaccaro & Marks, 1999), and creating psychological safety (Edmondson, 2003). Monitoring and backup behavior are the predominant mechanisms by which team members facilitate the timely mitigation of errors and adaptive recovery. Communication is important due to its role in the provision of feedback and updating of shared mental models and TSA. Finally, coordination is a team-level phenomenon that requires the team to organize the sequence and pacing of its actions (Marks et al., 2001).

The inputs that influence plan execution are teams' shared mental models and sense of psychological safety. Shared mental models are essential for effective monitoring, backup behavior, and communication. If team members do not have an adequate understanding of events, goals, or tasks, they cannot communicate with each other efficiently or provide and/or seek assistance when needed. An environment that is considered by teammates to be psychologically safe can facilitate acceptance of feedback, monitoring, and/or backup behavior. Thus, the emergent states from plan execution (i.e., shared mental models and psychological safety) contribute to the emergence of an adaptive team.

Returning to the cockpit crew example, the execution of the plan proposed during Phase 2 is set in motion. The well-functioning, adaptive team would utilize the individual behaviors necessary for plan execution on a consistent basis (i.e., monitoring, backup behavior, communication, and leadership). For example, the pilot assigned the task of monitoring any radio transmission to the engineer and the copilot noticed that the engineer was not performing the job in the way intended. By monitoring the performance of their teammate, the copilot has observed what could become an error. In order to help the engineer better understand his or her role, the copilot could engage in backup behavior and provide them with verbal feedback concerning performance expectations. This, in turn, would reemphasize the importance of communication of expectations. For most commercial cockpit crews, the pilot position is alternated (pilot–copilot) with each flight. Therefore, if the copilot was captaining this particular aircraft they would initially be in the leadership position unless they did not feel comfortable; in that case the leadership position would revert to the senior pilot. The role of the leader in this situation is crucial; he or she helps to establish the environment needed for successful plan execution.

Finally, coordinating all aspects of the individual behaviors will help to determine the sequence and pacing of the plan execution. Thus, in our example the crew would benefit from creating a prioritized list of actions for the problem at hand, hopefully having oxygen masks released as the most significant concern. They would also want to coordinate a timeline of how fast they are losing pressure, to give them an idea of how much time they have to find an appropriate location to land. Unlike previous stages that utilized only one or two of the emergent states, plan execution is impacted by the teams' shared mental models, TSA, and sense of psychological safety. All of these elements need to be maintained at high levels throughout plan execution.

After the plan has been enacted the team engages in the final segment, team learning. Team learning can be viewed as an iterative process by which relatively permanent changes occur as a result of group interaction though which members acquire, distribute, interpret, reflect on, change and store, and retrieve knowledge and information (Edmondson, 1999; Van Offenbeek, 2001). In order for a team to engage in these discussions, a sense of psychological safety must be established among the team members.

Our example cockpit crew would be more adaptive in the loss-of-pressure situation environment by utilizing team learning. After they have executed their plan of action for the situation at hand, the team would subsequently improve their collective understanding via discussion of actions. If each individual feels psychologically safe to talk about the situation at hand, they could possibly identify methods that might have prevented the escalation of the situation. In our example of the plane losing pressurization, perhaps one of the team members missed a cue that could have aided in identifying the problem before it became acute. Discussing the events could help this individual identify such cues in future situations, thus increasing his or her knowledge and as a result aiding the team in being more adaptive in future circumstances. However, if the team has not established a psychologically safe environment, the ability of team members to learn from each other might decrease, and subsequent adaptation might not be as successful as it might otherwise be.

Now that the key components of adaptive team performance have been briefly described, the next question is . . . "so what." What does this model of team adaptation mean for teams operating in "the wild"? What are the bottom-line practical implications? A few such implications are described next.

WHAT DOES THIS MEAN FOR TEAMS IN CONTEXTS THAT MANDATE ADAPTATION?

Working from the model depicted in Figure 23.1, adaptive team performance can be seen as a general process that underlies many team functions (e.g., dealing with performance hindrances, generating innovative solutions to problems, adopting new routines), and over time culminates in team adaptation. If one accepts the proposition that team adaptation is important to teams that operate in dynamic, mission-critical contexts, the next question becomes a practical one: How can team adaptation be facilitated?

One method by which to facilitate team adaptation is through training—team training. Team training is a set of theoretically derived techniques and methods that can be combined with a set of required competencies and training objectives to form an instructional strategy (Salas & Cannon-Bowers, 1997). Within the current context, such training would serve to target the team processes and the emergent states that facilitate team adaptation. Though team training can take many forms, we argue that training with the objective of promoting or facilitating team adaptation should be grounded within an approach known as scenario-based training (SBT). Next, this approach and why it is suited as a base method for training team adaptation is briefly described.

Scenario-Based Training

SBT differs from typical lecture-based training in that the curriculum it relies on is interactive lessons. Cannon-Bowers, Burns, Salas, and Pruitt (1998) describe SBT as "training that relies on controlled exercises or vignettes, in which the trainee is presented with cues that are similar to those found in the actual task environment and then given feedback regarding his or her responses" (p. 365). SBT seeks to present exercises that provide known opportunities to observe behaviors that have been targeted for training. Within this method, scenarios are scripted a priori and events are embedded such that explicit links are maintained between training objectives, exercise design, and performance assessment (Fowlkes, Dwyer, Oser, & Salas, 1998). This strategy focuses training on the objectives and incorporates deliberate guided practice so as to ensure practice results in learning the targeted objectives. Although the method of SBT is generic and can be applied to many types of skills and contexts, the training programs are highly context specific.

SBT provides an ideal grounding for training that seeks to promote adaptive team performance and the corresponding team adaptation, due to the fact that the training produced is highly context specific, in this case, specific to situations that require individual and team adaptation. The importance of context can be seen in Figure 23.1. Figure 23.1 depicts the first step in the adaptive cycle as the situation assessment process whereby a cue pattern is noticed. Cue patterns that trigger a recognition of the need for team adaptation are given meaning through what the individual and/or team understands about the current situation, seen through the lens of the team mental model and TSA. It is only through the assignment of this meaning that the decision can be made as to whether a change or modification is needed. Similarly, researchers have argued that adaptation "requires an understanding of deeper principles underlying the task, executive-level capabilities to recognize and identify changed situations, and knowledge of whether or not the existing repertoire of procedures can be applied" (Smith, Ford, & Kozlowski, 1997, p. 93). This type of understanding and the corresponding assignment of meaning rely heavily on task experience, and is the reason team experience is depicted as an antecedent to adaptive team performance within Figure 23.1. Given the importance of contextual expertise to the process of adaptive team performance and team adaptation, the use of realistic training environments is essential to build this context-dependent expertise. Training procedures grounded within SBT provide just such a setting.

SBT facilitates the transfer of experiential learning by basing scenarios on actual operational experiences. In addition, SBT allows the practicing and rehearsal of mission-critical events that may happen infrequently, be dangerous, or be expensive to train in any other manner. SBT has proven beneficial in many complex environments covering a broad range of domains (Salas, Priest, Wilson, & Burke, 2006).

Next, a subset of instructional features that can be embedded within the SBT framework to promote team adaptation are delineated. In doing so, their relation to the components comprising adaptive teams is described.

Instructional Feature 1: Variable Practice Scheduling. Traditionally, instructional models have argued for practice to proceed in a sequential fashion, yet research has begun to suggest that this may not always be the best approach within complex environments, especially in cases where skills and strategies might need to be reconfigured. Specifically, research has demonstrated that variable practice scheduling may produce better transfer and retention although initial acquisition/learning may suffer (De Croock, Van Merrienboer, & Pass, 1998; Sanders, Gonzalez, Murphy, Pesta, & Bucur, 2002). Variable practice scheduling has been defined as practice in which the order of trials is "randomized so that a given task [is] never practiced on successive trials" (Schmidt & Bjork, 1992, p. 210). It has been argued that variable practice scheduling "forces the learner to retrieve and organize a different outcome on each trial" (Schmidt & Bjork, 1992, p. 212). Thereby, use of this practice schedule can assist in building schemata that possess more breadth in their interconnections; thereby facilitating flexibility and reconfiguration of linkages.

Instructional Feature 2: Incorporate Cue-Recognition Training. In real-world decision-making environments, it is imperative that teams identify and attend to important environmental cues, thereby promoting accurate levels of situation awareness. A prerequisite to the development of accurate TSA is that individuals notice and interpret environmental cues within the situation assessment phase, which culminates in situation awareness. Not only must these cues be noticed and an accurate cue/pattern assessment made, but members must also be aware of the relationship between team process behaviors and cue/pattern assessments (Salas, Cannon-Bowers, Fiore, & Stout, 2001).

The challenge comes within the argument that although novices can be told about important environmental cues and provided practice in recognizing cues, this knowledge may not transferred to novel situations (Bransford, Franks, Vye, & Sherwood, 1989). Cue-recognition training is one instructional method that has been used to make important aspects of the environment more salient, thus increasing the likelihood that the team will attend to such aspects in dynamic situations (Salas et al., 2001). The use of perceptual contrasts during experiential learning is one manner in which cue-recognition training can be implemented. Specifically, the use of perceptual contrasts has been suggested as one manner by which to promote "noticing" as well as the enhancement of underlying conceptual knowledge, which is often slower to develop (Bransford et al., 1989; Lesgold et al., 1988). Training that incorporates perceptual contrasts by varying case features and correspondingly the range of typicality can be structured such that opportunities are provided whereby trainees assess how new situations are the same or different from situations previously encountered. This cognitive process is similar to that which is used by experts to determine whether situations are typical (i.e., recognition-primed decision making can be used) or atypical (i.e., involving the use of mental simulation).

Additional rationale for the use of perceptual contrasts in this situation is that instructional approaches that require learners to be actively involved in learning may be more effective in creating deeper and more acute understanding of instructional material than approaches that place learners in a more passive role (Schwartz & Bransford, 1998). Thus, research has begun to examine whether a variation of the perceptual contrast method as used by Schwartz and Bransford can be applied to teaching team members to adaptively apply teamwork behaviors, thereby facilitating transfer-appropriate processing (Fritzsche et al., 2004). Specifically, a series of video-based scripted events were created in which the adaptive application of various teamwork skills was targeted. During training, trainees were given subject-matter expert account of what should happen next for a team's action to be successful within the context of specific vignettes. Next, trainees were asked to identify the critical cues from the subject-matter expert's accounts. In other words, trainees were required to discover the similarities and differences in the subject-matter expert accounts and synthesize them. Conversely, the control group was provided with the same subject matter expert accounts already synthesized. During performance the participants were shown a variety of scenarios and asked to identify what should happen next. Preliminary results indicate that those who experienced perceptual contrasts were able to make more accurate decisions in regard to the adaptive application of teamwork behaviors as compared to the control group.

When training for cue recognition it is beneficial for the teammates to train in an environment where they are required to actively search for the relevant cues. If the team is simply informed of the relevant cues by an instructor or trainer, they may not develop the necessary skills to identify these same cues on their own because they did not experience the process by which a team recognizes an environmental cue. As a result of lack of experience in cue recognition, performance can suffer (Salas et al., 2001). Additionally, it is more beneficial for training to make contrasts that are grounded in context when identifying cues (e.g., explaining to a soldier that weapon *X* is more beneficial in combat and listing the specific reasons as opposed to simply informing them to use weapon *X*).

Instructional Feature 3: Incorporate Strategies That Encourage the Use of Metacognition. Metacognition can be conceptualized in terms of its executive roles as "both an awareness of and ability to regulate one's own cognitive processes" (Banks, Bader, Fleming, Zaccaro & Barber, 2001, p. 10). This instructional feature facilitates two aspects of the team adaptation model contained within Figure 23.1 (i.e., team learning and development of accurate shared mental models). Team learning has been argued to be a vital process in order for team adaptation to occur (Edmondson, 1999), as it is by this process and the corresponding feedback that knowledge is updated. Team learning is a process that is depicted as occuring at the end of the adaptive cycle, but may occur at any point in which the team has enough "downtime" to review its processes in some manner.

Metacognitive skills facilitate team learning and the corresponding updating of shared mental models by increasing the team's capability to engage in self-regulatory activities during all phases of the adaptation cycle. It is through self-regulatory processes (i.e., planning, monitoring, and adjustment of cognitive and task strategies) whereby cue–strategy linkages are explicitly monitored and discussed, allowing teams to gain awareness of the current state of their knowledge and the effectiveness of their strategies. To facilitate the use of metacognition, questions may be embedded within the context of SBT such that they cause team members to evaluate and discuss cue–response linkages, note discrepancies, and update their knowledge.

Team self-correction is another strategy that can be leveraged to facilitate metacognitive skill. Team self-correction relies on members monitoring team action and then using this information during periods of low workload to discuss and learn from past experiences. For training, prior to the implementation phase of the adaptive cycle team members should be taught techniques for monitoring and then categorizing their own behaviors as to their degree of effectiveness. This information can then be capitalized on during periods of low workload or during the debriefing sessions that occur after plan execution. In the debriefing session, members can discuss things that went well, as well as areas for improvement, and then discuss how to adjust their actions to respond in a more effective manner (see Smith-Jentsch, Zeisig, Acton, & McPherson, 1998). The feedback provided for team self-correction should be task focused and constructive, and should rely on members taking an active role during the process. Incorporating team-based strategies that facilitate metacognitive skill (e.g., team self-correction) ensures that team member expectations and interpretations of events remain calibrated and that this calibration is communicated to all team members. This, in turn, can promote appropriate adaptive team action.

Instructional Feature 4: Incorporate Anchored Instruction. Contextual knowledge is an important precursor to a team's ability to be adapt. Thus, SBT should be designed such that learning is couched in the specific context of interest where a link between information on the problem and the appropriate actions is maintained. The goal of anchored instruction is to "enable students to notice critical features of problem situations and to experience the changes in their perception and understanding of the anchor as they view the situation from new points of view" (Bransford et al., 1989, p. 14). Thus, by incorporating anchored instruction students will be more adaptive as a result of experiencing different points of view in decision-making and problem-solving environments, thereby, increasing the breadth of their mental models. It is hypothesized that by anchoring examples and instruction in the context of interest, team members will be better able to build the appropriate linkages that can result in improved strategic knowledge. This method is in contrast to instructional strategies where provided examples are not grounded in a specific context, but are presented in a broad, generalized manner. In turn, this knowledge builds the foundation for expertise and the recognition of cue–response discrepancies.

A RESEARCH AGENDA: DIRECTION FOR THE FUTURE

The aforementioned instructional features are means by which to aid teams at being more adaptive. In addition to instructing teams in how to be more adaptive, a map for future research endeavors that examine teams in a naturalistic environment is delineated.

Direction 1: Testing, Revision, and Expansion of Proposed Model of Team Adaptation. The model proposed here in Figure 23.1 is a first attempt to delineate a prescriptive model that illustrates the antecedents, processes, and emergent states that comprise adaptive team performance and lead to team adaptation. Though many of the individual links within this model have been tested in isolation, they remain to be tested as a set. In addition, there are some links within the model that make intuitive sense but have not been empirically tested. Finally, the multilevel aspects (i.e., individual and team-level constructs that combine to predict team-level performance) of the model remain to be tested as often this side of team research is ignored.

In addition to testing the proposed model there needs to be more focused examination of team adaptation in actual contexts—not only the moment-to-moment adaptation that is common within effective teams, but also longer-term team adaptation. Because this model is an initial attempt, we hope that it serves as food for thought, but are also confident that there will be expansion and revision of the model as more is learned about team adaptation in real-world environments.

Direction 2: Investigation of Appropriate Measurement Tools and Indices. Another aspect of adaptive teams that would benefit from further attention involves developing measures of adaptive team performance and team adaptation. Measuring team behavior has often proven to be cumbersome and difficult. Thus, it is of no surprise that developing and implementing measures of adaptive team behaviors is even more challenging. Current research has utilized the skills of subject-matter experts and self-report measures in order to quantify adaptive team performance and team adaptation. These methods require repeated interruptions in the data collection in order to attain an accurate picture of the adaptive process (Burke et al., 2006). These interruptions and inaccuracies, in turn, present multiple challenges in regard to assessing the validity of the measures. Therefore, it is imperative for future research to focus on finding better, less obtrusive measures for adaptive team performance.

Direction 3: Investigation of Communities of Practice as a Mechanism by Which to Promote Team Adaptation. Communities of practice (CoPs) have commonly been defined as joint enterprises that are continually renegotiated and produce a shared repertoire of resources that members have developed over time (Wenger, 1998). These joint enterprises typically represent an emergent property of a community or operational specialty that arises based on an unmet need. Although the predominant number of CoPs are not currently used to promote the competencies required for team adaptation, because of the repository of contextually based knowledge contained within them these enterprises could serve as a mechanism by which to examine and facilitate team adaptation. We argue that from a theoretical standpoint CoPs could be structured by those who create them in such a manner as to facilitate the growth of experiential learning and the development of breadth and flexibility within members' mental models. For example, CoPs pertaining to SASO could be designed such that the main page links to various themes (e.g., working with local populace, working within the political structure, working with joint operations) where users can submit questions, experiences, and answers with regard to a specific theme within the broader community or occupational specialty.

A second possibility would be to have even higher level categories (e.g., peacekeeping, humanitarian, peace enforcement) where experiences and outcomes might be posted. Site visitors could then look across situations and see different perspectives on similar problems, with the end goal of extracting common themes in terms of the cue–response contingencies. The presentation of information from multiple perspectives has been argued to facilitate breadth and flexibility within cognitive structures (e.g., mental models) (Spiro, Feltovich, Jacobson, & Coulson, 1992).

Finally, another possibility is for a community or occupational specialty to form a Web site where site visitors could post key information regarding essential coordination processes and workarounds. The site could be structured such that a team returning from deployment might pick several critical incidents and each member of the team would provide his or her perspective on that incident. This information is then available to others within that specialty. This might allow multiple perspectives to be examined at both an individual level (i.e., role) and team level. By examining various perspectives categorized around a common theme and extracting key cue–response linkages, this method might provide an avenue by which tacit and contextual knowledge is built.

Though the preceding suggestions may work in theory, from a practical standpoint CoPs are typically unstructured, representing informal learning. Therefore, research needs to be conducted to examine potential barriers to expanding this practice in the manner suggested earlier (i.e., climate, culture). Additionally, once identified, methods by which roadblocks can be mitigated are in need of investigation. It may be that in order to be utilized in the manner suggested previously the changes need to be transparent to the user. So the question becomes, how can the information that is typically provided be used to facilitate team adaptation and the development of expertise in a manner that is transparent?

Direction 4: Investigation of Best Strategies by Which to Facilitate Team Adaptation. In order to facilitate team adaptation, we must integrate current research on adaptive teams and determine optimal training strategies for adaptive teams. In the current chapter, we have argued for SBT as a methodology within which training for adaptive teams should be grounded. Additionally, we have proposed a number of instructional strategies that can be incorporated into such training and might impact various components within the team adaptation model. However, these suggestions are primarily based on theory and many have yet to be specifically applied to the area of *team* adaptation. Therefore, the proposed strategies need to be empirically evaluated within the context of team adaptation. It remains to be seen what the most efficacious and efficient strategy is with regard to team adaptation. Is there a "best" training strategy, or should training methodology depend on operational context?

Direction 5: Investigate How to Design Information Technologies That Support Performance and at the Same Time Facilitate Team Adaptation. Whereas this chapter has discussed instructional strategies that might facilitate team adaptation via their impact on various aspects of the team adaptation model proposed by Burke et al. (in press), there is another avenue that should be pursued—system design. Because teams are increasingly operating within technologically rich environments, the design of such environments and the technologies contained within them might be thought of as a mechanism by which to facilitate team adaptation. Through the merger of system design and training to facilitate adaptive capacity within teams, we might reach the goal of developing adaptive teams more quickly than is currently the case.

CONCLUSION

In this chapter, we have attempted to shed some light on the concept of team adaptation and its significance for teams' decision making. For better or for worse, expectations for teams have risen considerably. Teams are expected to: (a) survive and even thrive in dynamic, demanding surroundings, (b) interact with complicated technology and/or machinery, (c) take on multiple roles, (d) work under stress and time constraints, (e) work together and coordinate as a team, (f) work under unclear goals and objectives, and (g) work in dangerous situations where threats are unknown and/or ill-defined, creating situations in which team expertise is stretched. Obviously, all of this poses a monumental challenge to team members, researchers, and practitioners. We have discussed in some detail the nature of team adaptation and the need for adaptive teams in real-world environments, and we have summarized a model of team adaptation with an emphasis on the adaptive cycle. From this model, an instructional strategy and a series of instructional features were offered and a research agenda identified. It is our hope that this chapter will aid in broadening understanding concerning the nature and role of team adaptation in operational settings.

ACKNOWLEDGMENTS

This work was supported by funding from the Army Research Laboratory's Advanced Decision Architecture Collaborative Technology Alliance (Cooperative Agreement DAAD19-01-2-0009). The views expressed in this work are those of the authors and do not necessarily reflect official Army policy.

REFERENCES

Banks, D., Bader, P., Fleming, P., Zaccaro, S. J., & Barber, H. (2001, April). *Leader adaptability: The role of work experiences and individual differences.* Paper presented at the 16th annual meeting of the Society for Industrial and Organizational Psychology, San Diego, CA.

Bransford, J. D., Franks, J. J., Vye, N. J., & Sherwood, R. D. (1989). New approaches to instruction: Because wisdom can't be told. In S. Vosniadou & A. Ortony (Eds.), *Levels of processing and human memory* (pp. 470–497). Hillsdale, NJ: Lawrence Erlbaum Associates.

Burke, C. S., Fowlkes, J. E., Wilson, K. A., & Salas, E. (2003). A concept of soldier adaptability: Implications for design and training. In R. Hoffman & L. Tocarcik (Co-Chairs), *Innovative design concepts for the objective force* (pp. 143–148) (To appear in the *Proceedings of the Army Research Laboratory Collaborative Technology Alliance Annual Symposium*). U.S. Army Research Laboratory, Adelphi, MD.

Burke, C. S., Stagl, K., Salas, E., Pierce, L., & Kendall, D. (2006). Understanding team adaptation: A conceptual analysis and model. *Journal of Applied Psychology, 19*(6), 1189–1207.

Cannon-Bowers, J. A., Burns, J., Salas, E., & Pruitt, J. (1998). Advance technology in scenario-based training. In J. A. Cannon-Bowers & E. Salas (Eds.), *Making decisions under stress: Implications for individual and team training* (pp. 365–374). Washington, DC: American Psychological Association.

De Croock, M. B. M., van Merrienboer, J. J. G., & Paas, F. G. W. C. (1998). High versus low contextual interference in simulation-based training of troubleshooting skills: Effects on transfer performance and invested mental effort. *Computers in Human Behavior, 14,* 249–267.

Edmondson, A. C. (1999). Psychological safety and learning behavior in work teams. *Administrative Science Quarterly, 44,* 350–383.

Edmondson, A. C. (2003). Speaking up in the operating room: How team leaders promote learning in interdisciplinary action teams. *Journal of Management Studies, 40*(6), 1419–1452.

Endsley, M. R. (1995). Toward a theory of situation awareness in dynamic systems. *Human Factors, 37,* 32–64.

Fowlkes, J. E., Dwyer, D. J., Oser, R. L. & Salas, E. (1998). Event-based approach to training (EBAT). *International Journal of Aviation Psychology, 8,* 218–233.

Fritzsche, B. A., Burke, C. S., Flowers, L., Stagl, K., Wilson, K., & Salas, E. (2004, May). Improving team adaptability using contrasting cases training. In C. S. Burke & E. Salas (Co-Chairs), *Emerging theoretical and empirical approaches used in team effectiveness research.* Symposium conducted at the annual meeting of the American Psychological Society, Chicago.

Klein, G., & Pierce, L.G. (2001, June). *Adaptive teams.* Paper presented at the 6th International Command and Control Research and Technology Symposium. In *Proceedings of the 2001 6th International Command and Control Research and Technology Symposium.* Annapolis, MD: Department of Defense Cooperative Research Program. Retrieved May 16, 2003, from http://www. dodccrp. org/6thICCRts

Lesgold, A., Rubinson, H., Feltovich, P., Glaser, R., Klopfer, D., & Wang, Y. (1998). Expertise in a complex skill: Diagnosing x-ray pictures. In M. T. H. Chi, R. Glaser, & M. J. Farr (Eds.), *Nature of expertise* (pp. 311–342). Hillsdale, NJ: Lawrence Erlbaum Associates.

Marks, M. A., Mathieu, J. E., & Zaccaro, S. J. (2001). A temporally based framework and taxonomy of team processes. *Academy of Management Review, 26,* 356–376.

Orasanu, J. (1990). Shared mental models and crew decision making (Tech. Rep. No. 46). Princeton, NJ: Princeton University Cognitive Science Laboratory.

Salas, E., Cannon-Bowers, J. A., Fiore, S. M., & Stout, R. J. (2001). Cue-recognition training to enhance team situation awareness. In M. McNeese, E. Salas, & M. Endsley (Eds.), *New trends in cooperative activities: Understanding system dynamics in complex environments* (pp. 169–190). Santa Monica, CA: Human Factors and Ergonomics Society.

Salas, E., Priest, H. A., Wilson-Donnelly, K. A., & Burke, C. S. (2006). Scenario-based training: Improving mission performance and adaptability. In A. B. Adler, C. A. Castro, & T. W. Britt (Eds.), *Minds in the military: The psychology of serving in peace and conflict: Vol. 2. Operational stress* (pp. 32–53). Westport, CT: Praeger Security International.

Salas, E., Prince, C., Baker, D. P., & Shrestha, L. (1995). Situation awareness in team performance: Implications for measurement and training. *Human Factors, 37,* 123–136.

Sanders, R. E., Gonzalez, D. J., Murphy, M. D., Pesta, B. J., & Bucur, B. (2002). Training content variability and the effectiveness of learning: An adult age assessment. *Aging Neuropsychology and Cognition, 9*(3), 157–174.

Schmidt, R. A., & Bjork, R. A. (1992). New conceptualizations of practice: Common principles in three paradigms suggest new concepts for training. *Psychological Science, 3*(4), 207–217.

Schwartz, D. L., & Bransford, J. D. (1998). A time for telling. *Cognition & Instruction, 16*(4), 475–522.

Smith, E. M., Ford, J. K., & Kozlowski, S. W. J. (1997). Building adaptive expertise: Implications for training design strategies. In M. A. Quinones & A. Ehrenstein (Eds.), *Training for a rapidly changing workplace: Applications of psychological research* (pp. 89–118). Washington, DC: American Psychological Association.

Smith-Jentsch, K. A., Zeisig, R. L., Acton, B., & McPherson, J. A. (1998). Team dimensional training: A strategy for guided team self-correction. In J. A. Cannon-Bowers & E. Salas (Eds.), *Making decisions under stress: Implications for individual and team training* (pp. 271–297). Washington, DC: American Psychological Association.

Spiro, R. J., Feltovich, P. J., Jacobson, M. J., & Coulson, R. L. (1992). Cognitive flexibility, constructivism, and hypertext: Random access instruction for advanced knowledge acquisition in ill-structured domains. In T. M. Duffy & D. H. Jonassen (Eds.), *Constructivism and the technology of instruction: A conversation* (pp. 57–75). Hillsdale, NJ: Lawrence Erlbaum Associates.

van Offenbeek, M. (2001). Processes and outcomes of team learning. *European Journal of Work & Organizational Psychology, 10*(3), 303–317.

Wenger, E. (1998). Communities of practice: Learning as a social system. In *The Systems Thinker.* Retrieved September 22, 2005, from http://www.co-i-l.com/coil/knowledge-garden/cop/lss.shtml

Williams, D., & Chandrasekaran, R. (2003, June 19). U.S. troops getting frustrated in Iraq: Disillusioned soldiers unhappy about murky postwar role. *The Washington Post.* Retrieved June 19, 2003, from http://www. msn. com

Zaccaro, S., & Marks, M. (1999). The roles of leaders in high-performance teams. In E. Sundstrom et al. (Eds.), *Supporting work team effectiveness: Best management practices for fostering high performance* (pp. 95–125). San Francisco: Jossey-Bass.

Author Index

S

Subject Index